Protein-based Therapeutics

Dev Bukhsh Singh • Timir Tripathi
Editors

Protein-based Therapeutics

Springer

Editors
Dev Bukhsh Singh
Department of Biotechnology
Siddharth University, Kapilvastu
Siddharth Nagar, Uttar Pradesh, India

Timir Tripathi
Department of Biochemistry
North Eastern Hill University
Shillong, Meghalaya, India

Indira Gandhi National Open University
(IGNOU), Regional Center Kohima
Kohima, Nagaland, India

ISBN 978-981-19-8248-4 ISBN 978-981-19-8249-1 (eBook)
https://doi.org/10.1007/978-981-19-8249-1

This Springer imprint is published by the registered company Springer Nature Singapore Pte Ltd.
The registered company address is: 152 Beach Road, #21-01/04 Gateway East, Singapore 189721, Singapore

Preface

Protein-based therapeutics constitute an important class of drugs used for the treatment of multiple diseases, including cancers, genetic disorders, autoimmune diseases, and infectious diseases. In recent years, significant progress and advancements have been made in the techniques and approaches used for the production, modification, qualitative analysis, and delivery of therapeutic proteins. Novel design and delivery strategies of therapeutic proteins have improved significantly, triumphing many drawbacks, challenges, and issues. One of the critical challenges for protein-based therapeutics is poor membrane permeability, whereas another challenge is poor *in vivo* stability and short half-life. Advances in structural biology, recombinant biology, biochemistry, biophysics, drug design and discovery, synthetic biology, and pharmacogenomics have endowed new landmarks for peptide drug discovery, synthesis, and clinical applications. Today, a large number of approved therapeutic proteins are available in the market for clinical applications, and many are in the clinical, preclinical, or development phases. The global market and demand for therapeutic proteins are increasing rapidly, but the manufacturing and production of protein-based therapeutics are highly complex processes. A detailed comprehension of pathways targeted by therapeutic proteins and issues related to safety and efficacy are significant from an application point of view. This book provides a thorough and descriptive knowledge of various topics related to protein-based therapeutics, such as their clinical applications, methods and strategies to design, recombinant production, antibodies as protein therapeutics, success history of streptokinase, formulation and systemic delivery strategies, biochemical targets, pharmacogenetic biomarkers, immunogenicity, safety and efficacy issues, emerging trends and challenges in the field of protein therapeutics, biosimilar, biobetter, and biosuperior protein therapeutics, and therapeutic protein-based vaccines. The chapters discuss diverse aspects of protein-based therapeutics, their production, application, delivery, safety, efficacy, immunogenicity, and pharmacogenomic information.

The book has been written considering the demand for researchers and students looking to study disease treatment using proteins as therapeutic agents. It includes a large number of figures and illustrations for a better and clearer understanding. The collection will also help provide insights into therapeutic proteins for postgraduate and research students studying drug design, discovery, and development, pharmacology, pharmaceutics, pharmacogenomics, medicinal chemistry, biochemistry, structural biology, protein chemistry, etc. We are confident that the book will be beneficial for readers to understand broad aspects of protein-based therapeutics. We look forward to the valuable suggestions and feedback of readers related to the content of this book.

Siddharth Nagar, Uttar Pradesh, India Dev Bukhsh Singh
Shillong, Meghalaya, India Timir Tripathi

Contents

About the Editors

Dev Bukhsh Singh is an Assistant Professor at the Department of Biotechnology, Siddharth University, Kapilvastu, Siddharth Nagar, India. He also served as Assistant Professor at the Department of Biotechnology, Chhatrapati Shahu Ji Maharaj University, Kanpur, India. He received B.Sc. and M.Sc. from the University of Allahabad, an M.Tech from the Indian Institute of Information Technology, Prayagraj India, and a Ph.D. in Biotechnology from Gautam Buddha University, India. He has been actively involved in teaching and research since 2009 and is teaching bioinformatics, biostatistics, proteomics, genomics, and other courses to B.Sc. and M.Sc. Biotechnology students. His areas of research are molecular modeling, chemo-informatics, and drug designing. He authored numerous research articles and book chapters in the fields of computational biology, medicinal chemistry, molecular modeling, drug design, and systems biology. He is a member of various national and international academic bodies and is a reviewer for several international journals.

Timir Tripathi is the Regional Director of Indira Gandhi National Open University (IGNOU), Regional Centre Kohima, Nagaland, India. Earlier, he served as a Senior Assistant Professor and Principal Investigator at the Department of Biochemistry, North-Eastern Hill University, Shillong, India. He holds a Ph.D. from the Central Drug Research Institute, Lucknow, India. He was a visiting faculty at ICGEB, New Delhi, India, and Khon Kaen University, Thailand. He is known for his research in the fields of protein biophysics, biochemistry, structural biology, and drug discovery. He has over 17 years of experience in teaching and research on protein structure, function, and dynamics at the postgraduate and doctoral levels. He has developed and improved methods to investigate and analyze proteins. His research areas include protein interaction dynamics and understanding the roles of non-catalytic domains in regulating the catalytic activity of proteins. He has published over 100 research papers, reviews, commentaries, viewpoints, and editorial articles in international journals and published several book chapters. He has edited five books and authored a textbook on spectroscopic methods for undergraduate & postgraduate students. He has received several national and international awards.

Introduction to Protein Therapeutics

1

Monika Jain, Ankit Kumar, Rameez Jabeer Khan, Rajat Kumar Jha, Ekampreet Singh, Jayaraman Muthukumaran, and Amit Kumar Singh

Abstract

Proteins are a group of biomolecules that serve as functional and structural constituents of cells. The 20 standard amino acids combine to synthesize polypeptide chains, and their sequence in protein offers insights into their three-dimensional structure and biological roles, which depend on their physical interaction with other molecules. Proteins function as hormones, enzymes, and cytokines and play a crucial role in biological pathways and intercellular communication. Therapeutic proteins are drugs that are recombinantly modified versions of naturally existing human proteins. The complicated structure of proteins can be portrayed in multiple ways to aid scientists in studying their properties. The use of therapeutic proteins in treating cancer, HIV, and other disorders is common. Here, we introduce the functional aspects of therapeutic proteins and their applications. Moreover, we discuss the available computational resources for studying therapeutic proteins and their related information.

Keywords

Protein · Protein structures · Amino acid · Therapeutic proteins · Monoclonal antibodies · Immunogenicity · PEGylation · THBdB · Biosimilars

Monika Jain and Ankit Kumar contributed equally with all other contributors.

M. Jain · A. Kumar · R. J. Khan · R. K. Jha · E. Singh · J. Muthukumaran (✉) · A. K. Singh (✉)
Department of Biotechnology, School of Engineering and Technology, Sharda University, Greater University, Greater Noida, Uttar Pradesh, India
e-mail: j.muthukumaran@sharda.ac.in; amitk.singh@sharda.ac.in

D. B. Singh, T. Tripathi (eds.), *Protein-based Therapeutics*,
https://doi.org/10.1007/978-981-19-8249-1_1

1.1 Introduction

Proteins are the most innumerable biological macromolecules in all cells and subcellular organelles. They exhibit great diversity in terms of structure and function even within a single cell [1]. These complex biopolymers are comprised of simple monomeric units known as amino acids. The twenty universal amino acids are utilized inside a cell for protein synthesis. The amino acids are linearly joined via covalent peptide bonds to form proteins. By different combinations of the amino acid monomers, cells can generate a vast diversity in protein functionality [2, 3].

1.1.1 Structural Organization of Proteins

Proteins exhibit different levels of complexity in their structure. Protein architecture is categorized into four levels: primary, secondary, tertiary, and quaternary, with the intricacy of the structure increasing as we progress up the hierarchy (Fig. 1.1). The primary structure of proteins depicts the sequence of amino acids in which they are arranged in a particular protein. It describes the covalent bonds that link amino acids (peptide bonds). The secondary structure informs about the spatial arrangement of amino acids for a particular segment of protein. Interaction with other segments of proteins and positioning of side chains is not considered while describing the secondary structures. There are different types of secondary structures that are stable and are observed in proteins; these include α-helices, β-sheets, coils, turns, etc. [4, 5].

α-helix is the principal secondary structural element commonly found in almost all proteins. This structure is formed around an imaginary central axis running longitudinally. There are 3.6 amino acid residues in one turn of the helix, with one complete turn of the helix being 5.4 Å in length. The α-helix could be either right-

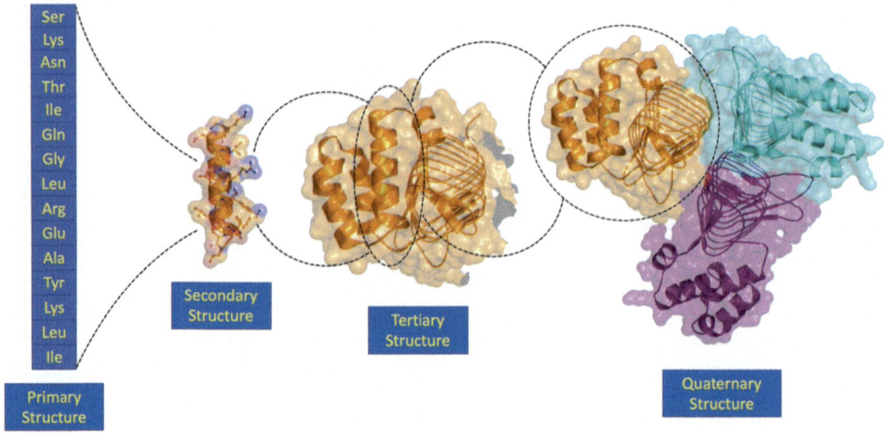

Fig. 1.1 Structural organization of proteins

handed or left-handed. However, the right-handed alpha helix is most common in proteins. The stability of α-helical structures is mainly contributed by intramolecular hydrogen bonding. The hydrogen bonds are formed between the hydrogen atom linked to the electronegative nitrogen atom of nth to the carbonyl group oxygen of the n+4th amino acid residue [4, 2]. Other helices sometimes observed in protein structures are π helix and 3_{10} helix. In the π helix, the hydrogen bond is formed between the first and fifth amino acid residues, while, in the 3_{10} helix, the first and the third residues are involved in hydrogen bonding [4, 3]. The β pleated sheets are formed when hydrogen bonding (intermolecular) forms between neighboring polypeptide chains. Here the backbone of the polypeptide acquires a zigzag shape, giving the overall structure a pleated (ridge-like) appearance. β sheets possess two different orientations of polypeptide strands in their structure, parallel and antiparallel [4, 3]. β turns are found in globular proteins with compact folded structures. Few residues are present in turns or loops where the polypeptide chain reverses its direction. β turns are the most common type of nonrepetitive structures and constitute, on average, 25% of the protein structure [6]. Protein stability, molecular recognition, and protein folding rely heavily on the formation of β turns. β turns are the linking components that connect consecutive helix or conformation turns. Four amino acid residues rotate 180° degrees in the turn structure. The first residue's carbonyl oxygen creates a hydrogen connection with the amino group hydrogen in the fourth. Interhydrogen bonding connection does not occur between the peptide groups of the two core residues [2, 7].

The tertiary structure of proteins is an all-inclusive three-dimensional silhouette of protein. This structure is the collective result of interactions among side chains of amino acid residues, hydrophobic interactions, hydrogen bonding, ionic interactions, and weak Van der Waals forces [8]. The hydrophobic interaction is a key player in the proper folding of protein to its tertiary structure. As discussed earlier, the amino acid side chains exhibit different properties; they may be polar, nonpolar, or charged. Being hydrophobic, nonpolar amino acids aggregate toward the interior of proteins, and the polar and charged residues, which are hydrophilic, accumulate toward the periphery. Inside the core of the tertiary structure, the nonpolar amino acids come close enough to provide an opportunity for weak van der Waals forces to play a role in protein stabilization. Hydrogen bonding and ionic interactions between polar and charged amino acids also play a part in structure stabilization. When proteins fold, cysteine residues are brought in close proximity to each other, allowing their sulfhydryl groups to form disulfide linkages. The disulfide bonds further enhance the overall stability of protein structure [9–11].

Quaternary structure is found in proteins that are composed of more than one polypeptide chain (subunits). These subunits interact in a specific geometric fashion. The conformational organization of multiple protein subunits in 3D space is called quaternary structure. The subunits can be either identical or different and can work independently or in conjunction to carry out a particular function [12].

1.2 Functions of Proteins

Proteins have a wide range of functionality inside cells. They are involved in almost every biochemical reaction occurring inside biological systems. Enzymes that act as biocatalysts and catalyze biological reactions inside the cell are proteins, for example, alcohol dehydrogenase (oxidoreductases), methyltransferase (transferases), aminopeptidase (hydrolases), aldehyde-lyases (lyases), glutamate racemase (isomerases), and DNA ligase (ligases) [2]. Proteins also function as hormones, which act as chemical messengers to regulate physiological processes. Insulin, glucagon, and growth hormone are some examples of peptide hormones [13]. Receptors present on the surface of cells that aid in cell-to-cell communication are also proteins, such as GPCRs [14], hormonal receptors, and cell-specific markers such as CD4 helper T cells and CD8 cytotoxic T cells [2]. Proteins also act as transporter molecules to aid in the movement of other molecules, atoms, or other chemical substances across the cell membrane, such as ABC transporters and ion channels [15]. Various biological roles of proteins are shown in Fig. 1.2. The biological functions are determined by how it interacts with other molecules. Ligands bind to protein receptors present on cells for cellular communication and play a vital part in biochemical processes and intercellular interactions. Proteins play an essential role in the protection against harmful substances and diseases. All life forms, including plants, bacteria, and mammals, possess these defense proteins.

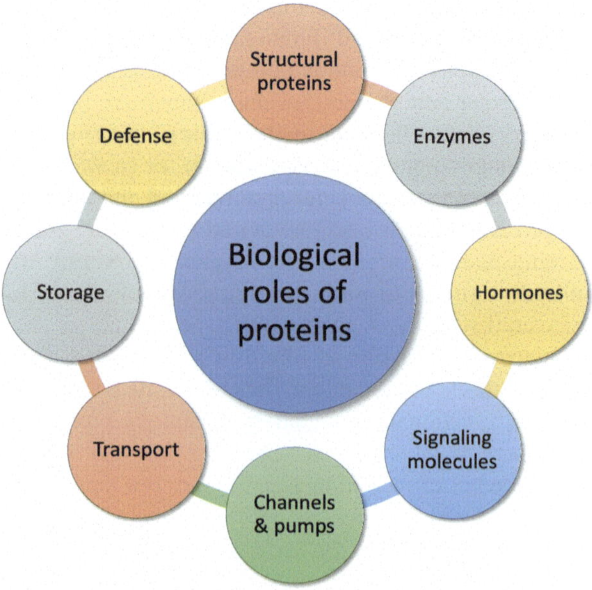

Fig. 1.2 Important biological roles of proteins

1.3 Therapeutic Proteins

Therapeutic proteins are medicines that are recombinantly modified versions of naturally existing human proteins. The three-dimensional structure of a protein can be used to study its interactions and functions. The use of therapeutic proteins in the treatment of cancer, HIV, and other disorders is common. Major therapeutic proteins include monoclonal antibodies, interferons, and cytokines. The US FDA has approved over 100 therapeutic proteins widely utilized to treat diseases such as cancer, diabetes, anemia, and infections. Engineered protein scaffolds, antibodies, bone, anticoagulants, blood factors, Fc fusion proteins, enzymes, hormones, interferons, morphogenetic proteins, interleukins, growth factors, and thrombolytics are among the molecular categories of therapeutic proteins [16]. Therapeutic proteins have transformed disease treatment as they are highly effective in vivo. Protein treatments aid a precisely targeted therapeutic process by correcting for a crucial protein deficiency, allowing for a more tailored approach to treatment. Protein therapeutics, compared to small molecule drugs, offer superior binding selectivity and specificity, allowing them to target specific phases in disease pathophysiology. Most drugs used to suppress the immune response in long-term inflammatory diseases were restricted to small molecule drugs like corticosteroids and cyclosporine A before the introduction of protein therapies. However, therapeutic proteins work broadly and indiscriminately to block both beneficial and detrimental immune responses, resulting in substantial adverse effects [17]. With the advent of recombinant DNA (rDNA) technology, various human proteins have been commercialized. In addition, many human proteins are now expressed heterologously in bacteria, yeast, or mammalian systems and are available to treat various diseases [18].

However, there are various advantages of protein therapeutics, such as (i) they have a sophisticated and highly specialized set of functions, (b) there is a lower risk of interfering with normal biological processes and causing negative consequences, and (c) immune responses are less likely to be elicited, and the drug is well tolerated, and (d) gene therapy is not required for effective replacement therapy. The mode of action of protein therapeutics depends on the pathology of the disease. First, if some unwanted extracellular molecule like cellular metabolite exists, the therapeutic protein (such as some enzyme) can degrade it. Secondly, if there is a protein deficiency in the body, external therapeutic protein can restore its deficiency and the health of the individual. Thirdly, if there are improper signaling or immune responses, therapeutic proteins can act as activators or inhibitors of receptors, triggering these responses. In contrast to small-molecule medications, therapeutic proteins cannot diffuse across cell membranes due to their large size. As a result, they are mostly used after cell surface receptors or extracellular substances.

Engineering therapeutic proteins involves various strategies considering their half-life in the body, shelf life, immunogenicity, pharmacokinetics, and pharmacodynamics. These involve the use of fusion proteins and posttranslational or chemical modifications. For example, glycosylation helps increase the size of protein, which in turn reduces renal clearance. It can also enhance the solubility and stability, thereby increasing the half-life of the protein. Similarly, chemical modification can be done by PEGylation, where polyethylene glycol is attached to drug molecules, therapeutic proteins, or vesicles, both covalently and noncovalently, to increase the half-life of protein therapeutics.

1.3.1 Classification of Therapeutic Proteins

Therapeutic proteins can be classified based on function, clinical applications, and pharmacological properties [17]. In general, the therapeutic proteins are classified into four groups, which are as follows:

1.3.1.1 Group I: Therapeutic Proteins with Enzymatic Activity

This includes the classic strategy of replacing the deficient enzyme or protein that causes a particular disorder or disease. Protein hormones, such as human growth hormone and insulin, are well-known examples of group I therapeutic proteins. Insulin is used in the treatment of diabetes mellitus Type I (insulin dependent) and Type II (insulin independent). There are various analogues of insulin available that are produced by genetic engineering. They differ in amino acid sequence and chemical modifications. Examples are long-acting insulin glargine, diarginyl insulin, and fast-acting insulin lispro [19]. Recombinant human growth hormone (rhGH) is used to treat various diseases resulting from its deficiency [20]. Other examples include lactase for patients lacking this enzyme in the gastrointestinal (GI) tract and blood clotting factors in hemophiliacs.

The second strategy in this group involves enhancing the magnitude or modulating the expression timing of a particular normal protein. For example, recombinant erythropoietin released by the kidney stimulates RBC production in the bone marrow, especially in chemotherapy-induced anemia patients. Similarly, patients with renal failure are also given this to help with anemia [21]. IVF treatments also use the recombinant follicle-stimulating hormone that leads to the maturation of a large number of follicles and recombinant HCG to promote follicle rupture [22]. Another example is Alteplase (recombinant tissue plasminogen activator; tPA) for the treatment of blood clots.

The third approach in the group I category is to add a novel function to the naturally occurring protein spatially and temporally. There are various examples of such protein therapeutics. For example, recombinant human deoxyribonuclease I (DNAse I) is used to clear leftover DNA in neutrophils in the respiratory tract of patients suffering from cystic fibrosis [23]. Other examples include enzymes from bacteria and yeast, such as collagenase, used in chronic dermal ulcers and severe burns [24]. Similarly, botulinum toxin is used for dystonia and cosmetic purposes [25]. A few group I therapeutic proteins are listed in Table 1.1 [17].

1.3.1.2 Group II: Therapeutic Proteins with Special Targeting Activity

Group II therapeutic proteins include two categories: (a) proteins that interfere with some other molecules in the body and (b) molecules that are delivered at a specific

Table 1.1 Few vital therapeutic proteins of the group I category

S. No.	Therapeutic protein	Role	Trade name	Clinical use
1.	Insulin	Replacement	Humulin, Glargine, NPH, Lispro, etc.	Diabetes mellitus
2.	Growth hormone	Replacement	Genotropin, Norditropin, Serostim, etc.	Growth failure due to GH deficiency, Turner syndrome
3.	Factor VIII	Replacement	Bioclate, Helixate, Recombinate, etc.	Hemophilia A
4.	Lactase	Replacement	Lactaid	Inability to digest lactose
5.	Human albumin	Replacement	Albumarc, Albutein, Plasbumin	Hypovolemia, hyperbilirubinemia
6.	β-Glucocerebrosidase	Replacement	Cerezyme	Gaucher's disease
7.	Pancreatic enzymes	Replacement	Cotazym, Pancrease	Cystic fibrosis, pancreatic insufficiency, etc.
8.	Immunoglobulins	Replacement	Octagam	Immunodeficiencies
9.	Erythropoietin	Augmenting of existing protein	Epogen	Anemia due to renal failure or chemotherapy
10.	Human follicle-stimulating Hormone (FSH)	Augmenting	Follistim	Assisted reproduction, IVF
11.	Human chorionic gonadotropin (HCG)	Augmenting	Ovidrel	Assisted reproduction, IVF
12.	Alteplase	Augmenting	Activase	Myocardial infarction
13.	Botulinum toxin type A	Augmenting	Botox	Dystonia, cosmetic uses
14.	Collagenase	Augmenting	Santyl, collagenase etc.	Dermal ulcers, burns, etc.
15.	Human deoxyribonuclease I	Augmenting	Pulmozyme	Cystic fibrosis, respiratory diseases

site in the body. Both these approaches are carried out by using monoclonal antibodies (mAbs). The mAbs are a type of protein therapy that has revolutionized preventive medicine and helped treat diseases like cancer, immune system disorders, and viral infections. They are based on natural antibodies produced against foreign particles and pathogens inside the body. Since the mAbs are made by a single clone of B cells, they are monospecific and homogeneous. Natural antibodies are used as antigen binders, and the binding domains of ligands are fused with the antibodies to create a target-specific binder that triggers the immune system to destroy specific cells or tissues. Many such mAbs have been constructed by rDNA technology. The mAbs are less likely to produce off-target effects due to their high affinity and selectivity for their molecular target. This makes them especially valuable in experimental models for determining the role of a target in disease pathogenesis. Currently, the majority of approved fusion proteins are Fc fusions [26]. Many mAbs are used in treating autoimmune disorders, such as Alemtuzumab for treating multiple sclerosis, Benralizumab for asthma, and Ocrelizumab for rheumatoid arthritis and lupus erythematosus. The mAbs can decrease the overactive immunological responses that characterize autoimmune disorders by targeting distinct components of the immune system. Many mAbs have been developed for the target-specific killing of cancer cells by blocking some receptors and ligands. The main targets for therapeutic mAbs are growth factor receptors that are upregulated in tumor cells, like epidermal growth factor receptor and human epidermal growth factor receptor 2 (HER). For example, Trastuzumab prevents dimerization and internalization of the HER2 receptor, resulting in receptor endocytic degradation and activation of the immune system in breast cancer. Cetuximab is a chimeric mAb used in the treatment of colorectal and head and neck cancer [27].

The mAbs are also used to selectively deliver toxic compounds inside cancer cells, for example, ibritumomab tiuxetan is an anti-CD20 mAb labeled with yttrium-90 or indium-111 and used to treat non-Hodgkin's lymphoma [28]. Some mAbs can target the tumor microenvironment, thus preventing metastasis and angiogenesis. Bevacizumab inhibits the binding of vascular endothelial growth factor (VEGF) to receptors on vascular endothelial cells, thus preventing angiogenesis. Some mAbs enhance antitumor immune response, such as ipilimumab and nivolumab. Many mAbs are used in the treatment of infectious diseases, such as those caused by cytomegalovirus, hepatitis A and B viruses, HIV-1, hepatitis B, Ebola, SARS-COV-2, etc. [29]. Many more mAbs have been produced for antiplatelet therapy, migraine prevention, treatment, and the production of immune complex vaccines for both preventive and therapeutic vaccination. Various applications of mAbs are shown in Fig. 1.3, and some examples of mAbs used as therapeutic proteins are given in Table 1.2 [17]. A wide range of mAb-based techniques has shown to be effective in treating the cancer patients. Examples include unlabelled IgG that attaches directly to cancer cells, mAb that affects the active host response to cancer, immunoconjugates that transport deadly moieties to cancer, and constructions that leverage the specificity of mAb to retarget cellular immunity toward the cancer cell [30].

1.3.1.3 Group III: Therapeutic Protein Vaccines

The group III category of therapeutic proteins includes three approaches: (a) protection against deleterious agents, (b) treatment of autoimmune disorders, and (c) cancer treatments. Adaptive immunity plays a vital role in the fight against foreign particles and cancer cells. The cell-mediated immune response includes the activation of T cells by antigen presentation. Various vaccines involve injecting killed or attenuated pathogens or their proteins inside the body to generate an immune response against them [31]. They have certain side effects, which can be avoided by using therapeutic protein vaccines, which are highly target-specific.

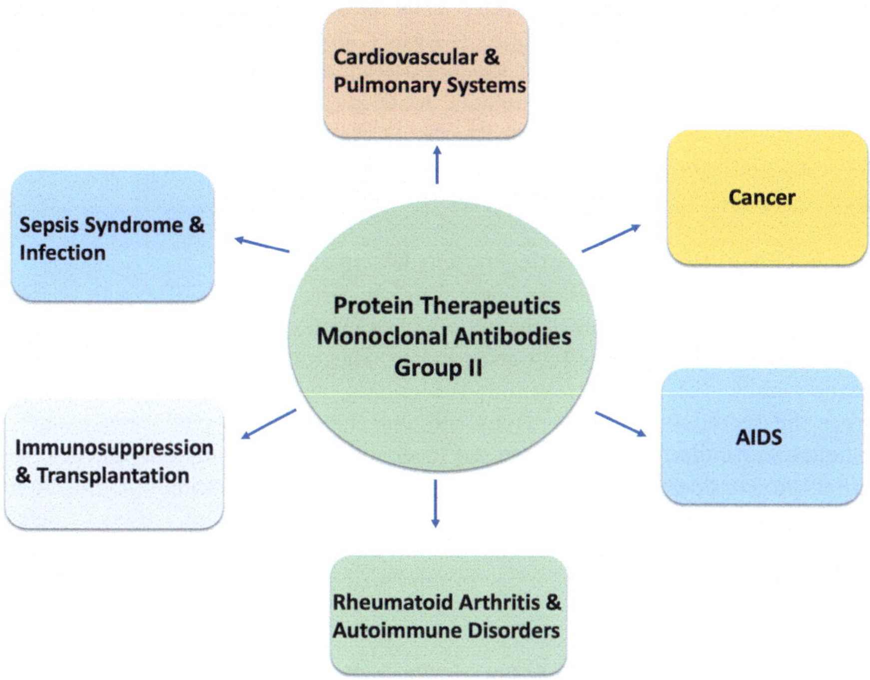

Fig. 1.3 Various clinical applications of monoclonal antibodies as protein therapeutics

Table 1.2 Few important therapeutic monoclonal antibodies

S. No.	Monoclonal antibody	Target	Clinical use
1.	Alemtuzumab	CD52	Multiple sclerosis, B-cell chronic lymphocytic leukemia
2.	Adalimumab	TNF-α	Rheumatoid arthritis
3.	Benralizumab	CD125	Asthma
4.	Ocrelizumab	CD20	Rheumatoid arthritis, lupus erythematosus
5.	Tocilizumab	IL-6 receptor	Rheumatoid arthritis
6.	Bevacizumab	VEGF	Colorectal cancer
7.	Trastuzumab	HER2	Breast cancer
8.	Cetuximab	EGFR	Head and neck cancer
9.	Ibritumomab tiuxetan	CD20	Non-Hodgkin's lymphoma
10.	Pembrolizumab	PD-1	Melanoma and other cancers
11.	Rituximab	CD20	Chronic lymphocytic leukemia
12.	Ibalizumab	CD4	MDR HIV
13.	Oblitoxaximab	*Bacillus anthracis*	Anthrax
14.	Ibalizumab	CD4	MDR HIV
15.	Rmab	Rabies virus glycoprotein	Rabies

Many such vaccines are still in clinical trials. A few important examples of such vaccines are the recombinant hepatitis B surface antigen (HBsAg) protein vaccine [17] and the human papillomavirus vaccine (HPV), responsible for causing cervical cancer and genital warts. Recombinant proteins protect against the overactive immune system; for example, anti-Rhesus D Ag immunoglobulin prevents the Rh-positive fetus in an Rh-negative mother. Many therapeutic protein vaccines for cancer treatment are in clinical trials [17].

1.3.1.4 Group IV: Therapeutic Protein Diagnostics

Many therapeutic proteins are used only for diagnostic purposes. They can be used in vivo as well as in vitro. One of the important examples of protein diagnostics is the purified protein derivative (PPD) test used to detect *Mycobacterium tuberculosis* infection cases [32]. Another example is human protein secretin, which is used to trigger the secretion of the pancreas and thus helps detect pancreatic exocrine dysfunction. Similarly, the TSH is used to diagnose residual thyroid cancer cells. Various protein diagnostics are used as imaging agents to detect multiple infections and pathological conditions. These imaging agents are frequently used to detect otherwise undetectable diseases. Diagnosis of cancer, myocardial injury, and location of occult infection sites are a few examples of protein-based imaging [33].

1.3.2 Challenges for Therapeutic Proteins

Therapeutic proteins are expressed in bacteria, yeasts, and mammalian cell cultures and are used to treat or prevent diseases. They are among the latest scientific innovations in pharmaceuticals. Unlike traditional chemically manufactured drugs, recombinant proteins are large and complex molecules with sophisticated and unique mode of action. Since the chemical production of proteins is challenging due to their size and complexity, they are made biologically using the protein synthesis machinery of cells. Protein solubility, delivery method, distribution, and stability are significant factors when using protein therapies [34]. The success of protein-based therapeutics depends on safety, efficacy, and quality. However, the protein therapeutics can have side effects due to their interactions with intended and unintended targets inside the body. The side effects can be suppressive or stimulatory to the immune system, which can lead to deleterious impacts [16].

1.3.2.1 Efficacy

Efficacy is an essential parameter for the approval of protein therapeutics by regulatory bodies. There is a requirement for gradual improvement in the efficacy of therapeutic protein-based therapies and to make them applicable to many more diseases. Another challenge involves the development of personalized treatments for diseases like cancer and autoimmune disorders and the prediction of toxicity levels.

1.3.2.2 Quality

Quality is also required for approval by the regulatory bodies. Developing protein therapeutics and delivery mechanisms involves complex steps and present significant challenges for quality control. For safety and approval issues, high-quality protein therapeutics with minimum heterogeneity and contamination are required. With the advent of computational technologies and high-throughput protein engineering experimental methods, there is ample scope for research and development in protein therapeutics, evaluating associated risks, and designing new technologies to limit them.

1.3.2.3 Stability

The stability of therapeutic proteins is also an important concern among researchers [35]. The stability of peptides and proteins over their shelf-life is a fundamental barrier in their preparation into effective dosage forms. A minimum shelf life of 1.5 or 2 years is desirable at room temperature or refrigeration storage. The stability issue can be associated with either physical instability or chemical instability. Physical stability is defined as the change in protein structure through denaturation, precipitation, and aggregation. Chemical instability can result from various pathways like deamidation, β-elimination, racemization, disulfide exchange, hydrolysis, and oxidation. Oxidation can degrade proteins in both solution and lyophilized states. The primary structure often indicates chemical degradation sites, while the physical destabilization is challenging to predict based on the primary structure. However, if the majority of the residues are hydrophobic, it implies a considerable susceptibility to adsorption and aggregation. Various genetic engineering approaches and computational protein design techniques are being used to improve the stability of therapeutic proteins [36].

1.3.2.4 Immunogenicity

Immunogenicity is the tendency to trigger an immune response. An immunological reaction is initiated if the body perceives the protein as foreign. The proteins derived from animal sources are potentially immunogenic. Even recombinant human proteins are generally immunogenic for various reasons, though to a lower extent. One reason is the presence of impurities like oxidized forms, aggregates, media, or residual host cell proteins (HCPs). Another reason for immunogenicity is the dose and route of administration. Subcutaneous routes are more immunogenic as compared to intramuscular and intravenous. T-cells have a role in the immunological response to protein-based therapeutics by activating B cells, which then generate antibodies, including those that obstruct protein therapeutics [37]. The amount and kind of post-translational modifications in a protein (such as glycosylation) can significantly affect the immunogenicity of the protein [38]. As a result, a protein expressed in *E. coli* that is not glycosylated is likely to elicit a different immunological response than a protein produced in a mammalian cell line because the latter is glycosylated. Antibodies raised against an immunogenic protein can sometimes reduce the protein's potency. These antibodies generally do not induce severe allergic or anaphylactic reactions, but if they deactivate the original protein, they can have fatal implications. Diverse analytical techniques are available to evaluate proteins' features correlating to their immunogenicity [39].

1.4 Methods for Production of Therapeutic Proteins

The production of next-generation protein therapeutics with improved efficacy, safety profiles, and delivery is a hot area of pharmaceutical research. Various techniques have been developed in the last two decades to design new protein therapeutics with excellent efficacy and safety. The two important approaches for protein engineering are directed evolution and rational design.

Directed evolution depends on mutant library generation with desired properties. Several systems have been constructed for high-throughput screening of vast libraries of protein mutants coupled to phage, bacteria, yeast, mRNA, or ribosomes. Directed evolution can be achieved by random mutagenesis, scanning mutagenesis, and block mutagenesis. A random mutagenesis is a powerful tool for identifying amino acid locations in a protein that are linked to its role in a particular type of activity or metabolic pathway. Similarly, scanning mutagenesis is a frequently used approach for replacing amino acids in a protein in a systematic manner to better understand the structure-function link. Block mutagenesis or simultaneous saturation mutagenesis of many adjacent residues can quickly access combinations of mutations that may elicit synergism [40].

The structural study of proteins is used in rational design to discover sites where mutations might be introduced to introduce or enhance desirable features. Several therapeutic enzymes with novel or enhanced activity have been successfully developed using rational design approaches. High-throughput computational design approaches are major contributors to the rational approach. Various characteristic features of protein therapeutics that can be modified, improved, or developed by protein engineering methods are affinity, specificity, cross-reactivity, half-life, stability, efficacy, and immunogenicity [41].

1.5 Delivery of Therapeutic Proteins

Thanks to improvements in rDNA technology and various protein engineering methodologies, many proteins and peptides have been commercialized as therapeutics. Protein-based treatments have shown to be beneficial in treating various cardiovascular, metabolic, and cancer diseases. The high specificity of protein therapeutics decreases their interference with normal biological processes and makes them less immunogenic. Despite these benefits, there are obstacles in administering these medications effectively compared to small molecule therapies delivered through oral or inhalation routes. When protein-based medicines are taken orally, they are degraded by enzymes in the Gastrointestinal tract (GI) tract and liver, lowering their therapeutic efficacy. The inhalation delivery avoids GI tract and liver breakdown; however, it still involves bypassing innate defense mechanisms of the lungs, such as

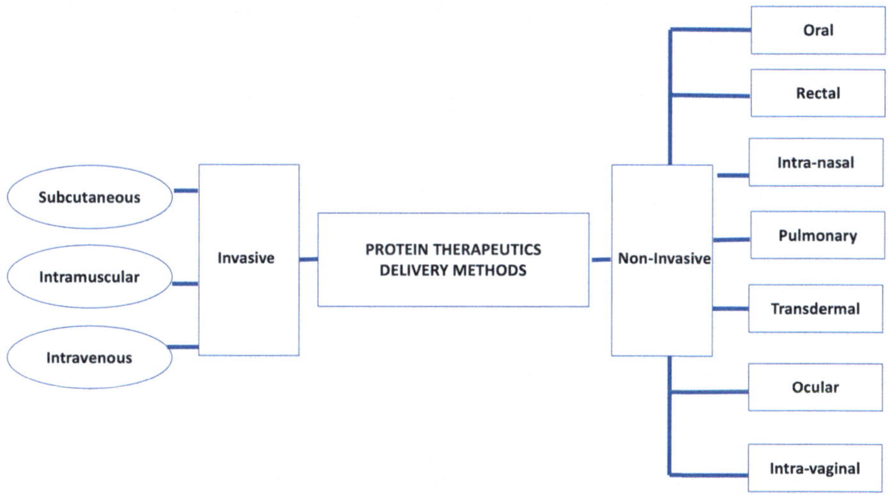

Fig. 1.4 Various methods of protein therapeutics delivery

macrophages and enzymes. Proteins are degraded less in the lungs than in the GI tract, yet protein therapeutics are too bulky to be absorbed in the lungs [42]. The various delivery mechanisms of protein therapeutics are shown in Fig. 1.4.

Due to the difficulties in oral and inhalation delivery, the injection has become the primary delivery route for protein therapies. The short biological half-lives, fragile structures, and specific physicochemical properties of therapeutic proteins affect their delivery. As a result, developing novel delivery strategies for protein-based treatments is in great demand, and researchers are exploring potential options. However, clinical acceptance sometimes becomes difficult due to their therapeutic efficacy and the targeted delivery, emphasizing the need for controlled-release delivery systems. The design of consistent delivery systems for the efficient administration of therapeutic proteins in steady and physiologically active forms has lately piqued interest [43].

1.5.1 Invasive Delivery Systems

When therapeutic protein is directly delivered into the blood, it is called the invasive delivery system. Invasive delivery routes usually include subcutaneous (SC), intramuscular (IM), and intravenous (IV). However, there are certain disadvantages of invasive delivery routes. These are (i) pain at the site of injection, (ii) possibility of infection at the site, (iii) hypersensitivity reactions, (iv) more time consuming with respect to expertise to inject and repeated visits to hospitals, and (v) complicating manufacturing process increasing the cost. Successful delivery of protein therapeutics via invasive mechanisms requires overcoming these disadvantages.

In recent years, various advancements have been made in the invasive delivery of therapeutic proteins. Chemical modification and/or colloidal carrier systems are the most commonly explored techniques [42]. The problems of instability and rapid clearance can be achieved by modifying the structure of protein therapeutics. PEGylation, which involves the covalent coupling of activated polyethylene glycol (PEG) with targeted proteins, has been intensively researched for the delivery of therapeutic proteins via invasive pathways [44]. PEG moieties are repeating units of ethylene glycol that are both neutral and amphiphilic. PEGylation also prevents protein degradation leading to decreased protein clearance [45]. Without biodegradation, intact PEG is excreted from the body. PEGylation can change a variety of physicochemical properties of the protein therapeutics, such as (i) increased size and molecular weight, (ii) modification in conformation, (iii) intermolecular interactions are hampered, (iv) enhanced hydrophilicity, (v) changes in binding properties, etc. All these significantly affect the pharmacological behavior of protein therapeutics. Because of a slower rate of renal removal and a reduction of proteolytic degradation and opsonization, PEGylation results in a longer circulation duration of conjugated molecules [46]. PEG is thought to improve the in vivo efficacy of protein therapeutics by regulating the pharmacokinetic and pharmacodynamic effects.

Glycosylation and mannosylation are other chemical changes that can be introduced into therapeutic proteins. Glycoengineering is the process of modifying the carbohydrates linked to the protein to change its pharmacokinetic properties. Invasive delivery of many therapeutic proteins has been examined using these methods. Proteins that have been glycosylated or mannosylated have better pharmacological characteristics than those that have been PEGylated [47].

Colloidal carrier systems are another area where therapeutic proteins have shown considerable promise, and one of the examples is lipid-mediated delivery. Liposomes, oily suspensions, lipid implants, submicron lipid emulsions, lipid microbubbles, solid lipid nanoparticles, lipid microtubules, inverse lipid micelles, and lipid microcylinders are examples of lipid-mediated delivery of protein therapeutics [48]. The key benefit of lipid drug delivery is the flexibility to build multiple kinds of lipid drug carriers based on the molecular architecture of the lipids employed in the formulation. Liposomes are bilayer vesicles formed by natural or artificial phospholipids [49].

1.5.2 Noninvasive Delivery Systems

Noninvasive delivery of therapeutic proteins has received much attention in recent years. However, due to their relatively high molecular weight and associated instability, efficient administration via noninvasive pathways is still a concern. There are many advantages of the oral route over invasive and other noninvasive delivery mechanisms of protein and peptide therapeutics; however, these therapeutics lose their bioavailability due to enzymatic degradation, pH variation, and barriers to GI tract and hepatic metabolism. Several methods have been established for increasing the systemic bioavailability of orally given therapeutic proteins. Some of the strategies are as follows:

(i) Cyclization of protein therapeutics structure prevents degradation by digestive enzymes in the GI tract.
(ii) PEGylation enhances the absorption and stability of proteins.
(iii) Protein lipidation also contributes to the stability and improved transportation of therapeutic proteins through the membranes of the GI tract.
(iv) Prodrug formation of therapeutic proteins, which are activated by some enzymatic degradation in the body; however, there are limited methodologies available for this procedure.
(v) Chemical modifications in protein structure.
(vi) Conjugation of therapeutic proteins with vitamins and transport carrier molecules enhances their absorption.
(vii) Coadministration of enzyme inhibitors along with protein therapeutics.
(viii) Coadministration of chemically inert, nontoxic, nonallergic absorption enhancers like chitosans, fatty acids, lectins, etc., reduces mucosal viscosity and causes a transient gap in the intestinal barrier and open gap junctions and other mechanisms that aid absorption. However, they also have certain side effects [42].

Besides these delivery approaches, a few other strategies are also used to effectively deliver protein therapeutics. Protein therapeutics have shown good absorption and stability when administered via mucosal, rectal, intranasal, pulmonary, transdermal, ocular, and intravaginal delivery systems. Pharmaceutical researchers have significant challenges in delivering therapeutic proteins to their target sites, limiting their application. Some of these challenges are listed in Table 1.3 [42]. Enzyme inhibitors, absorption stimulants, and structural alterations of therapeutic proteins have been used to boost the oral administration of therapeutic proteins [50]. Although some of these approaches are promising, delivering the requisite therapeutic protein

Table 1.3 Noninvasive route of administration of protein therapeutics and their challenges

S. No.	Route of administration	Challenges
1.	Ocular	Solution drainage Tear dilution Low tolerance
2.	Oral	pH variation Enzymatic degradation Less permeability
3.	Nasal	Small surface area Mucociliary clearance
4.	Pulmonary	Phagocytosis by macrophages Mucociliary clearance Necessity of complex device
5.	Transdermal	Less permeability Local allergic response
6.	Rectal	Variable absorption Interruption by defecation Low patient acceptability

to its site of action without causing significant adverse responses is still a task to complete [43].

1.6 Biosimilars

As defined by the FDA, biosimilars are products that are highly similar to the reference product without clinically significant differences in safety, purity, and potency. The definition of a biosimilar is based on three factors: (i) it must be a biological product, (ii) the reference product must be a previously approved biological product, and (iii) high resemblance in terms of safety, quality, and efficacy with the reference product must be substantiated. Biosimilars offer a way to reduce the cost of biological pharmaceuticals while also increasing patients' exposure to life-saving biological therapies. In contrast to generic small-molecule drugs, biosimilar development is more challenging due to fundamental distinctions between chemical drugs and biological products. Small molecule pharmaceuticals are manufactured using known chemical processes and assessed using established analytical methods, assuring that a generic version has the same active substance as the reference product. They are usually not immunogenic without coupling to a carrier protein due to their small size. On the other hand, biologics are proteins made from living sources such as bacteria, yeast, or mammalian cells and are significantly larger. Because of their size and complexity, biologics are recognized by the immune system and can cause a variety of immunological responses, some of which might have clinically severe repercussions, such as loss of efficacy, anaphylaxis, and infusion reactions. Biosimilars can be an alternate option for health-care institutions to control the rising cost of pharmaceuticals. However, the level of investigation required to evaluate biosimilar drugs for formulary inclusion is far more rigorous due to their intrinsic complexity and diversity [51].

1.7 Computational Resources for Therapeutic Proteins

Biological databases are collections of biological information obtained from various experimental studies. Based on the nature of biological databases, they are classified into (i) primary databases, (ii) secondary or derived databases, and (iii) composite databases (combination of both primary and secondary databases). Based on the content, biological databases can also be classified into several types, such as nucleotide sequence databases, protein sequence databases, protein structural databases, chemical structural databases, metabolic pathway databases, genome databases, SNP databases, databases of FDA-approved drug molecules, literature databases, gene expression databases, databases for mass spectroscopy data, databases for electrophoretic data, etc.

Several generalized protein-related databases are available for therapeutic proteins, such as NCBI-GenPept, UniProt, MIPS, Prosite, Blocks, Pfam, PDB, PDBSum, Alphafold, SCOP, CATH, String, etc. Figure 1.5 shows specialized

Fig. 1.5 Specialized therapeutic protein databases and tools

databases/tools mainly used to explore the sequence, structural, and functional aspects of therapeutic peptides and proteins. Moreover, the resources facilitate designing therapeutic proteins with various applications.

1.7.1 THBdB (Database of FDA-Approved Therapeutic Peptides and Proteins)

THBdB is a manually curated database of US FDA-approved therapeutic proteins and peptides (http://crdd.osdd.net/raghava/thpdb/). It has compiled information from ~1000 research articles, 70 patents, and other sources, particularly, DrugBank. As of 04/03/2022, the database has 852 items, including detailed information on 239 FDA-approved therapeutic proteins and peptides and their 380 drug variations. The database contains information on amino acid sequences, physio-chemical properties, amino acid residue composition, disease area, mode of action, pharmacological class, pharmacodynamics, administration route, toxicity, etc. Moreover, the structural information of therapeutic proteins and peptides is also available in this database.

1.7.2 TTDB (Therapeutic Target Database)

TTDB (http://db.idrblab.net/ttd) is a resource developed by the Innovative Drug Research and Bioinformatics Group (IDRB) at Zhejiang University, China, and the Bioinformatics and Drug Design Group at the National University of Singapore (NUS), Singapore. This database contains information on known and undiscovered therapeutic protein and nucleic acid targets, as well as the target disease, pathway, and drugs for each target. In addition, this database also provides comprehensive knowledge about target function, amino acid sequence, 3D structure, ligand binding site details, therapeutic class, clinical development status, etc.

1.7.3 SATPdb (Database of Structurally Annotated Therapeutic Peptides)

SATPdb (http://crdd.osdd.net/raghava/satpdb) is a collection of structurally and functionally annotated therapeutic peptides from 22 publicly available peptide databases and datasets. There are 19,192 experimentally verified therapeutic peptide sequences in the current edition (sequence length between 2 and 50 amino acid residues). These peptides have been organized into ten classes based on their primary function or therapeutic property.

1.7.4 CKTTD (Checkpoint Therapeutic Target Database)

CKTTD (http://www.ckttdb.org/) is the world's first comprehensive database of immunological checkpoint targets, including proteins, microRNAs, and long non-coding RNAs (LncRNAs), as well as their modulators. It could be used as a computational resource for cancer medication discovery and immunotherapy research.

1.7.5 HAPPENN (Novel Tool for Hemolytic Activity Prediction for Therapeutic Peptides)

HAPPENN (https://research.timmons.eu/happenn) is a novel web server for predicting hemolytic activity for therapeutic peptides and utilizes the concept of neural networks. This innovative classifier and web-server allow the research community to use it for computational screening of peptide drugs or lead candidates for therapeutics.

1.7.6 BioDADPep (Bioinformatics Database for Antidiabetic Peptides)

BioDADPep (http://omicsbase.com/BioDADPep/) is a web-based database of published literature about Type 1 and Type 2 diabetes mellitus peptides, their targets, and other relevant data.

1.8 Conclusions

Protein-based treatments have enjoyed much success in the clinic and are getting significant attention nowadays. Newly designed proteins, including selective mAbs and multispecific fusion proteins, mAbs coupled with small molecule medicines, and proteins with enhanced pharmacokinetics, etc., are being discovered rapidly. The small-molecule-based protein therapeutics cannot replace the precise and complicated set of functions; however, they have a low risk of interfering with regular biological processes and causing negative consequences. Protein-based therapeutics are frequently well tolerated and lower the risk of triggering immunological responses. However, along with many advantages, protein therapeutic approaches have certain limitations also. Therapeutic proteins can be used to treat various diseases and disorders, but some can trigger an immunological response, especially when given several doses over time. Also, the immunogenicity of proteins is influenced by several factors, including structural characteristics, preparation methods, contaminants or impurities in the preparation, dosing frequency, duration of therapy, mode of delivery, appropriate formulation, and patient genetic factors. Another vital area of research in protein therapeutics is the delivery mechanism. The stability of proteins in delivery devices and the design of appropriate target-specific protein carriers are two major research topics in protein delivery. For beneficial therapeutic results, many efforts have been made to deliver proteins/peptidic medicines through diverse modes of administration. In a nutshell, we introduced the structural and functional aspects of therapeutic proteins and their clinical applications. Additionally, we discussed key computational resources needed to investigate the sequence, structural, and functional features of therapeutic proteins and peptides to develop therapeutic proteins and peptides for various clinical purposes.

Acknowledgments Dr. Amit Kumar Singh thanks Indian Council of Medical Research (ICMR) and Indian National Science Academy (INSA), New Delhi, India. The authors thank Sharda University, Greater Noida, India, for support.

References

1. Whitford D (2013) Proteins: structure and function. John Wiley & Sons
2. Nelson DL, Lehninger AL, Cox MM (2008) Lehninger principles of biochemistry. Macmillan
3. Voet D, Voet JG (2010) Biochemistry, 4th edn. Wiley, Hoboken
4. Branden CI, Tooze J (2012) Introduction to protein structure. Garland Science
5. Tripathi, T. and Dubey, V.K. (2022) Advances in protein molecular and structural biology methods. Academic Press, USA [ISBN 978-032-39-0264-9]. pp 1–714. https://doi.org/10.1016/C2020-0-03170-1
6. Petersen B, Lundegaard C, Petersen TN (2010) NetTurnP–neural network prediction of beta-turns by use of evolutionary information and predicted protein sequence features. PLoS One 5(11):e15079
7. Fu H, Grimsley GR, Razvi A, Scholtz JM, Pace CN (2009) Increasing protein stability by improving beta-turns. Proteins Struct Funct Bioinf 77(3):491–498
8. Singh DB, Tripathi T (2020) Frontiers in protein structure, function, and dynamics. Springer, Singapore [ISBN 978-981-15-5529-9]. pp 1–452. https://doi.org/10.1007/978-981-15-5530-5
9. Rittle J, Field MJ, Green MT, Tezcan FA (2019) An efficient, step-economical strategy for the design of functional metalloproteins. Nat Chem 11(5):434–441
10. Snead D, Eliezer D (2019) Intrinsically disordered proteins in synaptic vesicle trafficking and release. J Biol Chem 294(10):3325–3342
11. Gligorijević N, Minić S, Robajac D, Nikolić M, Ćirković Veličković T, Nedić O (2019) Characterisation and the effects of bilirubin binding to human fibrinogen. Int J Biol Macromol 128:74–79
12. Janin J, Bahadur RP, Chakrabarti P (2008) Protein–protein interaction and quaternary structure. Q Rev Biophys 41(2):133–180
13. Møller N, Jørgensen JOL (2009) Effects of growth hormone on glucose, lipid, and protein metabolism in human subjects. Endocr Rev 30(2):152–177
14. Haglund K, Dikic I (2005) Ubiquitylation and cell signaling. EMBO J 24(19):3353–3359
15. Eide DJ (2004) The SLC39 family of metal ion transporters. Pflugers Arch 447(5):796–800
16. Dimitrov DS (2012) Therapeutic proteins. Methods Mol Biol (Clifton, N.J.) 899:1–26. https://doi.org/10.1007/978-1-61779-921-1_1
17. Leader B, Baca QJ, Golan DE (2008) Protein therapeutics: a summary and pharmacological classification. Nat Rev Drug Discov 7(1):21–39. https://doi.org/10.1038/nrd2399
18. Johnson-Léger C, Power CA, Shomade G, Shaw JP, Proudfoot AE (2006) Protein therapeutics--lessons learned and a view of the future. Expert Opin Biol Ther 6(1):1–7. https://doi.org/10.1517/14712598.6.1.1

19. Cheng A, Patel DK, Reid TS, Wyne K (2019) Differentiating basal insulin preparations: understanding how they work explains why they are different. Adv Ther 36(5):1018–1030. https://doi.org/10.1007/s12325-019-00925-6
20. Miller BS, Velazquez E, Yuen K (2020) Long-acting growth hormone preparations - current status and future considerations. J Clin Endocrinol Metab 105(6):e2121–e2133. https://doi.org/10.1210/clinem/dgz149
21. Corwin HL, Gettinger A, Pearl RG, Fink MP, Levy MM, Shapiro MJ, Corwin MJ, Colton T, EPO Critical Care Trials Group (2002) Efficacy of recombinant human erythropoietin in critically ill patients: a randomized controlled trial. JAMA 288(22):2827–2835. https://doi.org/10.1001/jama.288.22.2827
22. Van Wely M, Westergaard LG, Bossuyt PM, Van der Veen F (2003) Human menopausal gonadotropin versus recombinant follicle stimulation hormone for ovarian stimulation in assisted reproductive cycles. Cochrane Database Syst Rev 1:CD003973. https://doi.org/10.1002/14651858.CD003973
23. Fuchs HJ, Borowitz DS, Christiansen DH, Morris EM, Nash ML, Ramsey BW, Rosenstein BJ, Smith AL, Wohl ME (1994) Effect of aerosolized recombinant human DNase on exacerbations of respiratory symptoms and on pulmonary function in patients with cystic fibrosis. *The Pulmozyme Study Group*. N Engl J Med 331(10):637–642. https://doi.org/10.1056/NEJM199409083311003
24. Waycaster C, Carter MJ, Gilligan AM, Mearns ES, Fife CE, Milne CT (2018) Comparative cost and clinical effectiveness of clostridial collagenase ointment for chronic dermal ulcers. J Comp Eff Res 7(2):149–165. https://doi.org/10.2217/cer-2017-0066
25. Yoshida K (2021) Effects of botulinum toxin type A on pain among trigeminal neuralgia, myofascial temporomandibular disorders, and oromandibular dystonia. Toxins 13(9):605. https://doi.org/10.3390/toxins13090605
26. Lagassé HA, Alexaki A, Simhadri VL, Katagiri NH, Jankowski W, Sauna ZE, Kimchi-Sarfaty C (2017) Recent advances in (therapeutic protein) drug development. F1000Research 6:113. https://doi.org/10.12688/f1000research.9970.1
27. Castelli MS, McGonigle P, Hornby PJ (2019) The pharmacology and therapeutic applications of monoclonal antibodies. Pharmacol Res Perspect 7(6):e00535. https://doi.org/10.1002/prp2.535
28. Scott AM, Allison JP, Wolchok JD (2012) Monoclonal antibodies in cancer therapy. Cancer Immun 12:14
29. Flego M, Ascione A, Cianfriglia M, Vella S (2013) Clinical development of monoclonal antibody-based drugs in HIV and HCV diseases. BMC Med 11:4. https://doi.org/10.1186/1741-7015-11-4
30. Weiner GJ (2015) Building better monoclonal antibody-based therapeutics. Nat Rev Cancer 15(6):361–370. https://doi.org/10.1038/nrc3930
31. Kalita P, Tripathi T (2022) Methodological advancements in the design of peptide-based vaccines. Drug Discov Today 27(5):1367–1380
32. Campos-Neto A, Rodrigues-Júnior V, Pedral-Sampaio DB, Netto EM, Ovendale PJ, Coler RN, Skeiky YA, Badaró R, Reed SG (2001) Evaluation of DPPD, a single recombinant Mycobacterium tuberculosis protein as an alternative antigen for the Mantoux test. Tuberculosis (Edinb) 81(5–6):353–358. https://doi.org/10.1054/tube.2001.0311
33. Jowell PS, Robuck-Mangum G, Mergener K, Branch MS, Purich ED, Fein SH (2000) A double-blind, randomized, dose response study testing the pharmacological efficacy of synthetic porcine secretin. Aliment Pharmacol Ther 14(12):1679–1684. https://doi.org/10.1046/j.1365-2036.2000.00881.x
34. Nag N, Khan H, Tripathi T (2022) Strategies to improve the expression and solubility of recombinant proteins in *E. coli*. In: Tripathi T, Dubey VK (eds) Advances in protein molecular and structural biology methods. Academic Press, [ISBN: 978-032-39-0264-9]. pp. 1–12. https://doi.org/10.1016/B978-0-323-90264-9.00001-5

35. Tripathi T (2013) Calculation of thermodynamic parameters of protein unfolding using far-ultraviolet circular dichroism: A review. J Protein Proteomics 4(2):85–91
36. Nail SL, Jiang S, Chongprasert S, Knopp SA (2002) Fundamentals of freeze-drying. Pharm Biotechnol 14:281–360. https://doi.org/10.1007/978-1-4615-0549-5_6
37. Narasimhan C, Mach H, Shameem M (2012) High-dose monoclonal antibodies via the subcutaneous route: challenges and technical solutions, an industry perspective. Ther Deliv 3(7): 889–900. https://doi.org/10.4155/tde.12.68
38. Jenkins N, Curling EM (1994) Glycosylation of recombinant proteins: problems and prospects. Enzym Microb Technol 16(5):354–364. https://doi.org/10.1016/0141-0229(94)90149-x
39. Patten PA, Schellekens H (2003) The immunogenicity of biopharmaceuticals. Lessons learned and consequences for protein drug development. Dev Biol 112:81–97
40. Sinha R, Shukla P (2019) Current trends in protein engineering: updates and progress. Curr Protein Pept Sci 20(5):398–407. https://doi.org/10.2174/1389203720666181119120120
41. Lutz S, Iamurri SM (2018) Protein engineering: past, present, and future. Methods Mol Biol (Clifton, N.J.) 1685:1–12 https://doi.org/10.1007/978-1-4939-7366-8_1
42. Rehman K, Hamid Akash MS, Akhtar B, Tariq M, Mahmood A, Ibrahim M (2016) Delivery of therapeutic proteins: challenges and strategies. Curr Drug Targets 17(10):1172–1188. https://doi.org/10.2174/1389450117666151209120139
43. Pisal DS, Kosloski MP, Balu-Iyer SV (2010) Delivery of therapeutic proteins. J Pharm Sci 99(6):2557–2575. https://doi.org/10.1002/jps.22054
44. Roberts MJ, Harris JM (1998) Attachment of degradable poly(ethylene glycol) to proteins has the potential to increase therapeutic efficacy. J Pharm Sci 87(11):1440–1445. https://doi.org/10.1021/js9800634
45. Caliceti P, Veronese FM (2003) Pharmacokinetic and biodistribution properties of poly(ethylene glycol)-protein conjugates. Adv Drug Deliv Rev 55(10):1261–1277. https://doi.org/10.1016/s0169-409x(03)00108-x
46. Delgado C, Francis GE, Fisher D (1992) The uses and properties of PEG-linked proteins. Crit Rev Ther Drug Carrier Syst 9(3–4):249–304
47. Sinclair AM, Elliott S (2005) Glycoengineering: the effect of glycosylation on the properties of therapeutic proteins. J Pharm Sci 94(8):1626–1635. https://doi.org/10.1002/jps.20319
48. Rawat M, Singh D, Saraf S, Saraf S (2008) Lipid carriers: a versatile delivery vehicle for proteins and peptides. Yakugaku Zasshi: J Pharm Soc Jpn 128(2):269–280. https://doi.org/10.1248/yakushi.128.269
49. Chonn A, Cullis PR (1995) Recent advances in liposomal drug-delivery systems. Curr Opin Biotechnol 6(6):698–708. https://doi.org/10.1016/0958-1669(95)80115-4
50. Jitendra, Sharma PK, Bansal S, Banik A (2011) Non-invasive routes of proteins and peptides drug delivery. Indian J Pharm Sci 73(4):367–375. https://doi.org/10.4103/0250-474X.95608
51. Lucio SD, Stevenson JG, Hoffman JM (2013) Biosimilars: implications for health-system pharmacists. Am J Health Syst Pharm 70(22):2004–2017. https://doi.org/10.2146/ajhp130119

Clinical Applications of Protein-Based Therapeutics

2

Manoj Kumar Yadav, Ankita Sahu, Anu, Nehaa Kasturria, Anjali Priyadarshini, Archana Gupta, Kanika Gupta, and Anil Kumar Tomar

Abstract

Regarded as highly successful in their clinical applications, therapeutic proteins are now widely applied for the precise treatment of several diseases. Common forms of therapeutic proteins include enzymes, antibodies, and recombinant proteins. Here, we discuss different aspects of the clinical applications of protein-based therapeutics, including therapeutic proteins, their mechanisms and metabolism, challenges, precision medicine, and computer-aided drug designing. In addition, an overview of recently approved therapeutic proteins is provided. Conclusively, this chapter delivers comprehensive information on clinical applications of protein-based therapeutics, emerging trends, and challenges.

M. K. Yadav (✉) · Anu
Department of Biomedical Engineering, SRM University, Delhi-NCR, Rajiv Gandhi Education City, Sonepat, Haryana, India
e-mail: manoj.yadav@srmuniversity.ac.in

A. Sahu
Tumor Biology, ICMR-National Institute of Pathology, New Delhi, India

N. Kasturria · A. Priyadarshini
Department of Biotechnology, SRM University, Delhi-NCR, Rajiv Gandhi Education City, Sonepat, Haryana, India

A. Gupta
Department of Microbiology, SRM University, Delhi-NCR, Rajiv Gandhi Education City, Sonepat, Haryana, India

K. Gupta
Department of Computer Science, Jamia Millia Islamia, New Delhi, India

A. K. Tomar
Department of Biophysics, All India Institute of Medical Sciences, New Delhi, India

D. B. Singh, T. Tripathi (eds.), *Protein-based Therapeutics*,
https://doi.org/10.1007/978-981-19-8249-1_2

Keywords

Therapeutic proteins · Precision medicine · Computer-aided drug designing · Enzymes · Antibodies · Recombinant proteins

2.1 Introduction

Proteins, undoubtedly one of the most versatile macromolecules, have been extensively studied for their widespread biomedical and therapeutic applications [1]. Due to their properties like natural origin, biodegradability, biocompatibility, recognition by cells, reduced immunogenic potential, and natural bioactivity, therapeutic proteins offer several advantages compared to synthetic therapeutic molecules. In addition, they are easy to functionalize and engineered for specific locations or applications through alteration of their primary amino acid sequences. Presently, proteins form a dominating segment of the pharmaceutical industry as they have tremendous therapeutic potential against various diseases and syndromes. The demand for protein-based medicines has significantly augmented, mainly due to increased medical awareness and the prevalence of chronic diseases. From $140 billion in 2016, the global protein therapeutics market is projected to reach $566 billion by 2030 [2]. However, the high costs of therapeutic proteins and stringent government regulations remain significant challenges that negatively impact market growth.

Based on their molecular types, therapeutic proteins are classified into several types, such as antibodies, enzymes, fusion proteins, recombinant proteins, blood factors, anticoagulants, growth factors, interferon, hormones, etc. They act differently on biological or drug targets. Some common pharmacological activities of therapeutic proteins are replacing deficient or abnormal proteins, interfering with molecules or pathways, and delivering other molecules [3]. Protein-based therapeutics have evolved a lot with technological advances in the fields of drug discovery and protein engineering. For customized drug designing, the most critical aspects of protein-based molecules are understanding their mechanism of action and the structure-function relationship. Continuous improvements in traditionally existing therapies and methods to identify drug targets have resulted in developing drugs with better efficacy and targeted clinical applications [4]. This chapter provides a comprehensive overview of protein-based therapeutics, their mechanism, clinical applications, and challenges. A detailed illustration is provided about enzymes and antibodies as therapeutic proteins, followed by the introduction, applications, and prospects of precision medicine, an emerging, highly innovative, and targeted medicine approach that looks into an individual's genetics, environment, and lifestyle. The discovery and development of drugs consist of very complex and time-consuming processes. However, recent decades have seen much growth in this field due to the application of computer-aided drug design (CADD). Thus, we also briefly discuss CADD approaches in drug designing and their application in developing protein therapeutics. The chapter concludes with a discussion of recently approved therapeutic proteins, emerging trends, challenges, and opportunities in the field, followed by the safety and efficacy of protein-based drugs.

2.2 Enzyme as Biologics

In contrast to regular therapeutics, biologics are comparatively larger molecules having higher molecular weights. Most of these molecules are unstable at room temperature and require refrigeration for storage. Also, biologics are produced by complex processes, and at times, even slight changes in their formulation might lead to degradation of their efficacy in disease management. It is nearly impossible to produce an exact copy of biologics, and thus, nearly similar biologics, compared to the original one, are manufactured that are referred to as biosimilars.

In 1878, a German physiologist Wilhelm Kühne coined the word "enzyme" [5, 6]. The enzymes are responsible for biological catalysis and are also called biocatalysts. Biocatalysis is a remarkable property of enzymes to speed up the specific biological reaction in living organisms. The study of enzyme kinetics provides information about a diverse range of reactions, metabolism, cell regulation, and how poisons and drugs affect the enzymes [7, 8]. The first enzyme discovered was diastase (a mixture of amylases), which catalyzes the hydrolysis of starch into maltose. It has a wide range of clinical, food, forensic, biochemical, medicinal, pharmaceutical, and environmental applications [7]. Almost all enzymes are proteins, and a functional enzyme has different components, such as holoenzyme (functional unit of enzyme and conjugated protein), apoenzyme (polypeptide segment of the enzyme and inactive precursor), coenzyme (small organic moiety), or zymogens (simple protein enzymes, which are secreted in an inactive form). Enzymes are classified into six functional classes that catalyze a specific reaction, that is, oxidoreductases, transferases, hydrolases, lyases, isomerases, and ligases/ lyases [9]. This classification enables the identification and separation of diverse chemical reactions in living organisms. The functionality of an enzyme is intrinsically linked to its three-dimensional (3D) structure and is determined by the shape when it binds to the substrate that creates an ideal fit for catalysis.

2.2.1 Biological Process: How Enzymes Work?

A large number of biochemical reactions occur in the human body to carry out essential metabolic processes. Thousands of enzymes produced in the human body help to accelerate metabolism, growth, digestion, building muscle, healing, destroying toxins, reproduction, liver function, nerve function, and so on [8, 10]. The functions of enzymes are strongly affected by pH and temperature [7, 11]. The lock and key model related to substrate-enzyme interaction was postulated by Emil Fischer. The key signifies a substrate, and the lock and keyhole represent the enzyme and its active site. The shape and size of the substrate are complementary to the active site of the enzyme [5, 7]. The substrate perfectly binds at the active site and forms the enzyme-substrate complex, allowing rapid biochemical reaction.

2.2.2 Therapeutic Enzymes

The concept of therapeutic enzymes has been exploited for several decades [10]. Some enzymes are the preferred markers of various diseases, including cancer, infectious disease, myocardial infarction, clotting, pancreatitis, inherited diseases, neurodegenerative disorders, etc. These markers help in disease management via the diagnosis, prognosis, and assessment of responses to a therapeutic intervention [6]. Effective therapeutic solutions via various enzymes are helpful for the treatment of several diseases. The wide variety of uses of therapeutic enzymes is depicted in Table 2.1 and Fig. 2.1.

2.3 Antibodies as Biologics

Antibodies are the proteins produced to bind a specific antigen. The main task of antibodies is to get associated with the specific substance chemically considered alien by our body, such as bacteria, viruses, and other foreign substances in the blood, to neutralize them. Biochemically, antibodies are immunoglobulins, protective proteins produced by our immune system when it recognizes the presence of any foreign substance, commonly referred to as an antigen.

Biologics have the potential to mount an immune response against them, reducing their efficacy, and sometimes, it becomes life-threatening due to the generation of antibiologics antibodies. To compensate, other cotherapy options are used to treat a particular disease. It is always advisable to keep track of antibiologics antibodies in a patient's blood using therapeutic drug monitoring. It is one of the most prominent techniques that detects the presence of biologics and antibiologics antibodies in the blood [12]. If antibiologics antibodies are present above a permissible limit and the amount of biologics is insufficient to reduce the inflammation, immune-modulators are introduced into the treatment regimen [13]. Antibody-based biologics can be subdivided into three major categories: monoclonal antibody (mAb) products, non-mAb products, and vaccines [14]. The working mechanism of vaccines relies solely on mounting immune response and is mainly used for only prophylactic purposes rather than therapeutic uses and, thus, not elaborated here.

2.3.1 Monoclonal Antibody Products as Biologics

Monoclonal antibodies (mAbs) are immunoglobulin G (IgG) that imitate the natural IgG function within the body. Their role is to bind to the foreign particles to neutralize them. Fc Fusion proteins (FcFPs), consisting of the Fc receptor of the IgG, can also bind to a modified protein. Like natural IgGs, mAbs, and FcFPs bind to

Table 2.1 Enzymes and their therapeutic uses

Enzyme name	Therapeutic uses
Enzymes for the treatment of infectious diseases	
Lysozyme	Treatment of HIV infection, Rainbow trout, Barrett's oesophagitis, Chronic gastritis, Coeliac disease, Lymphocytic colitis, and Crohn's colitis
Chitinases	Allosamidin
Enzymes for the treatment of inherited diseases	
Alteplase; recombinant human tissue plasminogen activator	Used for the treatment of heart attacks
Pegademase bovine	For enzyme replacement therapy in severe combined immunodeficiency disease (SCID), caused by the chronic deficiency of ADA.
Imiglucerase Velaglucerase Taliglucerase	For replacement therapy in patients with Gaucher's disease type I, a lysosomal storage disease (LSD)
Phenylalanine hydroxylase (PAH)	Phenylketonuria (PKU)
Imiglucerase	Replacement therapy in patients with types I, II, and III Gaucher's disease
Sacrosidase	Treatment of congenital sucrase-isomaltase deficiency
Agalsidase-α Agalsidase-β	Treatment of Fabry disease
Cancer treatment	
PEGylated arginine deaminase, an arginine-degrading enzyme	Treatment of human melanoma and hepatocellular carcinomas
Rasburicase	Treatment of malignancy-associated or chemotherapy-induced hyperuricemia
Topoisomerase IIα	Breast cancer treatment
ASNase	Treatment of breast, rectal, acute lymphoblastic leukemia, and colon cancer
Other treatments	
Streptokinase (a nonenzymatic protein)	Treatment of thromboembolic diseases and heart attacks
Urokinase (UK)	Treatment of thrombotic disorders
L-asparaginase	Hodgkin's disease and melanosarcoma
Amylase, lipase, and protease	Diabetes treatment
Dornase α	Improves lung function in patients with cystic fibrosis (CF)
Agalsidase β	Treatment of Fabry's disease
Nattokinase	Treatment of cardiovascular diseases
Glutenase Prolyl endopeptidases (PEPs)	Treatment of celiac disease
Chymotrypsin	Treatment of pain relief and swelling
α-amylase	Treatment of type 2 diabetes mellitus
Collagenase	Treatment of Dupuytren's disease (DD)

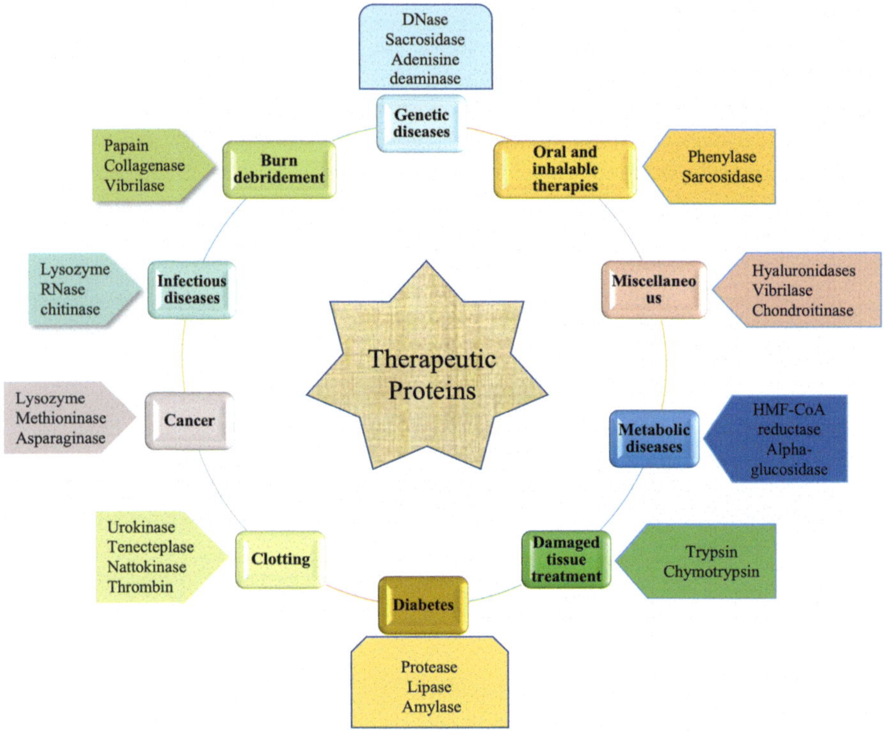

Fig. 2.1 Therapeutic enzymes used in the treatment of different diseases

extracellular targets, cells, or pathogens to neutralize them by disrupting their functions and removing them from circulation or modulating or imitating their activity. For example, inflammatory cytokines such as tumor necrosis factor-α (TNF-α) or interleukin-1b (IL-1b) neutralize the infected cells and result in immunosuppression [15]. The mAbs are usually derived from mice and rats and humanized to various degrees by engineering amino acid substitutions that make them similar to the human gene sequence through recombinant DNA (rDNA) technologies.

2.3.1.1 Biological Characteristics of mAbs

The mAbs are monospecific antibodies made from identical clones of a unique parent cell [16]. The essential biological characteristics of mAbs are listed below.

1. They show edacity effects against target cells.
2. They have the ability to obstruct protein-protein interactions with different targets like the serum, extracellular, and membrane-bound proteins.
3. They can mediate multiple processes like antibody-dependent cellular cytotoxicity (ADCC), complement-dependent cytotoxicity (CDC), antibody-dependent

cellular phagocytosis (ADCP), antibody-mediated immune complex formation with clearance, or a completely silent Fc with none of the above activities [17].
4. They can penetrate certain tissues.
5. The mAbs have minimal drug-drug interactions.
6. They show little or no non-mechanism-mediated toxicity.
7. They also have little or no off-target activity or drug-metabolism-related issues.

2.3.1.2 Applications of mAbs

Modifications in mAbs, such as Fab fragments, and bifunctional antibodies, are used to produce various biologics to treat a variety of diseases.

Fab Fragments

Fab fragments are the single binding site for the antigen. The important clinical applications of Fab fragments are as follows:

1. ***Caplacizumab*** is a humanized, bivalent, variable-domain-only fragment with a high affinity for the von Willebrand factor (VWF). The interaction between VWF and platelet plays a central role in microvascular thromboses in patients with thrombotic thrombocytopenic purpura (TTP). Caplacizumab disrupts the interactions between VWF multimers and platelets and is used to treat acquired TTP conditions [18].
2. ***Ranibizumab*** is a recombinant humanized Fab fragment that binds to and inhibits the human vascular endothelial growth factor A (VEGF-A) [19]. It inhibits the binding of VEGF to its receptors and slows down the related vision loss, and is used in treating age-related macular degeneration.
3. ***Abciximab*** is a Fab antibody fragment derived from a chimeric human-murine mAb (7E3) that binds to platelet IIb/IIIa receptors, resulting in steric hindrance and thus inhibiting the platelet aggregation [20]. Abciximab has been used in unstable angina and reduction of thrombosis in various coronary stenting procedures.

Bifunctional Antibodies

Bifunctional antibodies are antibodies with dual specificity. Both the immunoglobin chains are fused together to form a single antibody molecule. A few examples of bifunctional antibodies are as follows:

1. ***Emicizumab*** binds to two coagulation factors (factor IXa and factor X), taking the place of activated factor VIII (factor VIIIa) in the coagulation cascade [21]. The mAb is used for the prophylaxis of hemophilia patients.
2. ***Blinatumomab*** is a bispecific T cell and B cell engager molecule that binds to the cell surface proteins, CD3 present on T cells and CD19, present on precursor B-cell acute lymphoblastic leukemia (ALL) cells, and takes the site of cytotoxic T cells to recognize malignant B cells [22].
3. ***Catumaxomab*** is a bispecific trifunctional antibody that binds to the T-cell surface molecule CD3 and epithelial cell adhesion molecule (EpCAM), a tumor

cell surface marker, and receptor Fc region on dendritic cells [23]. This combination of antigen-binding helps in maintaining antitumor immune responses.

2.3.2 Non-Monoclonal Antibody Products as Biologics

The function of some natural proteins, enzymes, hormones, or peptides is disrupted in a healthy individual, resulting in the rise of physiological-function-related diseases. To counter this, non-monoclonal antibody (non-mAb) products with similar physiological effects are given to the patients [24]. Non-mAb therapy helps patients to recover by filling the physiological gaps. In general, the molecular weight of non-MAb products is more than 700 Da. These products are usually homogenous in nature and can be heterogenous only if they are glycosylated. Non-mAb products may include hormones, enzymes, interferons, interleukins, growth factors, or even natural or mimetic peptides [25]. The recent advancements in rDNA technology lead to the development of many hormones and non-MAbs under the biologics category to treat various diseases (Table 2.2).

2.4 Precision Medicine

The application of medicines is patient specific for treating a particular disease. It works better in certain patients compared to others, but the reason for this differential effectiveness was unknown a few decades back. Some patients face severe side effects, and others have fewer adverse effects when treated with anticancer drugs [36]. Over the past six decades, evidence has emerged indicating that a substantial portion of the variability in drug response is genetically determined as age, nutrition, health status, environmental exposure, and epigenetic factors play critical contributory roles [37]. The unique genetic constitution and differential gene expression in an individual is responsible for variation in drug responses. Precision medicine is an emerging practice that uses an individual's genetic profile to guide decisions regarding disease prevention, diagnosis, and treatment [38, 39]. The genetic profiling of every patient is necessary before treating them with a particular medication to increase the treatment efficacy with fewer side effects. One can take guidance from the knowledge of an individual's genome profile to preselect the treatment protocols that minimize adverse side effects or ensure more successful outcomes. After completing the human genome project, many advancements have occurred in the field of precision medicine. The individual genomic sequence data can indicate their susceptibility to certain diseases before they manifest, allowing physicians and patients to design a plan for monitoring and prevention [40]. The science of studying how the genetic variations affect drug responses in an individual is pharmacogenomics, an evolving field to better understand an individual's responses to different treatments.

Table 2.2 Some commercially available non-mAb biologics and their uses

Non-MAb biologics	Trade names	Mode of action	Conditions treated	Ref.
Insulin	Novolin R (Novo Nordisk) Humulin R (Eli Lilly, Indiana, USA)	Regulates carbohydrate and fat metabolism	Diabetes mellitus	[26]
Insulin-like growth factor 1	Increlex (Tercica/Ipsen, California, USA)	Stimulates growth in response to GH	Laron dwarfism	[27]
Growth hormone	Nutropin (Genentech), Genotropin (Pfizer, New York, USA)	Stimulates growth and cell reproduction and regeneration	Idiopathic short stature	[28]
Erythropoietin	Erythropoietin	Stimulates red blood cell production	Anemia	[29]
Granulocyte-colony-stimulating factor	Neupogen (Amgen, California, USA)	Stimulates granulocyte and stem cell production	Used to accelerate recovery after chemotherapy	[30]
Interferon α-2b	Intron-A (Schering-Plough, New Jersey, USA)	Antiviral activity	Hepatitis C, Hepatitis B	[31]
Interferon β-1a	Avonex (Biogen Idec, USA)	Anti-inflammatory; improves the integrity of the blood-brain barrier	Multiple sclerosis	[32]
Interferon β-1b	Betaferon (Bayer healthcare), Extavia (Novartis, Basal, Switzerland)	Anti-inflammatory; improves the integrity of the blood-brain barrier	Multiple sclerosis	[33]
Interleukin 2	Proleukin (Prometheus Laboratories Inc., California, USA)	Mediates the immune response	Malignant melanoma, renal cell cancer	[34]
Interleukin 11	Neumega (Pfizer, New York, USA)	Stimulates platelet production	Used to accelerate recovery after chemotherapy	[35]

2.4.1 Benefits of Precision Medicine

Though the current use of precision medicine is limited, it has the potential to offer a wide range of applications in the coming years.

1. ***Better medication selection:*** The adverse reactions to medicines, one of the drawbacks during treatment, leads to the death of many individuals. Although the Food and Drug Administration (FDA) approved drugs have to qualify the stringent parameters before coming to the market, there is either less or no information related to their response when given to certain individuals. The

medication may appear safe for a large population; however, some patients may experience harmful side effects due to genetic variations. The study of pharmacogenomics may help to predict how a particular group of individuals will be able to respond to one specific medication and can play a deciding role in its selection.

2. ***Safer dosing options:*** Following FDA approval, the standard dosage of a medication is decided based on factors such as liver or kidney function, weight, age, etc. However, these parameters might not be sufficient. Standard dosage may work well for one group of individuals but may be toxic for another group because of underlying genetic variations. Currently, clinicians generally decide which medication is appropriate for treating a particular disease based on their diagnosis. Once the field of pharmacogenomics matures, clinicians can directly consider an individual's genetic profile to decide the optimal medication dosage.
3. ***Improvements in drug development:*** Pharmaceutical companies often spend years conducting research and clinical trials of a new drug before it reaches the market. Diagnostic and device firms and pharmaceutical companies typically have to test a product in a large cohort to ensure its safety and efficacy. The study of pharmacogenomics may help these companies to ensure the efficacy of drug testing. For example, if a company has an advanced idea that the drug can optimally work in participants with a particular type of genetic variations, and may cause adverse reactions to others, then those participants having adverse reaction can be excluded from the clinical trials. This will speed up the whole clinical trial process, and a specific population can be treated with the same medication.

2.4.2 Applications of Precision Medicine

A promising application of precision medicine lies in the discovery and manipulation of potential drug targets for the treatment of cancer. Precision medicine is used to treat chronic myeloid leukemia (CML). However, the discovery of molecular predispositions, that is, the presence of genetic variants, in various diseases, such as CML, has made it possible to design and develop specific therapeutic agents against novel molecular targets. With this applicability, precision medicine has identified a novel molecular target, Bcr/Abl tyrosine kinase. This kinase is an oncoprotein expressed in more than 95% of CML patients, and administration of a competitive inhibitor helps to achieve almost 80% cytogenetic responses in newly diagnosed CML patients [40].

In 2017, FDA approved more number of precision medicines. One of the drugs approved was pembrolizumab [41], which was marked as the first robust cancer therapy approved for clinical use based on a specific biomarker rather than a tumor's location. Similarly, trastuzumab-dskt (Ogivri™) was approved as the first biosimilar agent that targets both stomach and breast tumors overexpressing the HER2 gene, facilitating competition and lowering healthcare costs [42]. Since these drugs are developed based on specific biomarkers, a need of companion genetic tests is needed to identify all the biomarkers. MSK-IMPACT™ (screens 468 genes) and Foundation One CdX™ (screens 324 genes) assays are examples of companion genetic tests. Both are solid tumor tests and massively parallel sequencing *in vitro* diagnostic tests [43]. These tests allow screening of multiple oncogenes to identify variants that might assist in the clinical management of the patients.

2.4.3 Future Prospects of Precision Medicine

In the past, physicians practiced intuition-based diagnosis and used their knowledge to provide medicine for the treatment of diseases. In the present time, clinicians rely on evidence-based diagnosis and treatment. They recommend the medicine based on evidence produced by scientific research, including clinical trials. In the future, precision medicines will be used according to algorithms that will consider the comprehensive information of an individual patient, including their genome, epigenetics, and lifestyle. Therefore, medicine in the twenty-first century must focus on attaining the four *P*'s: *prediction*, *prevention*, *personalization*, and *participation,* as stated by Dr. Leroy E. Hood [44]. Currently, patients are treated based on symptoms and diagnosis, which requires a transformation using precision medicine where the treatment is planned using the genetic profile of an individual. This evolution of medical treatment in the past, present, and future is summarized in Fig. 2.2.

2.5 Computer-Aided Drug Design

The discovery and development of new therapeutics is a complex and time-consuming process requiring much experimentation and research. Traditionally, a drug discovery takes an average of 10 to 15 years before it reaches the market for sale, with an estimated cost of 58.8 billion USD in 2015 [45, 46]. The high investment cost and failure rate of traditional methods prompted a need to utilize computational methods to aid drug discovery. Computer-aided drug design (CADD)

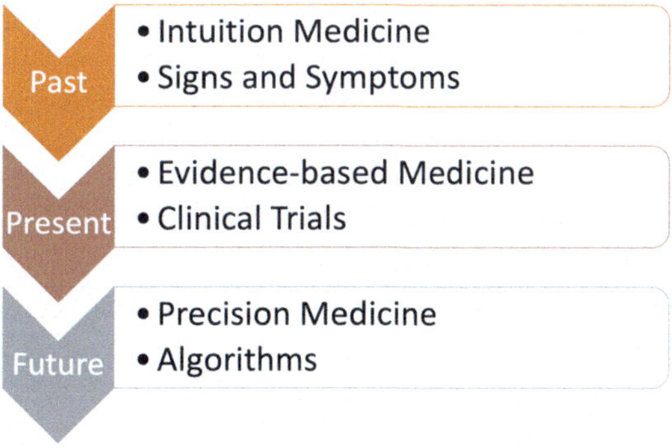

Fig. 2.2 Evolution in the field of medicine

refers to the use of computational tools and available data resources for designing, storing, analyzing, and modeling lead molecules to establish them as candidate drugs. This will facilitate studying chemical and biological interactions between the lead compounds and their biological targets. The technique systematically evaluates the potential lead candidates before their synthesis and *in vitro* and *in vivo* testing [47]. The computational resources comprise a screening process to select the best possible candidates and later estimate their physicochemical properties, such as absorption, distribution, metabolism, excretion, and toxicity (ADMET). CADD is routinely applied to discover and improve the quality of identified lead compounds quickly. Nowadays, the different applications of CADD techniques are used to speed up the drug design and discovery process.

2.5.1 Approaches of CADD in Designing Protein-Based Therapeutics

CADD strategies rely on the accessibility and availability of the 3D structure information of biological target and candidate molecules. This technique can be broadly divided into structure-based drug design (SBDD) and ligand-based drug design (LBDD), based on the information available for a protein receptor and ligands, respectively. The availability of the 3D structure of a receptor leads to the implementation of structure-based drug design methods. If only the ligand information is known, then ligand-based drug design can be adopted.

2.5.1.1 Structure-Based Drug Design Approach

This approach can only be used in the drug design process if the structure information of the protein receptor target is available. One can identify the active site and analyze the key amino acid residues responsible for its biological functions using the 3D structure. This information can then be used to create protein-based therapeutics that can outcompete the natural ligands, thereby interfering with the biological pathways to prevent the disease. The foundation for structure-based drug design was laid by the easy identification of binding cavities due to the availability of 3D structures of a large number of therapeutically important proteins. It is a precise, efficient, and rapid process, because it involves the 3D structure of a protein and knowledge about the disease at the molecular level [48]. SBDD is a multicycle process that leads to the development of potential lead candidates for clinical trials. The most notable success story involves FDA-approved drugs that inhibit the human immunodeficiency virus (HIV)-1, such as amprenavir, an inhibitor of HIV protease discovered through protein structure modeling and molecular dynamics (MD) simulations [49].

Along with success stories, some failures have also been reported. For example, RPX00023 was claimed to be an antidepressant with agonistic activity toward receptor 5-HT1A, but it behaved as an inhibitor of the receptor. Such failures highlight the limitations of SBDD strategies. To overcome these limitations, continuous improvements and developments have been incorporated into SBDD techniques, but still, consistent solutions need to be developed.

2.5.1.2 Ligand-Based Drug Design Approach

Ligand-based drug design (LBDD) is another method where information on ligand molecules is essential to use on previously unknown drug targets. LBDD methods are used when the experimental 3D structure of a receptor is not available. The structural and physicochemical properties of the known ligands that bind to the known drug target are analyzed to study their desired pharmacological activity [50]. The relationship between physicochemical properties and drug activity is known as a structure-activity relationship (SAR), which can be used to optimize known drugs or help design new drugs with improved activity [51]. LBDD methods also include substrate analogues that interact with the target molecule to produce the desired pharmacological effect.

The preparation of small-molecule libraries is the initial step of LBDD, where chemical structures of different compounds are created, processed, and analyzed in the form of molecular graphs. A molecular graph comprises a network of nodes and edges, in which atoms are represented as nodes and bonds between different atoms as edges. The molecular graphs communicate by using connection tables and their linear notations. The different sections and sub-sections of a connection table contain information related to atoms, atom types, connection types, and their coordinate positions in the 3D or 2D space. Specific file formats are used to store the ligand information, such as .mol2, .sdf, .pdb, etc. Simplified molecular-input line-entry specification (SMILE) and Wiswesser line notation are examples of linear notations where alphanumeric characters are used to store the ligand information. Linear notation is preferred for storing or transferring millions of small molecules due to its compactness compared to connection tables [52]. The quantitative structure-activity relationship (QSAR) and pharmacophore modeling concepts are used for designing drugs based on LBDD approaches.

2.5.2 Quantitative Structure-Activity Relationship

The quantitative structure-activity relationship (QSAR) is a computational method for determining the relationship between the structural properties of chemical compounds and their biological activities [47]. It is based on the principle that different structural properties yield different biological activities [53]. Structural properties include physicochemical properties, whereas biological activities correspond to pharmacokinetics, that is, ADMET, of drug molecules. The development of a QSAR model begins with recognizing a group of chemical entities or lead molecules that exhibit the desired biological activity. Then, suitable molecular descriptors are identified that are associated with various structural and physicochemical properties of the molecules of interest. Molecular descriptors are mathematical representations of molecular properties generated by associated algorithms. Finding the set of molecular descriptors is a significant step in constructing QSAR models. It helps reduce computational time, improve prediction performance, and better understand data in machine learning [54]. Further, statistical methods are employed to derive a quantitative relationship between molecular descriptors and their associated biological activity. Finally, the developed QSAR model is validated and tested for structural stability and predictive power. A QSAR model helps predict the movements of untested chemicals and aids in rational drug design through computer-aided simulation, molecular modeling, and virtual screening of suitable compounds.

2.5.3 Applications of CADD in Protein-Based Therapeutics

Therapeutic proteins are genetically engineered proteins that substitute abnormal or malfunctioned human proteins to cure a disease. In CADD, protein-drug interactions are simulated to determine their binding affinities. Virtual database screening helps screen large libraries efficiently to identify potent drugs that are likely to have high binding affinities to the target. Target may be any enzyme or protein linked to a specific disease. Structural information about the target is also required to learn about its functions. One can harness the structural information of proteins already available in the PDB database. The missing structural information can be predicted using bioinformatics approaches, such as homology modeling, threading, or *ab initio* predictions. Dhanavade et al. predicted the 3D structure of cysteine protease using molecular modeling, which degrades amyloid-β peptide, a major cause of Alzheimer's disease (AD) [55–57]. In recent years, CADD has successfully identified potential drugs for treating several neurodegenerative disorders. ROCK-I and NOX2 are two of the most promising potential therapeutic targets for various neurodegenerative disorders [58, 59]. Inhibition of these two enzymes can help manage neurodegenerative disorders like autism spectral disorder, AD, and fragile X syndrome. Utilizing this information, Alokam et al. identified chemical entities that behave as dual inhibitors of these enzymes using a combination of pharmacophores and the molecular docking approach of CADD [60]. Also, *in vitro* validation demonstrated their inhibitory potentials to ROCK-I and NOX2.

In the COVID-19 pandemic, CADD served as a powerful tool for identifying therapeutic proteins against rapidly mutating SARS-CoV-2 [61]. The main protease (M^{pro}) enzyme is crucial for the survival of pathogen as it is involved in replication and maturation. The structure-based virtual screening successfully identified four compounds having the ability to disrupt the normal functioning of M^{pro} protein. Later, ADMET analysis, molecular docking, and MD simulations were applied to explore their binding conformational stability at the active site of M^{pro} protein. The study identified crucial ligand amino acid residues, such as GLN189, SER10, GLU166, ASN142, PHE66, and TRP132, that participate in stabilizing the protein-ligand interaction of SARS-CoV-2 M^{pro} [62]. Nowadays, machine learning approaches are used in conjunction with CADD to identify repurposed therapeutics [63]. Thus, CADD is serving as a rapid and promising technology in the development of protein-based therapeutics [64, 65].

2.6 Overview of Recently Approved Protein Therapeutics for Clinical Applications

Recombinant human proteins can be used as therapeutics for treating many illnesses such as diabetes mellitus and multiple sclerosis. The production of high-quality and functional recombinant proteins is crucial in drug therapy. Although many applications of recombinant proteins exist as potent therapeutics, the production of antidrug antibodies (ADAs) is a matter of concern that limits its use. The aggregates formed during the formulation of recombinant proteins lead to the breakage of immune system tolerance and result in the production of ADAs. Various strategies are applied to minimize the aggregation and reduce immunogenicity to make protein therapeutics safer and more efficient.

2.6.1 Diabetes

Purified from the porcine and bovine pancreas in the 1920s, insulin was the first therapeutic protein discovered. Since it had a nonhuman origin, its immunogenicity was expected. The patients suffering from diabetes mellitus were treated with insulin for over 80 years [66]. In addition to the source being nonhuman, early purification methods were also not up to the mark, resulting in the development of anti-insulin antibodies in most patients. To overcome the issue, therapeutic insulin is now mainly produced as recombinant human protein, and advanced purification methods take care of purity. Although recombinant insulin is a safer drug, it has been reported that ADAs, including subclasses of immunoglobulins, developed in about 50% of diabetic patients treated with recombinant human insulin [67, 68]. The presence of insulin autoantibodies in diabetes Type I patients hampers the interpretation of clinical data [69]. Several theories have been put forward to explain why ADAs develop against insulin in many patients. One of the most promising theories suggests the involvement of insulin ADAs with themselves and the formation of aggregates as insulin antibodies have a high tendency to self-associate. These aggregates contribute to high immunogenicity [70].

The glucagon-like peptide-1 (GLP-1) receptor agonist lixisenatide (lyxumia 1) was approved for the treatment of type 2 diabetes [71]. The GLP-1 receptor agonist is compared to other antidiabetic drugs, that is, exenatide, insulin glargine, metformin, sitagliptin, liraglutide, or placebo in type 2 diabetes patients [72]. It is linked with other biomolecules like lipids, carbohydrates, polyethylene glycol, or proteins to increase its efficacy. The GLP-1 receptor agonist, along with these conjugates, acts as cell-targeting peptides or cell-penetrating peptides. It induces insulin release and suppresses glucagon release in type 2 diabetes [73]. Another study showed that the C-peptide activates the phosphorylation of insulin receptor tyrosine kinase and glycogen synthase kinase 3 and results in the mobilization of insulin-responsive glucose transporter, increased amino acid uptake, and glycogen synthesis. This suggests that C-peptide signaling may cross-talk with the insulin pathway at the level of the insulin receptor [51]. The clinical studies indicate that the replacement of

C-peptides in type 1 diabetic patients shows advantageous effects on somatic and autonomic diabetic peripheral neuropathy (DPN). Apart, the C-peptide also reduces the diabetes-induced glomerular hyperfiltration and, thus, decreases the excretion of urinary albumin [74].

2.6.2 Interferon-β

Relapsing-reemitting multiple sclerosis is generally treated by interferon-β (IFN-β) or recombinant human IFN-β (Rhu IFN-β). Though these are the most promising and efficient anti-inflammatory drugs for treating multiple sclerosis, many patients do not respond to them [75]. As suggested, this can be attributed to the production of neutralizing antibodies (nAbs) against the IFN-β [76]. IFN-β 1a (Avonex®, Rebif®) and IFN-β 1b (Betaseron®) products are available in the market that differ in their source of production, glycosylation pattern, amino acid sequences, and degree of aggregation [77]. These drugs can mount different levels of immunogenicity in patients as their formulation sources, administered routes, dosage, and frequency regimes differ.

Interestingly, patients with a history of developing nAbs, when treated with IFN-β, result in the disappearance of nAbs. This indicates that the production of nAbs does not form the memory and, thus, possibly does not involve in the classical immune response. Besides that, there is an increasing research interest in investigating and characterizing the formation of aggregates in IFN-β formulations and their potency in eliciting an immune response in patients by breaking immune tolerance [75]. One study reported that multiple sclerosis patients who received IFN-β 1b developed more nAbs than those who received IFN-β 1a [78]. This observation is most probably correlated to the levels of aggregates as IFN-β 1b formulation shows a higher degree of aggregation than IFN-β 1a [79]. The self-binding characteristic of IFN-β 1b is high, due to which they cluster together. Also, the lack of glycosylation in these molecules promotes aggregation [80]. The exact cause of the formation of ADAs is not well understood. However, aggregation is considered an essential contributory factor for immunogenicity in almost all cases, which requires comprehensive and exploratory studies to identify and validate the causes of their formation.

2.6.3 Cancer

An array of peptide-based therapeutics has been tested in preclinical models to check their efficacy in curing cancer. Therapeutics are developed based on a synthetic polymeric carrier elastin-like polypeptide (ELP), which can be synthesized in variable sequences and sizes to stabilize the therapeutic peptide and avoid crossing the placental interface to prevent fetal exposure and potential developmental effects [81]. The therapeutic peptides possess a targeting delivery feature to recognize cancer cells effectively. These peptides increase the specificity and efficacy of drug delivery with minimal side effects [82]. The cyclic peptide, cCPGPEGAGC (PEGA), is a homing peptide that can identify cancer cells. In conjugation with the cell-penetrating peptide pVEC, this peptide was selectively taken up by different breast cancer cells [83]. Another peptide, D2A21, and its gel formulations have been used in wound-healing products to treat infected burns, wounds, and several types of cancer. A TAT peptide derived from the N-terminus of p53, fused with a peptide derived from the VHL tumor-suppressor gene, inhibits insulin-like growth factor I receptor (IGF-IR) signaling in renal cell carcinomas [83].

2.7 Emerging Issues and Developments in Proteins-Based Therapeutics

Both native and recombinant therapeutic proteins are an essential class of medicines developed to treat a wide variety of diseases. Therapeutic proteins, including vaccines, antigens, or hormones, are produced using rDNA technology and protein purification methods. The drug developers apply protein engineering to achieve desirable molecular characteristics to make these therapeutic proteins safe and effective. Drug targeting is an important aspect of therapeutics to treat several diseases. So, it is essential to devise better drug targeting and delivery methods to have improved potency and functionality. The knowledge of the mechanism of action and structure-function relationship of a protein is essential for engineering its activity or introducing new desired activities. The customization of existing proteins or the generation of novel therapeutics having specific clinical applications is a developing field in drug design. Besides protein engineering, technological advancements in genetic engineering are also used to develop therapeutic proteins to tackle a wide range of life-threatening conditions. However, there are challenges and limitations associated with the use of therapeutic proteins to combat life-threatening conditions [84], which include (i) optimal utilization of therapeutic proteins and peptides via the oral route, (ii) extensive hepatic first-pass metabolism, (iii) degradation in the gastrointestinal tract, and (iv) large molecular size and poor permeation.

2.7.1 Issue of Demand and Supply

Therapeutic protein development projects are time consuming and budget extensive. Also, the associated development processes have various intricacies of cellular metabolism, pharmacokinetics, and pharmacodynamics, making their development task more difficult. To reduce the complexity and overcome related limitations, pharmaceutical scientists do a lot of preresearch and testing to select only those molecules with a maximum chance of success in clinical trials. It is a well-known and documented fact that from discovery to the pharmaceutical market, a new drug molecule takes more than ten years, and yet, its success rate is not guaranteed

[17, 85]. Therefore, much research needs to be done to consider selective molecules to maximize the chances of success.

2.7.2 Issues Related to Immunogenicity

Immunogenicity is a challenging aspect of disease management. A particulate matter, however small in amount, can lead drastically enhance immunogenicity [86, 87]. Protein-based therapeutics develop either from nonhuman or human sources and have chances of producing neutralizing and/or nonneutralizing antibodies [88]. In some cases, using protein therapeutics may also lead to an array of adverse immune reactions from mild inflammation to severe anaphylaxis. The therapeutic proteins given to patients may also neutralize endogenous proteins in some cases and, thus, lead to adverse effects [89]. T cells are a critical arm of the immune system, and their activity is regulated via T cell receptor interactions. Therefore, the prior knowledge of all the T cell epitopes present on the surface of therapeutics can enhance the immune tolerance level and, thus, minimize the unwanted immunogenic responses [90].

The immunogenicity of protein therapeutics can be reduced by designing depleting T cell epitopes (deimmunization process) [91]. This idea led to many deimmunized therapeutic proteins in clinical trials. The quality and quantity of T cell epitopes are measured by using T-cell-based assays. These *in vitro* methods, along with computational techniques, facilitate the identification and removal of T cell epitopes. The desired mutations can be incorporated into the peptide sequences using *in silico* tools, which are later implemented into deimmunized T cell epitope protein sequences. The resultant peptides have limited capacity for MHC binding and produce decreased immune responses. Antibodies are mainly deimmunized protein therapeutics. The rise in the unwanted level of immunogenicity is diminished by using deimmunized antibodies. The deimmunized antibodies specific for prostrate membrane antigen have passed different stages of clinical trials and are approved for clinical use. In clinical trials, these antibodies are conjugated with a radioactive probe and do not show antitherapeutic immune responses [92–95]. So, T cell epitopes are one of the crucial factors taken into consideration to control antitherapeutic antibody responses. Thus, deimmunized protein-based therapeutics may provide a safe class of new biologics.

2.7.3 Issue of Protein Stability

The prolonged stability of therapeutic proteins in a clinical setting is a desirable but challenging trait. It is one of the limitations in making them ideal clinical therapeutics [96]. The possible aggregation of therapeutic proteins increases if they are stored in the high concentrations required for using them on a large-scale [97, 98]. The aggregation decreases their overall activity and results in immunological reactions [97]. This problem can be overcome by spatial aggregation propensity, which identifies aggregation-prone regions in a protein sequence, and then those regions are mutated to engineer stable antibodies [99, 100]. High temperature also influences the stability of therapeutic proteins. The proteins lose their activity and structural integrity when stored at room temperature. The best practice is to keep the purified therapeutic proteins at or below 4 °C for an extended period. During the release of therapeutic proteins, it may also form particles that compromise their stability and induce an immune response in patients when administered [97]. The common strategies to enhance protein stability include (i) the inclusion of desired mutations in the protein [101, 102], (ii) optimization of the formulation of therapeutic proteins [103], (iii) use of thermosensitive polymers [96], (iv) encapsulation of therapeutic proteins, (v) use of biodegradable polymers for the delivery of therapeutic substances, and (vi) use of nontoxic nanostructured materials.

2.7.4 Issues of Metabolism and Elimination

The metabolism and elimination of protein-based therapeutics, such as those used for hepatic diseases (like liver cirrhosis), poses a significant hurdle in their successful clinical uses. The noninvasive administration of protein-based therapeutic by using alternative routes can possibly solve the issue of hepatic metabolism. Moreover, the hepatic first-pass metabolism may also be overcome by using invasive delivery of therapeutic proteins. Another problem with therapeutic proteins is that most have a short half-life. To overcome this issue, therapeutic proteins are encapsulated and/or conjugated with biocompatible polymers [104]. Nowadays, the half-life of these therapeutics is enhanced by using existing fusion protein technology.

The hurdles in developing and delivering therapeutic proteins may be overcome by studying their pharmacokinetic properties and pharmacodynamic effects. Complete identification and analysis of their pharmacokinetic parameters are required for predicting the biodisposition of these agents. Though recent advancements in applied technologies have solved such problems to some extent, some unknown factors are responsible for creating hindrances to efficiently using much-needed therapeutics. For instance, poor intestinal absorption and intestinal first-pass metabolism significantly impact the clearance of protein-based therapeutics if they are given via the oral route. Therefore, in-depth knowledge of the routes of their administration and the underlying mechanism of metabolism is needed to tackle the issues of early clearance. Several such protein-based therapeutics have recently been developed, such as Oral Recosulin, Octreolin®, Sandimmune®, etc., and many are in the clinical stages of development.

2.8 Conclusion and Future Prospects

Protein-based therapeutics are engineered drugs with a wide range of clinical applications. Rapid progress has been made in the last decade toward developing engineered proteins to treat several life-threatening conditions. Clinical safety and efficacy are some of the essential features to overcome by working on various factors such as disease biology, individual genetic profile, and selection of target population on the patient side, while safety margin, route of delivery, half-life, stability, and solubility on the protein-based therapeutics side. The mode of drug delivery is a concern for increasing the efficacy. The oral delivery of therapeutic proteins is one of the efficient ways to replace the invasive routes, only if the problems of poor absorption and intestinal first-pass metabolism are handled. The recent advancements in several cross-cutting technologies have made the oral delivery of therapeutic proteins possible. Apart from that, the stability of protein during its formulation and decreasing its development cost remain a significant challenge in front of research communities. These problems can be handled by parallel use of advanced *in vitro*, *in vivo*, and *in silico* techniques. Many protein-based therapeutics are either FDA approved or in the final stages of approval, with many reaching the global market, and hundreds are in preclinical studies and clinical development. These therapeutics have been successful in treating a variety of conditions, from diabetes mellitus to cancers. The design and development of therapeutic proteins considering novel scaffolds with superior biochemical and physiological activities will be primary areas of research in the coming decades.

References

1. Singh DB, Tripathi T (2020) Frontiers in protein structure, function, and dynamics. Springer Nature, Singapore
2. Sumant OSS (2017) Protein Therapeutics Market, Market Research Report, Allied Market Research, p 211
3. Dimitrov DS (2012) Therapeutic proteins. Methods Mol Biol 899:1–26
4. Tripathi T, Dubey VK (2022) Advances in protein molecular and structural biology methods, 1st edn. Academic Press, Cambridge
5. Heckmann CM, Paradisi F (2020) Looking back: a short history of the discovery of enzymes and how they became powerful chemical tools. ChemCatChem 12(24):6082–6102
6. Hemalatha T, UmaMaheswari T, Krithiga G, Sankaranarayanan P, Puvanakrishnan R (2013) Enzymes in clinical medicine: an overview. Indian J Exp Biol 51(10):777–788
7. Robinson PK (2015) Enzymes: principles and biotechnological applications. Essays Biochem 59:1–41
8. Tandon S, Sharma A, Singh S, Sharma S, Sarma SJ (2021) Therapeutic enzymes: discoveries, production and applications. J Drug Deliv Sci Technol 63:102455
9. Martínez Cuesta S, Rahman SA, Furnham N, Thornton JM (2015) The classification and evolution of enzyme function. Biophys J 109(6):1082–1086
10. Meghwanshi GK, Kaur N, Verma S, Dabi NK, Vashishtha A, Charan PD, Purohit P, Bhandari HS, Bhojak N, Kumar R (2020) Enzymes for pharmaceutical and therapeutic applications. Biotechnol Appl Biochem 67(4):586–601
11. Vimal A, Kumar A (2019) Chapter 35 - transforming the healthcare system through therapeutic enzymes. In: Kuddus M (ed) Enzymes in food biotechnology. Academic Press, pp 603–625
12. Van Stappen T, Vande Casteele N, Van Assche G, Ferrante M, Vermeire S, Gils A (2018) Clinical relevance of detecting anti-infliximab antibodies with a drug-tolerant assay: post hoc analysis of the TAXIT trial. Gut 67(5):818
13. Dalal SR, Cohen RD (2015) What to do when biologic agents are not working in inflammatory bowel disease patients. Gastroenterol Hepatol (N Y) 11(10):657–665

14. Kalita P, Tripathi T (2022) Methodological advances in the design of peptide-based vaccines. Drug Discov Today 27(5):1367–1380
15. Kany S, Vollrath JT, Relja B (2019) Cytokines in inflammatory disease. Int J Mol Sci 20(23): 6008
16. Ndoja S, Lima H (2017) Monoclonal antibodies. In: Thomaz-Soccol V, Pandey A, Resende RR (eds) Current developments in biotechnology and bioengineering. Elsevier, pp 71–95
17. Strohl WR, Knight DM (2009) Discovery and development of biopharmaceuticals: current issues. Curr Opin Biotechnol 20(6):668–672
18. Scully M, Cataland SR, Peyvandi F, Coppo P, Knöbl P, Kremer Hovinga JA, Metjian A, de la Rubia J, Pavenski K, Callewaert F, Biswas D, De Winter H, Zeldin RK (2019) Caplacizumab treatment for acquired thrombotic thrombocytopenic purpura. N Engl J Med 380(4):335–346
19. Campochiaro PA, Heier JS, Feiner L, Gray S, Saroj N, Rundle AC, Murahashi WY, Rubio RG (2010) Ranibizumab for macular edema following branch retinal vein occlusion: six-month primary end point results of a phase III study. Ophthalmology 117(6):1102–1112.e1
20. Tcheng JE, Kandzari DE, Grines CL, Cox DA, Effron MB, Garcia E, Griffin JJ, Guagliumi G, Stuckey T, Turco M, Fahy M, Lansky AJ, Mehran R, Stone GW (2003) Benefits and risks of abciximab use in primary angioplasty for acute myocardial infarction: the controlled abciximab and device investigation to lower late angioplasty complications (CADILLAC) trial. Circulation 108(11):1316–1323
21. Blair HA (2019) Emicizumab: a review in Haemophilia A. Drugs 79(15):1697–1707
22. Burt R, Warcel D, Fielding AK (2019) Blinatumomab, a bispecific B-cell and T-cell engaging antibody, in the treatment of B-cell malignancies. Hum Vaccin Immunother 15(3):594–602
23. Linke R, Klein A, Seimetz D (2010) Catumaxomab: clinical development and future directions. MAbs 2(2):129–136
24. Strohl WR, Strohl LM (2012) Introduction to biologics and monoclonal antibodies, therapeutic antibody engineering. Woodhead Publishing, pp 1–595
25. Pradhananga S, Sayers JR (2012) Natural synthesis: biologics, biosimilars and biobetters in protein hormone therapy. Biochemist 34(1):10–15
26. Kowarski CR, Lin SS, Michalek LM, Kowarski AA (1988) Comparing the biological activity of humulin R and novolin R using an intravenous bioassay. Pharm Res 5(4):245–246
27. Latrech H, Simon A, Beltrand J, Souberbielle JC, Belmejdoub G, Polak M (2012) Postprandial hyperglycemia corrected by IGF-I (Increlex®) in Laron syndrome. Horm Res Paediatr 78(3): 193–200
28. Lal RA, Hoffman AR (2018) Long-acting growth hormone preparations in the treatment of children. Pediatr Endocrinol Rev 16(Suppl 1):162–167
29. de Las Cuevas Allende R, Díaz de Entresotos L, Conde Díez S (2021) Anaemia of chronic diseases: pathophysiology, diagnosis and treatment. Med Clin (Barc) 156(5):235–242
30. Dale DC, Crawford J, Klippel Z, Reiner M, Osslund T, Fan E, Morrow PK, Allcott K, Lyman GH (2018) A systematic literature review of the efficacy, effectiveness, and safety of filgrastim. Support Care Cancer 26(1):7–20
31. Swartz ML (1991) Intron A (interferon-alpha-2b recombinant) for injection/Schering-Plough Corporation. Gastroenterol Nurs 14(1):40–43
32. Kleinschnitz C, Niemczyk G, Rehberg-Weber K, Wernsdörfer C (2015) Interferon Beta-1a (AVONEX®) as a treatment option for untreated patients with multiple sclerosis (AXIOM): a prospective, observational study. Int J Mol Sci 16(7):15271–15286
33. Hadjigeorgiou G, Dardiotis E, Tsivgoulis G, Doskas T, Petrou D, Makris N, Vlaikidis N, Thomaidis T, Kyritsis A, Fakas N, Treska X, Karageorgiou C, Sotirli S, Giannoulis C, Papadimitriou D, Mylonas I, Kouremenos E, Vlachos G, Georgiopoulos D, Mademtzoglou D, Vikelis M, Zintzaras E (2014) Observational study assessing demographic, economic and clinical factors associated with access and utilization of health care services of patients with multiple sclerosis under treatment with interferon beta-1b (EXTAVIA). PLoS One 9(11):e113933

34. Clark JI, Wong MKK, Kaufman HL, Daniels GA, Morse MA, McDermott DF, Agarwala SS, Lewis LD, Stewart JH, Vaishampayan U, Curti B, Gonzalez R, Lutzky J, Rudraptna V, Cranmer LD, Jeter JM, Hauke RJ, Miletello G, Milhem MM, Amin A, Richart JM, Fishman M, Hallmeyer S, Patel SP, Van Veldhuizen P, Agarwal N, Taback B, Treisman JS, Ernstoff MS, Perritt JC, Hua H, Rao TB, Dutcher JP, Aung S (2017) Impact of sequencing targeted therapies with high-dose Interleukin-2 immunotherapy: an analysis of outcome and survival of patients with metastatic renal cell carcinoma from an on-going observational IL-2 clinical trial: PROCLAIM(SM). Clin Genitourin Cancer 15(1):31–41.e4
35. Kaye JA (1998) FDA licensure of NEUMEGA to prevent severe chemotherapy-induced thrombocytopenia. Stem Cells 16(Suppl 2):207–223
36. Vogenberg FR, Isaacson Barash C, Pursel M (2010) Personalized medicine: part 1: evolution and development into theranostics. P T 35(10):560–576
37. Ginsburg GS, Phillips KA (2018) Precision medicine: from science to value. Health Aff (Millwood) 37(5):694–701
38. Anna M, Nurrani Mustika D, Andi W (2016) Personalized medicine: the future of health care. Indones Biomed J 8(3):46–127
39. Mathur S, Sutton J (2017) Personalized medicine could transform healthcare. Biomed Rep 7(1):3–5
40. Gameiro GR, Sinkunas V, Liguori GR, Auler-Júnior JOC (2018) Precision medicine: changing the way we think about healthcare. Clinics (Sao Paulo) 73:e723
41. Bradford D, Demko S, Jin S, Mishra-Kalyani P, Beckles AR, Goldberg KB, Lemery S, Ward A, Keegan P, Pazdur R (2020) FDA accelerated approval of pembrolizumab for recurrent locally advanced or metastatic Merkel cell carcinoma. Oncologist 25(7):e1077–e1082
42. Miller EM, Schwartzberg LS (2019) Biosimilars for breast cancer: a review of HER2-targeted antibodies in the United States. Ther Adv Med Oncol 11:1758835919887044–1758835919887044
43. Bilkey GA, Burns BL, Coles EP, Mahede T, Baynam G, Nowak KJ (2019) Optimizing precision medicine for public health. Front Public Health 7:42
44. Schmidt C (2014) Leroy Hood looks forward to P4 medicine: predictive, personalized, preventive, and participatory. J Natl Cancer Inst 106(12)
45. Hay M, Thomas DW, Craighead JL, Economides C, Rosenthal J (2014) Clinical development success rates for investigational drugs. Nat Biotechnol 32(1):40–51
46. Mullard A (2016) Biotech R&D spend jumps by more than 15. Nat Rev Drug Discov 15(7): 447
47. Song CM, Lim SJ, Tong JC (2009) Recent advances in computer-aided drug design. Brief Bioinform 10(5):579–591
48. Lionta E, Spyrou G, Vassilatis DK, Cournia Z (2014) Structure-based virtual screening for drug discovery: principles, applications and recent advances. Curr Top Med Chem 14(16): 1923–1938
49. Wlodawer A, Vondrasek J (1998) Inhibitors of HIV-1 protease: a major success of structure-assisted drug design. Annu Rev Biophys Biomol Struct 27:249–284
50. Güner O, Clement O, Kurogi Y (2004) Pharmacophore modeling and three dimensional database searching for drug design using catalyst: recent advances. Curr Med Chem 11(22): 2991–3005
51. Yu W, MacKerell AD Jr (2017) Computer-aided drug design methods. Methods Mol Biol 1520:85–106
52. Sharma V, Wakode S, Kumar H (2021) Chapter 2 - Structure- and ligand-based drug design: concepts, approaches, and challenges. In: Sharma N, Ojha H, Raghav PK, Goyal RK (eds) Chemoinformatics and bioinformatics in the pharmaceutical sciences. Academic Press, pp 27–53
53. Verma J, Khedkar VM, Coutinho EC (2010) 3D-QSAR in drug design--a review. Curr Top Med Chem 10(1):95–115

54. Gramatica P (2007) Principles of QSAR models validation: internal and external. QSAR Comb Sci 26(5):694–701
55. Dhanavade MJ, Jalkute CB, Barage SH, Sonawane KD (2013) Homology modeling, molecular docking and MD simulation studies to investigate role of cysteine protease from Xanthomonas campestris in degradation of Aβ peptide. Comput Biol Med 43(12):2063–2070
56. Tripathi T, Kalita P (2019) Synergistic effect of amyloid-β and tau disrupts neural circuits. ACS Chem Neurosci 10(3):1129–1130
57. Tripathi T, Khan H (2020) Direct interaction between the β-amyloid core and tau facilitates cross-seeding: a novel target for therapeutic intervention. Biochemistry 59(4):341–342
58. Labandeira-Garcia JL, Rodríguez-Perez AI, Villar-Cheda B, Borrajo A, Dominguez-Meijide-A, Guerra MJ (2015) Rho kinase and dopaminergic degeneration: a promising therapeutic target for Parkinson's disease. Neuroscientist 21(6):616–629
59. Sorce S, Krause KH (2009) NOX enzymes in the central nervous system: from signaling to disease. Antioxid Redox Signal 11(10):2481–2504
60. Alokam R, Singhal S, Srivathsav GS, Garigipati S, Puppala S, Sriram D, Perumal Y (2015) Design of dual inhibitors of ROCK-I and NOX2 as potential leads for the treatment of neuroinflammation associated with various neurological diseases including autism spectrum disorder. Mol BioSyst 11(2):607–617
61. Padhi AK, Rath SL, Tripathi T (2021) Accelerating COVID-19 research using molecular dynamics simulation. J Phys Chem B 125(32):9078–9091
62. Mohan A, Rendine N, Mohammed MKS, Jeeva A, Ji HF, Talluri VR (2021) Structure-based virtual screening, in silico docking, ADME properties prediction and molecular dynamics studies for the identification of potential inhibitors against SARS-CoV-2 M(pro). Mol Divers:1–17
63. Yadav MK, Ahmad S, Raza K, Kumar S, Eswaran M, Pasha Km M (2022) Predictive modeling and therapeutic repurposing of natural compounds against the receptor-binding domain of SARS-CoV-2. J Biomol Struct Dyn:1–13
64. Shukla R, Tripathi T (2021) Molecular dynamics simulation in drug discovery: opportunities and challenges. In: Singh SK (ed) Innovations and implementations of drug discovery strategies in rational drug design. Springer Nature, Singapore, pp 295–316
65. Shukla R, Tripathi T (2020) Molecular dynamics simulation of protein and protein-ligand complexes. In: Singh DB (ed) Computer-aided drug design. Springer Nature, Singapore, pp 133–161
66. Sauerborn M, Brinks V, Jiskoot W, Schellekens H (2010) Immunological mechanism underlying the immune response to recombinant human protein therapeutics. Trends Pharmacol Sci 31(2):53–59
67. Greenfield JR, Tuthill A, Soos MA, Semple RK, Halsall DJ, Chaudhry A, O'Rahilly S (2009) Severe insulin resistance due to anti-insulin antibodies: response to plasma exchange and immunosuppressive therapy. Diabet Med 26(1):79–82
68. Radermecker RP, Renard E, Scheen AJ (2009) Circulating insulin antibodies: influence of continuous subcutaneous or intraperitoneal insulin infusion, and impact on glucose control. Diabetes Metab Res Rev 25(6):491–501
69. Jaeger C, Winter S, Eckhard M, Hardt P, Brendel MD, Bretzel RG (2008) Binding characteristics and crossreactivity of insulin autoantibodies and insulin antibodies directed to three different insulin molecules. Acta Diabetol 45(3):191–194
70. Fineberg SE, Kawabata TT, Finco-Kent D, Fountaine RJ, Finch GL, Krasner AS (2007) Immunological responses to exogenous insulin. Endocr Rev 28(6):625–652
71. Kaspar AA, Reichert JM (2013) Future directions for peptide therapeutics development. Drug Discov Today 18(17–18):807–817
72. Keservani RK, Sharma AK, Jarouliya U (2015) Protein and peptide in drug targeting and its therapeutic approach. Ars Pharmaceutica (Internet) 56:165–177
73. Nauck MA, Meier JJ (2005) Glucagon-like peptide 1 and its derivatives in the treatment of diabetes. Regul Pept 128(2):135–148

74. Luppi P, Cifarelli V, Wahren J (2011) C-peptide and long-term complications of diabetes. Pediatr Diabetes 12(3 Pt 2):276–292
75. Bermel RA, Rudick RA (2007) Interferon-beta treatment for multiple sclerosis. Neurotherapeutics 4(4):633–646
76. Farrell R, Kapoor R, Leary S, Rudge P, Thompson A, Miller D, Giovannoni G (2008) Neutralizing anti-interferon beta antibodies are associated with reduced side effects and delayed impact on efficacy of interferon-beta. Mult Scler 14(2):212–218
77. Kagawa Y, Takasaki S, Utsumi J, Hosoi K, Shimizu H, Kochibe N, Kobata A (1988) Comparative study of the asparagine-linked sugar chains of natural human interferon-beta 1 and recombinant human interferon-beta 1 produced by three different mammalian cells. J Biol Chem 263(33):17508–17515
78. Bertolotto A, Deisenhammer F, Gallo P, Sölberg Sørensen P (2004) Immunogenicity of interferon beta: differences among products. J Neurol 251:ii15–ii24
79. Hermeling S, Jiskoot W, Crommelin D, Bornaes C, Schellekens H (2005) Development of a transgenic mouse model immune tolerant for human interferon Beta. Pharm Res 22(6): 847–851
80. Runkel L, Meier W, Pepinsky RB, Karpusas M, Whitty A, Kimball K, Brickelmaier M, Muldowney C, Jones W, Goelz SE (1998) Structural and functional differences between glycosylated and non-glycosylated forms of human interferon-beta (IFN-beta). Pharm Res 15(4):641–649
81. Bidwell GL 3rd, George EM (2014) Maternally sequestered therapeutic polypeptides - a new approach for the management of preeclampsia. Front Pharmacol 5:201–201
82. Zhang XX, Eden HS, Chen X (2012) Peptides in cancer nanomedicine: drug carriers, targeting ligands and protease substrates. J Control Release 159(1):2–13
83. Regberg J, Srimanee A, Langel U (2012) Applications of cell-penetrating peptides for tumor targeting and future cancer therapies. Pharmaceuticals (Basel) 5(9):991–1007
84. Muheem A, Shakeel F, Jahangir MA, Anwar M, Mallick N, Jain GK, Warsi MH, Ahmad FJ (2016) A review on the strategies for oral delivery of proteins and peptides and their clinical perspectives. Saudi Pharm J 24(4):413–428
85. Shih H (2013) Discovery process for antibody-based therapeutics, development of antibody-based therapeutics. Translat Consider 9–32
86. Carpenter JF, Randolph TW, Jiskoot W, Crommelin DJ, Middaugh CR, Winter G, Fan YX, Kirshner S, Verthelyi D, Kozlowski S, Clouse KA, Swann PG, Rosenberg A, Cherney B (2009) Overlooking subvisible particles in therapeutic protein products: gaps that may compromise product quality. J Pharm Sci 98(4):1201–1205
87. Rosenberg AS (2006) Effects of protein aggregates: an immunologic perspective. AAPS J 8(3):E501–E507
88. Anjum R, Zahra N, Rehman K, Alam R, Parveen A, Tariq M, Akash MSH (2013) Comparative analysis of serum lipid profile between normotensive and hypertensive. J Mol Genetic Med 7:64
89. Schellekens H (2005) Factors influencing the immunogenicity of therapeutic proteins. Nephrol Dial Transplant 20(Suppl 6):vi3–vi9
90. De Groot AS, Moise L, McMurry JA, Wambre E, Van Overtvelt L, Moingeon P, Scott DW, Martin W (2008) Activation of natural regulatory T cells by IgG Fc-derived peptide "Tregitopes". Blood 112(8):3303–3311
91. Baker MP, Reynolds HM, Lumicisi B, Bryson CJ (2010) Immunogenicity of protein therapeutics: the key causes, consequences and challenges. Self Nonself 1(4):314–322
92. Galsky MD, Eisenberger M, Moore-Cooper S, Kelly WK, Slovin SF, DeLaCruz A, Lee Y, Webb IJ, Scher HI (2008) Phase I trial of the prostate-specific membrane antigen-directed immunoconjugate MLN2704 in patients with progressive metastatic castration-resistant prostate cancer. J Clin Oncol 26(13):2147–2154
93. Morris MJ, Divgi CR, Pandit-Taskar N, Batraki M, Warren N, Nacca A, Smith-Jones P, Schwartz L, Kelly WK, Slovin S, Solit D, Halpern J, Delacruz A, Curley T, Finn R,

O'Donoghue JA, Livingston P, Larson S, Scher HI (2005) Pilot trial of unlabeled and indium-111-labeled anti-prostate-specific membrane antigen antibody J591 for castrate metastatic prostate cancer. Clin Cancer Res 11(20):7454–7461
94. Pandit-Taskar N, O'Donoghue JA, Morris MJ, Wills EA, Schwartz LH, Gonen M, Scher HI, Larson SM, Divgi CR (2008) Antibody mass escalation study in patients with castration-resistant prostate cancer using 111In-J591: lesion detectability and dosimetric projections for 90Y radioimmunotherapy. J Nucl Med 49(7):1066–1074
95. Vallabhajosula S, Goldsmith SJ, Kostakoglu L, Milowsky MI, Nanus DM, Bander NH (2005) Radioimmunotherapy of prostate cancer using 90Y- and 177Lu-labeled J591 monoclonal antibodies: effect of multiple treatments on myelotoxicity. Clin Cancer Res 11(19 Pt 2): 7195s–7200s
96. Akash MS, Rehman K, Chen S (2013) IL-1Ra and its delivery strategies: inserting the association in perspective. Pharm Res 30(11):2951–2966
97. Jiskoot W, van Schie RMF, Carstens MG, Schellekens H (2009) Immunological risk of injectable drug delivery systems. Pharm Res 26(6):1303–1314
98. Shire SJ, Shahrokh Z, Liu J (2004) Challenges in the development of high protein concentration formulations. J Pharm Sci 93(6):1390–1402
99. Chennamsetty N, Voynov V, Kayser V, Helk B, Trout BL (2009) Design of therapeutic proteins with enhanced stability. Proc Natl Acad Sci U S A 106(29):11937–11942
100. Voynov V, Chennamsetty N, Kayser V, Helk B, Trout BL (2009) Predictive tools for stabilization of therapeutic proteins. MAbs 1(6):580–582
101. Lawrence T, Fong C (2010) The resolution of inflammation: anti-inflammatory roles for NF-kappaB. Int J Biochem Cell Biol 42(4):519–523
102. Strickler SS, Gribenko AV, Gribenko AV, Keiffer TR, Tomlinson J, Reihle T, Loladze VV, Makhatadze GI (2006) Protein stability and surface electrostatics: a charged relationship. Biochemistry 45(9):2761–2766
103. Schneider CP, Trout BL (2009) Investigation of cosolute-protein preferential interaction coefficients: new insight into the mechanism by which arginine inhibits aggregation. J Phys Chem B 113(7):2050–2058
104. Tariq M, Akash MSH, Rehman K, Shuqing C (2015) Development of therapeutic proteins: advances and challenges. Turk J Biol 39

3 Protein Engineering Methods to Design Protein Therapeutics

Priyanka Narad, Romasha Gupta, Isha Gupta,
and Abhishek Sengupta

Abstract

The expeditious growth of modern medicine has proven to be indispensable for protein therapeutics. In some instances, protein therapeutics are the only known cure. Continuous efforts have been put in to enhance their production and efficacy. The advancement of protein engineering methods has aided in developing effective protein therapeutics with increased affinity, pharmacokinetics, pharmacodynamics, immunogenicity, and productivity. In terms of novel protein therapeutic platform technologies, different synthetic protein scaffolds are in the early phases of clinical trials. Protein therapeutics have immense potential to improve human health. As protein engineering efforts have grown and diversified, understanding the biophysical and biochemical properties of proteins and applying this information to developing effective pharmacological medications has become increasingly relevant. The overall purpose of the various methodologies for designing protein therapeutics is to better understand their role in medicine for proper care and diagnosis, leading to further innovation and development.

Keywords

Protein therapeutics · Pharmacodynamics · Pharmacokinetics · Protein engineering · Affinity · Immunogenicity · Protein scaffolds · Pharmacological medicines

---

P. Narad · R. Gupta · I. Gupta · A. Sengupta (✉)
Amity Institute of Biotechnology, Amity University Uttar Pradesh, Noida, Uttar Pradesh, India
e-mail: asengupta@amity.edu

D. B. Singh, T. Tripathi (eds.), *Protein-based Therapeutics*,
https://doi.org/10.1007/978-981-19-8249-1_3

3.1 Protein Engineering: A Realm for Biological Therapeutics

Protein engineering approaches for designing proteins with different activities, particularly in the case of protein therapeutics, are evolving rapidly. Natural proteins may be adequate for some purposes, such as enzyme replacement therapy, yet designing unique stimulus-responsive proteins is much more anticipated. Naturally powerful proteins are closely regulated in the body and are frequently posttranslationally controlled. Many are produced as inactive zymogens and subsequently removed or neutralized when no longer required. Protein engineering is used to create biochemically innovative therapeutics personalized to patients' needs. Designing protein treatments to boost responsiveness for designated initiation and balance and utilizing and developing these medications have tremendous potential for improvement. Current studies into protein molecular recognition, allostery, and catalysis and the advent of computational protein design tools will hasten this transformation.

3.1.1 Introduction

Current drug therapies have a wide range of outcomes; however, combining safety and efficacy with a short therapeutic window becomes difficult. Drug systems that respond to stimuli could be a solution to these problems, allowing for improved control of drug metabolism and action. These smart stimulus-responsive drugs may be more successful and have fewer adverse effects than current treatments, because they respond correctly to varied physiological and pathological signals. Hydrogels, for example, are nonprotein components of innovative medication systems and are currently being developed, but protein function control will be critical for reaching sophisticated in vivo systems for smart drugs. Proteins are ideal for critical biosensing and functional responses, because they can respond to stimuli with precise sensitivity [1].

Although protein engineering is frequently described as essential for future therapeutics, its role is commonly mentioned in a limited context, such as different scaffolds that act as antibody substitutes, fusion proteins for increased half-life, covalent conjugation with polyethylene glycol (PEG), that is, PEGylation, glycosylation, or mutagenesis for reduced immunogenicity, etc. In a practical sense, protein engineering will have a considerably broader influence. Protein engineering currently allows for tuning and creating protein stimulus-responsive activity, with the potential to change the landscape of drug metabolism. When combined with effective anticancer therapies, the use of modified proteins for targeted drug activation or neutralization can improve pharmacokinetics and pharmacodynamics while also reducing adverse effects.

Many recent breakthroughs in related fields such as biosensing, biocatalysis, and synthetic biology can be used in the emerging field of modified drug metabolism. Protein engineering includes three primary approaches: knowledge-based mutagenesis (KBM), computational protein design (CPD), and directed evolution (DE), in addition to the alteration of proteins by derivatization. Knowledge-based mutagenesis, the most basic kind of rational protein engineering, uses broad biochemical principles and prior knowledge to guide the mutagenesis of native proteins to generate enhanced or unique structure and/or functional features. Molecular modeling algorithms are used in computational protein design to determine amino-acid sequences that fold into the desired structure. This serves as a thorough examination of knowledge of the structure-function link. Computational protein design frequently comprises developing protein design candidates by altering residues on a high-resolution structure, then energetically analyzing the designs to find variants that are best for specific physicochemical properties such as protein stability or enzymatic activity. Random mutation or gene recombination is used in directed evolution to impart desirable traits into proteins [2]. Functional variants with specified features are subsequently found from these libraries through screening or selection. In directed evolution, in its most basic form, some of the de novo possibilities of computer design are missing. On the other hand, this approach can be used on a protein without having a complete understanding of its structure or the molecular process that allows it to operate. KBM, CPD, and DE techniques are often combined for researchers to quickly bestow desired physicochemical features and speed up discovery [Fig. 3.1]. These methods, when combined, provide a formidable toolkit for manipulating an enzyme's input and output sensitivities by altering substrate selectivity, binding affinity, or generating novel activity.

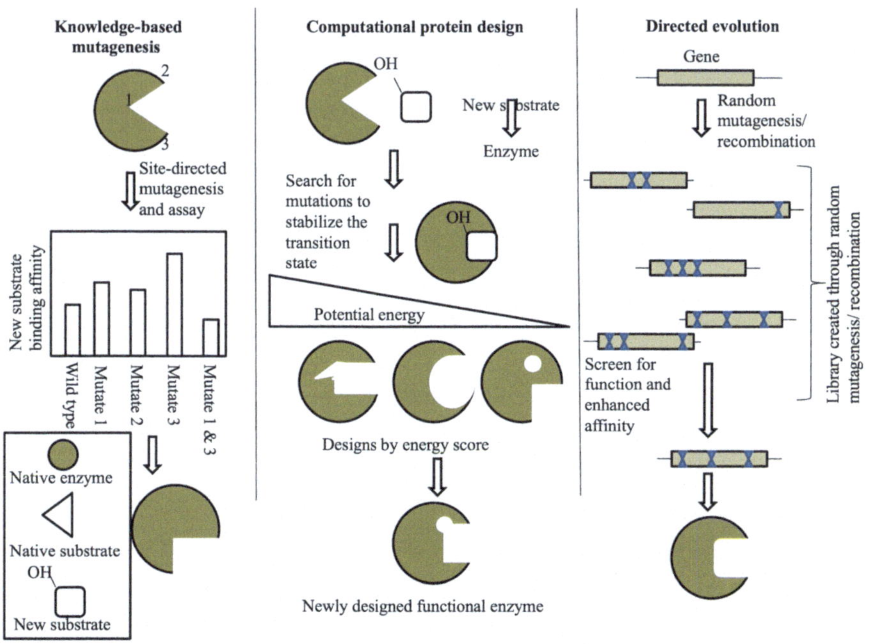

Fig. 3.1 Knowledge-based mutagenesis, computational protein design, and directed evolution are shown here for applying protein engineering to modify the substrate specificity of an enzyme

Medicine, notably pharmacology, is on the verge of a paradigm change from small-molecule chemical therapy to therapeutic intervention, which encompasses protein, cellular, and gene therapy. Since proteins are vital to biological processes, protein engineering is a common and essential method for these new biologics. Controlling half-life and immunogenicity in protein therapies is critical for clinical usage, and protein engineering will remain vital in these domains [3]. However, the ultimate promise of protein engineering for future treatments is significantly greater and is oriented on protein function and regulation, especially stimulus-responsive activity. Protein design constraints associated with traditional pure protein delivery methods will likely be reduced or eliminated when other intracellular protein delivery methods and gene therapy are developed.

3.1.2 Protein Structure Modeling and Prediction

All cells in our bodies and all living creatures from all kingdoms are made up of proteins. Although the deoxyribonucleic acid (DNA) encodes the information required for life to continue, proteins are responsible for the dynamic processes of replication, defense, life maintenance, and reproduction. There are 20 natural amino acids that occur more frequently than other unique amino acids with distinct functions. These 20 amino acids can be linked together in different combinations of sequence and length to form polypeptide chains, or proteins, depending on the genetic code and stereochemical properties. In terms of their functions, proteins may have a constitutive or transitory cell expression. It is also worth noting that efforts are being made to create proteins from nonnatural amino acids.

A protein, or a protein complex, performs biochemical events such as respiration, transport of gases, food absorption, energy metabolism and storage, heat or cold physiological reactions, or any other life process that can be imagined. Proteins evolve under natural selection pressures and play a specific role in every reaction in a living organism [4]. Protein function can be seen in a variety of ways. The fact that all of the activities they may accomplish are dependent on a single principle, the 20 amino acids that make up a protein, is remarkable. As a result, studying proteins, their composition, shape, dynamics, and function becomes crucial [5]. The function of protein is determined by its structure, which is influenced by physical and chemical factors. Another significant element to consider when examining these molecules is that traditional chemical, biological, mathematical, physical, and informatics sciences have collaborated in a new field known as bioinformatics to provide a new level of understanding about life organization.

3.1.2.1 Protein Structure

Protein structure refers to the three-dimensional (3D) arrangement of atoms in an amino acid chain. Proteins are made up of L-amino acids polymerized through peptide bonds. However, this holds for any polypeptide or short peptide. To understand what a protein is, consider that, unlike other polymers formed of L-amino acids, they result from biological evolution. They serve a biological purpose that is

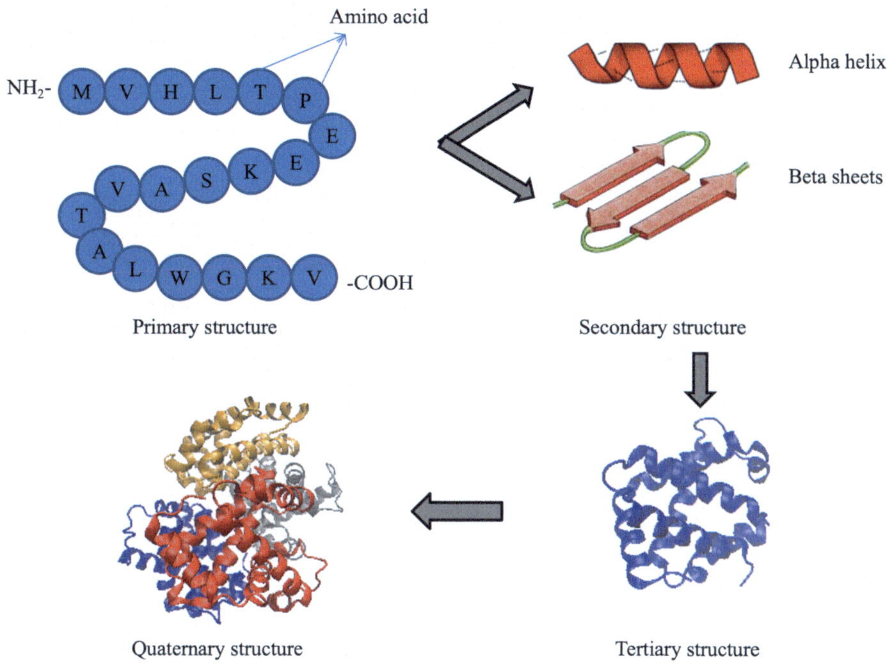

Fig. 3.2 Diagrammatic representation of the four levels of protein structure viz. primary, secondary, tertiary, and quaternary structures

defined by their distinctive 3D structures. The primary, secondary, tertiary, and quaternary structures of a protein are the four levels that are commonly utilized [Fig. 3.2]. Supersecondary structures and domains, which are intermediate levels between secondary and tertiary structures, are frequently present in proteins with specific functional roles.

Primary Structure

The primary structure of a protein is defined by the linear amino-acid sequence linked together by peptide bonds. In short, the sequence determines the structure and function of proteins. The fundamental structure is depicted from the N-terminus to the C-terminus [6]. The genetic code table can be used to infer it from the deoxyribonucleic acid (DNA) or ribonucleic acid (RNA) sequence. The most prevalent amino acid residue on the N-terminal extremity of the primary structure in the natural proteins is methionine, coded by the ATG start codon. However, in bacteria, instead of methionine, formyl-methionine is used as the first residue [7]. A polypeptide chain is a one-dimensional (1D) heteropolymer of amino acid residues. Although stop codons can incorporate two more amino acids (selenocysteine and pyrrolysine) in unusual instances, only 20 naturally occurring amino acids are directly encoded by the appropriate codons [8]. Nature can construct the enormous

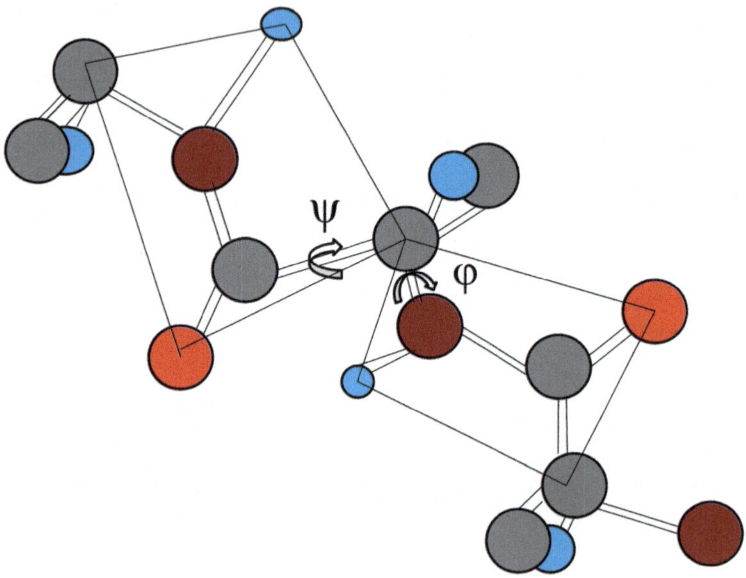

Fig. 3.3 A small segment of polypeptide chain with planar peptide groups (CONH bond) and the torsion angles ϕ and ψ are shown

variety of activities that proteins execute in living beings with the 20 different amino acids.

Due to the molecular nature of the amide bond, which takes on partial double bond properties, the peptide bond between the two amino acids is prone to resonance [Fig. 3.3]. The C-N distance is shorter than a single bond but larger than a double bond. Furthermore, the structure becomes rigid, because the linkage is no longer freely rotatable. As a result, the α-carbon atoms of two neighboring amino acids and the peptide carbonyl (C=O) and NH groups reside within the same plane [9]. The angle of torsion (dihedral) associated with this connection must be either 180° (*trans*) or 0° to be planar (*cis*). Whenever the amino acid proline occurs on the bond's C-terminal side, it forms a *cis* peptide bond. This is because of the unique character of the proline side chain, which creates a covalent bond link with its main chain nitrogen, resulting in a closed ring. As a result, proline is classified as a secondary amino acid, and neither the *trans* nor the *cis* forms are particularly beneficial in terms of energy. The torsional rotation around the polypeptide's single bonds is the polypeptide's only source of conformational freedom. Since the peptide bond is not freely rotatable, it is effectively stuck in the *trans*-state. Only two single bonds exist per residue throughout the main chain, and the torsion angles phi (ϕ) and psi (ψ) can be derived from these bonds, which are found to be 0° in the *trans* conformation. The confirmation of a specific residue's main chain can thus be defined in terms of these two parameters, which can be represented in a two-dimensional (2D) coordinate system (the Ramachandran plot), with both ϕ and ψ varying between −180° and +180°.

Secondary Structure

As protein synthesis progresses, interactions among nearby amino acid residues start to occur, resulting in secondary conformations or local patterns. The well-known α-helices and β-strands are examples of secondary structures. The consequence of systematic repeats of the ϕ and ψ angles for all residues along the polypeptide is invariably a helix. Based on the combination of ϕ and ψ, an infinite number of helix forms are theoretically feasible. However, only a tiny percentage of all theoretically possible helices are found in protein structures. Regarding globular proteins, the most important ones correspond to the α-helix and the β-strand [10, 11] [Fig. 3.4].

The α-helix is coiled with 3.6 amino acids per helical turn (5 helical turns = 18 amino acids) with a spiral shape. Helices are generally right-handed; left-handed helices only appear in exceptional circumstances, such as sequences with a lot of glycines. The primary forces that stabilize the α-helix are hydrogen bonding. The terms rise, repeat, and pitch are used to distinguish the parameters of any helix. The number of residues in a helix before it begins to repeat itself is known as the repeat. The repetition for an α-helix is 3.6 amino acids per turn per helical turn, and with each residue added, the helix elevates a certain distance (0.15 nm per amino acid). The pitch is the distance between the helix's complete turns (0.54 nm for an α-helix).

α-helices have several physicochemical and structural features in common. They are usually amphipathic, with the hydrophobic facing toward the protein core and a hydrophilic facing the water. Competing water and the amide groups of the protein chain strain the hydrogen bonds on the hydrophilic face causing helix curvature. Due to the arrangement of the individual dipoles of the peptide units, the macroscopic

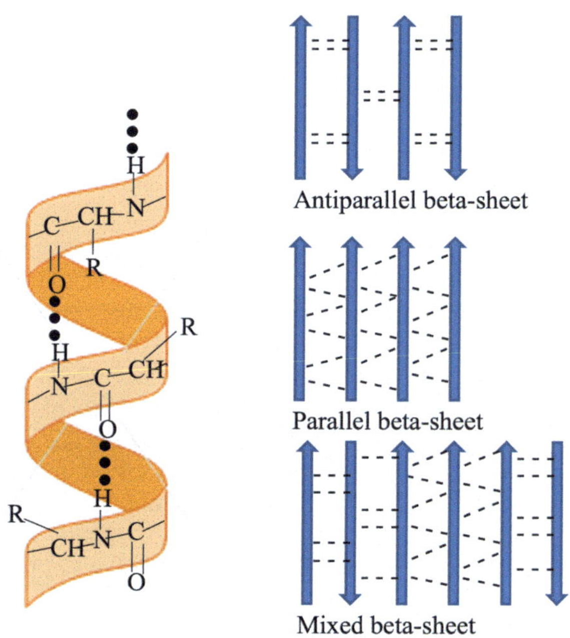

Fig. 3.4 Common secondary structures: α-helix and the β-pleated sheets. The strands in parallel β sheets run in the same direction but in opposite directions in antiparallel sheets. In mixed sheets, few strands run parallel, while others run antiparallel

dipole associated with the whole helix is generally of the order of 0.5 units of positively charged ions at the N-terminus and 0.5 units of negatively charged ions at the C-terminus [12]. Because of the free NH groups within the first turn, the N-termini of α-helices are suitable binding sites for anions like phosphate. When such an anion is absent, the side chain of the first helical residue (asparagine or aspartic acid) is commonly used to cap the N-terminus (the N-cap). Since proline is the most abundant amino acid at the second position, the sequence Asn-Pro may theoretically constitute a powerful helix initiation signal.

A β-strand is commonly described as a 2D flattened helix. Instead of coils, β-strands have bends, often referred to as pleats. β-strands can be arranged to create intricate structures like sheets, barrels, and other arrangements. Higher-order β-strand structures are sometimes called supersecondary structures, because they involve interactions between amino acids that are not close in the primary sequence. Strands can be placed parallel or antiparallel in a higher-order structure. Internal hydrogen bonds cannot be formed by β-strands [13]. The creation of hydrogen bonds between strands allows them to organize into sheets. Based on the relative orientation of the strands that make up the sheet, it might be parallel, antiparallel, or mixed. In β-sheets, the ϕ and ψ angles vary depending on the shape, and thus, β-sheets can have several topologies. Almost all β-sheets are twisted, which is a typical property. The sheet has a left-handed twist when viewed perpendicular to the strands and a right-handed twist when viewed parallel. Given the left-handed twist of the individual strands, this may appear to be a contradiction, but it is only because every second residue forms hydrogen bonds to the same end of the strand.

Tertiary Structure

Although the structure is required for protein function, the amino-acid sequence determines all the features of a protein. As protein synthesis progresses, distinct secondary structure components begin interacting with one another, resulting in folds that bring amino acids closer than they would be in the primary structure. Interactions between the R-groups of amino acids in proteins and the interactions between the polypeptide backbone and amino acid side groups have a role in folding at the tertiary level of structure [Fig. 3.5]. The complete 3D structure of the polypeptide is the result of the folding process. Nonfibrous proteins fold into unique 3D structures known as globular proteins. The same forces that drive the formation of a globular protein also keep it stable, such as hydrophobic forces, ionic interactions, hydrogen bonding, ionic bonds, disulfide bonds, and metallic bonds. Detergents, heat, pH fluctuations, urea, and reducing agents destabilize the structure of a protein, resulting in the unfolding and loss of its structure and function.

There are several tools to study the secondary structure and protein folds. Structural Classification of Proteins (SCOP) (https://scop.mrc-lmb.cam.ac.uk/) [14] and Class, Architecture, Topology, and Homology Protein Structure Database (CATH) (https://www.cathdb.info/) [15] employ hierarchical methods, which are two of the most extensively used classifications of protein folds. Folds are classified at the top of the hierarchy depending on the composition of their secondary structure. There is a widespread belief that the total number of folds is minimal and that our

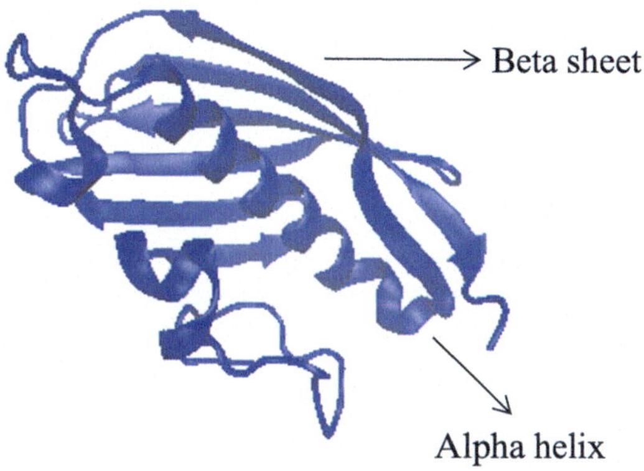

Fig. 3.5 Tertiary structure of a protein, the polypeptide chain's total 3D arrangement in space (PDB code: 1Q4S)

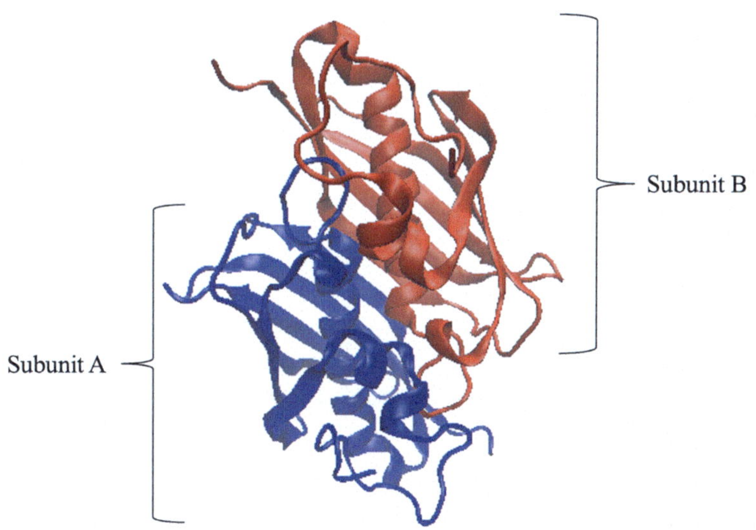

Fig. 3.6 The quaternary structure of a protein is the organization of multiple protein chains or subunits in a tightly packed pattern (PDB code: 1Q4S)

understanding of all folds that soluble proteins employ will be complete within the next few years due to structural genomics initiatives. It is evident that some folds are more prevalent than others, and some authors have labeled them superfolds, for example, the triose phosphate isomerase (TIM) barrel, the jelly-roll, and the up-and-down 4-helix bundle [16].

Quaternary Structure

The protein subunits interact via noncovalent interactions to form quaternary structures or oligomers in a specific arrangement of the subunits [Fig. 3.6]. Only a small percentage of proteins are made up of a single polypeptide chain. The remaining proteins form oligomers, macromolecular assemblages, and polymers as homo- or hetero-oligomers [17]. When two polypeptide chains combine to form an interface, they form isologous or heterologous interfaces. The interaction in the first scenario results in the formation of an asymmetric dimer with a two-fold axis, in which if region A on the first subunit interacts with a region B on the second, region A of the second must interact with region B on the first. This is not the case in the case of heterologous interactions. They can be symmetric or asymmetric, resulting in ring-like structures with higher-order (>2) symmetry axes. Internal symmetry exists in almost all oligomeric formations. Because the individual subunits are asymmetric and chiral, symmetry can only be achieved through the coupling of subunits around pure rotation axes.

The amino-acid sequences of most proteins are thought to include all the data required to fold them into their 3D structure. However, complementary "patches" on the surfaces of proteins are required to form the quaternary structure. As a result, the patches buried in the subunit interfaces play an essential role in protein oligomerization.

3.1.2.2 Protein Structure Prediction

Computational approaches for protein structure prediction from their amino-acid sequence have blossomed over the last few decades. From a physical standpoint, the amino-acid sequence specifies the core structure of the protein, with the native form as the most stable conformation with the minimum free energy. Though we know that various physical principles drive protein folding, fully describing such a complicated macromolecule is incredibly difficult. On the other hand, searching the massive quantity of conformational space is still challenging. Protein structure prediction can be performed in a variety of ways. First-principles methods with database information, first-principles methods without database information, comparative modeling, and fold recognition are the four structure prediction approaches. As prediction methods have evolved, the distinctions between these categories have blurred [18]. Despite the hazy differences, categorizing the approaches is helpful.

Comparative Modeling

The structure of a protein is predicted using comparative modeling by comparing its amino-acid sequence to sequences whose native 3D structure is known. Comparative modeling or homology modeling is based on the idea that structure similarity follows sequence similarity. Examples of homology modeling tools include the SWISS-MODEL (https://swissmodel.expasy.org/), MODELLER (https://salilab.org/modeller/), etc. However, the degree of sequence similarity significantly impacts the accuracy of comparison modeling predictions. Predictions are usually of good quality when the target and template share more than 50% sequence identities. More than 80% of the C-atoms may lie within 3.5 Å of their actual positions if the sequence is 30–50% identical. However, if the sequence identity is less than 30%, the prediction will likely contain many mistakes. Comparative modeling generally entails selecting templates from a database, aligning them to the target sequence, and fine-tuning side-chain geometry and low-identity regions. The first phase, template identification, has well-established techniques and implementations. Basic Local

Alignment Search Tool (BLAST) (https://blast.ncbi.nlm.nih.gov/Blast.cgi) and its improvements, such as Position-Specific Iterative Basic Local Alignment Search Tool (PSI-BLAST), have become well established and are used as benchmarks for any new method development [19].

Fold Recognition

Fold recognition methods aim to predict the 3D folded structure of amino-acid sequences when comparative modeling methods fail. Examples of fold recognition tools include 3D-pssm (http://www.sbg.bio.ic.ac.uk/3dpssm/index2.html), PDBeFold (https://www.ebi.ac.uk/msd-srv/ssm/), etc. The idea behind fold recognition algorithms is that the protein structure is more evolutionarily conserved than sequence. The number of unique folds has been estimated by categorizing the structures in the protein data bank (PDB) into families [20]. Since there are more sequences than folds, fold identification systems attempt to find a model fold for a certain target sequence from the available folds even though no sequence similarity is established. Fold recognition techniques employ advanced sequence comparison methods. However, the distinction between fold recognition and homology modeling methods is hazy, especially in approaches based on hidden Markov models (HMM). As noted earlier, comparative modeling fold identification techniques include models and location-specific iterated BLAST searches. Another method for fold recognition is to predict and compare secondary structures [21]. The fact that sequences with less than 10% similarity can also have more than 80% secondary structure similarity underpins this subclass. However, any such method is only as efficient as the secondary structure prediction method on which it is based. Finally, one of the most important fold recognition methods is threading, which involves comparing sequences to a known 3D fold [22]. Threading algorithms try to match a target sequence to a known structure in a fold library. Threading-based approaches are well known for their high computational costs. Several threading algorithms neglect the interaction between residues in pairs, and thus, the threading problem is significantly simplified, which can be handled using dynamic programming.

Prediction Based on First Principles Using Data from Database

The term ab initio refers to structure prediction approaches that do not rely on experimentally established structures. For example, trRosetta (https://yanglab.nankai.edu.cn/trRosetta/help/), I-TASSER (https://zhanggroup.org/I-TASSER/), etc. Since the emergence of fragment-based approaches, ab initio methods have grown increasingly ambiguous. These methods compare fragments of a target to fragments of known structures retrieved from the PDB [20]. After finding the correct fragments, they are put together into a structure, often using scoring systems and optimization algorithms. The method can be compared to physics-based first principle methods, since scoring functions are analogous to energy functions, and fragment assembly with optimization algorithms is akin to free energy optimization. The first principle methods compatible with free energy minimization can be called first principle structure prediction methods; however, the fragment assembly methods do not hold the same. Local interactions induce a bias but do not uniquely characterize the local structure, according to fragment assembling approaches. Nonlocal interactions between nonneighboring residues are projected to fix local degrees of freedom, resulting in a compact overall design. By pooling across observable fragment configurations in known protein structures, fragment-based approaches mimic this structural bias. Compact structures are formed by randomly mixing pieces and simulating annealing when acceptable fragments have been found. Following that, scoring functions generated from conformational statistics of known proteins are used to evaluate the fitness of a conformation. Improvement in fragment assembling methods may be attributed to Baker et al. Notably, in the existing critical assessment of structure prediction (CASP) studies, fragment assembly strategies performed reasonably well across target classes [23].

Prediction Based on First Principles Without Using Data from Database
The first principle structure prediction is an essential supplement to any knowledge-based methodology for various reasons. These methods use Anfinsen's thermodynamic theory by attempting to find the minimal free energy of a protein in a particular surrounding. Examples of such predictions include CrySPY (https://tomoki-yamashita.github.io/CrySPY_doc/), ASTRO-FOLD (https://www.ncbi.nlm.nih.gov/pmc/articles/PMC1303441/), etc. Sometimes, even a remotely related structure homologue may not be available. In these instances, the only option is to use first-principle methods. Additionally, new structures are continually being discovered that cannot be detected using methods that rely on the comparison to known structures. Third, knowledge-based strategies have been criticized for accurately predicting protein structures without requiring a comprehensive understanding of structural formation mechanisms and driving forces. In contrast, first principle structure prediction methods rely on physical models of these systems to make predictions. As a result, they can assist in distinguishing between right and wrong modeling assumptions and deepen our understanding of protein folding mechanisms. Only physically relevant potentials and atom representations can be used to apply such approaches to any given target sequence. Because of the large range of targets and the difficulty of directly or indirectly employing database knowledge, such approaches are meant for the most challenging protein structure prediction methods [24].

3.1.3 Protein and Metabolite Identification

Protein identification is one of the most basic tasks that mass spectrometry (MS) can undertake. In over 90% of samples for MS analysis, protein identification is anticipated. Peptide mapping and tandem MS are the two main types of MS techniques used to identify proteins. Protein identification is performed using either a bottom-up or top-down approach. Identifying proteins that are predigested into peptide fragments is known as bottom-up MS, while the intact protein is detected using top-down MS. The bottom-up approach is more prevalent than the top-down, which still faces substantial technological challenges before becoming widely used in MS laboratories.

3.1.3.1 Peptide Mapping

Peptide mapping is performed on a single protein or a combination of proteins. Before MS analysis, the protein must be digested into peptides for identification via peptide mapping. After the peptides are digested, their molecular weight is determined. Peptide masses produced from a library of in silico digested proteins or translated nucleic acid sequences are compared to experimental peptide masses [Fig. 3.7]. If several experimental masses meet values for a particular protein in the database within a given mass tolerance, the protein can be identified. Since the accuracy and precision between the experimental and in silico masses are essential for proper protein identification, generating the peptide map on equipment with high mass measurement accuracy is preferred; the more matches between the experimental and database peptide mass, the more confident the protein sequence identification. Specific software is also required to translate raw MS data and identify a protein. Several free software applications are available for examining peptide mapping data [Table 3.1]. At the absolute least, the programs require an input that includes the experimental peptide ion list and a database against which it may be compared.

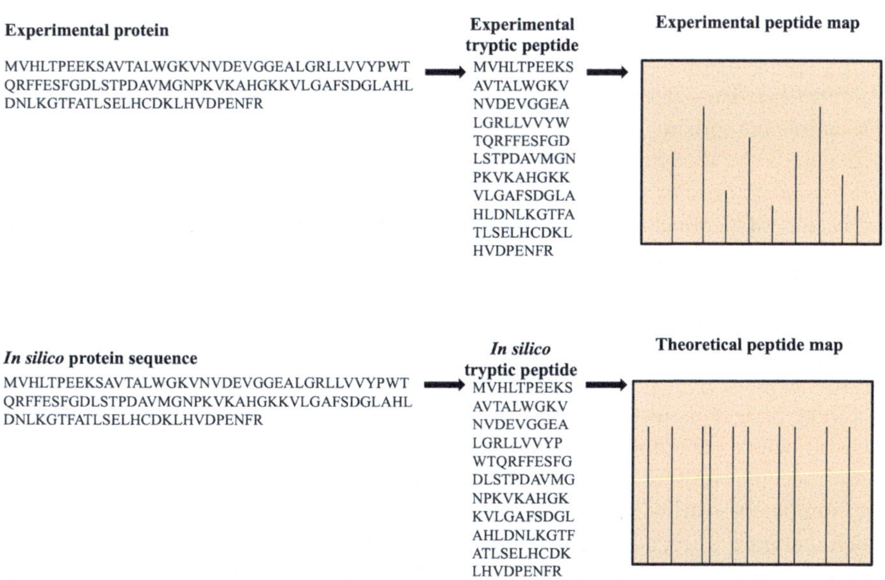

Fig. 3.7 Mass spectrometry (MS) peptide mapping for protein identification. In the initial stage, the protein is proteolytically digested, and the experimental masses of the peptides are determined using MS. Depending on the specificity of the enzymes utilized, the sequences of the proteins in the specified database are digested in silico. The masses of the peptides are estimated, and theoretical mass spectra are constructed. To identify the correct protein, the best match among experimental and theoretical mass spectra is employed

Table 3.1 Protein identification software based on peptide mass fingerprinting and tandem mass spectrometry data

Technique	Software Program	URL
Peptide mass fingerprinting	Mascot	http://www.matrixscience.com
	MS-Fit	http://prospector.ucsf.edu
	MultiIdent	http://web.expasy.org
	MassWiz	http://masswiz.igib.res.in
	Protein Lynx	http://www.matrixscience.com
	ProFound	http://prowl.rockefeller.edu
	PepMAPPER	http://www.nwsr.manchester.ac.uk
	Request	http://www.thermo.com
Tandem mass spectrometry	Crux	http://noble.gs.washington.edu/proj/crux
	MS-Tag	http://prospector.ucsf.edu
	Pep-Frag	http://prowl.rockefeller.edu
	X! Tandem	http://www.thegpm.org/tandem
	Sonar MS/MS	http://hs2.proteome.ca/prowl/sonar/sonar_cntrl.html
	OMSSA	http://pubchem.ncbi.nlm.nih.gov/omssa

3.1.3.2 Tandem Mass Spectrometry

Only single proteins or a simple mixture of two or three proteins can be identified using peptide mapping. MS2 is necessary for a complicated sample with multiple proteins [25]. This approach can distinguish between isolated proteins and proteins in a sample comprising up to 100,000 species. Bottom-up protein identification with MS2 entails the digestion of proteins into peptides. Collision-induced dissociation (CID) is a process in which peptides collide with inert gas and fragment into a succession of peptide ladders. Fortunately, CID causes peptide ions to fragment largely along their backbone.

3.1.3.3 Protein Databases

In addition to a mass spectrometer and tools for translating raw MS data to identify peptide sequences, a protein database is required. Although de novo sequencing should be able to identify MS2 spectra in the future, this process will take years of research before becoming widespread. Amino acid or nucleotide sequences that are transcribed into protein sequences can be found in protein identification databases. The number of entries in these databases has exploded in tandem with the rate at which genome sequencing has progressed over the years. There are numerous publicly accessible sequence databases. The UniProt knowledgebase (https://www.uniprot.org/), the national center for biotechnology information (NCBI) nonredundant protein database (https://www.ncbi.nlm.nih.gov/refseq/) [26], or the International Protein index database (http://www.ebi.ac.uk/IPI) are used to validate most experimental MS data. These databases are updated regularly to ensure that proteome researchers have access to the most up-to-date sequence information.

3.1.3.4 Bottom-Up and Top-Down Mass Spectrometry

The majority of proteins identified by MS adopt a bottom-up strategy. Proteins are digested into peptides in this approach, and identifying these peptides is used as a substitute for protein identification. The fundamental issue in this technique is that it lacks clear proof of important biological features such as alternate splice variants, various modifications, and distinct protein cleavage sites. Bottom-up aggregation of all discovered peptides into a single protein sequence makes it impossible to discriminate between active and inactive populations. Top-down proteomics measures the intact proteins, allowing each changed variation to generate its signal [27]. As a result, numerous altered protein versions may be detected. The primary advantage of top-down MS is that the molecular weight of the entire protein is determined experimentally, allowing the contributions of each amino acid to be recorded as a single signal. Even with today's high mass measurement accuracy instruments, knowing only the total mass of a protein is inadequate for its identification. The protein must be fragmented in the same way as peptides examined with MS2. Unfortunately, top-down techniques for recognizing complete proteins have lower sensitivity than bottom-up approaches for peptides.

3.1.3.5 Metabolite Identification

The most extensively utilized techniques for metabolome analysis are mass spectrometry (MS) and nuclear magnetic resonance (NMR) spectroscopy. Global metabolomics is a research strategy that involves identifying, characterizing, and estimating as many metabolites in biological samples as analytically possible to find changes that can be used to distinguish between two samples.

MS Metabolite Identification

In gas chromatography-mass spectrometry (GC/MS) and liquid chromatography-mass spectrometry (LC/MS), metabolites are identified and confirmed using fragmentation pattern, molecular weight, and retention time or by comparing the metabolites' spectra and pattern of fragmentation to those of pure substances. There is no single screening strategy for nontargeted metabolomics that can provide complete coverage because of weak ionization, limited analyte concentration, and poor chromatographic retention. The knowledge and expertise, as well as available tools, are used to perform ab initio identification of small molecules. Organic compounds that have the same molecular weight but exist in different forms make the identification process difficult. In this case, the isomers are resolved by chromatography before being identified by MS. Due to the linear and repeating nature of their building blocks, MS makes it easier to identify proteins, peptides, DNA, and RNA than it is to identify low-molecular-weight substances [28]. To aid in the identification of unknown metabolites, several accessible and valuable databases have been created [Table 3.2].

A targeted metabolomics analysis concentrates on a specific group of molecules, such as the principal metabolites of a potential treatment candidate or a known metabolic pathway that leads to drug toxicities or detoxification. The identification method requires analyzing experimental data, such as mass-to-charge ratio (m/z) and

Table 3.2 Metabolic pathway websites that help in the identification of unknown metabolites

Pathway	Description	URL
KEGG	A database of pathways that shows how genes or molecules are connected.	https://www.genome.jp/kegg/
Reactome	An open-source curated database with analytic tools for pathways and networks.	http://www.reactome.org
BioCyc	A repository of organism-specific pathway and genome databases.	http://biocyc.org
ExPASy	It is a bioinformatics portal that's both extensible and integrative.	http://www.expasy.org
SGD	*Saccharomyces* genome database contains information on the yeast *Saccharomyces cerevisiae's* molecular biology and genetics.	http://www.yeastgenome.org
PMN	A curated database of plant metabolic networks.	http://www.arabidopsis.org

retention duration, with pure standards if the goal is to locate known metabolites. If somehow the metabolic compounds can be predicted, metabolite identification requires finding representative standards and analyzing experimental data for the expected metabolites. Metabolite identification becomes substantially more difficult when the metabolites cannot be predicted or are unknown, as in the case of biomarker discovery, and may require numerous chromatographic separations with varying chemistries and mobile phases. A high-resolution mass spectrometer for accurate mass analysis, statistical analytic tools for data mining, metabolite databases, and access to hundreds of pure compounds for confirmation would also be required.

NMR Metabolite Identification

The core strength of NMR spectroscopy is metabolite identification, either by structure confirmation or de novo. The NMR frequency of a nucleus is influenced by the nuclear isotope, the field strength of the NMR magnet, and the local electronic environment [29]. The overall structure is determined using scalar couplings, chemical shifts, molecular connections, and spatial data gained from a set of 1D and 2D experiments. The chemical shift scale, which is independent of the magnetic field intensity, is commonly used to express NMR peak frequencies in parts per million (ppm). Chemical shifts are common among several functional groups, and this insight is crucial for determining structure.

3.1.4 Folding of Proteins

When it comes to multidomain proteins, folding is a difficult task. Multidomain proteins are those that have more than 200 amino acids in them, with each domain having independent folding. The presence of multiple domains aids in the stability and flexibility of the protein. In contrast, the folding rate is fast when a single domain is examined. The rate-limiting phase in the folding process of multidomain proteins is the formation of interfacial interactions between distinct domains. The folding kinetics is divided into three stages: an initial fast phase; an intermediate state that lacks native-like characteristics, is more prone to proteolysis, and lacks catalytic properties; and a final stage with a relatively slow folding rate. During the final phase, the already folded domains of a protein are paired together. The rate of folding is inversely proportional to the solvent viscosity. High protein concentrations, such as those seen in the cellular environment, aid in domain pairing, but they tend to encourage the formation of dimers rather than monomers. In many cases, it has been found that ligand binding is one of the factors responsible for correct protein folding. Efficient protein folding is required for normal cell functioning, which is dependent on the coordinated actions of chaperones, chaperonins, and auxiliary cofactors [30]. When this system malfunctions, produced proteins are unable to fold correctly, resulting in their aggregation. Protein misfolding is the failure of a given protein to return to its original and active state. Due to the loss of normal physiological functioning, protein misfolding has a wide range of pathogenic ramifications (Fig. 3.8).

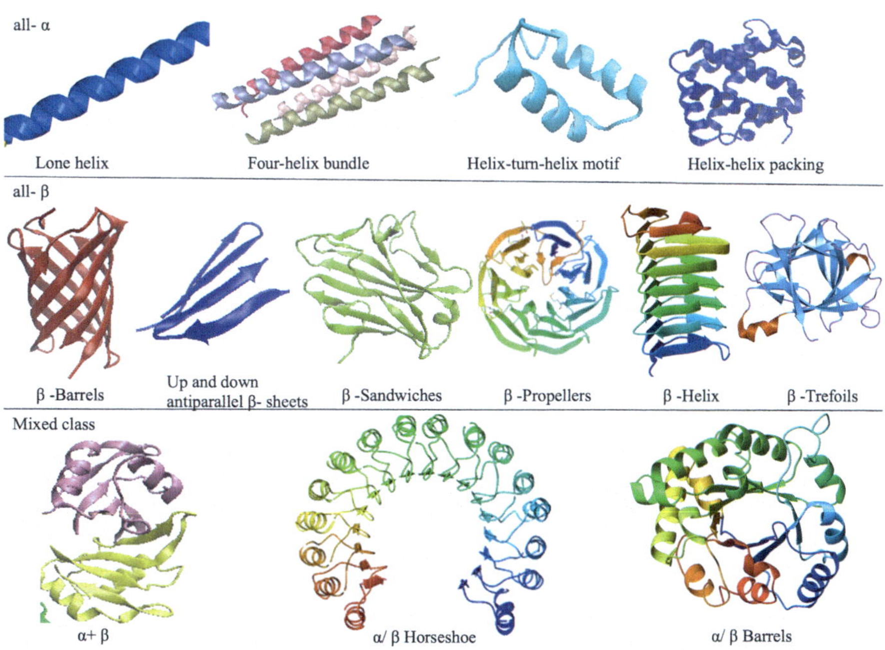

Fig. 3.8 Protein models from the all-α, all-β, and mixed classes are represented structurally

3.1.4.1 Structural Classes of Proteins

(a) **All-α Proteins**

(i) **Lone helix:** Lone helix proteins, such as alamethicin, are small proteins with only one helix.

(ii) **Helix-turn-helix motif:** Two antiparallel helices are joined by a short loop in the helix-turn-helix motif, as seen in the RNA-binding protein Rop.

(iii) **Four-helix bundle:** A bundle of four helices connected by three loops. The interfaces of helices are hydrophobic, while the surface is hydrophilic. They are found in photosynthesis reaction centers, G-protein-coupled receptor membrane-spanning regions, cytokines, DNA-binding proteins, etc. They are also seen in a globin fold with two bundles, each with four helices.

(iv) **Helix-helix packing:** α-helices are packed so that their interfacial regions are complementary, and their side chains are buried.

(b) **All-β Proteins**

(i) **β-sandwiches:** Also known as immunoglobulin folds, they are two sheets of immunoglobulin packed together like a sandwich. Two sheets are either aligned or orthogonal in their packing. The two sheets are connected by residues that are not in the β-sheet conformation.

(ii) **β-barrels:** A protein domain with an antiparallel β-sheet and no fixed strand arrangement.

(iii) **Up and down antiparallel β-sheets:** Antiparallel strands that make up the β-sheet are joined by loops of adjacent strands that resemble Greek keys. Hairpins join three up-and-down strands, with the fourth strand resting close to the first.

(iv) **β-propellers:** This fold is a super barrel with six four-stranded antiparallel sheets arranged in an up-down topology.

(v) **β-trefoils:** Has a three-fold axis of symmetry, for example, cytokinin interleukin-1.

(vi) **β-helix:** With β-strands looped around the structure, this fold resembles a helical topology.

(c) **Mixed Class**

(i) **α/β:** This fold is found in proteins that contain repetitive β-α-β supersecondary units (right-handed) with an outer layer of α-helices and an inner core of parallel β-sheets. The β-strands and α-helices are parallel to each other, while the α-helices are antiparallel to the strands. This fold is also known as the Rossmann fold and is found in many glycolysis enzymes, cytosolic proteins, and nucleotide-binding proteins.

(ii) **α/β horseshoe:** They resemble an open horseshoe with α-helices at the curve's surface and a curve produced by repeated units α/β with a parallel β-sheet. The β-strands run parallels to the center axis and are somewhat inclined to one another.

(iii) **α/β barrels:** This is an eight β-α sequence barrel-like structure, with the first strand of hydrogen linked to the last strand. The fold is closed like a barrel, with the α-helices on only one side of the β-sheet.

(iv) **α + β:** They possess α and β secondary structural features but no particular topology.

3.1.4.2 Protein Misfolding

With the help of chaperones, multidomain proteins assume their native state in vivo shortly after ribosome synthesis. In vitro protein folding has gained importance due to its association with various exciting facts, such as how the chain attains the most stable shape within milliseconds to seconds and the prediction of 3D shape from the proteins' amino-acid sequence. Proteins can assume zillions of different conformations, and if the proper stable conformation is found, a one degree deviation can dramatically raise the stability of the chain by several orders of magnitude. A unique folding mechanism likely drives the unfolded protein, allowing the process to occur rapidly. According to molecular simulations using lattice models of protein chains, protein folding is triggered by the formation of the nucleus, with the folding rate based on protein size. The entire process seems to be under thermodynamic control [31].

If the appropriate folding of the 3D structure does not happen correctly and the protein undergoes erroneous folding, it is referred to as protein misfolding. Amyloid is a term used to describe protein aggregation caused by misfolding. The deposition of amyloid has been linked to a large number of human diseases. Parkinson's disease, Alzheimer's disease, Huntington's disease, and spongiform encephalopathy are neurodegenerative disorders that damage the central nervous system.

3.1.5 Motif, Domains, and Scaffolds

In the PDB, there are iterations and laws of structural biochemistry that can be discovered using machine learning (ML) approaches. Generalized patterns can help with data retrieval and integration by establishing a theoretical framework to which additional structural data can be linked. Because they often represent common 3D structural properties, they can also be utilized to predict molecular conformation from the topological structure. In molecular biology, similar research into the links between protein structure and amino-acid sequence is a prominent topic. Before protein activity and function can be understood at the molecular level, the 3D structure of the protein must be understood.

A protein fragment is an amino-acid sequence that has been observed, such as a portion of the primary or tertiary structure. A protein motif is made up of one or more fragments that have been assembled into a single entity [32]. There are four different types of protein motifs. Sequence motifs are linear sequences of residue identifiers that are topologically arranged. Sequence-structure motifs are sequence motifs with secondary structure identifiers connected to one or even more residues in the motif. The corresponding structure is thought to be predicted by the sequence. Structure motifs are 3D structural items defined by the placements of residue objects in 3D Euclidean space. Even though most studies focus on the consistency of structure motif components, structure motifs lack sequence information. Finally, structure-sequence motifs combine 1D and 3D structures to connect sequence data to a structural motif [33]. When employing a structure sequence motif, the direction of inference among structure and sequence does not need to be set.

3.1.5.1 Machine Discovery of Protein Motifs

Finding various forms of protein motifs by machines and ML approaches are currently at the forefront of molecular biology. Empirical machine discovery refers to the theories and automated processes that are involved in finding novel regularities, conceptions, or dependencies in data. It is much simpler to link a protein motif to a concept with both a formal and an extensional description. When using concept discovery approaches, precise mathematical semantics for a motif are required to determine if an observation fits inside that concept's extension. Many studies have been done on machine discovery of protein motifs [34]. Most of the work is based on numerical clustering: fragments are described by a set of numeric parameters, a parameter is calculated based on these specifications, and motifs arise as clustering centroids throughout the clustering process.

Sequence and Sequence-Structure Motifs

Finding sequence motifs using structurally similar proteins or their fragments is an integral part of figuring out the structure of proteins and the sequence connections. Protein sequence motifs can help collect data and retrieve information bases structured by sequence similarity. To find sequence motifs, a maximum alignment with one or even more protein sequences can be employed, followed by the removal of residues at matched places. Identical residues are those that are conserved at matching alignment places. Because nonhomologous proteins rarely have long interconnected sequences of conserved regions, histograms of residue distributions at aligning locations are necessary to construct a consensus sequence motif. Most work on protein secondary structure prediction is based on the a priori interpretation of sequence motifs that are indicative of a specific type of secondary structure identification [35].

Structure Motifs

The discovery and use of structural elements that are less coarse than standard secondary structures, coupled with the packaging of secondary structures into three dimensions, has lately acquired popularity. This progress can be attributed to three factors. First, different approaches for distinguishing secondary structure from tertiary structure vary significantly. A prediction system that uses a single assignment technique for training and evaluation models its specific properties to some extent. Second, although these areas are neither randomized nor indefinable, unidentified folding patterns are called random coils. Finally, packing secondary structure elements is a complex process that relies on reliable secondary structure predictions. On the other hand, structure motifs are building pieces that can be utilized to precisely characterize a new protein's tertiary structure. Numerical clustering is commonly used to find structure motifs. This method needs a predetermined range of numerical values and a distance measure of multidimensional vectors to characterize observations [36].

Structure-Sequence Motifs

Both the amino-acid sequence and three-dimensional (3D) coordinate information are assigned to residues by the structure-sequence motifs. This motif differs from the sequence-structure motif, because it necessitates the presence of a 3D structure. In molecular biology, structure-sequence motifs are already receiving considerable interest. The inverse structure prediction problem, wherein a sequence is predicted for a given structure, is solved using a compact library containing structure-sequence motifs. The sequences of these motifs can be conserved sequences or probabilistic, with propensities for distinct amino acids at every sequence specified [37]. Each sequence to structure motif discovery has a limitation: it can only associate a structure with a single feature or conserved sequence.

3.1.6 Current and Future Protein Therapeutics

The first protein therapeutics were recombinant analogues of natural proteins. Drugs that replace a missing or abnormal protein, enhance an existing pathway, provide a novel function, intrude with a compound or organism, or deliver a payload such as a cytotoxic drug, or protein effectors, are regarded as protein therapeutics based on their pharmacological activity. Some molecular categories of protein treatments include antibodies, anticoagulants, blood factors, engineered protein scaffolds, enzymes, growth factors, hormones, interferons, interleukins, and thrombolytics. With 24 commercialized antibody medications and over 240 more in clinical development in the USA, antibody-based drugs are the most common and rapidly expanding type of protein treatment [38]. The development of enhanced protein therapeutics, such as antibodies, has resulted from a convergence of clinical, commercial, scientific, and technical forces highlighting major unmet requirements and providing solutions to fulfill them. Protein treatments have a bright future ahead. They can make a difference in terms of efficacy, safety, and immunogenicity. These goals will likely be aided by several new technologies for manufacturing and developing protein therapeutics.

3.2 Therapeutic Protein Engineering

3.2.1 Protein Engineering Techniques

Due to the rapid advancements, particularly in recombinant DNA (rDNA) technology, a variety of protein engineering approaches are now available. The most familiar method in protein engineering is the rational design technique, which includes site-directed mutagenesis (SDM) of proteins [39]. When the structure and function of the protein of interest are known, rational structure-based protein design is a promising strategy. However, there is a scarcity of information on protein structures for protein engineering. For the desired protein properties, evolutionary techniques were presented as an alternative [40]. Peptidomimetics is another approach that is used in protein engineering. The design and manufacture of biologically stable peptide analogues comprise imitating or suppressing the functioning of enzymes or endogenous peptides [41]. In vitro protein evolution systems and other protein engineering techniques are equally relevant. They are predicated on the premise of gene hierarchy evolution. Following combinatorial and hierarchical processes, modern genes are assumed to have arisen from the genotypic analysis of classical genes [42].

In contrast to rational methods where site-directed mutagenesis is used, random mutagenesis is used in random methods, and in evolutionary approaches, DNA shuffling is employed. In the DNA shuffling approach, a set of genes containing double-stranded DNA and identical sequences are obtained through various species or produced using error-prone polymerase chain reaction (PCR). DNaseI is used to degrade these genes, resulting in randomly cleaved small fragments that are purified and reassembled using DNA polymerase. The pieces are used as PCR primers, which align and cross-prime each other. As a result, an rDNA is formed that contains parts of multiple parent genes. Alternatives to DNaseI include using a mixture of restriction endonucleases rather than DNaseI and the staggered extension procedure, which would not involve parental gene fragmentation.

Cell surface libraries coupled with fluorescence-activated cell sorting (FACS) or phage display technology are also used to screen large libraries of proteins/enzymes [43]. A substrate scissile bond is used in the technique, which can be either cleaved or not by a surface-displayed enzyme. A fluorophore and a quencher are linked via a scissile bond on the specified substrate. The fluorescence emission is suppressed by the quenching fluorophore if the surface-displayed enzyme fails to cleave the substrate's scissile bond. As a result, no fluorescence is produced. The fluorophore and quenching fluorophore are separated when the enzyme breaks the scissile bond of the substrate, which results in fluorescence. Phage display technology is another excellent method for screening large protein libraries. Degenerate reverse primers are used in a PCR for random mutagenesis of the initial cDNA throughout the whole target region. The products from the PCR are then subcloned into a bacteriophage vector that encodes the phage coat protein. Each phage in the mutant pool expresses a unique protein that is visible on the phage's coat protein on the surface. Elution tests help identify variants that adhere strongly to a target substrate and can be used to screen them. The variants that have been discovered are then purified and sequenced as a consequence. In protein engineering, flow cytometry, a powerful technique for single-cell research, is widely used [44]. In antibody and peptide surface display research and enzyme engineering of intra- and extracellular enzymes, there are various examples of sorting based on ligand binding. Several random mutagenesis procedures utilized in protein engineering have also been evaluated and compared for their benefits and drawbacks.

Another set of valuable tools in protein engineering and production are cell-free translation systems [45]. They can be used instead of in vivo protein expression.

When template DNA or messenger RNA (mRNA) is given to a reaction mixture and allowed to incubate without the cells, proteins are generated. The ribosomal protein system of cells is used in cell-free translation systems, which is acquired as a supernatant after centrifugation at 30000 g of a cell extract from *E. coli* or other bacteria. Ribosomes, aminoacyl-tRNA synthetases, tRNAs, and translation factors are all found in this supernatant, which is essential for protein synthesis. The synthesis of artificial proteins, physiologically active proteins, and membrane proteins are possible applications. With more research, cell-free translation systems can be a potential substitute for in vivo protein expression due to a high level of controllability and simplicity. In addition, the disadvantages of recombinant protein expression in living cells, including protein degradation and aggregation, will be avoided [46].

Another vital protein engineering strategy for changing enzymatic functions is designed divergent evolution. Divergent molecular evolution hypotheses underpin the method. According to these theories, enzymes with more specific and catalytic functions developed from those with more permissive functions. Second, this process is driven by a few amino acid alterations, and third, the effects of double or multiple mutations are typically cumulative. As a result, the method allows for identifying mutation combinations that introduce the necessary functions into enzymes [47]. Either naturally occurring peptides or systems that have been intentionally designed make up stimulus-responsive peptide systems. The capacity of proteins and peptides to modify conformation in response to an external stimulus such as pH, temperature, or specific chemicals is used in these systems [48].

Protein engineering studies benefit from the receptor-based quantitative structure-activity relationship (QSAR) approaches. The development of these strategies relies on a computational blend of structure-activity correlation and structure-based design [49]. They provide crucial pharmacological information about therapeutic targets. For example, comparative binding energy (COMBINE) research examines how amino acid variations and structural alterations in a protein affect the activity in a group of homologous protein receptors [50].

3.2.1.1 Rational Protein Design

The rational design approach for protein engineering is site-directed mutagenesis of proteins [51] [Fig. 3.9]. Site-directed mutagenesis allows a specific gene sequence corresponding to a particular amino acid(s) to be introduced inside a target gene. For site-directed mutagenesis, there are two methods. One is the overlap extension method. This method uses two primer pairs, one of which carries the mutant codon along with the mismatched sequence. The PCR reaction consists of two PCRs and the creation of two double-stranded DNA (dsDNA) products and uses these four primers. Two heteroduplexes are generated during denaturation and annealing, and each strand of the heteroduplex contains the desired mutagenesis codon. The nonmutated primer set is then used for amplifying the mutagenic DNA, and the filling of the overlapping ends of each heteroduplex at the 3′ and 5′ end is done by DNA polymerase. The "whole plasmid single round PCR" method is another site-directed mutagenesis technique. The commercially available

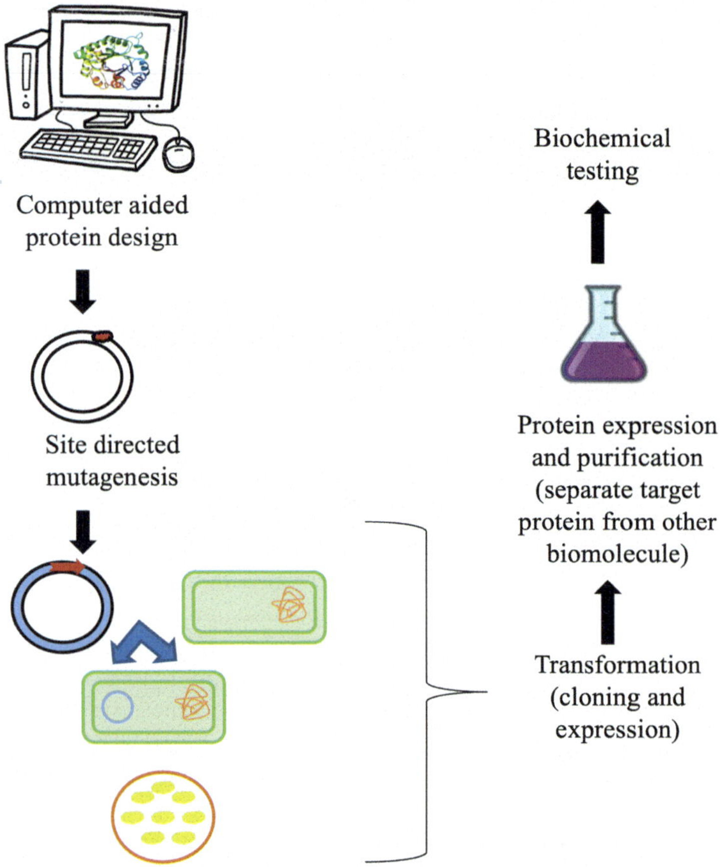

Fig. 3.9 Diagrammatic representation of protein engineering by rational design

QuikChange Site-Directed Mutagenesis Kit (Agilent Technologies) is based on this method. It requires two oligonucleotide primers with the desired mutations complementary to the opposite strands of a dsDNA plasmid template. PCR is performed with DNA polymerase, and both the template strands are reproduced without the primers being dislocated, yielding a modified plasmid with nonoverlapping breaks. After that, DpnI methylase is employed to selectively digest the vector to generate a circular, nicked vector containing the mutant gene. When the nicked vector is transformed into competent cells, the nick within the DNA is repaired, yielding a circular, altered plasmid.

3.2.1.2 Random Mutagenesis

In many cases of protein engineering, there is limited information on the protein structure and function. Evolutionary techniques such as random mutagenesis and selection were developed as a substitute to design new proteins with the desired properties. Random mutagenesis could be a helpful technique, especially when there is little knowledge about the structure and function of a protein. Saturation mutagenesis is a simple and widespread approach to random mutagenesis [52]. It comprises substituting each of the native amino acids for a specific amino acid in a protein, resulting in all possible variants at that location.

Localized or region-specific random mutagenesis is another way for protein engineering that blends rational and randomized methods. It entails simultaneously replacing a small number of amino acid residues at a specific point to produce proteins with new functions. The approach uses overlap extension and whole-plasmid, single-round PCR mutagenesis, just like site-directed mutagenesis. Because the amino acid codons are randomized, a variety of 64 specific forward and reverse primers entrenched on a statistical mixing of four bases and three nucleotides in a randomized codon can be used.

3.2.1.3 Combinatorial Protein Design

In rational protein engineering, changes are generally made in or near the region of proteins that are directly relevant to the function, such as the active site of enzymes or the interaction surfaces of binding proteins. Using rational approaches, specific substitutions are impossible to foresee [53]. In the pursuit of better protein variations, many researchers were compelled to look at a larger sequence space, which led to the creation of technologies for random mutagenesis, in which genes or gene fragments are genetically diversified at random (Fig. 3.10).

The discovery of methods for random mutagenesis, in which genes or gene segments are genetically diverse in a random manner, is a vital component of the objective, which is to examine a broader sequence space in their hunt for superior protein variants. Random or semirandom nucleotide substitution and random recombination of gene segments are common methods of such diversification. Because these methods frequently create a high number of protein variants, effective functional analysis is required through powerful screening methods. Using synthetic oligonucleotides during gene assembly or PCR amplification is the most popular approach to introducing genetic diversity through genetic engineering [55].

Error-Prone PCR

Using the DNA polymerase's inherent error rate as a starting point, nucleotide alterations can potentially be inserted into a DNA fragment during PCR amplification. The Taq DNA polymerase from *Thermus aquaticus* is the most often used thermally stable enzyme for PCR amplification of DNA [56]. This enzyme has no proofreading function, resulting in a mutation rate of about 5.5×10^{-4} per nucleotide. The amplified DNA fragment must be long to ensure that mutations are introduced under conventional PCR settings with this relatively low error rate [57]. As a result, numerous ways to increase the error rate have been developed, such as changing the buffer composition or using a high concentration of DNA polymerase, a significant number of PCR cycles, and a less amount of template. The fact that various sections of a gene have varied error rates is a disadvantage of these PCR-based approaches. The mutants that result tend to favor specific substitutions.

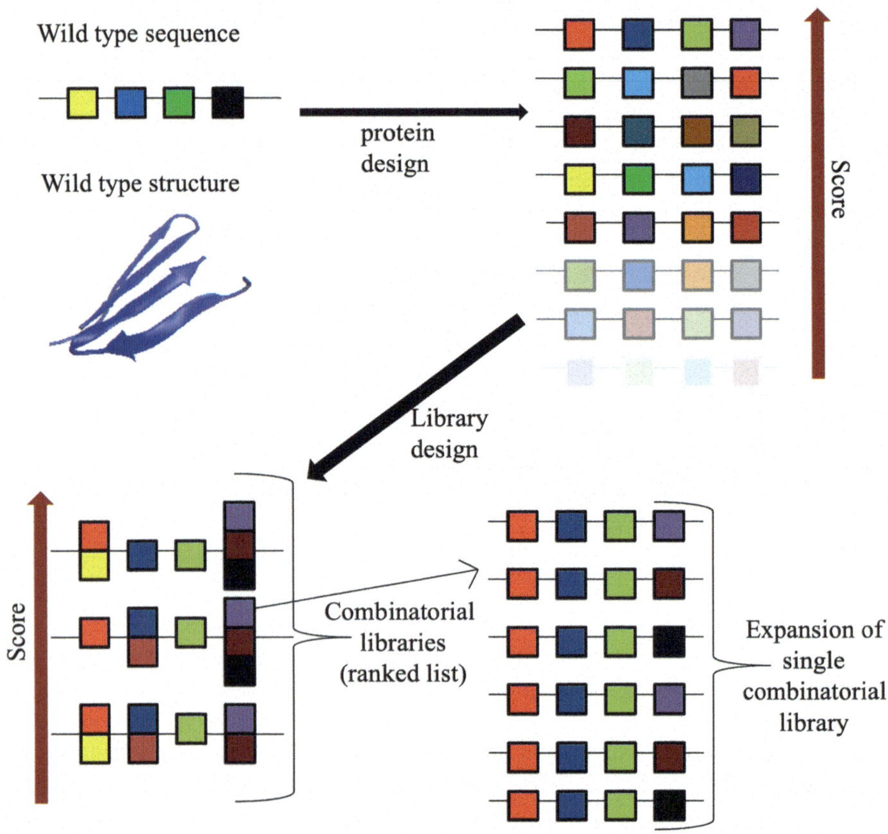

Fig. 3.10 A general scheme is employed to construct combinatorial mutation libraries based on computational protein design calculations. Each box represents a position in the protein chain, and a protein sequence is shown as a line of boxes. Different colored boxes indicate different amino acids. The far-right set of sequences corresponds to a combinatorial library's extension into the set of sequences it represents. The energy of the expansion sequences is used to choose which combinatorial library should be tested experimentally [54]

DNA Shuffling

The ability to make better protein variations and novel proteins have been revolutionized by DNA shuffling [58]. A pool of homologous genes, either natural variants of a gene or a pool generated through genetic engineering, serves as the beginning material. In most cases, DNAseI is used to digest input genes randomly, and fragments of a particular length or a range of lengths are purified from a gel after electrophoretic separation. The isolated fragments are assembled by thermal cycling with a DNA polymerase, and then the full-length fragments are amplified using standard PCR. When a pool of homologous genes encoding functional variants of a protein is used as the starting material for DNA shuffling, the genetic variants that are recombined are often associated with functional variations, avoiding the introduction of deleterious mutations and speeding up the search for variants with desired traits. A set of homologous genes originating from different species, or genes that have already been changed and screened for the desired feature, such as binding to a specific epitope, is used as the beginning material for such tasks.

3.2.1.4 Knowledge-Based Protein Design

A knowledge-based technique could be used to model the tertiary structures of proteins. This is dependent on finding similarities between a protein's secondary structure, motifs, domains, or ligand interactions with those of the known 3D structures [Fig. 3.11]. The approach picks fragments from structurally conserved regions based on sequence homology while simultaneously aligning the known tertiary structures. The alignment is done using either "average structure" or "framework," which is built on loops from homologous proteins, replaces sidechains, and reduces the energy of the final model. Knowledge-based techniques have been used

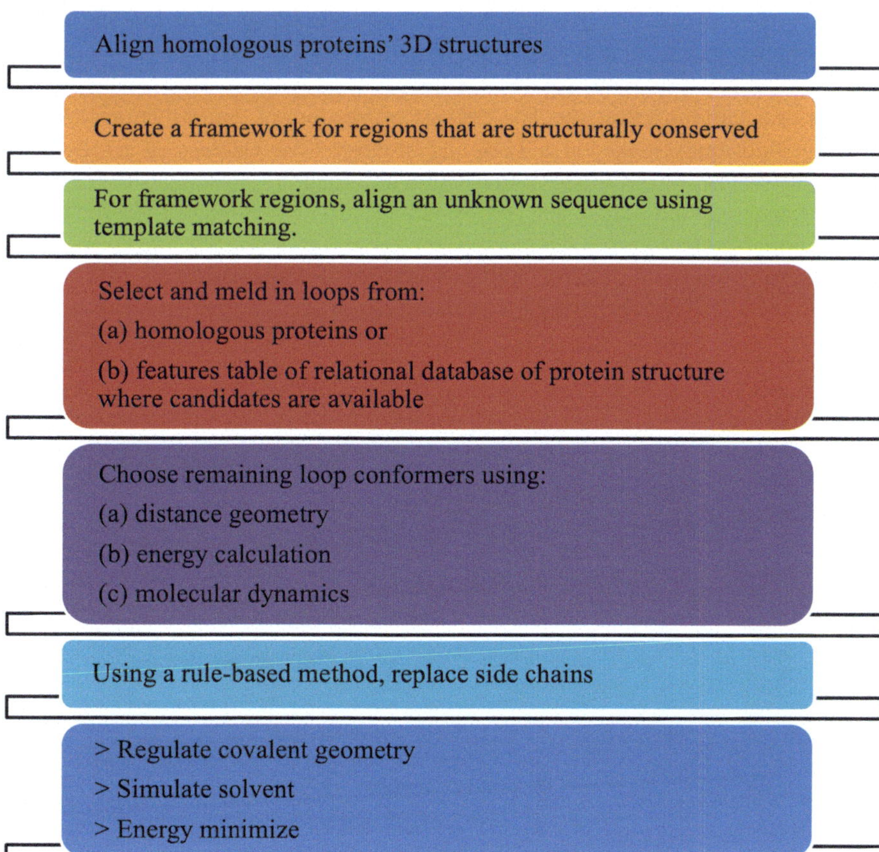

Fig. 3.11 Process of applying a knowledge-based approach to model proteins

to create surface tissue peptide superstructures, transmembrane-binding peptides, and protein crystals [59].

3.2.1.5 Computational Protein Design

A well-folded protein structure is always in the lowest free energy state, according to the fundamental understanding of computational protein design [60]. As a result, finding the lowest-energy amino-acid sequence for a particular target structure with specific functionalities is the most crucial goal of protein design. Packing side chains for a large number of candidate sequences and calculating the free energy of each sequence with the target structure is the most common method for designing proteins. As a result, in the last three decades, to improve the capability and efficiency of computational protein design, the majority of efforts have been focused on two goals: designing sampling strategies to explore sequence space and free-energy minimization [Fig. 3.12]. In the computational protein design process, selection pressures are required, just as they are during natural protein evolution. The direction of evolution is dictated by selection pressure, which is a function of free energy. As a result, the desired results can only be reached by employing a free energy function that accurately describes the nature of interactions inside or between proteins. Another important aspect of computational protein design is sampling methods or a sampling strategy. The degrees of freedom of conformations and the sampling algorithms make up a sampling strategy. The degrees of freedom of conformations encompass the total translation and rotation movements of a protein and the backbone and side-chain conformations of each residue. The efficacy of computational protein design depends on the accuracy of scoring functions, given the enormous number of possible outputs with diverse sequences.

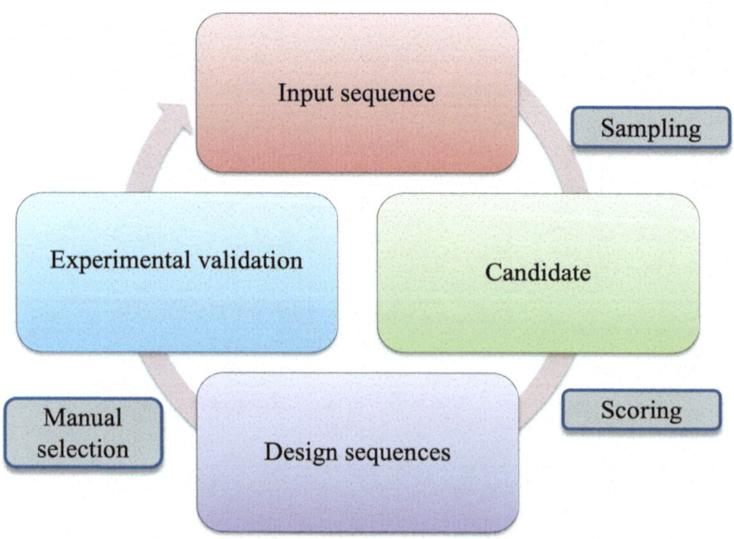

Fig. 3.12 A flowchart depicting the major steps in the computational protein design process

3.2.1.6 Directed Evolution

Directed evolution is one of the most widely utilized protein engineering techniques [61]. Because directed evolution is designed to replicate natural evolution, there are several in vitro methods for generating diversity. In vitro mutagenesis, homologous recombination, and nonhomologous recombination are the most common ways to introduce variations. After creating a random mutant library, compounds with better attributes are identified via high-throughput screening and selection procedures [Fig. 3.13]. The most often utilized high-throughput method for protein evolution is fluorescence-activated cell sorting (FACS), because it automatically detects and isolates cells with desired gene variations without using a diffuse fluorescent reporter [62]. Phage, ribosome, mRNA, and cell surface display, auxotroph complementation, organismal fitness, compartmentalized partnered replication (CPR), in vitro compartmentalization (IVC), and phage-assisted continuous directed evolution (PACE) are various selection strategies. Screening and selection strategies have several elements in common [63]. In the following selection stage, the selected progeny are used as parents. The process continues until the protein has the desired properties.

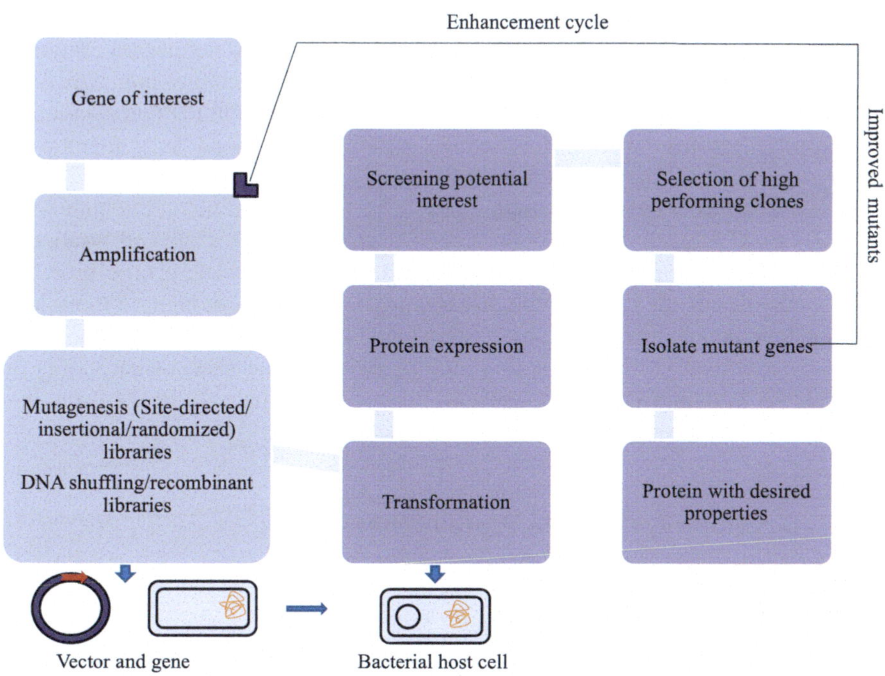

Fig. 3.13 Systematic representation of the key stages in the directed evolution cycle

3.2.2 Source, Targets, and Mechanism of Action of Protein Therapeutics

The human immune system has become increasingly sophisticated, which aids in the prevention and management of diseases. Any component of the immune system's molecular machinery that functions insufficiently, ineffectively, or improperly causes diseases to varying degrees. Extrinsic immune regulation using natural human immune regulators is an intriguing strategy for treating diseases. Human genome data has emerged as a promising source of information by providing a massive database for finding therapeutic targets. Before the invention of rDNA technology, therapeutic proteins like growth hormone and follicle-stimulating hormone were extracted directly from the human body. Hormones, recombinant antibodies, cytokines, enzymes, and interferons of human origin are now available on the market due to advances in rDNA technology. They are produced industrially in mammalian, yeast, or bacterial expression systems.

In addition to human immunoregulatory proteins, many viruses can manipulate the human immune system to aid their proliferation in the host [64]. Viruses encode and produce a diverse spectrum of immunomodulatory proteins that target cytokine-mediated signaling, antigen presentation, and antibody response controlled by the major histocompatibility complex (MHC) to avoid or subvert host immune detection and destruction. After generations of co-evolution, these virus-engineered immunomodulatory proteins demonstrate incredible potency and specificity unmatched by commercial pharmaceuticals, laying a solid platform for protein therapies research. As a result, 'virome' mining for humans has been proposed as a method of identifying additional therapeutic possibilities [65].

Protein treatments have three main modes of action depending on the pathology of the disease. Extracellular compounds such as cell metabolites and cell lysate can be destroyed using enzyme treatments. Second, if a shortage of specific proteins, such as enzymes, is the cause of the disease, protein therapy can be used to replace them and help people recover. Third, protein therapies work as cell surface receptors inhibitors or activators in disorders involving inappropriate immune responses or dysregulated signaling pathways, such as chronic inflammatory diseases, autoimmune diseases, viral diseases, and malignancies [66].

3.2.3 Developing Effective Protein Therapeutic

Although the human genome has many potentially therapeutic proteins, these proteins were not developed in the body with therapeutic goals in mind; hence, they lack the optimal specificity, stability, and activity for disease treatment. Protein treatments face several problems, including protein stability and immunogenicity. Proteins have limited stability both physically and chemically. This has numerous consequences, including a short half-life within the body, resulting in limited efficacy and frequent administration; hence, they are also difficult to manufacture. Both of these contribute to their high cost to commercialize as pharmaceuticals.

They must be administered via injection, since they are quickly processed in the intestines if taken orally, impacting patient compliance and pharmaceutical commercialization costs. As a result, protein treatments must be developed with enhanced stability, effectiveness, pharmacokinetics, pharmacodynamics, and expression productivity [67]. Only protein therapeutics face the issue of immunogenicity. The human immune system recognizes pathogens by their proteins or processed protein products. As a result, immunogenicity testing is critical in the development of protein therapeutics [68].

3.2.3.1 Improving Pharmacokinetics and Reducing Immunogenicity

Chemical modification, glycosylation, and fusions are methods to improve the efficacy of therapeutic proteins in the human body. Recombinant DNA technology allows a protein of interest to be fused to an endogenous human protein, increasing its size and decreasing clearance in the kidneys, which occurs when the molecular weight is less than 70,000 Da. Glycosylation, which involves the addition of sugars to a protein's surface, reduces renal clearance by increasing protein size. Furthermore, it improves protein solubility, stabilizes it against heat and free radical damage, and protects it from proteolysis and immunological surveillance, thereby contributing to a prolonged serum half-life. Glycoengineering is the process of inserting glycosylation patterns into the primary sequence of a therapeutic protein [69]. Finally, PEGylation is a common approach for enhancing the serum half-life of protein therapies. The bulky PEG molecule acts as a steric barrier, preventing therapeutic proteins from being broken down by proteases, and the increased size of the protein-PEG conjugation prevents its clearance from the kidneys [70]. In general, immunogenicity reduction entails modifying protein therapies to be immune surveillance-free. In addition, changes to the primary sequence of a therapeutic protein can also assist in minimizing its immunogenicity.

3.2.4 Examples of Protein Therapeutics

Various protein treatments have benefited from protein engineering and design approaches to improve their stability, activity, specificity, affinity, pharmacodynamics, pharmacokinetics, productivity, and immunogenicity [71].

3.2.4.1 Extracellular Targets of Protein/Enzyme Therapeutics

Besides binding proteins that target soluble chemicals and membrane-bound receptors, protein/enzyme therapies can also target additional compounds in the extracellular environment. In the therapeutic realm, one example is the use of amino-acid-degrading enzymes as an anticancer therapy. Rapidly expanding tumor cells may be auxotrophic for specific metabolites, which can be depleted in the plasma to specifically suppress tumor growth, in contrast to healthy cells. Enzyme therapies have also been discovered to be useful in treating cystic fibrosis. Frequent bacterial infections cause neutrophils in the lungs to accumulate and eventually lyse, releasing extracellular DNA and resulting in unusually sticky mucus. Dornase alfa

(Pulmozyme®, Genentech), also known as recombinant human DNase I, is an inhalable aerosol that reduces lung infections and improves lung function and quality of life [72]. Dornase alfa is manufactured using Chinese hamster ovary (CHO) cells, which produce a glycosylated protein with limited immunogenic factors, leading to its high cost.

3.2.4.2 Monoclonal Antibodies as Therapeutics

Antibodies have been propelled by mAbs and antibody engineering to become a rapidly emerging class of treatments for various human diseases. Antibodies have become a rapidly emerging class of treatments due to their precise specificity, ease of designing fragments using display technologies, and chimerization or humanization to improve stability [73]. Two potential areas to be explored are synthesizing multiantibody combinations for synergistic effects and pairing antibodies to immunotoxic medicines to induce tumor death.

3.2.4.3 Protein Therapeutics as Replacements for Proteins That Are Defective or Insufficient

The deficit or complete lack of an endogenous protein can be treated with replacement therapy in various medical conditions. Protein hormones, such as insulin and human growth hormone, are the most well-known protein treatments. Globally millions of people suffer from Type I and Type II diabetes, and many rely on insulin injections. Numerous short- and long-acting insulin analogues have been developed through protein engineering to minimize meal-time glucose peaks and meet the daily insulin requirement of the body. The second most well-known protein hormone, recombinant human growth hormone (rhGH), has been used to treat growth hormone deficiency in humans for decades [74]. Growth failure in children caused by various diseases and human immunodeficiency virus (HIV)-related wasting and lipodystrophy are treated with rhGH. Transfusion of human plasma-derived coagulation factors is the standard treatment for bleeding issues, and handling inherited or acquired coagulation factor deficiency is another long-standing area of protein replacement therapy [75]. Recombinant production has become a reality because of breakthroughs in mammalian cell culture and molecular biology approaches, as well as the sequencing of the human coagulation factor gene. The access to a safe, pathogen-free coagulation cascade has also enabled the off-label testing of these medicines in individuals who are not hemophiliacs. The third type is enzyme replacement therapy (ERT), which addresses acquired or inherited enzyme deficiencies. ERT has transformed the treatment of lysosomal storage disorders and other uncommon genetic diseases [76]. For best efficiency, an exogenously given replacement enzyme should be directly delivered to the specific cell types.

3.2.4.4 Protein Therapeutics Using Cytokines and Their Receptors

Regulatory proteins such as cytokines and their receptors are produced by WBCs and various other cell types. They are important targets and tools in developing protein therapies [77]. Cytokines are endogenous proteins that include tumor necrosis factors (TNFs), interleukins, colony-stimulating factors (CSFs), interferons, epidermal growth factors (EGFs), etc. Clinical trials include recombinant cytokines, anticytokine antibodies, and soluble receptors, as cytokines have been shown to fight against and contribute to disease. Erythropoietin, a CSF that promotes red blood cell proliferation, is perhaps the most well-known cytokine treatment due to its use as a performance-enhancing drug [78]. Recombinant human erythropoietin is used to treat anemia in individuals undergoing chemotherapy or with renal failure. Interferons are a family of secreted proteins that control normal and malignant cell survival, increase innate and acquired immune responses, and regulate viral infection resistance; for a few cases, the use of interferon therapy in humans has been approved.

3.3 De Novo Protein Design for Biotechnological Purposes

Traditional protein engineering strategies relied solely on altering existing proteins with a function similar to the desired function or a suitable structure and adequate stability for tolerating alterations. Before the availability of de novo protein design, it was necessary to add the desired functionality. On the other hand, de novo computational protein design removes this barrier by allowing access to a nearly limitless number of protein forms that can be used to build functions. Interest in engineering proteins at the sequence level to produce new folds and functions has risen as our understanding of the link among sequence, structure, and function has progressed. Protein structure prediction attempts to correctly predict the 3D structure based on its amino-acid sequence alone. When only low homology templates are available, structure prediction becomes difficult. The inverse challenge is de novo protein design, which involves determining a sequence that will fold into a stable structure. Because multiple sequences can fold into the same structure, there is degeneracy within the protein design space. The availability and precision of protein structure templates for protein design can significantly impact success. As a result, the ability to build effective protein templates via protein structure prediction is vital for protein design and progress in biotechnology and drug discovery [79].

3.3.1 Physical Principles of Protein Design

Hydrophobic residues, which are distant from the solvent and present in the protein core, are critical factors in protein folding [80]. To counteract polar functional groups that interact with the solvent in the unfolded state but hide during protein folding, intraprotein hydrogen bonds must be formed; otherwise, the significant energy expense of removing water will prevent folding. Nonpolar side chains that fit together like jigsaw pieces to create tightly packed cores contribute to the characteristics of globular protein structures, in which the polar carbonyl and amide groups of the polypeptide backbone can form hydrogen bonds. Hydrogen bonding at the termini of α-helices and torsional and steric stresses alter the free energy of folding, favoring specific backbone topologies while inhibiting others.

Proline, for example, has a rigid interior ring and can only be used with a select few backbones, whereas glycine, on the other hand, lacks a side-chain, allowing it to bend firmly in loops between secondary structures [81]. The interaction of atoms in proteins with one other and the solvent are captured by an energy function representing protein folding. Using such an energy function to predict and design protein structures necessitates methods of finding structures and sequences with very low energy by sampling different backbone and side-chain conformations. Backbone and side-chain sampling are done in different ways. In side-chain sampling, discrete combinatorial optimization is used to locate amino acids and side-chain conformations, resulting in low-energy, closely packed protein cores. If the amino-acid sequence is already known, the search encompasses the separate rotameric states of each side chain, as in the case of protein structure prediction [82].

3.3.2 De Novo Protein Design

The term de novo protein design refers to the process of creating novel proteins that are unrelated to those found naturally, using physical principles of intramolecular and intermolecular interactions. The de novo design starts with the discovery of a primary sequence that can fold into the appropriate structural topology [Fig. 3.14]. In general, the design is completed in three steps: the construction of a protein backbone, sequence optimization, and computational and experimental validation of the chosen sequence structure match. Both the sequencing and the specific structure of the backbone are unknown in a fundamental protein design (de novo) problem. Since only a limited percentage of backbone conformations permit nearly ideal core packaging and hydrogen bonding between buried hydrogen bond donors and acceptors, design simulations begin with a massive number of alternatives. These starting backbones can be made by synthesizing short peptide segments or parametrically defining the structure with algebraic equations [83]. For each backbone conformation, combinatorial sequence optimization algorithms are used to identify the lowest-energy conformation. The proposed structure is then placed through ab initio structure prediction calculations to see if it is the lowest-energy state in the specified sequence [84].

Although new structures are the focus of most current contributions to de novo design, the field is focusing on inventing new biological functions and their applications. Designer proteins are starting to impact biomedical and synthetic biology research. Immune system modulators, viral infection inhibitors, protein logic gates, and self-assembling biomaterials are just a few fascinating discoveries in recent years.

3.3.3 Design of Proteins and Peptides for Therapeutic Applications

Computational methods have recently been employed to develop new proteins and peptides for medical purposes. The elucidation of the structures, sequences, and

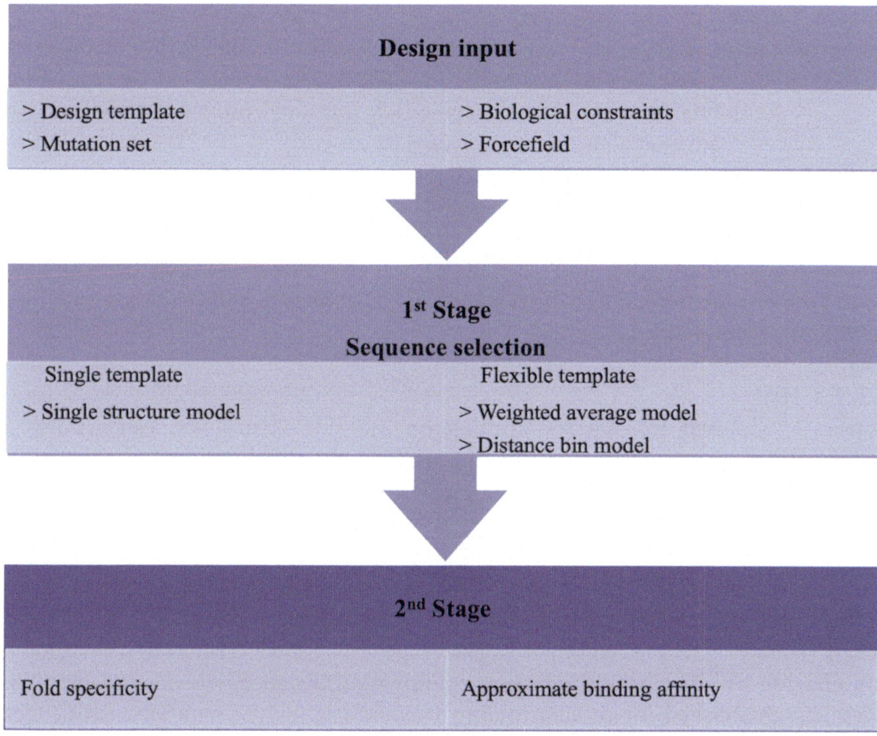

Fig. 3.14 A general flowchart of de novo design strategy. The de novo design approach includes design inputs, sequence selection, and fold validation stages

interaction patterns of numerous disease-related proteins has permitted the use of computational approaches for peptide therapy design [85]. The use of peptides as drugs will become increasingly common, although there will be hurdles in achieving the goal.

3.3.3.1 Alzheimer's Disease

Eisenberg et al. used computational protein design to predict and experimentally test a peptide inhibitor of tau protein fibrillation linked to Alzheimer's disease and an inhibitor of an amyloid fibril that enhances sexual transmission of HIV [86]. The designs inhibit elongation by adhering to the steric zipper's end. Depending on the tau protein inhibitor technique, they flipped the chirality of the design target to allow the use of the Rosetta toolkit for rotameric, fixed-backbone sequence optimization. They designed L-amino-acid sequences that effectively interact with the D variant of the scaffold. D-amino-acid-containing peptides were used as experimental inhibitors once the scaffold was returned to its original L-form. The shape complementarity of the D-peptides was then checked. Inhibition is achieved by creating a tight-binding interface and collisions that prevent a cascade of amyloid-forming sequences from propagating [87].

3.3.3.2 Cancer

Therapeutic proteins/peptides can disrupt signaling pathways, block the cell cycle by altering cyclin-dependent kinase activity, or directly cause apoptosis by regulating apoptosis-regulating proteins [85]. Cysteine-rich intestinal protein 1 (CRIP1) is a breast cancer biomarker that can be detected at an early stage. To locate CRIP1-binding peptide sequences, Hao et al. used phage display [88]. Cosic et al. employed the resonant recognition model (RRM) to create a short medicinal peptide with anticancer and cytotoxic properties against the myxoma virus. A protein sequence is represented by RRM in the form of a succession of numbers that may be examined using Fourier transformation, which results in a discrete spectrum with a strong link to biological activity.

3.3.3.3 HIV

Correia et al. used Rosetta for grafting on the side chains and establishing a computational approach for transplanting a 4E10, a continuous structural epitope, into the scaffold proteins for immunological presentation and conformational stabilization [89]. The approach created designs with epitopes that bind to a monoclonal antibody (mAb) 4E10 better than 4E10 alone, preventing HIV-positive sera from neutralizing them. The structure of the C14linkmid peptide and the hydrophobic core of gp41 was used by Floudas et al. to create HIV-1 entry inhibitors [90]. Cross-linked peptide gp41's C-terminal heptad repeat is called C14linkmid. A global optimization-based sequence selection employing a distance-dependent force field originally established for protein folding was used to choose candidate sequences from the vast combinatorial space. Fold-specificity computations were used to rerank these sequences, which examines conformations close to the template structure with modifications defined by freshly produced sequences. Its purpose is to determine how well a new sequence fits into the fold of the design template. Calculations of estimated binding affinity, which approximate the binding equilibrium constant, were utilized to examine a subset of the highest-ranking sequences discovered during the fold-specificity step. The optimum design for several HIV-1 donors and mutations had an IC_{50} ranging from 29 to 253 mm. Protein WISDOM, an interactive web interface, was created using this de novo design approach [91].

3.3.3.4 Antibody Therapeutics

Gray et al. employed Rosetta to integrate a noncanonical amino acid (NCAA) into the antibody CDR, with the optimal design experimentally crosslinking 52% of the accessible antigen [92]. Ellington et al. devised a method of supercharging for replacing numerous surface residues in proteins with charged amino acids, which they used to create an antibody with a 30-fold increase in affinity and better thermal inactivation resistance. Pantazes and Maranas developed OptCDR to build antibodies that attach to a specific antigen epitope [93]. They used it to make antibodies targeting a peptide from the fluorescein, hepatitis C capsid, and vascular endothelial growth factor (VEGF), and they used computational metrics and binding energies to verify the process. The templates are prototype structures from the database's random diversity (D), variable (V), and joining (J) sections, generating gene combinations with the fewest amino acid changes from the target. Using this information, they could predict antibody tertiary structures with an average all-atom root-mean-square deviation (RMSD) of 1.9 Å on a testing set of 260 antibodies. When a target structure is accurately predicted, antibodies can be computationally affinity generated using the iterative protein redesign and optimization (IPRO) methodology. IPRO is an iterative framework that uses a mixed-integer optimization model to optimize side-chain replacements in user-defined design positions, followed by local minimizations that allow the backbone of the protein to adapt to the new side-chains [94].

3.3.4 Designing Repeat Proteins

Proteins having repeating sequence patterns fold with structural motifs are well defined, and in contrast to other proteins, they provide more information on the sequence-structure relationship. The evolutionary process that gave rise to these repeat proteins is rather remarkable. A structurally compatible section of the sequence is repeated in tandem, and the connecting segments diverge to take novel functions within the structural compatibility. The repeating modularity indicates that structural elements are consistent from one segment to the next, enabling each unit to connect to the one before it while forming the same interactions for the one after it [95]. The bulk of these units have 5 to 50 residues and fold into a single domain when the surrounding units are present. If only a few repeated units are present, these structures can be very similar to other globular proteins. However, as the repeating unit number rises, nonglobular shapes emerge. For this type of repeat protein, where the associated modules are required for folding, the whole structure twists further into a superhelix, and the secondary structural pieces generate a base coil with either right- or left-handed spiral handedness. The hydrophobic residues that link the modular units together and a few topologically crucial residues that define each unique shape are usually conserved traits for a repeat protein family. Repeat proteins have distinct properties that make them appealing to designers. The structural and sequence similarity between recurrent units within a protein suggests that an idealized unit with family qualities could be developed. Limiting repeat protein connections to only surrounding repetitions also simplifies that long-range interactions are ubiquitous and typically irregular in the formation of globular proteins. Repeat units can be inserted, withdrawn, or replaced without affecting the overall structure as long as the relevant interfaces between units are present. The goal of producing repeat proteins is to create modular systems to build unique scaffolds for multiple applications. Most repeat protein engineering efforts have focused on creating idealized copies of natural repetitive units with stable, homogeneous, and modular structures [96].

3.3.5 Designing Recombinant Therapeutics

Drug discovery is entering a new era in which disease management decisions are made at the genetic and protein level, and here protein therapeutics are becoming increasingly important. Several Food and Drug Administration (FDA)-approved technical treatments involve recombinant human proteins. The future potential for such treatments is enormous, in a way that hundreds of proteins are produced by the human body, and millions are made by other organisms. Furthermore, recombinant proteins can be combined with small molecule drugs to have additive or synergistic benefits [97]. Patients with colorectal cancer who receive a combination of the small molecule drug irinotecan and the recombinant mAb cetuximab have a higher chance of surviving. Irinotecan with cetuximab has therapeutic synergy, because both drugs target the same epidermal growth factor receptor (EGFR) signaling pathway, with cetuximab blocking a downstream target and irinotecan decreasing the pathway's initiation [98]. The early success of recombinant insulin production in the 1970s created an atmosphere of enthusiasm and hope, which was unfortunately followed by disappointment in the 1980s when vaccine attempts, nonhumanized mAbs, and cancer trials all failed. Despite these obstacles, there has been substantial development recently. In medicine, protein treatments will continue to play an essential role for years to come, given the vast number of protein therapies in use and clinical trials for a variety of ailments.

3.3.6 Advances and Challenges

Protein therapies are an important development in modern medicine; in some circumstances, they are the only effective medicines available. Protein engineering and design generate effective protein therapeutics with improved specificity, affinity, activity, stability, pharmacokinetics, pharmacodynamics, productivity, and immunogenicity. When developing new protein treatments, drawing on expertise from various fields, including pathology, immunology, molecular biology, and nanotechnology, is essential. The introduction of the function necessitates the modification of a large number of amino-acid residues, which will alter structural elements. Crystal structures of engineered enzymes with unexpected loop reconfigurations demonstrate this. Sequence alterations can cause unfolding or aggregation in native proteins, which are frequently marginally stable. De novo generated proteins should have a higher level of stability, making them more reliable starting molecules for developing novel functions [3]. There are numerous examples of proteins being employed successfully for medicinal purposes. Nonetheless, due to many obstacles that must be overcome in developing and using protein therapies, potential protein treatments that have failed vastly outnumber those that have succeeded thus far. A few of the major limitations and obstacles are listed below:

1. Protein solubility, administration mechanism, transport, and stability are crucial factors that can prevent a protein therapy from being properly implemented.

2. The therapeutic protein may trigger an immune response in the body, which is a second significant hurdle. This immune response can occasionally neutralize the protein, causing the patient adverse reactions. On the other hand, immune responses to proteins that are not of human origin are more common. Immune responses against therapeutic proteins are rapidly developing and had previously hampered the extensive clinical deployment of mAbs. The quest for antibody treatments that do not require immune monitoring and response has been a driving force behind the evolution of antibody production technologies.
3. For a protein to be physiologically active, posttranslational modifications (PTMs) such as glycosylation, phosphorylation, and proteolytic cleavage are typically required. These criteria require certain cell types to produce and adequately modify the protein. Furthermore, for large-scale synthesis, the production of recombinant proteins involves the use of genetically modified cell types. Not only must the host system create physiologically active protein, but it must also make enough of it to address clinical demand. The technology must also allow for the purification and long-term preservation of a therapeutically active protein.
4. The cost of the production of protein treatments is another significant challenge. Switching from tedious purification of placentally derived protein to recombinant methods, for example, has made it possible to produce enough β-glucocerebrosidase to treat Gaucher's illness in many humans [99]. Despite this, the cost of the recombinant protein per patient per year exceeds $100,000.
5. Gaucher's illness also highlights another concern that is ethics. For example, the promise of effective but costly protein therapy for small but severely ill patient populations, such as Gaucher's disease patients, can create difficulty in terms of how health-care organizations allocate their financial resources.

3.4 Unleashing the Potential of Therapeutic Proteins

3.4.1 Advanced Protein Engineering Reinforcing Next-Generation Therapeutics

Protein engineering can modify an already existing protein to enhance its qualities. It is a critical technology for improving enzyme properties for pharmaceuticals, green chemicals, and biofuels, as well as expanding our fundamental understanding of how enzymes work and have evolved. The method has already achieved several commercial successes. Proteases developed by the Biotech company Genencor can survive the bleach in laundry detergents and help clean clothes. For the engineering of enzymes to manufacture a pharmaceutical intermediate, Codexis received the 2006 Green Chemistry Award. Verenium created enzymes for enhanced oil recovery and cellulosic ethanol generation. Metabolix developed enzymes that aid in the manufacturing of polyesters.

There are numerous phases in the protein engineering process where strategy must be devised. Individual situations can help clarify the benefits or motivations for choosing a specific technique for a particular challenge. The larger pattern, on the other hand, may not be as straightforward. Protein engineering methodologies nowadays vary greatly and are based on frequently contradicting concepts. This is attributed to a lack of critical methodological comparison [100]. Making a comparison like this will aid future protein engineering, because it will allow them to learn from their success and errors. To achieve faster progress in protein engineering, researchers must first have a better understanding of how protein structure influences protein function as well as a thorough evaluation of the various protein engineering approaches. Many tests involve numerous protein modifications occurring simultaneously, obscuring the factors that led to success. Related to these, two suggestions have been made:

1. Every innovation in protein engineering should be backed by an empirically proven molecular foundation hypothesis. Protein engineering problems will become easier to address as more instances and assumptions are accumulated.
2. All protein engineering technologies should be studied and tested thoroughly against other methods using quantitative comparisons. Authors should give information such as the proportion of variations consistently with enhanced characteristics and the degree of these enhancements when publishing a discovery. Details of the dimensions of the theoretical and experimental libraries, and the proportion of the total library that was screened, should be provided. They should also inform that compared to other strategies, how did their strategy rate. Rather than depending on published studies, these comparisons should be conducted in the same laboratory with the same problem. Similar comparisons should be included in computational approaches.

Protein engineering will move toward rational design in the future as the concepts of protein engineering are defined [101]. Organic synthesis is complex, as there are some principles to follow, such as constructing the carbon framework first and changing functional groups to ensure that stereochemistry is always maintained. Chemists know that convergent synthesis is preferable to sequential synthesis and that protecting groups should be avoided wherever possible. Finding the best protein engineering technique and understanding how mutations work will aid in the faster advancement of protein engineering. Future solutions will take less effort if methods are compared. Because each challenge has a different purpose, the quantity of information accessible, and protein specifics, it is doubtful that one strategy will win out. However, some techniques will outperform others. The similarities will also aid in the development of protein engineering concepts and our understanding of enzyme function. The rational design will become more trustworthy, and the path to solutions will be sped up even more.

3.4.2 Genetic Engineering

The three primary genetic engineering approaches, that is, directed evolution, rational design, and semirational design, have been proved useful in creating new

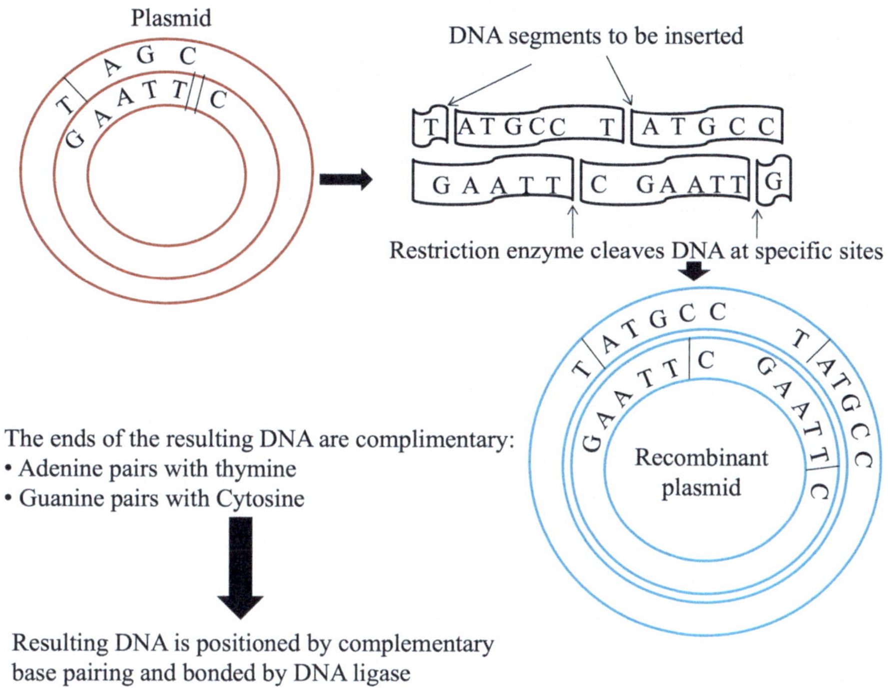

Fig. 3.15 Genetic engineering for the production of therapeutic proteins

proteins with improved physical and chemical properties and/or unique functions [102]. However, these well-established approaches are facing new challenges due to additional characteristics like pharmacokinetics, pharmacodynamics, and immunogenicity. The engineering of protein therapeutics has benefited from various rational and semirational design methodologies [Fig. 3.15]. Rational or computation design methods are preferred to improve the stability and solubility and anticipate and minimize immunogenicity of protein therapies. The main disadvantage of the rational design is that it necessitates some understanding of protein structure, mechanism, and structure-function correlations. High-throughput selection or screening processes are highly desired for directed evolution. A variety of library selection and screening processes have been developed for diverse uses. Among the library selection and screening approaches, display methods have been widely used and proven particularly useful for creating protein therapies with better affinity and specificity. The common goal of all display technologies is to establish a physical connection between the genotype and the protein displayed on the platform, allowing for the selection and recovery of a library of target protein variants for further engineering. This connection is directly accessible to binding analysis. Typically, a cell-surface display library is made by converting cells using DNA variations and FACS to look for mutants with the desired phenotype. FACS enables quantitative high-throughput enrichment of positive clones; however, it is not ideal for phage display libraries due to the small size of phage particles. Yeast, bacteria, insect, and mammalian cells have all been studied for their potential to display protein libraries [103]. Of all these platforms, yeast display has attracted the most attention. The advantage of yeast display is that it has posttranslational processing pathways that allow complex human proteins to be folded and glycosylated [104]. Using the same library as phage display, yeast display was shown to extensively investigate the immunological antibody repertoire, resulting in the selection of twice as many new antibodies. The cell-free display is a new approach that has proven effective in identifying and engineering high-affinity therapeutic proteins. The main benefit of cell-free display systems is that the transcription and/or translation processes are completed entirely in vitro, eliminating the requirement to insert DNA into host cells, which limits the size of the library that may be employed in other ways of display. The libraries created by in vitro display systems are of a larger magnitude than those produced by conventional display technologies.

3.4.3 Antibody and Therapeutic Protein Engineering

Many attempts have been made to improve the engineering technologies, safety, and efficacy of therapeutic antibodies. Stability is one of the most important functional requirements for using antibodies in therapeutic and diagnostic applications. Antibody/therapeutic protein engineering strives to create clinically useful proteins with higher efficacy and safety, lower immunogenicity, and better delivery. The most prevalent and fastest-growing class of protein therapeutics is antibody-based medicines, such as therapeutically effective antibody-drug conjugates and bispecific antibodies [105]. Besides, clinical trials for synthetic protein scaffolds are still in the early phases, and they may have certain advantages over traditional antibodies. Concerns of pharmacokinetics and dynamics must be addressed carefully while creating protein therapeutics. Proteins and peptides are becoming more popular as therapeutic instruments for treating several disorders, although they have some drawbacks. Because of the distinctive structure of this organ and the presence of many barriers that protect the internal structure of the eye, administering drugs based on proteins can be challenging in the case of eye disorders. Nanoencapsulation of peptides and proteins has helped partly resolve this problem, which provides several benefits for ocular delivery, including drug protection following topical or intravitreal injection, from metabolic activity, regulated and sustained release, and enhanced drug bioavailability [106]. The administration of protein therapeutics to the brain is another example. Because of their poor bioavailability and the presence of the blood-brain barrier (BBB), proteins cannot be delivered to the brain via parenteral administration. Even when proteins are supplied by nonparenteral means, other challenges remain. Chemical modification of proteins and particle-based carriers has been used to improve such delivery in relevant central nervous system (CNS) disease models, and a few have reached clinical proof of concept. As a

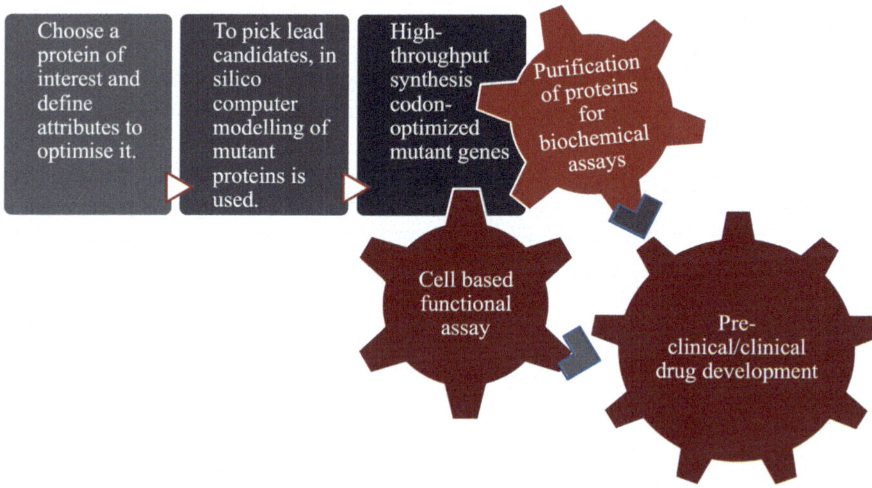

Fig. 3.16 Diagram for engineering and validation of new therapeutic proteins

result, we require a more thorough comprehension of the variables that affect the properties of therapeutic proteins to produce them more effectively. One method uses integrated exposure and efficacy/toxicity data to model pharmacokinetics and dynamics throughout the preclinical period. These preclinical models are extremely helpful in refining experimental designs for future preclinical studies. Furthermore, when properly simplified and statistically modified, these models provide the framework for future clinical evaluation. Finally, these models improve the translation of protein treatments from preclinical to clinical trials. Protein therapeutic synthesis has progressed at a breakneck pace [Fig. 3.16]. In contrast to traditional approaches that solely make proteins, emerging methods can use the ability of proteins to attach to nanoparticles (NPs) to form protein-NP conjugates with varying biodistribution, clearance, and toxicity features.

3.4.4 Exon Shuffling as a Method of Protein Evolution

Exon shuffling is a chemical mechanism that allows new genes to be formed. It is a technique for ectopically joining two or more exons from different genes, or duplicating the same exon, to create a new exon-intron structure. Exon shuffling occurs via various mechanisms, including crossover during parental genome sexual recombination, transposon-mediated exon shuffling, and illegitimate recombination [107]. A set of splice frame rules governs exon shuffling. By introducing a sequence between two successive codons, introns can cause a gene's reading frame to be disrupted (phase 0 introns), between the first and second nucleotides of a codon (phase 1 introns), or between the second and third nucleotides of a codon (phase

2 introns). Exons are likewise classified into nine groups according to the phase of the flanking introns.

Introns disrupt the coding sequence in eukaryotic genes. After transcription, splicing removes the introns, leaving only the coding sequences (exons). One folding domain is encoded by one exon in many genes, suggesting that intron recombination might combine separate exons into genes to produce new proteins, a process known as exon shuffling, resulting in fast protein evolution. Many instances, such as blood-clotting proteins, support the notion of exon shuffling for protein evolution. According to the introns-early version of the exon-shuffling theory, exon shuffling has developed in eukaryotic cells to increase the genetic plasticity of these creatures. Regardless, exon shuffling is a critical factor for protein formation, suggesting that this natural evolutionary process could be explored for developing novel enzymes. Exon shuffling needs to establish spliceosomal introns before it can start to play a role in protein evolution. The self-splicing introns of the RNA world were not suitable for exon recombination via intronic recombination. These introns were necessary for the function of the gene and could not be recombined. Furthermore, there is much evidence that spliceosomal introns represent a recent evolution with a limited evolutionary distribution [108]. Likewise, to understand when exon shuffling became relevant in eukaryotes, researchers looked at the evolutionary distribution of modular proteins that emerged via this strategy in different animals. These studies discovered an inverse association between genome compactness and the proportion of intronic and repetitive sequences.

3.4.5 Site Saturation Mutagenesis Methods and Applications in Protein Engineering

The amino acids encoding the proteins can be modified through PTMs at specific sites. Recombinant DNA technology allows polypeptides to be easily manufactured and manipulated, and various mutagenesis procedures have been created. Site-directed mutagenesis (SDM) is a technique for studying the function of a single amino acid in a protein. When an alanine is replaced with another in SDM, the process is known as alanine-scanning mutagenesis.

Site-saturation mutagenesis (SSM) approach adds a layer of complexity by allowing an amino acid to be replaced with any of the other 19 potential substituents. As a result, the outcome of mutagenesis is a group of clones, each of which has a different codon in the targeted area, which is why it is referred to as "saturated." This has a benefit over SDM in that all possible substitutions can be made with the same amount of work, enabling a more in-depth analysis of the original amino acid's function at the targeted site. The design of the SSM experiment varies depending on the application [109] [Fig. 3.17].

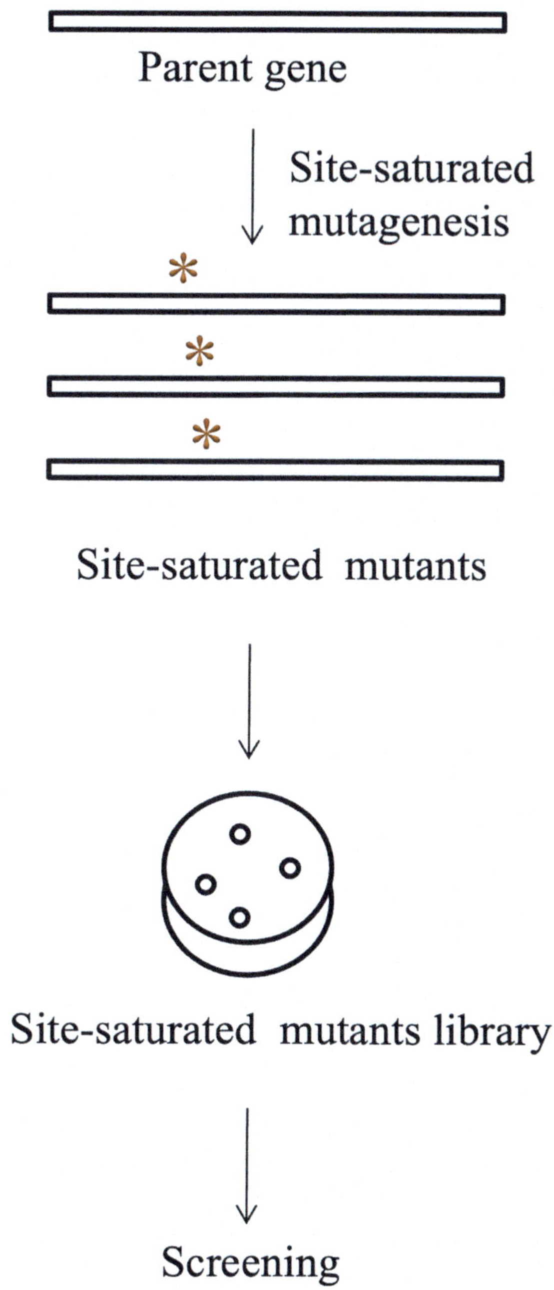

Fig. 3.17 The major experimental phases of a site saturation mutagenesis-based in vitro directed evolution approach

3.5 Applications of Protein Engineering in Therapeutics

Protein engineering has a wide range of medical uses. Protein engineering as a tool for cancer treatment research is a prime topic. Pretargeted radioimmunotherapy is a promising cancer treatment [110]. As protein engineering and rDNA technology advance, pretargeted radioimmunotherapy is expected to become increasingly common. The ability of antibodies to identify antigens selectively and with greater affinity is used to develop new antibodies as anticancer medicines, and protein engineering approaches are employed for altering antibodies to target cancer cells.

The phrase "modular protein engineering" has recently been coined for potential cancer therapeutics. Antibodies have a wide range of uses in protein engineering. Antibody engineering is now possible due to improvements in rDNA technology [111]. To produce antibodies with desired antigen-binding properties, combinational techniques such as bacteriophage display libraries are developed as a potential alternative to hybridoma technology. Phage display has proven effective in various domains, including immunology, protein engineering, and oncology. Antibodies are increasingly being used as molecular imaging tools. Protein engineering methods have been used to enhance the pharmacokinetic properties of antibodies. Specialized imaging probes for target organs, variants of an antibody of different sizes, and sites that bind to antigens have been created.

3.6 Case Study: Protein Engineering for Cardiovascular Therapeutics

Clinical studies of cell-based cardiovascular therapy have received much attention, and preliminary results are promising, especially given the surge in interest in new heart-repair therapeutics. On the other hand, the mode of action of cell-based therapeutics is unknown, and many specifics of characteristics, quality control, and cell administration must be evaluated before widespread therapeutic use. Regardless, the success of cell-based cardiac treatments in human research has inspired curiosity in new therapeutic development [112].

The discovery of paracrine-acting protein release as a mechanism for improving cardiac function via cell-based treatments in the heart has received interest in protein-based therapy approaches for heart repair. Despite their rapid adoption in various fields such as cancer and inflammatory illnesses, protein therapies have low penetration in the cardiovascular market. Protein therapy development represents a promising field in cardiovascular medicine as the majority of new medication approvals move to proteins. Improvements in heart microvasculature design have taken a lot of time and effort. The ability of proteins to drive cardiomyocyte proliferation has recently been discovered, raising hopes that protein-based approaches to heart regeneration would be successful. However, engineered protein therapies have yet to become a significant component of cardiologists' toolset.

A recent study provides an overview of the use of protein therapy approaches in the heart, focusing on treatments that have progressed to the clinical trial stage [112]. It explains how new findings in molecular cardiology, when paired with new and promising protein engineering approaches, can potentially pave the way for a new generation of heart-repair treatments. Much of the effort in therapeutic development has focused on two pathways that contribute to myocardial repair and regeneration: cardiomyogenesis and vascularization. The molecular foundations of cardiomyogenesis are still a matter of discussion, although the fundamental processes of vascularization, such as arteriogenesis, angiogenesis, and vasculogenesis, are well defined.

3.7 Conclusion

Proteins have long been recognized as a clinically and financially relevant therapeutic class. To fully exploit the potential of protein therapeutics, ongoing efforts are necessary to improve their efficacy and discover new protein-based medicines. Protein engineering is a widely used and vital technology that aids in developing novel biologics such as protein therapeutics, cellular therapy, and gene therapy. Engineering protein therapeutics is an emerging pharmaceutical industry with tremendous growth potential. Current research into allostery, catalysis, molecular recognition of proteins, and the advent of computer protein design tools will hasten this transformation. In the future, there will be more synergy between protein engineering and design efforts, thanks to a better knowledge of interactions between protein structure and function and advancements in bioinformatics and systems biology techniques. To develop novel protein therapeutics, researchers draw on expertise from various fields, including molecular biology, immunology, pathology, and nanotechnology [113–115]. Protein treatments have given researchers insight into their strengths, limitations, and prospective enhancements. The success of protein treatments in healthcare provides a strong incentive to develop better proteins and new approaches, and new platform technologies are expected to make this possible. Considering these advancements, protein therapeutics can be confidently assumed to be the future of human medicine.

References

1. Tobin PH, Richards DH, Callender RA, Wilson CJ (2014) Protein engineering: a new frontier for biological therapeutics. Curr Drug Metab 15(7):743–756
2. Kuchner O, Arnold FH (1997) Directed evolution of enzyme catalysts. Trends Biotechnol 15(12):523–530
3. Carter PJ (2011) Introduction to current and future protein therapeutics: a protein engineering perspective. Exp Cell Res 317(9):1261–1269
4. Lesk AM (2001) Introduction to protein architecture: the structural biology of proteins. Oxford University Press, Oxford
5. Breda A, Valadares NF, de Souza ON, Garratt RC (2007) Protein structure, modeling and applications. In: Bioinformatics in tropical disease research: a practical and case-study approach [Internet]. National Center for Biotechnology Information (US)

6. Kopp J, Schwede T (2004) Automated protein structure homology modeling: a progress report. Pharmacogenomics 5(4):405–416
7. Piatkov KI, Vu TT, Hwang CS, Varshavsky A (2015) Formyl-methionine as a degradation signal at the N-termini of bacterial proteins. Microbial Cell 2(10):376
8. Zhang Y, Baranov PV, Atkins JF, Gladyshev VN (2005) Pyrrolysine and selenocysteine use dissimilar decoding strategies. J Biol Chem 280(21):20740–20751
9. MacArthur MW, Thornton JM (1991) Influence of proline residues on protein conformation. J Mol Biol 218(2):397–412
10. Karpen ME, De Haseth PL, Neet KE (1992) Differences in the amino acid distributions of 310-helices and α-helices. Protein Sci 1(10):1333–1342
11. Salemme FR (1983) Structural properties of protein β-sheets. Prog Biophys Mol Biol 42:95–133
12. Hol WGJ, Van Duijnen PT, Berendsen HJC (1978) The α-helix dipole and the properties of proteins. Nature 273(5662):443–446
13. Wang H, Logan DT, Danielsson J, Oliveberg M (2020) Exposing the distinctive modular behavior of β-strands and α-helices in folded proteins. Proc Natl Acad Sci 117(46): 28775–28783
14. Murzin AG, Brenner SE, Hubbard T, Chothia C (1995) SCOP: a structural classification of proteins database for the investigation of sequences and structures. J Mol Biol 247(4):536–540
15. Orengo CA, Michie AD, Jones S, Jones DT, Swindells MB, Thornton JM (1997) CATH–a hierarchic classification of protein domain structures. Structure 5(8):1093–1109
16. Wierenga RK (2001) The TIM-barrel fold: a versatile framework for efficient enzymes. FEBS Lett 492(3):193–198
17. Yu X, Wang C, Li Y (2006) Classification of protein quaternary structure by functional domain composition. BMC Bioinf 7(1):1–6
18. Deng H, Jia Y, Zhang Y (2018) Protein structure prediction. Int J Mod Phys B 32(18):1840009
19. Altschul SF, Madden TL, Schäffer AA, Zhang J, Zhang Z, Miller W, Lipman DJ (1997) Gapped BLAST and PSI-BLAST: a new generation of protein database search programs. Nucleic Acids Res 25(17):3389–3402
20. Berman HM, Westbrook J, Feng Z, Gilliland G, Bhat TN, Weissig H et al (2000) The protein data bank. Nucleic Acids Res 28(1):235–242
21. Klepeis JL, Floudas CA (2003) ASTRO-FOLD: a combinatorial and global optimization framework for ab initio prediction of three-dimensional structures of proteins from the amino acid sequence. Biophys J 85(4):2119–2146
22. Godzik A (2003) Fold recognition methods. Methods Biochem Anal 44:525–546
23. Lee J, Kim SY, Joo K, Kim I, Lee J (2004) Prediction of protein tertiary structure using PROFESY, a novel method based on fragment assembly and conformational space annealing. Proteins Struct Funct Bioinf 56(4):704–714
24. Floudas CA, Fung HK, McAllister SR, Mönnigmann M, Rajgaria R (2006) Advances in protein structure prediction and de novo protein design: A review. Chem Eng Sci 61(3): 966–988
25. McLafferty FW (1981) Tandem mass spectrometry. Science 214(4518):280–287
26. Pruitt KD, Tatusova T, Maglott DR (2005) NCBI reference sequence (RefSeq): a curated non-redundant sequence database of genomes, transcripts and proteins. Nucleic Acids Res 33 (suppl_1):D501–D504
27. Tran JC, Zamdborg L, Ahlf DR, Lee JE, Catherman AD, Durbin KR et al (2011) Mapping intact protein isoforms in discovery mode using top-down proteomics. Nature 480(7376): 254–258
28. Gingras AC, Aebersold R, Raught B (2005) Advances in protein complex analysis using mass spectrometry. J Physiol 563(1):11–21
29. Berger S, Braun S (2004) 200 and more NMR experiments. Wiley-Vch, Weinheim, pp 305–307

30. Dwevedi A (2015) Classes of protein and their folding. In: Protein folding. Springer, Cham, pp 13–26
31. Dobson CM (2003) Protein folding and misfolding. Nature 426(6968):884–890
32. Bhattacharyya RP, Reményi A, Yeh BJ, Lim WA (2006) Domains, motifs, and scaffolds: the role of modular interactions in the evolution and wiring of cell signaling circuits. Annu Rev Biochem 75:655–680
33. Chang DTH, Chien TY, Chen CY (2009) seeMotif: exploring and visualizing sequence motifs in 3D structures. Nucleic Acids Res 37(suppl_2):W552–W558
34. Conklin D (1995) Machine discovery of protein motifs. Mach Learn 21(1):125–150
35. Bystroff C, Baker D (1998) Prediction of local structure in proteins using a library of sequence-structure motifs. J Mol Biol 281(3):565–577
36. Watkins AM, Wuo MG, Arora PS (2015) Protein–protein interactions mediated by helical tertiary structure motifs. J Am Chem Soc 137(36):11622–11630
37. Dym O, Eisenberg D (2001) Sequence-structure analysis of FAD-containing proteins. Protein Sci 10(9):1712–1728
38. Dabhole MP (2014) Recent innovations in therapeutic recombinant protein. Innov Recent Protein Ther Recomb Pharma Bio World 10–18
39. Arnold FH (1993) Engineering proteins for nonnatural environments. FASEB J 7(9):744–749
40. Brindha J, Balamurali MM, Chanda K (2019) Evolutionary approaches in protein engineering towards biomaterial construction. RSC Adv 9(60):34720–34734
41. Venkatesan N, Kim B (2002) Synthesis and enzyme inhibitory activities of novel peptide isosteres. Curr Med Chem 9(24):2243–2270
42. Lipovsek D, Plückthun A (2004) In-vitro protein evolution by ribosome display and mRNA display. J Immunol Methods 290(1–2):51–67
43. Antikainen NM, Martin SF (2005) Altering protein specificity: techniques and applications. Bioorg Med Chem 13(8):2701–2716
44. Mattanovich D, Borth N (2006) Applications of cell sorting in biotechnology. Microb Cell Factories 5(1):1–11
45. Shimizu Y, Kuruma Y, Ying BW, Umekage S, Ueda T (2006) Cell-free translation systems for protein engineering. FEBS J 273(18):4133–4140
46. Jackson SE, Craggs TD, Huang JR (2006) Understanding the folding of GFP using biophysical techniques. Expert Rev Proteomics 3(5):545–559
47. Yoshikuni Y, Keasling JD (2007) Pathway engineering by designed divergent evolution. Curr Opin Chem Biol 11(2):233–239
48. Chockalingam K, Blenner M, Banta S (2007) Design and application of stimulus-responsive peptide systems. Protein Eng Des Sel 20(4):155–161
49. Fang C, Xiao Z (2016) Receptor-based 3D-QSAR in drug design: methods and applications in kinase studies. Curr Top Med Chem 16(13):1463–1477
50. Lushington GH, Guo JX, Wang JL (2007) Whither combine? New opportunities for receptor-based QSAR. Curr Med Chem 14(17):1863–1877
51. Marshall SA, Lazar GA, Chirino AJ, Desjarlais JR (2003) Rational design and engineering of therapeutic proteins. Drug Discov Today 8(5):212–221
52. Georgescu R, Bandara G, Sun L (2003) Saturation mutagenesis. In: Directed evolution library creation. Humana Press, pp 75–83
53. Moore GL, Maranas CD (2004) Computational challenges in combinatorial library design for protein engineering. AICHE J 50(2):262–272
54. Allen BD, Nisthal A, Mayo SL (2010) Experimental library screening demonstrates the successful application of computational protein design to large structural ensembles. Proc Natl Acad Sci 107(46):19838–19843
55. Weber H, Khorana HG (1972) CIV. Total synthesis of the structural gene for an alanine transfer ribonucleic acid from yeast. Chemical synthesis of an icosadeoxynucleotide corresponding to the nucleotide sequence 21 to 40. J Mol Biol 72(2):219–249

56. Tindall KR, Kunkel TA (1988) Fidelity of DNA synthesis by the Thermus aquaticus DNA polymerase. Biochemistry 27(16):6008–6013
57. Pritchard L, Corne D, Kell D, Rowland J, Winson M (2005) A general model of error-prone PCR. J Theor Biol 234(4):497–509
58. Stemmer WP (1994) Rapid evolution of a protein in vitro by DNA shuffling. Nature 370(6488):389–391
59. Poole AM, Ranganathan R (2006) Knowledge-based potentials in protein design. Curr Opin Struct Biol 16(4):508–513
60. Samish I, MacDermaid CM, Perez-Aguilar JM, Saven JG (2011) Theoretical and computational protein design. Annu Rev Phys Chem 62:129–149
61. Arnold FH (1998) Design by directed evolution. Acc Chem Res 31(3):125–131
62. Chen W, Georgiou G (2002) Cell-surface display of heterologous proteins: from high-throughput screening to environmental applications. Biotechnol Bioeng 79(5):496–503
63. Packer MS, Liu DR (2015) Methods for the directed evolution of proteins. Nat Rev Genet 16(7):379–394
64. Wen F, Rubin-Pitel SB, Zhao H (2009) Engineering of therapeutic proteins. Protein Eng Design 75:153
65. Santiago-Rodriguez TM, Hollister EB (2019) Human virome and disease: high-throughput sequencing for virus discovery, identification of phage-bacteria dysbiosis and development of therapeutic approaches with emphasis on the human gut. Viruses 11(7):656
66. Carton JM, Strohl WR (2013) Protein therapeutics (introduction to biopharmaceuticals). In: Introduction to biological and small molecule drug Research and Development. Elsevier, pp 127–159
67. Lagassé HD, Alexaki A, Simhadri VL, Katagiri NH, Jankowski W, Sauna ZE, Kimchi-Sarfaty C (2017) Recent advances in (therapeutic protein) drug development. F1000Research, 6
68. De Groot AS, Scott DW (2007) Immunogenicity of protein therapeutics. Trends Immunol 28(11):482–490
69. Sinclair AM, Elliott S (2005) Glycoengineering: the effect of glycosylation on the properties of therapeutic proteins. J Pharm Sci 94(8):1626–1635
70. Fontana A, Spolaore B, Mero A, Veronese FM (2008) Site-specific modification and PEGylation of pharmaceutical proteins mediated by transglutaminase. Adv Drug Deliv Rev 60(1):13–28
71. Leader B, Baca QJ, Golan DE (2008) Protein therapeutics: a summary and pharmacological classification. Nat Rev Drug Discov 7(1):21–39
72. Guichard MJ, Patil HP, Koussoroplis SJ, Wattiez R, Leal T, Vanbever R (2017) Production and characterization of a PEGylated derivative of recombinant human deoxyribonuclease I for cystic fibrosis therapy. Int J Pharm 524(1–2):159–167
73. Dirks NL, Meibohm B (2010) Population pharmacokinetics of therapeutic monoclonal antibodies. Clin Pharmacokinet 49(10):633–659
74. Romer T, Peter F, Saenger P, Starzyk J, Koehler B, Korman E et al (2007) Efficacy and safety of a new ready-to-use recombinant human growth hormone solution. J Endocrinol Investig 30(7):578–589
75. Kunert R, Reinhart D (2016) Advances in recombinant antibody manufacturing. Appl Microbiol Biotechnol 100(8):3451–3461
76. Concolino D, Deodato F, Parini R (2018) Enzyme replacement therapy: efficacy and limitations. Ital J Pediatr 44(2):117–126
77. Pestka S, Krause CD, Walter MR (2004) Interferons, interferon-like cytokines, and their receptors. Immunol Rev 202(1):8–32
78. Viscor G, Javierre C, Pagès T, Ventura JL, Ricart A, Martin-Henao G et al (2009) Combined intermittent hypoxia and surface muscle electrostimulation as a method to increase peripheral blood progenitor cell concentration. J Transl Med 7(1):1–6
79. Huang PS, Boyken SE, Baker D (2016) The coming of age of de novo protein design. Nature 537(7620):320–327

80. Fleming PJ, Rose GD (2005) Do all backbone polar groups in proteins form hydrogen bonds? Protein Sci 14(7):1911–1917
81. Marcos E, Chidyausiku TM, McShan AC, Evangelidis T, Nerli S, Carter L et al (2018) De novo design of a non-local β-sheet protein with high stability and accuracy. Nat Struct Mol Biol 25(11):1028–1034
82. Bower MJ, Cohen FE, Dunbrack RL Jr (1997) Prediction of protein side-chain rotamers from a backbone-dependent rotamer library: a new homology modeling tool. J Mol Biol 267(5): 1268–1282
83. Kuhlman B, Dantas G, Ireton GC, Varani G, Stoddard BL, Baker D (2003) Design of a novel globular protein fold with atomic-level accuracy. Science 302(5649):1364–1368
84. Pan X, Thompson MC, Zhang Y, Liu L, Fraser JS, Kelly MJ, Kortemme T (2020) Expanding the space of protein geometries by computational design of de novo fold families. Science 369(6507):1132–1136
85. Pirogova E, Istivan T (2013) Toward development of novel peptide-based cancer therapeutics: computational design and experimental evaluation. In: Bioinformatics of human proteomics. Springer, Dordrecht, pp 103–126
86. Sievers SA, Karanicolas J, Chang HW, Zhao A, Jiang L, Zirafi O et al (2011) Structure-based design of non-natural amino-acid inhibitors of amyloid fibril formation. Nature 475(7354): 96–100
87. Khoury GA, Smadbeck J, Kieslich CA, Floudas CA (2014) Protein folding and de novo protein design for biotechnological applications. Trends Biotechnol 32(2):99–109
88. Hao J, Serohijos AW, Newton G, Tassone G, Wang Z, Sgroi DC et al (2008) Identification and rational redesign of peptide ligands to CRIP1, a novel biomarker for cancers. PLoS Comput Biol 4(8):e1000138
89. Correia BE, Ban YEA, Holmes MA, Xu H, Ellingson K, Kraft Z et al (2010) Computational design of epitope-scaffolds allows induction of antibodies specific for a poorly immunogenic HIV vaccine epitope. Structure 18(9):1116–1126
90. Bellows ML, Taylor MS, Cole PA, Shen L, Siliciano RF, Fung HK, Floudas CA (2010) Discovery of entry inhibitors for HIV-1 via a new de novo protein design framework. Biophys J 99(10):3445–3453
91. Smadbeck J, Peterson MB, Khoury GA, Taylor MS, Floudas CA (2013) Protein WISDOM: a workbench for in silico de novo design of biomolecules. J Vis Exp 77:50476
92. Xu J, Tack D, Hughes RA, Ellington AD, Gray JJ (2014) Structure-based non-canonical amino acid design to covalently crosslink an antibody–antigen complex. J Struct Biol 185(2): 215–222
93. Pantazes RJ, Maranas CD (2010) OptCDR: a general computational method for the design of antibody complementarity determining regions for targeted epitope binding. Protein Eng Des Sel 23(11):849–858
94. Saraf MC, Moore GL, Goodey NM, Cao VY, Benkovic SJ, Maranas CD (2006) IPRO: an iterative computational protein library redesign and optimization procedure. Biophys J 90(11): 4167–4180
95. Tripp KW, Barrick D (2004) The tolerance of a modular protein to duplication and deletion of internal repeats. J Mol Biol 344(1):169–178
96. Parmeggiani F, Huang PS (2017) Designing repeat proteins: a modular approach to protein design. Curr Opin Struct Biol 45:116–123
97. Graddis TJ, Remmele RL Jr, McGrew JT (2002) Designing proteins that work using recombinant technologies. Curr Pharm Biotechnol 3(4):285–297
98. Xie YH, Chen YX, Fang JY (2020) Comprehensive review of targeted therapy for colorectal cancer. Signal Transduct Target Ther 5(1):1–30
99. Deegan PB, Cox TM (2012) Imiglucerase in the treatment of Gaucher disease: a history and perspective. Drug Des Devel Ther 6:81
100. Gebauer M, Skerra A (2020) Engineered protein scaffolds as next-generation therapeutics. Annu Rev Pharmacol Toxicol 60:391–415

101. Fersht AR, Serrano L (1993) Principles of protein stability derived from protein engineering experiments. Curr Opin Struct Biol 3(1):75–83
102. Azhar A, Ahmad E, Zia Q, Owais M, Md Ashraf G (2017) Recent updates on molecular genetic engineering approaches and applications of human therapeutic proteins. Curr Protein Peptide Sci 18(3):217–232
103. Daugherty PS, Chen G, Olsen MJ, Iverson BL, Georgiou G (1998) Antibody affinity maturation using bacterial surface display. Protein Eng 11(9):825–832
104. Van Antwerp JJ, Wittrup KD (2000) Fine affinity discrimination by yeast surface display and flow cytometry. Biotechnol Prog 16(1):31–37
105. Sandhu JS (1992) Protein engineering of antibodies. Crit Rev Biotechnol 12(5–6):437–462
106. Pescina S, Sonvico F, Santi P, Nicoli S (2015) Therapeutics and carriers: the dual role of proteins in nanoparticles for ocular delivery. Curr Top Med Chem 15(4):369–385
107. Kolkman JA, Stemmer WP (2001) Directed evolution of proteins by exon shuffling. Nat Biotechnol 19(5):423–428
108. Patthy L (1999) Genome evolution and the evolution of exon-shuffling—a review. Gene 238(1):103–114
109. Siloto RM, Weselake RJ (2012) Site saturation mutagenesis: methods and applications in protein engineering. Biocatal Agric Biotechnol 1(3):181–189
110. Lam L, Liu XY, Cao Y (2003) Pretargeted radioimmunotherapy, a potential cancer treatment. Drugs Future 28(2):167–173
111. Huang CJ, Lowe AJ, Batt CA (2010) Recombinant immunotherapeutics: current state and perspectives regarding the feasibility and market. Appl Microbiol Biotechnol 87(2):401–410
112. Jay SM, Lee RT (2013) Protein engineering for cardiovascular therapeutics: untapped potential for cardiac repair. Circ Res 113(7):933–943
113. Saudagar P, Tripathi T (2023) Advanced spectroscopic methods to study biomolecular structure and dynamics, 1st edn. Academic Press, Cambridge, MA
114. Singh DB, Tripathi T (2020) Frontiers in protein structure, function, and dynamics. Springer Nature, Singapore
115. Tripathi T, Dubey VK (2022) Advances in protein molecular and structural biology methods, 1st edn. Academic Press, Cambridge, MA

Recombinant Production of Therapeutic Proteins

4

Parvez Singh Slathia, Sagrika, Era Sharma, Ishfaq Ahmad Khan, Randeep Singh Thakur, and Preeti Sharma

Abstract

Therapeutic proteins are gaining importance in disease therapy because of their specificity, efficiency, safety, and reduced side effects. With the dawn of recombinant DNA technology, the genes for the therapeutic proteins can be cloned into various expression systems, thus eliminating the earlier practice of obtaining such proteins from animal or human sources. Various expression systems like bacterial, yeast, mammalian, and plant hosts have been successfully used to recombinantly produce therapeutic proteins. Each expression system has its own benefits and limitations, thus making the expression of all proteins in a single system impossible. Prokaryotic systems like *E. coli* are well established and widely used for production; however, when it comes to glycosylated proteins, the lack of a secretory system in prokaryotes makes them ineffective. For producing such proteins, eukaryotic systems, particularly mammalian expression systems, are better suited. We discuss the methods for recombinant production of major therapeutic proteins using different expression systems.

Keywords

Recombinant proteins · Bacterial expression system · Animal cell lines · Enzymes · Cytokines · Hormones · Clotting factors

---

P. S. Slathia (✉) · Sagrika · E. Sharma · I. A. Khan · R. S. Thakur · P. Sharma
School of Biotechnology, Shri Mata Vaishno Devi University, Kakrial, Katra, Jammu and Kashmir, India
e-mail: parvez.singh@smvdu.ac.in

D. B. Singh, T. Tripathi (eds.), *Protein-based Therapeutics*,
https://doi.org/10.1007/978-981-19-8249-1_4

4.1 Introduction

Proteins are the workhorses carrying out most, if not all, functions in the cell. Their diversity makes it feasible to perform both structural and functional roles [1]. Functionally, they perform their role in tandem, thus forming an extensive network of interconnected biochemical cycles. Sometimes slight changes in critical proteins of these biochemical cycles could lead to diseases. Not only this, but variation in the concentration of proteins can also lead to diseases. Proteins as therapeutic agents have been employed in a broad spectrum of diseases and syndromes, which reveals their effective and operative nature. Therapeutic proteins are divided into various categories based on their chemical composition, origin, mechanism of activity, and pharmacological action. Proteins and peptides have tremendous therapeutic potential and have time and again exploited for the benefit of mankind. Most of the therapeutic proteins were initially derived from animals, which raised significant concerns, and led to the use of recombinant DNA technologies to discover effective humanized proteins that had better half-life and met the high demands.

One of the most significant and fastest-growing categories of the biopharmaceutical business is therapeutic recombinant proteins. Since 1980, the burgeoning industry of therapeutic proteins has seen a paradigm change with improved efficacy, higher safety, and lower immunogenicity due to a convergence of clinical, scientific, technical, and commercial drives that have identified unmet requirements. Since the first protein therapies were authorized two decades ago, the discipline has evolved from discovering naturally occurring proteins to creating molecules tailored for optimal target recognition, pharmacokinetics, biodistribution, and therapeutic action [2]. Human proteins produced using genetic engineering rather than extracted from tissue samples have emerged as a significant therapeutic medicine area. Examples of therapeutic proteins are antibodies, anticoagulants, blood factors, engineered protein scaffolds, enzymes, Fc fusion proteins, growth factors, hormones, interferons, interleukins, and thrombolytics [3].

In the last few years, bacteria, yeast, filamentous fungi, insect cells, mammalian cells, transgenic animals, and transgenic plants have been used to produce recombinant proteins [4]. Human proteins will continue emerging as an essential treatment alternative for several human disorders as manufacturing technology and pharmacological knowledge improve [5]. Human insulin (Humulin), produced by Eli Lilly and authorized by the US Food and Drug Administration (FDA) in 1982, was the first human protein therapy produced from recombinant DNA technology [6]. In 1986, the FDA approved the first medicinal use of a recombinant protein generated in mammalian cells (human tissue plasminogen activator, tPA, Genentech). Therapies based on recombinant proteins have revolutionized the treatment of several diseases, including cancer, rheumatic disorders, etc. Furthermore, the recombinant protein market has expanded at a 35% annual average rate since 2001, showing that the biopharmaceutical business has a bright future [2]. Therapeutic proteins are used to treat a variety of life-threatening diseases, including diabetes (insulin), viral hepatitis (interferon (IFN), end-stage renal disease (erythropoietin), cancer treatment-related neutropenia (G-CSF), clotting disorders (Factor VII, VIII, IX), inborn metabolic errors (lysosomal enzymes), etc. [3]. Pharmaceutical companies increasingly rely on protein-based research to hunt for new and effective treatments. Biotechnology advancements have boosted and facilitated the manufacturing of therapeutically essential proteins to battle a variety of terminal illnesses. Although protein-based treatments have taken center stage in drug research and have improved human safety, several hurdles remain, such as safety and immunogenicity concerns, protein stability, and degradation issues [7]. Proteins have emerged as a critical class of medicines in recent decades, with more than 200 protein-based drugs now on the market, 90% of which are employed as treatments. Though non-monoclonal antibodies (mAbs) recombinant proteins like insulin, erythropoietin (EPO), interferons (INF), interleukins (ILs), and somatotropin (hGH) are in high demand for therapy, mAbs are the fastest expanding category in the therapeutic industry [8]. The development and production of therapeutic proteins are highly complicated procedures. Considering the complexity of therapeutic proteins in the context of their large molecular size, post-translational modifications (PTMs), and the variety of biological materials used in their manufacturing process, it is essential to improve specific functional attributes while maintaining product safety and efficacy.

Since 2011, the Centre for Drug Evaluation and Review (CDER) and the Centre for Biologics Evaluation and Review (CBER) of the US FDA have authorized 62 recombinant therapeutic proteins. Monoclonal antibodies consist of over half of the 62 therapeutic proteins (48%), followed by coagulation factors (19%) and replacement enzymes (11%) in all cases. Fusion proteins, hormones, growth factors, and plasma proteins received the remaining approvals (22%) [9]. Many host systems, spanning from prokaryotic to eukaryotic species, have expressed diverse heterologous proteins with the help of recombinant DNA (rDNA) technology.

For the last 30 years, recombinant microbial techniques (mostly bacteria *Escherichia coli*) have been used, and they constitute a significant source of industrial and therapeutic proteins [10]. Over a third of all approved recombinant therapeutic proteins are produced by *E. coli*. Because of its well-characterized genetics, fast growth, and high yield production, *E. coli* has been a preferred alternative and workhorse for expressing non-glycosylated proteins in the biotech sector. A plethora of knowledge and extensive tools are ideally matched for commercial applications of *E. coli* systems, such as expression vectors, production strains, protein folding, and fermentation technologies. The advancement of the systems continues to suit contemporary industrial demands, as seen by the FDA's recent approval of *E. coli*-produced antibody fragments and Fc-fusion proteins. However, due to poor product quality, several proteins cannot be generated in microbial systems, particularly if (PTMs) are needed for the stability and activity of proteins [11].

The need for recombinant proteins has risen dramatically in the last 20 years. Recombinant proteins have been produced extensively using mammalian cell lines. Mammalian cells have the cellular machinery to promote the secretion of the recombinant product and (PTMs), such as glycosylation, which is present in many marketed recombinant therapeutic proteins, so this expression system has several advantages over microbial systems [12]. For efficient manufacturing, human cell lines have emerged as a new and potent option. These cells can make recombinant

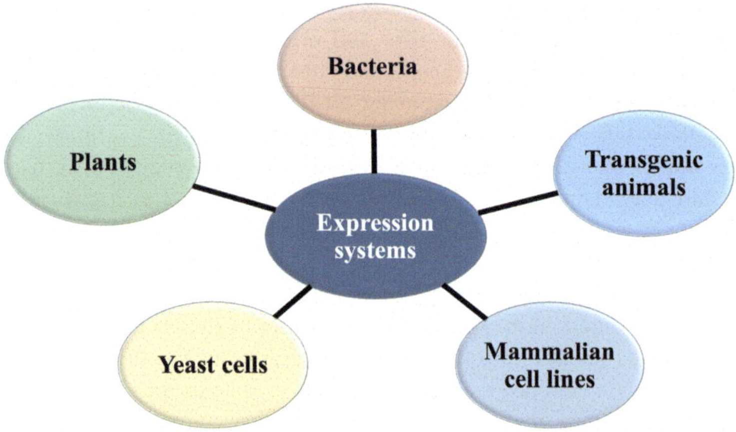

Fig. 4.1 Expression systems for recombinant therapeutic protein production

proteins with (PTMs) comparable to their native counterparts, resulting in proteins with a human-like glycosylation pattern that prevents immunogenic responses.

Plants have been created as recombinant protein production systems in a technology known as molecular plant farming over the last decade [13]. Transgenic plants and plant cell cultures are used as production methods for recombinant proteins having commercial, industrial, or medicinal uses in this technology. Plant molecular farming provides several benefits over traditional expression methods. Plants can create eukaryote-like PTMs, which allow the generation of recombinant humanized molecules that are highly comparable or even identical to the parent protein [14]. Furthermore, developing plant expression systems enables precise and regulated changes in recombinant proteins, which is beneficial for plant-based products that cannot be duplicated in bacterial or yeast expression systems [15]. Plants are exceptionally well adapted for generating therapeutic recombinant products like antibodies and vaccines, as they do not typically carry human diseases that can be linked with mammalian expression systems [16, 17]. Differences in the glycosylation pathway, and incredibly complicated processing of glycan side-chains, including specific host-specific alterations that do not occur in humans and vice versa, are the fundamental limits of plant expression systems. Glycoengineering, whose purpose is to produce recombinant proteins with homogeneous glycosylation that closely mimic the original system, can somewhat compensate for these flaws [10]. The different expression systems used for recombinant production of therapeutic proteins are shown in Fig. 4.1.

Immunogenicity is a challenging problem for all recombinant protein-based therapies. It is reasonable to suppose that all biotechnologically generated therapies will display immunogenicity at some point in the future. Antibody responses can result in decreased therapeutic protein effectiveness, neutralizing antibodies, inactivating native endogenous proteins, or inducing anaphylactic reactions. The Eprex instance is a well-known case [3]. Between 1998 and 2002, an extremely high percentage of individuals with chronic renal disease who had subcutaneously injected Eprex, a particular recombinant human erythropoietin preparation, were reported to have pure red cell aplasia. The decline in endogenous and therapeutically given erythropoietin produced by the therapy's neutralizing antibodies resulted in pure red cell aplasia. The presence of leachates in the formulation, which came from an interaction between polysorbate 80 and the uncoated rubber stopper, was most likely responsible for the antigenicity [3]. This emphasizes how difficult it is to forecast the safety of novel therapeutic protein products. On the one hand, we have observed that combining the recombinant protein with formulation/packaging ingredients might be more important for safety than alterations to the recombinant protein itself. It has been observed that combining the recombinant protein with formulation/packaging ingredients might be more important for safety than alterations to the recombinant protein itself. However, some molecular alterations in permitted biopharmaceuticals compared to the natural ones have been proven to be relatively safe and thus can be considered for therapeutic usage though this may lead to immunogenicity in the context of safety concerns. Drug inventors and manufacturers may now fine-tune and exploit beneficial functional features of proteins of interest while preserving product safety, effectiveness, or both, thanks to recent breakthroughs in protein-engineering technology. Protein engineering and design have been shown to help develop effective protein therapeutics with improved affinity, specificity, activity, stability, pharmacokinetics, pharmacodynamics, immunogenicity, and productivity.

4.2 Methods for the Production of Recombinant Therapeutic Proteins

Today various methods for producing recombinant therapeutic proteins like bacterial, fungal, plant, mammalian cell cultures, transgenic animals, etc., are available [18]. There are about 165 approved biopharmaceutical products, most of which are proteins. Therapeutic proteins are primarily derived from the blood tissues of humans or animals in a few cases [19]. The benefit of recombinant proteins is that it makes a platform for manufacturing more advanced and effective pharmaceutical products with high therapeutic profiles like safety, less immunogenicity, and increased half-life with better bioavailability. An overview of recombinant protein production is depicted in Fig. 4.2.

4.2.1 Production of Recombinant Proteins in *E. coli*

The bacterial expression system has been extensively used for rapid growth and high-quality yields as it is easy to control the expression of a prokaryotic system [20]. These systems are used for the production of non-glycosylated proteins [21]. The most widely used bacterial system for the production of recombinant proteins is *Escherichia coli*. *E. coli* has proved a boon in the biotech industry

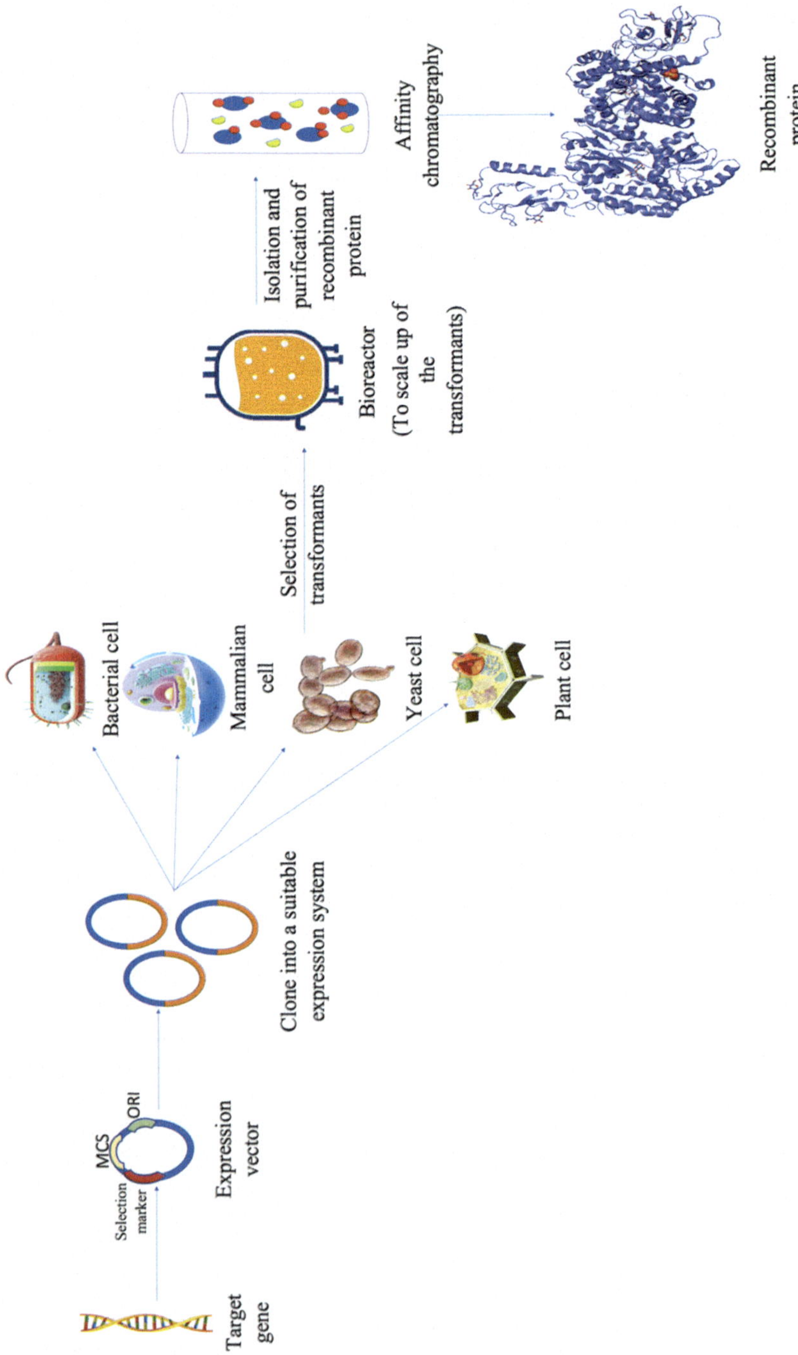

Fig. 4.2 General outline of production of recombinant proteins

because of its growth, high yield, and simple genetic system. It is always preferred in the biotech industry to manufacture non-glycosylated proteins. The first recombinant protein insulin, approved by the FDA, was expressed in *E. coli* which opened a new path for the production of different recombinant therapeutic proteins [11]. One major reason for the production of proteins by *E. coli* is that it can express a good quantity of proteins at a low culture media cost. The expression of proteins in *E. coli* requires consideration of various aspects like the host cells, vectors, and media composition for an efficient production [22, 23]. The general scheme for the expression of therapeutic recombinant proteins in *E. coli* is depicted in Fig. 4.3.

4.2.1.1 Host

In the biotech industry, *E. coli* K12, RV308, and W3110 strains are most commonly used for producing therapeutic proteins. K12 strains have been genetically engineered to minimize acetate production during cell development, further improving recombinant protein synthesis. Acetate is an unwanted by-product of fermentation that has a detrimental effect on protein production and cell growth when it accumulates in larger concentrations [11]. Other host variants are also engineered to meet particular needs, like the extraction of periplasmic therapeutic proteins.

4.2.1.2 Vector

The copy number of the plasmid and structural stability of the vector dramatically affect the productivity of recombinant proteins. The higher copy number leads to an increased mRNA pool which enhances productivity and vice versa. Antibiotics are added to growth media to balance the plasmid's copy number; however, they impose metabolic burdens on cells. Several vectors have been developed for protein production in *E. coli,* such as pUC (500–700 copies, a high copy number plasmid), and pBR322 (14–20 copies, a low copy number plasmid).

4.2.1.3 Media Preparation and Fermentation

The media plays a vital role in obtaining cell yield and high production of recombinant proteins. Generally, three types of media are used, i.e., chemically defined, semi-defined, and complex media. The media concentration must be analyzed very carefully, and it should contain all crucial components in optimal concentration to provide essential nutrients for cell growth. The most commonly used growth media are semi-defined and complex media because of their flexibility and ability to generate more cell density and protein yield. Following media preparation, fermentation is done, which increases the cell density and protein yield in a controlled physical, chemical, and biological environment based on parameters like pH, temperature, dissolved oxygen (D.O) level, and nutrient availability.

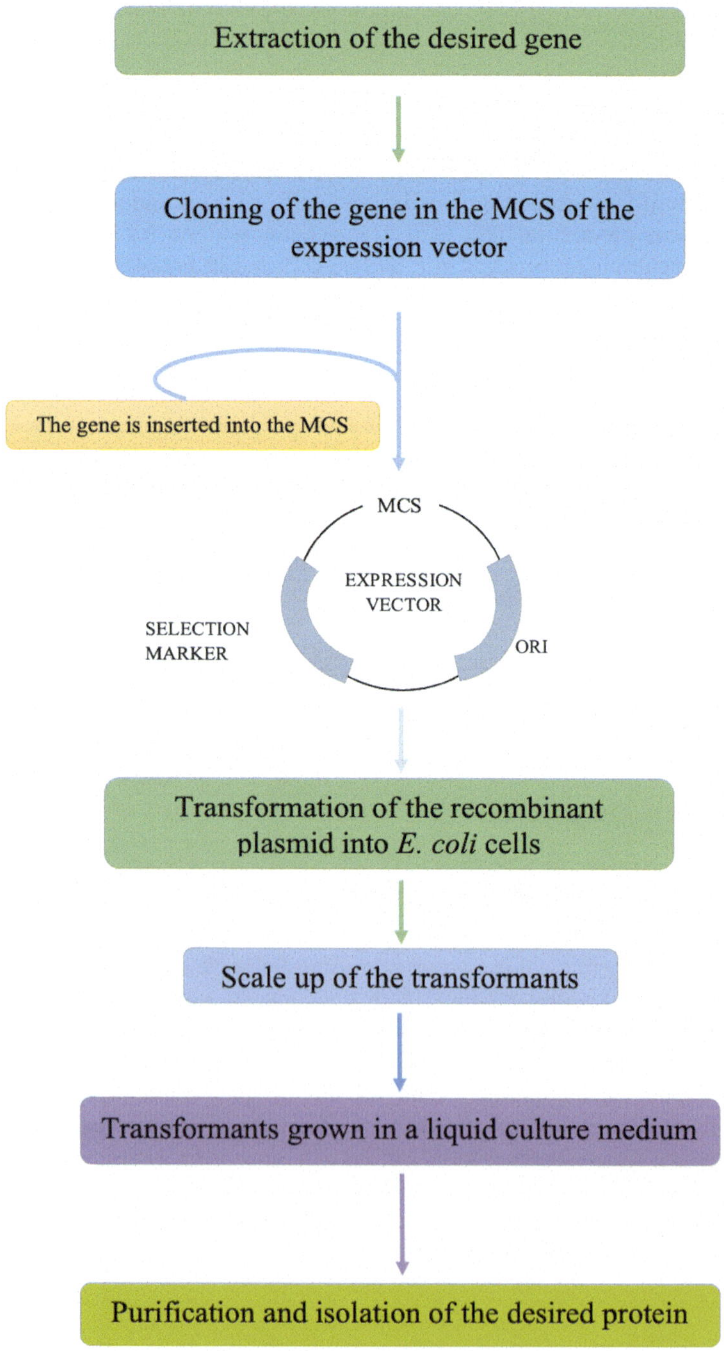

Fig. 4.3 Therapeutic protein production in *E. coli* expression system

4.2.2 Therapeutic Protein Production in Bacterial Cultures

4.2.2.1 Therapeutic Enzymes

L-asparaginase is obtained from submerged fermentation of bacteria *Erwinia chrysanthemi* and *E. coli* [24] and used for treating leukemia. This enzyme hydrolyzes L-asparagine to form aspartic acid and ammonia, which inhibits protein synthesis in leukemia cells [25]. There have been minor incidents of hypersensitivity to L-asparaginase; hence, a recombinant alternative was introduced known as Pegaspargase under the trade name Oncaspar in which the enzyme is linked to monomethylpolyethylene glycol. This enzyme is synthesized using an *E. coli* expression system [25]. Glucarpidase, also known as carboxypeptidase-G2, is an enzyme that metabolizes circulatory methotrexate (MTX) and is used in patients with renal problems who are receiving methotrexate to treat specific forms of cancer. The enzyme is produced in *E. coli* cells in the recombinant form [26]. Pegloticase is recombinant porcine-like uricase (urate oxidase) conjugated to PEG and is used to treat chronic gout refractory. It decreases uric acid levels and inhibits crystal formation by converting uric acid to allantoin, which is excreted by the kidneys [25]. It consists of modified mammalian urate oxidase, produced in *E. coli*, covalently conjugated to monomethoxy-polyethylene glycol [25].

4.2.2.2 Therapeutic Hormones

Humulin N (recombinant human insulin) is a man-made insulin product used for glucose control in patients with diabetes. Earlier, the purified insulin was obtained from pig or cow pancreas, which differs slightly in chemical composition from human insulin. The decrease in the supply of animal pancreas led to extensive use of rDNA technology and the production of humulin which was safe and efficacious. The human insulin gene was isolated and inserted into the plasmid of the *E. coli,* and then this recombinant bacterial cell produces insulin which can be easily harvested and stored for further use [27]. Insulin promotes cell division and growth through its mitotic effect, regulates carbohydrate, lipid, and protein metabolism, facilitates cellular glucose uptake, and maintains a normal blood sugar level [28].

Glucagon is a 29-amino acid peptide hormone secreted from pancreatic alpha cells in the islet of Langerhans [29]. Recombinant DNA technology has been efficiently utilized to produce glucagon by *E. coli* [30]. This hormone helps raise blood sugar levels by working in coordination with the liver. Hence it works best in case of hypoglycemia, a condition in which the blood sugar level is below normal. In 2020, FDA approved the generic glucagon for the treatment of hypoglycemia.

Metreleptin is a recombinant human leptin analogue used to treat complications of congenital or acquired generalized lipodystrophy. It is a 16 kDa protein that differs from endogenous human leptin in having an amino acid terminal methionine residue. It is also produced by utilizing the bacterial *E. coli* cells. Metreleptin, as a human leptin analogue, binds and activates the leptin receptor, imitating the physiological actions of endogenous leptin. It is sold under the trade name Myalept and improves LD-related metabolic problems such as glycemic control, hypertriglyceridemia, and insulin sensitivity [31]. It may disrupt the events that lead to lipotoxicity and show improvements in the hepatic disease parameters such as intrahepatic lipid content, liver volume, and non-alcoholic steatohepatitis (NASH) [32].

Calcitonin is a thyroid hormone that decreases calcium content in the blood when it is higher than usual. This polypeptide hormone contains 32 amino acids and is mainly generated by the thyroid gland's parafollicular cells (also known as C cells). Calcitonin is produced by cells of the ultimobranchial body in fish, birds, and other non-mammalian animals. Salmon calcitonin is a 32 amino acid alpha-helical polypeptide that varies significantly from human calcitonin in amino acid residues ranging from 10 to 27. These amino acid sequence variations boost the efficacy of salmon-derived calcitonin. The FDA authorized it for the treatment of bone ailments such as postmenopausal osteoporosis, Paget disease, and hypercalcemia. Recombinant calcitonin is synthesized using either rDNA technology or chemical peptide synthesis. A chemically synthesized salmon calcitonin gene was incorporated into the plasmid and introduced into the *E. coli* cells to produce peptide substrate for *in vitro* amidation [33]. The hormone also limits calcium absorption from the intestine and enhances calcium absorption by the kidney [34].

Parathyroid hormone (PTH) secreted from the parathyroid glands regulates calcium levels in the blood when they are too low. The hormone helps treat a rare condition of calcium metabolism called hypoparathyroidism, which can be chronic or resolve transiently. This condition is traditionally treated with calcium and vitamin D analogues that boost blood calcium, but this bypass the significant calcium reabsorption effects of PTH on the kidney, resulting in hypercalciuria and an increased risk of nephrocalcinosis and renal insufficiency. Recombinant human PTHs [PTH (1–84) and PTH (1–34)] are hence used for the treatment of hypoparathyroidism. Natpara (trade name) administration normalizes plasma calcium and phosphate levels resulting in the reduction in the doses of calcium and vitamin D analogues. The FDA has authorized teriparatide or human PTH (1–34) or Forteo (commercial name) for the treatment of osteoporosis in postmenopausal women and men who are at high risk of fracture [35]. These recombinant PTHs are produced in *E. coli* [36], and it has been shown that biosynthetic teriparatide (1–34) could be successfully secreted and expressed in *Bacillus subtilis* [37].

Growth hormone (GH), also called somatotrophin or human growth hormone, is a peptide hormone secreted by the anterior lobe of the pituitary gland. GH plays a vital role in regulating somatic development, body composition, and intermediate muscle and bone metabolism, particularly throughout adolescence. Initially, cadaveric GH was utilized in treatment, but it was phased out and replaced by recombinant human GH (rhGH), which is identical to human GH but contains an extra methionine [38]. Growth hormone treatment is mainly used to treat children with idiopathic short stature who cannot attain a normal adult height. Growth requires the combined work of thyroid and gonadal hormones, as well as GH, especially around puberty. Treatment with rhGH is effective in children with chronic kidney disease (CKD), which is linked with severe alterations in the GH-IGF-1 axis, resulting in growth retardation. The human *GH* gene is usually inserted into plasmids of *E. coli* bacteria to manufacture the rhGH. The production of rhGH could also be done in Chinese hamster ovary (CHO) cells, which have proven to be more advantageous than *E. coli* as the rhGH is secreted into protein-free production media, allowing for easier purification and avoiding inclusion body re-solubilization and protein refolding [39].

Insulin-like growth factor 1 (IGF-1), also called somatomedin C, is a hormone similar in structure to insulin and can improve insulin sensitivity while lowering glucose levels. IGF-1 is currently being researched as a therapeutic agent for cancer treatment, tissue repair, and the treatment of insulin-resistant diabetes. It is the direct mediator of GH's effect on statural growth and must be present for the normal growth of bones, cartilage, and organs in children. Hence it is used to treat children with short stature [40]. Recombinant human IGF-I has been created in a variety of host systems. To date, hIGF-1 has been produced using several expression platforms, including *E. coli*, yeast, a cell-free system, and transgenic plants [41].

4.2.2.3 Recombinant Cytokine Production

Interferon-alpha (IFN-α) was the first approved antitumor application for hairy cell leukemia in 1986 [42]. IFN-α (trade name Infergen) is an approved treatment for chronic hepatitis C virus (HCV) infection [43]. The clinical application of IFN-α is limited as it has a short life *in vitro* and shows severe toxicity at therapeutic doses, so a recombinant IFN-α, modified with the addition of polyethylene glycol (PEG) moiety is currently used as a treatment option for HCV along with certain other drugs like ribavirin. Two forms of pegylated interferon (PEG-IFNs) were developed in the early 2000s—PEG-IFN-alpha2a and PEG-IFN-alpha2b [44]. They have pro-apoptotic/anti-proliferative activity on tumor cells mediated by the activation of immune system cells and display a potent anti-angiogenic activity [45]. The FDA has authorized two kinds of interferon as adjuvant treatments for individuals with high-risk melanoma: interferon alfa-2b (Intron A) and peginterferon alfa-2b (Sylatron). Pegylated formulations of IFN-α is effectively used for the treatment of AIDS-associated Kaposi sarcoma [46]. It is most often used in treating chronic myeloid leukemia (CML) in combination with tyrosine kinase inhibitors such as imatinib, as it reduces the growth and division of leukemia cells [47]. Interferons have significant use in treating condyloma acuminata or genital warts [48]. IFN-n3 (Alferon N) injection made from human leukocytes has a spectrum of IFNs and is only licensed for treating adult patients with refractory or recurring condyloma acuminata. Recombinant interferons are commercially produced mainly in bacteria *E. coli* (IFN-α, IFN-β1b, IFN-γ) and rarely in a mammalian cell line CHO (IFN-β1a).

Interferon-beta (IFN-β) is secreted by many cells like lymphocytes (B cells, T cells, and NK cells), macrophages, fibroblasts, endothelial cells, and osteoblasts in response to pathogens. They induce immunoregulatory, antitumor, antiviral, and anti-inflammatory responses by activating the natural killer (NK) cells and macrophages [49]. Recombinant IFN-β has been authorized for the treatment of multiple sclerosis (MS), including rhIFNβ-1a (trade name Avonex) and rhIFNβ-1b (trade name Betaseron). In MS, the immune system attacks the protective covering or nerves, causing them to deteriorate [50]. The use of IFNβs may help in the reduction of MS relapses. rhIFN-β1a is synthesized using eukaryotic expression systems (CHO cells) and has an identical amino acid sequence to naturally occurring hIFN-β, while IFN-β1b is produced by the prokaryotic expression system in *E. coli* [51]. Synthetic PEGylated IFN-β1a is also widely used as a therapeutic agent with many perks like increased half-life and reduced glomerular filtration rate.

Interferon-gamma (IFN-γ) is a homodimer formed of polypeptide subunits. It is released by the activated lymphocytes such as CD4 T Th1 cells and CD8 CT cells, T cells, natural killer cells, B cells, and antigen-presenting cells (APCs). IFN-γ regulates inflammation, apoptosis, and cell cycle, among other biological responses. It is primarily engaged in host defense and immune surveillance, but it also aids in the development of adaptive immunity, regulation of inflammation, apoptosis, and cell cycle [52]. IFN-γ is used to treat a genetic disorder called chronic granulomatous disease, in which phagocytes are unable to kill certain types of bacteria and fungi. IFN-γis a safe and effective medication for lowering the frequency and severity of viral infections [53]. The FDA has approved recombinant IFN-γ (trade name Actimmune) for the treatment of malignant osteopetrosis, which is strikingly evident by increased bone density as a result of a failure in bone reabsorption by cells known as osteoclasts. IFN-gamma promotes osteoblast formation and maturation [54] and increases bone resorption [55]. Recombinant IFN-γ is commonly expressed in *E. coli* [56].

Interleukin-2 (IL-2) helps NK cells and T lymphocytes to grow. As a result, it is commonly employed in adoptive transfer protocols to grow lymphocytes in culture and extend the survival of transferred cells in cancer patients. Aldesleukin (recombinant IL-2) is being used in cancer therapy as a biological response modifier to strengthen the immune system and kill the cancer cells; hence IL-2 (trade name Proleukin) has been approved for the treatment of metastatic renal cell carcinoma and metastatic melanoma [57]. An *E. coli* expression system is used to produce the recombinant IL-2.

Interleukin-11 (IL-11) is a non-glycosylated protein belonging to the family of IL-6, produced by fibroblasts, neuronal cells, epithelial cells, and endothelial cells. This multifunctional anti-inflammatory cytokine promotes hematopoiesis, regulates T cell activity, activates macrophage proliferation and differentiation, and protects the mucosa [58]. It has also been found to enhance T cell polarization toward Th2, promote B-cell IgG production, accelerate osteoclast bone absorption, protect endothelial cells from oxidative stress, and control epithelial proliferation and apoptosis. The pharmacologic agent IL-11 was the first to be licensed for the treatment of chemotherapy-induced thrombocytopenia, which is the deficiency of platelets in blood [59]. Recombinant IL-11, a commercially available growth factor, increases the platelet count and is produced in *E. coli* [60].

4.2.2.4 Recombinant Toxins

Botulinum toxins are used widely in the cosmetic industry to provide transient anti-aging effects like reducing facial fine lines and wrinkles. Botulinum toxin type A and type B are proteases that inactivate SNAP-25, which is essential in synaptic vesicle fusion [61]. These toxins are produced by strains of *Clostridium botulinum*. Serotype A is commercially available for clinical use. Approved under the trade name Botox in 1989, this toxin is used for the treatment of strabismus, hemifacial spasms, and blepharospasm [62]. Full-length recombinant attenuated botulinum neurotoxins were expressed in *E. coli* and purified via an amino-terminal 6X-polyhistidine affinity tag [63].

4.3 Production of Recombinant Therapeutic Proteins in Mammalian Cell Cultures

Mammalian systems for the production of therapeutics were developed because of their ability to perform complex PTMs that are difficult to achieve by non-mammalian expression systems. Therefore, most recombinant therapeutic proteins are produced in mammalian expression systems, as shown in Fig. 4.4. The recombinant gene of interest is incorporated into the vector, which is then introduced into the desired cell line. The recombinant gene contains necessary transcriptional regulatory elements for the transcription, and in addition, a selection system is also incorporated. Glutamine synthetase (GS) and dihydrofolate reductase (DHFR) are the common markers for selecting transformed cells. These selective markers allow transformed cells to grow in a medium lacking glutamine and hypoxanthine/thymidine. Cells that survive are used to produce a clonal population. The most stable population with the appropriate growth and productivity characteristics is analyzed for recombinant protein expression [64, 65].

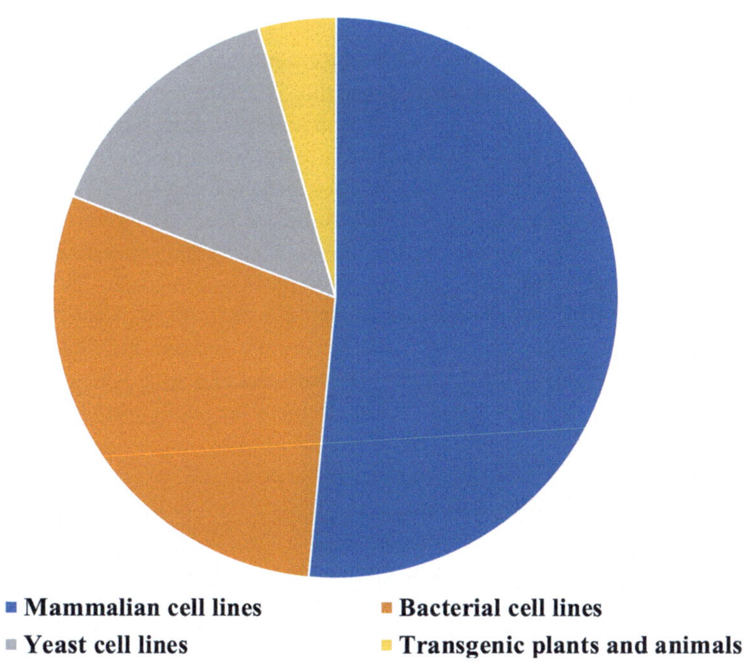

Fig. 4.4 Therapeutic proteins produced in different expression systems

The two primary cell culture systems used to produce recombinant therapeutic proteins in mammalian cells are adherent cell culture and cell suspension culture. Adherent cell culture includes seeding cells into roller bottles, filling the bottles with media, and harvesting cell culture fluids that are extracted within air filtered environment. Cultivation is also possible on microcarriers that are kept in suspension in stirred tank bioreactor. CHO cell lines cultivated on microcarriers are used for generating follicle-stimulating hormone (FSH). Suspension culture is the most commonly used culture because of its ability to support single-cell suspension growth. CHO cells, BHK (baby hamster kidney) cells, and human retina-derived PER-C6 cells grow well in suspension systems [65].

4.3.1 Therapeutic Proteins Expressed in Chinese Hamster Ovary (CHO) Cell Lines

The CHO cell lines are the most employed mammalian expression system for recombinant protein production due to their ability to generate complex PTMs easily [66]. In addition, CHO cell lines have high endurance for deviations in pH, pressure, temperature, and oxygen levels [67] and allow specific productivity [21, 67]. Therapeutic protein production in CHO cells is depicted in Fig. 4.5.

4.3.1.1 Enzyme Production

Asfotase alfa (trade name Strensiq) is a recombinant tissue-nonspecific alkaline phosphatase-Fc-deca-aspartate fusion protein that is incorporated in an engineered CHO cell line [68]. It is used in enzyme replacement therapy of hypophosphatasia [69]. Hypophosphatasia is a genetic disorder that is a rare metabolic bone disease characterized by reduced serum alkaline phosphatase and defective development of bones and teeth [70]. Agalsidase alfa enzyme is synthesized in CHO cultures with a plasmid having DNA sequences that code for α-galactosidase, a protein whose deficiency causes Fabry disease. It is a rare X-linked recessive glycosphingolipid storage disorder [71] that results in the absence of α-galactosidase A due to mutations in the *α-galactosidase A* gene. The enzyme Alglucosidase alfa is produced in CHO cells to express human acid α-glucosidase enzyme, deficiency of which causes glycogen storage Pompe disease [72]. Pompe disease is a fatal inherited disorder caused by the mutation in a gene that forms acid α-glucosidase that breaks glycogen. Mutations in this gene cause defects or deficiency of the acid α-glucosidase enzyme [73].

The enzyme imiglucerase is glucocerebrosidase produced in CHO cells, and after production, the exoglycosidase enzyme cleaves the carbohydrate moieties to expose mannose residues [74]. This enzyme is used for treating Gaucher disease, an inherited metabolic disorder characterized by accumulation of glycolipid glucocerebroside due to deficiency in enzyme glucocerebroside resulting from a mutation in the *GBA1* gene [75].

Elosulfase alfa is expressed in CHO cell lines incorporated with recombinant human N-acetylgalactosamine 6-sulfatase containing expression plasmid leading to

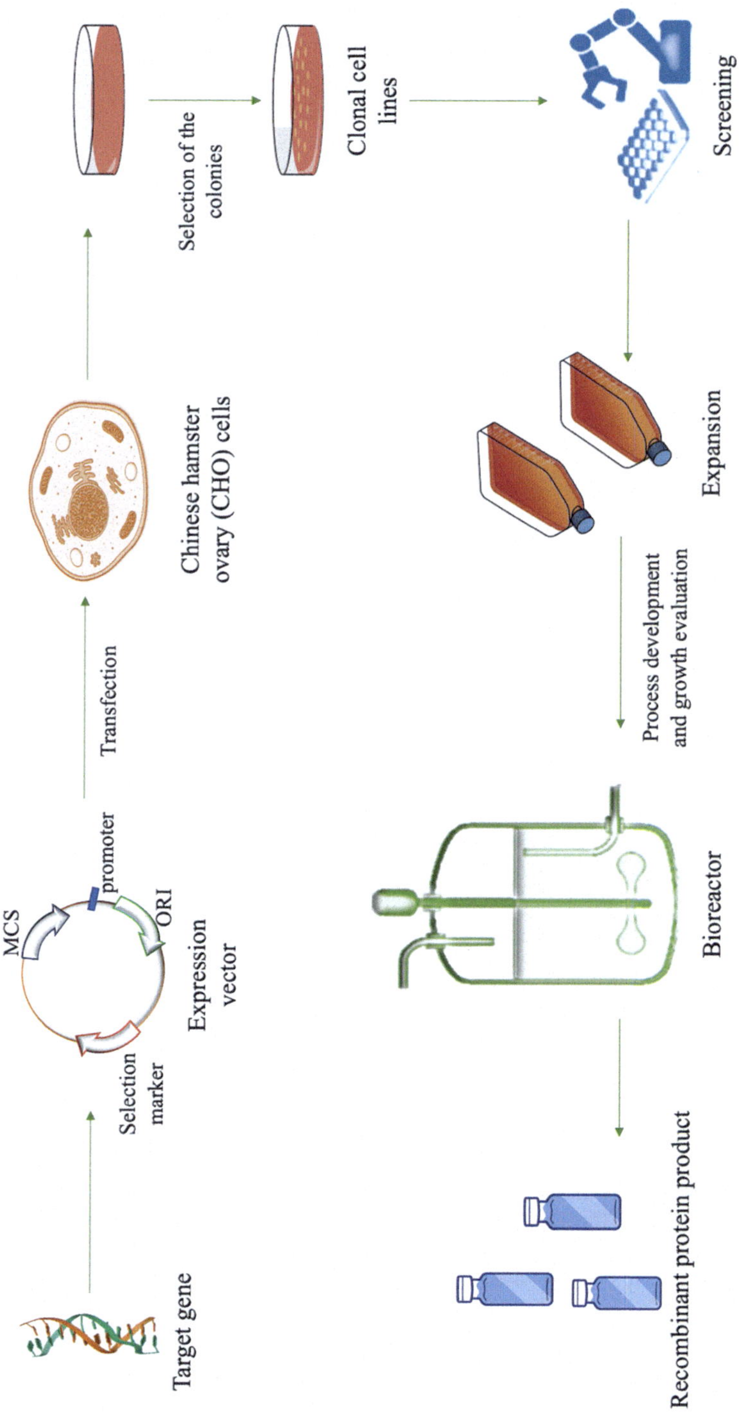

Fig. 4.5 Production of therapeutic proteins in CHO cell line

the generation of clones that produce the enzyme. It is used for the treatment of mucopolysaccharidosis type IVA [76]. It is a mucopolysaccharide storage disease that develops due to the deficiency of N-acetyl-galactosamine-6-sulfatase, leading to the accumulation of mucopolysaccharides in the body.

Galsulfase is a recombinant variant form of N-acetylgalactosamine 4-sulfatase enzyme used for treating mucopolysaccharidosis VI [25]. It is synthesized in the CHO cell line via the perfusion process. The enzyme N-acetylgalactosamine 4-sulfatase also known as arylsulfatase B is encoded by the *ARSB* gene. The lack of activity of this enzyme leads to the build-up of glycosaminoglycans in the body, thus leading to mucopolysaccharidosis VI, also known as the Maroteaux-Lamy syndrome [77].

Laronidase is a recombinant form of human enzyme α-L-iduronidase [25] produced in CHO cells consisting of six asparagine-linked glycosylation sites, two of them being mannose-6-phosphorylated. Iduronidase degrades glycosaminoglycans like dermatan sulfate and heparan sulfate. The deficiency of this enzyme leads to the accumulation of dermatan sulfate and heparan sulfate in the lysosomes and causes mucopolysaccharidosis I.

4.3.1.2 Clotting Factor Production

Recombinant human thrombin is obtained from a truncated transcript of the human prothrombin gene, which is then incorporated into CHO cells. The product is secreted into the culture media and is activated by proteolysis [78]. Thrombin is a type of protease formed from prothrombin, which converts fibrinogen to fibrin and thus aids in hemostasis. A recombinant form of factor IX under the trade name Rixubis is BAX362 produced using CHO cells grown in suspension culture. The CHO line releases the recombinant factor IX into a defined cell culture medium which is used for purification [79]. Factor IX is a part of the blood coagulation system. The activated factor IXa, with factor VIIIa, and phospholipids in the presence of Ca^{2+}, activates factor X [80]. The deficiency of factor IX causes hemophilia B, a hereditary bleeding disorder.

4.3.1.3 Production of Hormones

Human chorionic gonadotropin (HCG) is a 237 amino acid glycoprotein hormone produced primarily by syncytiotrophoblastic cells of the placenta during pregnancy. Recombinant DNA technology has been used to create rHCG from genetically modified CHO cells [81]. HCG is essential for the success of the pregnancy. It boosts fertility in females and is used to initiate mid-cycle follicular maturation and ovulation of women undergoing *in vitro* fertilization (IVF) treatment. An HCG dose helps the eggs develop before they are fertilized in a lab for those who use IVF. It is widely utilized in all types of reproductive treatments, including IVF and intrauterine insemination (IUIs).

Follicle-stimulating hormone (FSH), another gonadotropin, regulates the growth, sexual maturation, and reproductive development in both females and males. FSH is used in the treatment of infertility and hypogonadism along with human chorionic gonadotropin (hCG), which treats secondary hypogonadism [82]. Adding hCG to rFSH is associated with better quality embryos and higher pregnancy rates [83]. Recombinant FSH is also effective for the treatment of oligozoospermia [84]. The CHO cell line is used for recombinant FSH production [85].

Luteinizing hormone (LH) is a pituitary hormone that stimulates ovarian follicle maturity by interacting with receptors on ovarian follicles. A spike of LH in the middle of the menstrual cycle causes ovulation and the creation of progesterone by the corpus luteum, which is required for the development of the uterine endometrium and the implantation of the fertilized egg. LH stimulates the testes to produce testosterone in males. It is clinically utilized to enhance ovarian follicle development in assisted reproduction methods (ART) and in vitro fertilization (IVF). Lutropin alfa is the first and only recombinant human form of LH and is produced using genetically engineered CHO cells [86].

The pituitary gland produces and releases thyroid-stimulating hormone (TSH) into the bloodstream. The major functions of pituitary TSH are maintaining the secretion and biosynthesis of the thyroid hormones. The FDA approved a recombinant human TSH (thyrotropin alfa) for diagnostic use in patients with thyroid cancer. It works by causing thyroid cells to produce thyroglobulin (Tg), absorb radioactive iodine, and eliminate thyroid tissue that remains after surgery [87]. The rhTSH is produced in CHO cells.

4.3.1.4 Cytokine Production

Interleukin-12 (IL-12) is an immunoregulatory protein generated by macrophages, B cells, mononuclear cells, keratinocytes, and dendritic cells, with early undifferentiated pluripotent hematopoietic stem cells as its target. Recombinant IL-12, secreted by CHO cells, has the potential to kill tumor cells by limiting blood flow to them and activating white blood cells to kill them [88]. RhIL-12 restores the hematopoietic function and also improves the immune function. It has therapeutic and protective effects on radiation-related problems (which are prevalent after the therapy), such as blood cell loss, myelosuppression, and immune function decline or imbalance, indicating promising development and application prospects. It also plays a major role in inducing protective immunity against infectious agents like viruses [89].

4.3.2 Other Mammalian Cell Lines Used for Therapeutic Protein Production

Like CHO, other expression systems have also been used to produce protein therapeutics. The trend follows using recombinant tools for synthesizing structurally similar and/or human-like therapeutics in human embryonic kidney 293, fibrosarcoma HT-1080 cell lines, etc. [21]. Human cell lines as expression systems have several advantages, including rapid production, no immunogenic PTMs due to human-compatible glycosylation, and access to many transfection methods [21].

The first recombinant enzyme introduced and approved by the FDA in 1987 was a recombinant human tissue plasminogen activator (tPA) Alteplase (trade name Activase) [90]. It is a fibrinolytic agent that converts plasminogen to plasmin, which dissolves fibrin and fibrinogen, thus aiding cardiovascular diseases like acute myocardial infarction, ischemic stroke, pulmonary embolism, etc. It was produced by establishing a cDNA library of isolated human tissue-type plasminogen activator from a human melanoma cell line [91]. Other recombinant tissue plasminogen activators, tenecteplase, and reteplase were synthesized in the same cell lines. While tenecteplase is a mutated variant of tPA alteplase with 6 mutated amino acids (out of 527) [92], reteplase is a single-chain deletion mutant of tPA containing 355 amino acids [93].

Idursulfase alfa is a recombinant iduronate-2-sulfatase enzyme generated using a human HT-1080 fibrosarcoma cell line [94]. Its production is based on gene activation, which consists of incorporating a DNA promotor upstream of an endogenous gene [20]. The resultant enzyme has a similar structure to the endogenous enzyme. It is used to treat mucopolysaccharidosis type II, also called Hunter syndrome, a lysosomal storage disease characterized by defects in the functioning of lysosomal enzymes [94].

Another enzyme produced in fibroblast carcinoma human cell, velaglucerase alfa, uses a mannosidase I inhibitor to obtain the necessary glycosylation, producing high mannose-containing oligosaccharides that are used in enzyme replacement therapy for Gaucher disease [74]. Gaucher disease is an inherited metabolic disorder resulting from the absence of enzyme glucocerebrosidase, leading to the deposition of certain fats like glycolipid glucocerebroside.

Factor VIIa available for commercial use is a recombinant factor VIIa produced by transfecting the human factor VII gene into BHK cells cultured in bovine albumin. The product is secreted into the media and later purified by a chromatographic process [95]. It is used to treat factor FVII deficiency, an inherited coagulation disorder previously treated with prothrombin complex concentrates or plasma-derived factor VII concentrates. Being a recombinant product, there is no risk of blood-transmitted disease as it is not derived from human or animal plasma [96].

Factor VIII (FVIII) is a blood-clotting protein that also functions as an anti-hemophilic factor (AHF) in blood coagulation. It is a cofactor for factor IXa, which forms a complex with Ca2+ and phospholipids to convert factor X to the active form Xa. Recombinant variants are available in two forms: full-length in human serum albumin or formulated without human serum albumin but containing sucrose as the primary stabilizer [97]. Recombinant FVIII (rFVIII) is produced in mammalian cell culture. rFVIII (trade name Kogenate) is generated in the BHK21 cell line. The full-length FVIII coding sequence was assembled in the plasmid and introduced in BHK21 cells. rFVIII (trade name Recombinate) uses CHO cells as the expression vector. A variant, ReFacto, does not use full-length FVIII but lacks the b-region and has a truncated FVIII molecule with a peptide linker. In contrast, Kogenate combines full-length FVIII with an albumin-free formulation [97].

4.4 Production of Recombinant Therapeutic Proteins Using Plant Cell Cultures

The agricultural or plant industry is a well-settled and rapid growing industry, as crop growing and harvesting is simple, easy, and economical with the latest infrastructure. The most typical growing conditions like sunlight, water, inexpensive, and a wide range of fertilizers make this industry more profitable. Various plant expression systems are available, such as whole species, gene constructs, or individually targeting protein. These forms are chosen accordingly to the type of protein to be produced. For expressing genes in plants, a variety of approaches can be followed. The nuclear expression can be done by integrating transgenes into the nuclear genome of the plant. These transgenes undergo nuclear transcription and cytoplasmic translation. The genes for therapeutic protein can also be expressed in the tumor-inducing (Ti) plasmid of *Agrobacterium tumefaciens*, a well-established system for transgene expression in plants. The protein targeting can be achieved by using signal peptides for organellar storage or secretion, which localizes the protein to a specific organelle or targets it for secretion into media [98]. For chloroplast expression of the protein, the insertion of the transgene to the chloroplast genome can be done. The precise insertion of the gene prevents it from being placed in a poorly transcribed region of the genome, resulting in high levels of expression [99]. Suspension cell cultures can be used for the production of proteins that possess many advantages, such as sterility, confinement, and well-defined downstream processing techniques [100].

Plants are potential expression systems that can produce cheap and safer proteins than mammalian cells [101]. Different plant-based platforms can be used, including cells and field crops, *in vitro* culture systems (like plant cell suspensions and hairy roots), leaf-based, or seed-based systems [102]. Chlamydomonas algal-based expression is being explored by Phycobiologics, USA, to develop vaccines, growth factors, and enzymes [102]. Moss-derived enzyme α-galactosidase for Fabry disease and β-glucocerebrosidase for Gaucher disease are under phase I or/and preclinical trials [102]. Tobacco has been suggested as an effective plant expression system for the production of recombinant IFNs [103]. Gastric lipase, produced in maize by Meristem therapeutics for treating pancreatic insufficiency, has completed phase II trials [104].

Enzyme taliglucerase alfa was the first FDA-approved recombinant plant-derived therapeutic aiming to store vacuoles for producing glycans in the carrot cell expression system for human recombinant β-glucocerebrosidase for treating Gaucher disease [104]. Isolated carrot root cells were incorporated with *A. tumefaciens* binary Ti plasmid possessing acid β-glucosidase cDNA and a kanamycin resistance gene NPT II. The transformed cells were selected through antibiotic resistance, followed by clonal selection by evaluating the protein expression levels in transgenic callus. The selected clones were grown in suspension culture to create a master cell bank, and the cells were cultured in suspension in a closed bioreactor [105]. The schematic representation of taliglucerase alfa production in carrot root cells is shown below in Fig. 4.6.

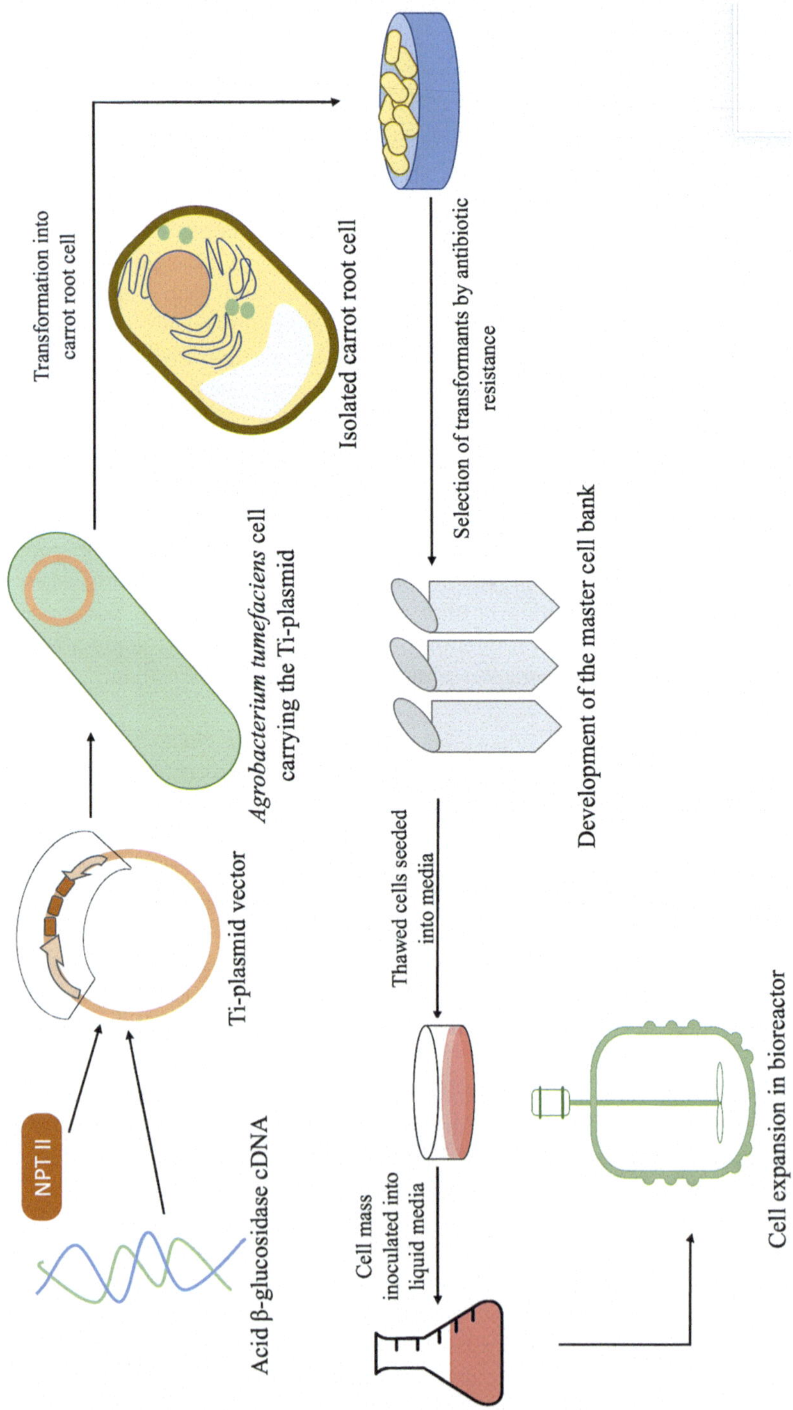

Fig. 4.6 Production of taliglucerase alfa in carrot root cells

4.5 Production of Recombinant Proteins in Transgenic Animals

Recombinant proteins nowadays have been synthesized using transgenic animals. Such animals carry a transgene coding recombinant protein incorporated into their genome, making them eligible to pass it to the offspring. The major advantage of using transgenic animals is the production of high product quality [106]. A suitable example would be a recombinant C1 esterase inhibitor. It is a serine protease inhibitor that inhibits various complement proteases and contact protease like kallikrein. An insufficient amount of C1 inhibitor can lead to cleavage of kininogen by protease kallikrein, which leads to the production of excess bradykinin. Bradykinin is a vasodilator that increases permeability and edema, leading to hereditary angioedema. It is a rare autosomal disorder caused by the deficiency of C1 esterase inhibitor [107]. Recombinant C1 esterase inhibitor consists of human plasma-derived C1 esterase inhibitor under the trade names Cinryze and Berinert. Another preparation is recombinant human C1 esterase inhibitor, a single-chain glycoprotein obtained from the milk of transgenic rabbits, which undergoes a three-step purification process [107]. The resultant recombinant protein is obtained by producing an expression vector incorporated with the gene that encodes for the desired protein fused to milk-specific regulatory elements. This gene is then introduced into the germline of the species. Alternatively, DNA microinjections or transposons can be used to incorporate the gene into a transposon and then into the pronucleus [108]. Recombinant protein human antithrombin III has been produced using the same method [108]. This drug is produced by microinjecting human antithrombin III genes into the embryos of goats and is sold under the trade name ATryn [109]. Roslin innovation center produced proteins with therapeutic potential in chicken eggs. These proteins are IFN-α 2a having antiviral and anticancer properties, and macrophage colony-stimulating factor (CSF) that enhances the repair of damaged cells and tissues [109]. The enzyme sebelipase alfa (trade name Kanuma) [110] is produced in the egg white of transgenic chicken eggs [109]. Sebelipase alfa is recombinant human lysosomal acid lipase involved in lipid metabolism; its deficiency causes defects in the breakdown of fats and cholesterol in the body.

4.6 Production of Recombinant Proteins in Yeast

All expression systems have their own advantages and limitations as *E. coli* is inexpensive, quick, and well established, but it frequently fails to produce proteins of the appropriate quality, particularly PTM proteins, whereas cell culture manufacturing is time-consuming and costly. An alternate option for the production of recombinant proteins is yeast. Yeasts combine ease of genetic manipulation with high productivity and eukaryotic protein processing machinery. In addition, the process of scaling up to large-scale production is well established. *Saccharomyces cerevisiae* fermentations have been done from antiquity in baking and wine production. With recent technological advancements, there has been a considerable improvement in the alcohol fermentation industry of yeast. However, for recombinant protein production, factors like plasmid instability, less yield of protein, and hyper glycosylation of proteins limit the use of *S. cerevisiae.* Many other strains were engineered with high humanized N-glycosylation, which was incorporated in *Pichia pastoris*, *Yarrowia lipolytica,* and *Schizosaccharomyces pombe*, which have improved the efficiency of producing recombinant proteins by using yeast strains. However, the protocols are the same as the other methods and consist of

i. Choosing a host strain that allows for appropriate folding and PTMs
ii. Selecting a relevant vector (integrative or episomal) with a suitable promoter and selection marker
iii. Codon optimizing the gene
iv. Choosing the appropriate signal sequence to target the recombinant protein (intracellular or extracellular)
v. Avoiding product proteolytic cleavage
vi. Selecting the correct medium for fermentation (nitrogen source, carbon source, and induction conditions)
vii. Setting parameters for bioprocessing (temperature, pH, O_2 transfer, etc.) [111]

The selection of the host strain is critical to the overall success rate of the process. The similarity of the host's biochemical conditions with the ability to initialize and translate RNA transcript, as well as its ability to change and sustain the protein that is translated, plays an essential role in selecting the strain. Both are critical factors in choosing the best cell for protein production. Factors like vector designing, culture medium, and composition also play an important role. Temperature, pH, aeration rate, agitation rate, feeding, and induction practices are bioprocess operations that significantly impact product manufacturing and quality [111].

The enzyme urate oxidase or rasburicase (trade name Elitek) is a recombinant enzyme expressed in genetically modified *S. cerevisiae* [112]. The coding sequence for rasburicase is derived from cDNA strains of *Aspergillus flavus.* This enzyme balances uric acid levels in plasma during cancer therapy [25]. It is a urate-lowering drug that breaks down the uric acid that accumulates in the blood when the tumor disrupts during cancer therapy. Ocriplasmin enzyme is a protease that cleaves peptide bonds after a lysine or arginine residue. It can cleave fibronectin, fibrinogen, collagen, and laminin and is a truncated form of human plasmin. It is expressed in methylotrophic yeast *P. pastoris* as inactive zymogen precursor microplasminogen. Catridecacog is a recombinant factor XIII (FXIII), whose subunit A is expressed in *S. cerevisiae* and is used for treating congenital factor XIII deficiency [113]. It is structurally identical to the human FXIII A-subunit. Factor XIII is an endogenous coagulation factor and the final enzyme of the blood coagulation cascade. It has clot stabilizing properties and anchors α2 antiplasmin to fibrin clot and fibrinogen [113]. Glucagon, a secretion of pancreatic α-cells in the islet of Langerhans and its recombinant form, has been effectively produced in the *S. cerevisiae* expression system [29, 30]. A comprehensive list of therapeutic proteins produced in different expression systems is shown in Table 4.1.

Table 4.1 List of recombinant therapeutic proteins obtained from different expression systems

Bacterial cells				
E. coli	*Clostridium botulinum*	Mammalian cells	Yeast cells	Plant cells
Humulin N Pegaspargase L-Asparaginase Pegloticase Glucarpidase Recombinant human growth hormone Calcitonin Metreleptin Glucagon PEG-IFN-alpha2a and PEG-IFN-alpha2b Insulin-like growth factor 1 (IGF-1) Recombinant hPTHs [PTH (1–84) and PTH (1–34)] Recombinant IL-11 Recombinant IFN-gamma rhIFNβ-1a and rhIFNβ-1b Aldesleukin	Botulinum toxin type A Botulinum toxin type B	Asfotase alfa Agalsidase alfa Alglucosidase alfa Imiglucerase Elosulfase alfa Galsulfase Laronidase Recombinant human Thrombin Recombinant form of factor IX Rh-Human chorionic gonadotropin (hCG) Rh-FSH Lutropin alfa Thyrotropin alfa Recombinant interleukin-12 Alteplase Idursulfase alfa Velaglucerase alfa Factor VIIa	Rasburicase Ocripasmin Factor XIII Glucagon	Taliglucerase alfa Gastric lipase Rh human IGF-I

4.7 Conclusions

Medicine is on the verge of a paradigm shift where the treatment and management decisions are determined at the genomic and proteomic levels. Recombinant human proteins account for the bulk of FDA-approved biotechnology medications today, and the future potential for these therapies is enormous. The methods for producing recombinant therapeutic proteins are still in their infancy and will reach their full potential in the coming years. High yield expression systems, targeting of proteins into specific cell organelles, novel protein purification strategies, etc., hold the promise for producing cost-effective therapeutic proteins. There is also a need for improving the bioprocesses for production, particularly in the case of animal and plant cells, to enhance protein yields. Protein engineering has been widely employed to ameliorate any immunogenicity or pharmacokinetics difficulties, and it is a valuable tool for improving treatments. We may expect more synergy between protein engineering and design efforts in the future, thanks to improved knowledge of protein structure-function correlations and rapid development of *in silico* bioinformatics and systems biology approaches. While developing new protein treatments, drawing on expertise from various fields, including molecular biology, pathology, immunology, and nanotechnology, is essential. The "Golden Age" of protein treatments can be achieved with concerted efforts. Finally, a convergence of protein-engineering, computational, and high-throughput experimental methodologies and "off-the-shelf" platform technologies has ushered in unprecedented prospects to build safe, effective, and more convenient protein therapies. These potentials come with hazards, but rapid breakthroughs in new technology and underlying research indicate that these risks can be controlled. Given all of these breakthroughs, it is reasonable to say that protein therapies are the way of the future for human medicine.

References

1. Singh DB, Tripathi T (2020) Frontiers in protein structure, function, and dynamics. Springer Nature, Singapore
2. Dabhole M (2014) Recent innovations in therapeutic recombinant protein. Pharma Bio World
3. Dingermann T (2008) Recombinant therapeutic proteins: production platforms and challenges. 3:90–97
4. Owczarek B, Gerszberg A, Hnatuszko-Konka KJBRI (2019) A brief reminder of systems of production and chromatography-based recovery of recombinant protein biopharmaceuticals 2019
5. Hajare A, Dange A, Shetty YJIJOPE (2008) Research, therapeutic protein production and delivery: an overview. 42:104–112
6. Riggs ADJER (2021) Making, cloning, and the expression of human insulin genes in bacteria: the path to Humulin. 42:374–380
7. Akash MSH, Rehman K, Tariq M, SJTJoB Chen (2015) Development of therapeutic proteins: advances and challenges. 39:343-358
8. Premo PM, A. I SR, A.J.C. ISHEMIJA, medicinska revija medical review, 381 929
9. Lagassé HD, Alexaki A, Simhadri VL, Katagiri NH, Jankowski W, Sauna ZE, Kimchi-Sarfaty CJF (2017) Recent advances in (therapeutic protein) drug development 6
10. Vukušić K, Šikić S, Balen BJPb (2016) Recombinant therapeutic proteins produced in plants: towards engineering of human-type O-and N-glycosylation. 118:75-90
11. Huang CJ, Lin H, Yang X (2012) Industrial production of recombinant therapeutics in *Escherichia coli* and its recent advancements. J Ind Microbiol Biotechnol 39:383–399
12. Picanco-Castro V, Tage Biaggio R, Tadeu Cova D, Swiech KJP, Letters P (2013) Production of recombinant therapeutic proteins in human cells: current achievements and future perspectives 20:1373–1381
13. Schillberg S, Raven N, Spiegel H, Rasche S, Buntru MJFips (2019) Critical analysis of the commercial potential of plants for the production of recombinant proteins. 10:720
14. Desai PN, Shrivastava N, Padh HJBa (2010) Production of heterologous proteins in plants: strategies for optimal expression 28:427–435
15. Dirisala VR, Nair RR, Srirama K, Reddy PN, Rao K, Satya Sampath Kumar N, Parvatam GJApp (2017) Recombinant pharmaceutical protein production in plants: unraveling the therapeutic potential of molecular pharming. 39:1–9
16. Sethi L, Kumari K, Dey NJMb (2021) Engineering of plants for efficient production of therapeutics. 63:1125–1137
17. Kalita P, Tripathi T (2022) Methodological advances in the design of peptide-based vaccines. Drug Discov Today 27:1367–1380
18. Tripathi T, Dubey VK (2022) Advances in protein molecular and structural biology methods, 1st edn. Academic Press, Cambridge, MA

19. Grillberger L, Kreil TR, Nasr S, Reiter M (2009) Emerging trends in plasma-free manufacturing of recombinant protein therapeutics expressed in mammalian cells. Biotechnol J 4:186–201
20. Swiech K, Picanço-Castro V, Covas DT (2012) Human cells: new platform for recombinant therapeutic protein production. Protein Expr Purif 84:147–153
21. Dumont J, Euwart D, Mei B, Estes S, Kshirsagar R (2016) Human cell lines for biopharmaceutical manufacturing: history, status, and future perspectives. Crit Rev Biotechnol 36:1110–1122
22. Nag N, Khan H, Tripathi T (2022) Strategies to improve the expression and solubility of recombinant proteins in *E. coli*. In: Tripathi T, Dubey VK (eds) Advances in protein molecular and structural biology methods. Academic Press, New York, pp 1–12
23. Chhetri G, Kalita P, Tripathi T (2015) An efficient protocol to enhance recombinant protein expression using ethanol in *Escherichia coli*. Methods X 2:385–391
24. El-Naggar NE-A, Moawad H, El-Shweihy NM, El-Ewasy SM, Elsehemy IA, Abdelwahed NAM (2019) Process development for scale-up production of a therapeutic L-asparaginase by Streptomyces brollosae NEAE-115 from shake flasks to bioreactor. Sci Rep 9:13571
25. Baldo BA (2015) Enzymes approved for human therapy: indications, mechanisms and adverse effects. BioDrugs Clin Immunotherapeut Biopharmaceut Gene Ther 29:31–55
26. Rattu MA, Shah N, Lee JM, Pham AQ, Marzella N (2013) Glucarpidase (voraxaze), a carboxypeptidase enzyme for methotrexate toxicity. PT 38:732–744
27. Landgraf W, Sandow J (2016) Recombinant human insulins – clinical efficacy and safety in diabetes therapy. Eur Endocrinol 12:12–17
28. Wilcox G (2005) Insulin and insulin resistance. Clin Biochem Rev 26:19–39
29. Rix I, Nexøe-Larsen C, Bergmann NC, Lund A, Knop FK (2000) Glucagon physiology. In: Feingold KR, Anawalt B, Boyce A, Chrousos G, de Herder WW, Dhatariya K, Dungan K, Hershman JM, Hofland J, Kalra S, Kaltsas G, Koch C, Kopp P, Korbonits M, Kovacs CS, Kuohung W, Laferrère B, Levy M, McGee EA, McLachlan R, Morley JE, New M, Purnell J, Sahay R, Singer F, Sperling MA, Stratakis CA, Trence DL, Wilson DP (Eds) Endotext, MDText.com, Inc. Copyright © 2000–2022, MDText.com, Inc., South Dartmouth (MA)
30. Mollerup I, Kornfelt T, Oreskov P, Tøttrup HV (1996) Industrial production of glucagon. In: Lefèbvre PJ (ed) Glucagon III. Springer, Berlin, pp 1–10
31. Brown RJ, Valencia A, Startzell M, Cochran E, Walter PJ, Garraffo HM, Cai H, Gharib AM, Ouwerkerk R, Courville AB, Bernstein S, Brychta RJ, Chen KY, Walter M, Auh S, Gorden P (2018) Metreleptin-mediated improvements in insulin sensitivity are independent of food intake in humans with lipodystrophy. J Clin Invest 128:3504–3516
32. Deeks E (2019) Metreleptin in lipodystrophy: a profile of its use. Drugs Ther Perspect 35: 201–208
33. Ray MV, Van Duyne P, Bertelsen AH, Jackson-Matthews DE, Sturmer AM, Merkler DJ, Consalvo AP, Young SD, Gilligan JP, Shields PP (1993) Production of recombinant salmon calcitonin by in vitro amidation of an *Escherichia coli* produced precursor peptide. Biotechnology (Nature Publishing Company) 11:64–70
34. McLaughlin MB JI (2021) Calcitonin. [Updated 2021 Sep 28
35. Brixen KT, Christensen PM, Ejersted C, Langdahl BL (2004) Teriparatide (biosynthetic human parathyroid hormone 1–34): a new paradigm in the treatment of osteoporosis. Basic Clin Pharmacol Toxicol 94:260–270
36. Gardella TJ, Rubin D, Abou-Samra AB, Keutmann HT, Potts JT Jr, Kronenberg HM, Nussbaum SR (1990) Expression of human parathyroid hormone-(1–84) in *Escherichia coli* as a factor X-cleavable fusion protein. J Biol Chem 265:15854–15859
37. Karimi M, Behzadian F, Rouhaninejad H, Yari S (2018) A feasibility study to evaluate Bacillus subtilis as a host for producing recombinant human parathyroid hormone. Avicenna J Med Biotechnol 10:147–151

38. Flodh H (1986) Human growth hormone produced with recombinant DNA technology: development and production. Acta Paediatr Scand Suppl 325:1–9
39. Catzel D, Lalevski H, Marquis CP, Gray PP, Van Dyk D, Mahler SM (2003) Purification of recombinant human growth hormone from CHO cell culture supernatant by Gradiflow preparative electrophoresis technology. Protein Expr Purif 32:126–134
40. Rosenbloom AL (2008) Insulin-like growth factor-I (rhIGF-I) therapy of short stature. J Pediat Endocrinol Metab JPEM 21:301–315
41. Ranjbari J, Babaeipour V, Vahidi H, Moghimi H, Mofid MR, Namvaran MM, Jafari S (2015) Enhanced production of insulin-like growth factor I protein in *Escherichia coli* by optimization of five key factors. Iran J Pharm Res 14:907–917
42. Assanto GM, Riemma C, Malaspina F, Perrone S, De Luca ML, Pucciarini A, Annechini G, D'Elia GM, Martelli M, Foà R, Tiacci E, Pulsoni A (2021) The current role of interferon in hairy cell leukaemia: clinical and molecular aspects. Br J Haematol 194:78–82
43. Parekh PJ, Shiffman ML (2014) The role of interferon in the new era of hepatitis C treatments. Expert Rev Gastroenterol Hepatol 8:649–656
44. Aghemo A, Rumi MG, Colombo M (2010) Pegylated interferons alpha2a and alpha2b in the treatment of chronic hepatitis C. Nat Rev Gastroenterol Hepatol 7:485–494
45. Bekisz J, Baron S, Balinsky C, Morrow A, Zoon KC (2010) Antiproliferative properties of type I and type II interferon. Pharmaceuticals (Basel) 3:994–1015
46. Rokx C, van der Ende ME, Verbon A, Rijnders BJA (2013) Peginterferon Alfa-2a for AIDS-associated Kaposi Sarcoma: experience with 10 patients. Clin Infect Dis 57:1497–1499
47. Talpaz M, Hehlmann R, Quintás-Cardama A, Mercer J, Cortes J (2013) Re-emergence of interferon-α in the treatment of chronic myeloid leukemia. Leukemia 27:803–812
48. Browder JF, Araujo OE, Myer NA, Flowers FP (1992) The interferons and their use in condyloma acuminata. Ann Pharmacother 26:42–45
49. Abdolvahab MH, Mofrad MRK, Schellekens H (2016) Chapter eight – Interferon beta: from molecular level to therapeutic effects. In: Jeon KW, Galluzzi L (eds) International review of cell and molecular biology. Academic Press, pp 343–372
50. Ghasemi N, Razavi S, Nikzad E (2017) Multiple sclerosis: pathogenesis, symptoms, diagnoses and cell-based therapy. Cell J 19:1–10
51. Pal D, Tripathy RK, Teja MS, Kumar M, Banerjee UC, Pande AH (2018) Antibiotic-free expression system for the production of human interferon-beta protein. 3 Biotech 8:36–36
52. Castro F, Cardoso AP, Gonçalves RM, Serre K, Oliveira MJ (2018) Interferon-gamma at the crossroads of tumor immune surveillance or evasion. Front Immunol 9:847–847
53. Errante PR, Frazão JB, Condino-Neto A (2008) The use of interferon-gamma therapy in chronic granulomatous disease. Recent Pat Antiinfect Drug Discov 3:225–230
54. Tang M, Tian L, Luo G, Yu X (2018) Interferon-gamma-mediated osteoimmunology. 9
55. Key LL, Rodriguiz RM, Willi SM, Wright NM, Hatcher HC, Eyre DR, Cure JK, Griffin PP, Ries WL (1995) Long-term treatment of osteopetrosis with recombinant human interferon gamma. 332:1594–1599
56. Razaghi A, Owens L, Heimann K (2016) Review of the recombinant human interferon gamma as an immunotherapeutic: impacts of production platforms and glycosylation. J Biotechnol 240:48–60
57. Jiang T, Zhou C, Ren S (2016) Role of IL-2 in cancer immunotherapy. Onco Targets Ther 5: e1163462–e1163462
58. Srinivasan L, Harris MC, Kilpatrick LE (2017) 128 – Cytokines and inflammatory response in the fetus and neonate. In: Polin RA, Abman SH, Rowitch DH, Benitz WE, Fox WW (eds) Fetal and neonatal physiology, 5th edn. Elsevier, pp 1241–1254.e1244
59. Li J, Shen L, Li Y, Zhang X, Li J, Gong J, Deng W (2007) Recombinant human interleukin-11 for treatment of chemotherapy-induced thrombocytopenia in patients with gastrointestinal cancer. Chinese-German J Clin Oncol 6:450–452
60. Miao J, Wang J, Peng S, Tang P, Zou M, Duan J, Zhao C, Ma X (1995) Expression of human interleukin-11 cDNA in *E. coli*. Sci China Ser B Chem Life Sci Earth Sci 38:1202–1209

61. Dimitrov DS (2012) Therapeutic proteins. Methods Mol Biol (Clifton, NJ) 899:1–26
62. Nigam PK, Nigam A (2010) Botulinum toxin. Indian J Dermatol 55:8–14
63. Webb RP (2018) Engineering of botulinum neurotoxins for biomedical applications. Toxins (Basel) 10:231
64. O'Flaherty R, Bergin A, Flampouri E, Mota LM, Obaidi I, Quigley A, Xie Y, Butler M (2020) Mammalian cell culture for production of recombinant proteins: a review of the critical steps in their biomanufacturing. Biotechnol Adv 43:107552
65. Wurm FM (2004) Production of recombinant protein therapeutics in cultivated mammalian cells. Nat Biotechnol 22:1393–1398
66. Bailey LA, Hatton D, Field R, Dickson AJ (2012) Determination of Chinese hamster ovary cell line stability and recombinant antibody expression during long-term culture. Biotechnol Bioeng 109:2093–2103
67. Lai T, Yang Y, Ng SK (2013) Advances in Mammalian cell line development technologies for recombinant protein production. Pharmaceuticals (Basel) 6:579–603
68. Daniel LH, Aikaterini A, Simhadri VL, Katagiri NH, Wojciech J, Chava K-SJF (2017) Recent advances in (therapeutic protein) drug development. 6
69. Tournis S, Yavropoulou MP, Polyzos SA, Doulgeraki A (2021) Hypophosphatasia. 10:5676
70. Mornet E, Nunes ME (1993) Hypophosphatasia. University of Washington, Seattle, Seattle (WA)
71. Keating GM, Simpson D (2007) Agalsidase Beta. Drugs 67:435–456
72. Beck M (2009) Alglucosidase alfa: long term use in the treatment of patients with Pompe disease. Ther Clin Risk Manag 5:767–772
73. Katzin LW, Amato AA (2008) Pompe disease: a review of the current diagnosis and treatment recommendations in the era of enzyme replacement therapy. J Clin Neuromuscul Dis 9:421–431
74. Tekoah Y, Tzaban S, Kizhner T, Hainrichson M, Gantman A, Golembo M, Aviezer D, Shaaltiel Y (2013) Glycosylation and functionality of recombinant β-glucocerebrosidase from various production systems. Biosci Rep 33:e00071
75. Stone WL, Basit H, Master SR (2017) Gaucher disease
76. Sanford M, Lo J (2014) Elosulfase alfa: first global approval. Drugs 74
77. Harmatz P, Shediac R (2017) Mucopolysaccharidosis VI: pathophysiology, diagnosis and treatment. Front Biosci (Landmark edition) 22:385–406
78. Ratner M (2008) Recombinant thrombin approved. Nat Biotechnol 26:250–250
79. Windyga J, Solano Trujillo MH, Hafeman AE (2014) BAX326 (RIXUBIS): a novel recombinant factor IX for the control and prevention of bleeding episodes in adults and children with hemophilia B, vol 5. Ther Adv Hematol, pp 168–180
80. Taran LD (1997) Factor IX of the blood coagulation system: a review. Biochem Biokhimiia 62:685–693
81. Farrag A, Costantini A, Manna C, Grimaldi G (2008) Recombinant HCG for triggering ovulation increases the rate of mature oocytes in women treated for ICSI. J Assist Reprod Genet 25:461–466
82. Fink J, Schoenfeld BJ, Hackney AC, Maekawa T, Horie S (2021) Human chorionic gonadotropin treatment: a viable option for management of secondary hypogonadism and male infertility. Expert Rev Endocrinol Metab 16:1–8
83. Partsinevelos GA, Antonakopoulos N, Kallianidis K, Drakakis P, Anagnostou E, Bletsa R, Loutradis D (2016) Addition of low-dose hCG to rFSH during ovarian stimulation for IVF/ICSI: is it beneficial? Clin Exp Obstet Gynecol 43:818–825
84. Foresta C, Selice R, Ferlin A, Garolla A (2009) Recombinant FSH in the treatment of oligozoospermia. Expert Opin Biol Ther 9:659–666
85. Prevost RR (1998) Recombinant follicle-stimulating hormone: new biotechnology for infertility. Pharmacotherapy 18:1001–1010
86. Kaufmann R, Dunn R, Vaughn T, Hughes G, O'Brien F, Hemsey G, Thomson B, O'Dea LS (2007) Recombinant human luteinizing hormone, lutropin alfa, for the induction of follicular

development and pregnancy in profoundly gonadotrophin-deficient women. Clin Endocrinol 67:563–569
87. Emerson CH, Torres MS (2003) Recombinant human thyroid-stimulating hormone: pharmacology, clinical applications and potential uses. BioDrugs Clin Immunotherapeut Biopharmaceut Gene Ther 17:19–38
88. Guo N, Wang W-Q, Gong X-J, Gao L, Yang L-R, Yu W-N, Shen H-Y, Wan L-Q, Jia X-F, Wang Y-S, Zhao Y (2017) Study of recombinant human interleukin-12 for treatment of complications after radiotherapy for tumor patients. World J Clin Oncol 8:158–167
89. Komastu T, Ireland DD, Reiss CS (1998) IL-12 and viral infections. Cytokine Growth Factor Rev 9:277–285
90. Vellard M (2003) The enzyme as drug: application of enzymes as pharmaceuticals. Curr Opin Biotechnol 14:444–450
91. Lohani P (2018) Biotechnological production of recombinant tissue plasminogen activator protein, its application. Int J Bio-Technol Res 8:9–20
92. Sidawy AN (2018) Rutherford's vascular surgery and endovascular therapy. Elsevier [Place of publication not identified]
93. Mohammadi E, Seyedhosseini-Ghaheh H, Mahnam K, Jahanian-Najafabadi A, Sadeghi HMM (2019) Reteplase: structure, function, and production. Adv Biomed Res 8:19–19
94. Whiteman DA, Kimura A (2017) Development of idursulfase therapy for mucopolysaccharidosis type II (Hunter syndrome): the past, the present and the future. Drug Des Devel Ther 11:2467–2480
95. Castillo B, Dasgupta A, Klein K, Tint H, Wahed A (2018) Chapter 11 - Pharmacologic agents in transfusion medicine. In: Castillo B, Dasgupta A, Klein K, Tint H, Wahed A (eds) Transfusion medicine for pathologists. Elsevier, pp 177–193
96. Dutta TK, Verma SP (2014) Rational use of recombinant factor VIIa in clinical practice. Indian J Hematol Blood Transfus 30:85–90
97. Boedeker BG (2001) Production processes of licensed recombinant factor VIII preparations. Semin Thromb Hemost 27:385–394
98. Shahid N, Daniell H (2016) Plant-based oral vaccines against zoonotic and non-zoonotic diseases. Plant Biotechnol J 14:2079–2099
99. Daniell H, Lin C-S, Yu M, Chang W-J (2016) Chloroplast genomes: diversity, evolution, and applications in genetic engineering. Genome Biol 17:134
100. Twyman RM, Stoger E, Schillberg S, Christou P, Fischer R (2003) Molecular farming in plants: host systems and expression technology. Trends Biotechnol 21:570–578
101. De Muynck B, Navarre C, Boutry M (2010) Production of antibodies in plants: status after twenty years. Plant Biotechnol J 8:529–563
102. Xu J, Towler M, Weathers PJ (2018) Platforms for plant-based protein production. Bioprocess Plant In Vitro Syst:509–548
103. Budzianowski J (2014) Tobacco – a producer of recombinant interferons. Przeglad lekarski 71:639–643
104. Gomes C, Oliveira F, Vieira S, Duque S (2019) Prospects for the production of recombinant therapeutic proteins and peptides in plants: special focus on Angiotensin I-Converting Enzyme Inhibitory (ACEI) peptides
105. Grabowski G, Golembo M, Shaaltiel Y (2014) Taliglucerase alfa: an enzyme replacement therapy using plant cell expression technology. Mol Genet Metab 112
106. Tripathi NK, Shrivastava A (2019) Recent developments in bioprocessing of recombinant proteins: expression hosts and process development 7
107. Riedl M (2015) Recombinant human C1 esterase inhibitor in the management of hereditary angioedema. Clin Drug Investig 35:407–417
108. Houdebine L-M (2009) Production of pharmaceutical proteins by transgenic animals. Comp Immunol Microbiol Infect Dis 32:107–121
109. Hunter P (2019) The prospects for recombinant proteins from transgenic animals. 20:e48757

110. Shirley M (2015) Sebelipase alfa: first global approval. Drugs 75:1935–1940
111. Kunze G, Kang HA, Gellissen G (2009) Hansenula polymorpha (Pichia angusta): biology and applications. In: Satyanarayana T, Kunze G (eds) Yeast biotechnology: diversity and applications. Springer, Dordrecht, pp 47–64
112. Waller DG, Sampson AP (2018) 31 – Hyperuricaemia and gout. In: Waller DG, Sampson AP (eds) Medical pharmacology and therapeutics, 5th edn. Elsevier, pp 385–389
113. Korte W (2014) Catridecacog: a breakthrough in the treatment of congenital factor XIII A-subunit deficiency? J Blood Med 5:107–113

Antibodies as Therapeutic Agents

5

Ravindra Singh Rawat and Sanjit Kumar

Abstract

Antibodies are immune system components secreted by B-cells. They have a propensity to bind foreign particles in the body. Antibodies have a Y shaped structure and bind to and kill pathogens such as viruses, bacteria, and parasites. In the past three decades, there has been a considerable increase in the number of diagnostic and therapeutic procedures that use monoclonal (mAbs) and polyclonal antibodies (pAbs). In the treatment of cancer, autoimmune diseases, and a variety of neurological disorders, mAbs are more effective than conventional antibodies. The high cost and poor efficacy of mAbs have now been overcome by antibody fragments like Fab, ScFv, and VHH with high binding affinity and ease of production. This chapter describes the basics of antibody structure and function and its use as a therapeutic molecule.

Keywords

Antibody · Fab region · Humanized mAbs · Complementarity determining regions · CDRs

5.1 Structure of Antibodies

Antibodies, which are produced by B-lymphocyte cells, are a component of the adaptive immune response system. Antibodies are active glycoprotein molecules that bind with very high specificity to any molecule our body recognizes as foreign

R. S. Rawat · S. Kumar (✉)
Centre for Bioseparation Technology (CBST), Vellore Institute of Technology, Vellore, Tamil Nadu, India

D. B. Singh, T. Tripathi (eds.), *Protein-based Therapeutics*,
https://doi.org/10.1007/978-981-19-8249-1_5

[1–3]. Higher mammals have antibodies categorized into five classes based on the type of heavy chains they contain: IgG, IgM, IgA, IgE, and IgD antibodies [4]. Each category of antibody is distinct from one another on the basis of physiological function, molecular weight, and glycosylation [4, 5].

Antibodies are made up of two polypeptide chains: one light chain and one heavy chain. Light chains are approximately 25 kDa in size, as opposed to heavy chains, which are about 50 kDa in size [4]. The light and heavy polypeptide chains fold into quaternary structures to form domains. These domains are stabilized by van der Waals' interactions and are further strengthened by the interchain and intrachain disulfide linkages [4]. Together, these domains form three regions: antigen-binding region, constant region, and the hinge region. The antigen-binding region is composed of variable light chain (V_L) and variable heavy-chain (V_H) domains [2]. This region is rich in hypervariable loops, also known as complementarity determining regions (CDRs), that bind to the antigen and are supported by a relatively rigid framework region formed by β-sheet structures [6]. The hinge region joins the antigen-binding fragment (Fab) to the constant region (Fc) in an antibody and provides flexibility to the immunoglobulin (Ig). Moreover, it links two heavy chains through disulfide linkages [7]. The size of the hinge region differs among different classes or subclasses of Ig molecules. While the hinge region is absent in IgE and IgM, it is present in IgG, IgA, and IgD [8]. In the case of IgD, the hinge region is very long, whereas the IgA hinge region is short but with heavy glycosylation in the IgA1 subclass [8]. The constant regions are made up of heavy-chain domains and are highly glycosylated. The constant regions of Igs are associated with effector functions such as complement activation [7, 9]. At times misperception of our own body molecules as foreign molecules under certain circumstances or due to predisposition to certain clinical conditions leads to autoimmune diseases like rheumatoid arthritis, myasthenia gravis, multiple sclerosis, celiac disease, etc. [10–13].

5.2 Functions of Antibodies

Antibodies can act against a pathogen via different modes of action. As a result, even for a particular pathogen, the antibody response cannot be delineated in a predictable manner. For instance, when antibodies encounter a pathogen, antibodies have the ability to neutralize the infection. Further, bound antibodies can initiate other effector functions that are detrimental to the pathogen. Functions of antibodies can be presented by different modes, which have different consequences at the cellular level. Complement-mediated lysis of pathogen or cells infected with the pathogen, phagocytosis, antibody-dependent cellular cytotoxicity (ADCC), and opsonization are a few modes by which antibodies can function [14, 15].

5.2.1 Effector-Independent Functions

The effector-independent functions help reduce the severity of infection by directly binding to the pathogenic cells; this is called neutralization. The binding of

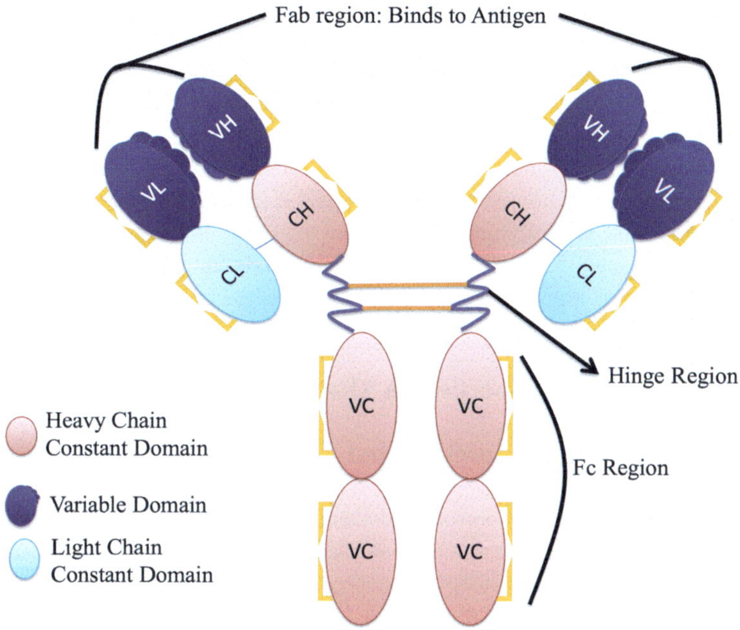

Fig. 5.1 Diagrammatic representation of the structure of an antibody

antibodies directly to the pathogens prevents the pathogen from adhering to the cell surfaces, which reduces the infectivity of pathogens. However, direct neutralization of a pathogen is not exclusively capable of completely removing the infection, and it is accompanied by effector functions of antibodies that enhance the response of our immune system.

Direct neutralization of pathogen by antibodies is governed by the antigen-binding region of the antibodies, as shown in Fig. 5.1. The antigen-binding (Fab) region is made up of variable light-chain (V_L) and variable heavy-chain (V_H) domains [2, 6]. These domains contain hypervariable regions, and antibodies originating from different B-cell clones can even differentiate epitopes on the same antigen [2]. Recently, the antigen-binding region has been a subject of extensive research for its application as a therapeutic molecule. A single-chain variable fragment (scFV) is one example of utilizing an antigen-binding region for therapeutic purposes. The details of scFV are discussed in a later part of this chapter.

5.2.2 Effector-Dependent Functions

The Fc (crystallizable fragment) region governs both the isotype and the effector functions of an antibody [16]. Fc is a part of the heavy-chain constant (C_H) region. C_H is a multi-domain region, and the number of domains varies in different isotypes [7]. For instance, IgD, IgA, and IgG C_H consist of three domains: C_{H1}, C_{H2}, and C_{H3}, whereas IgE and IgM possess C_{H4} as an additional domain.C_{H1} is a part of $F_{V,}$ and the rest of the domains constitute the F_C portion of the constant region. Glycosylation is an essential aspect of the Ig C_H region [4, 9]. It facilitates Igs to produce their effector function through binding to the F_C receptors on the cells [17]. Glycosylation provides stability, helps keep the protein in soluble form, and eases the transportation of Ig through mucosal membranes [5, 17, 18]. It is evident from structural studies that the glycosylation pattern, as well as the structure of the oligosaccharide, is well defined. The non-covalent interactions of these structured oligosaccharide moieties with the protein component in Igs help shape its structure and interaction with other proteins to a great extent [9, 19, 20].

5.3 Formats of Antibodies as Therapeutics

5.3.1 Polyclonal Antibodies (pAbs)

In nature, passive immunization already exists and occurs via the transfer of antibodies from the mother to the fetus through the placenta wall or through milk to a newly born infant. The concept of passive immunization was vague, while that of antibody was completely unknown until 1890. During the 1890s, the administration of sera from cured animals to infected children by Von Behring and colleagues was the first of its kind experiment to use anti-sera (now known as an antibody) as a therapeutic measure for diphtheria and tetanus toxin [21]. The following 20 years were an expansion of this anti-sera therapy for various other infections and clinical emergencies. For instance, anti-sera against meningococcal meningitis, botulism, measles, pneumonia, diphtheria, anthrax, and scarlet fever were developed [22–25]. The list of currently marketed pAbs is shown in Table 5.1.

5.3.1.1 Production of pAbs

In the modern world, the generation of pAbs is accomplished by immunizing either animals or human donors with the antigen and harvesting the produced antibodies [38]. The animals used for the production of pAbs are strictly screened for the presence of any infections, such as spongiform encephalopathies (TSEs) [39]. To ensure this, the manufacturers select animals from regions with a prolonged absence of TSEs. Moreover, animals are sourced from suppliers approved by concerned authorities, and the flocks of animals are regularly tested for infections [38].

The first step for the production of therapeutic pAbs is to immunize animals with the antigen, followed by regular screening for antibody titers in the animal serum. Subsequently, upon achieving a desirable titer, the serum is collected from the animals, and total pAbs are purified by precipitation or affinity chromatography [40]. If the Fab fragments are the desired product, the purified pAbs are subjected to cleavage by protease like pepsin [41] papain [42]. The Fab fragments thus obtained are further purified to remove other cleavage products. Like in any other therapeutic production process, the production of pAbs-based therapeutics also is constantly monitored for contamination with bacteria, bacterial components, and viruses

Table 5.1 Currently marketed polyclonal antibodies

Type	Product trade name	Therapeutic application	Manufacturing/ distributor company	Formulation	Viral inactivation/ reduction method	Reference
Complete polyclonal antibody	Gamunex®	Primary humoral immunodeficiency disease (PIDD)	Grifols Therapeutics Inc., USA	of 9%–11% protein in 0.16–0.24 M glycine	Caprylate, pH 4.25	[26]
	Flebogamma® DIF	PIDD	Instituto Grifols, Spain	5% sorbitol	Pasteurization, PEG	[27]
	Gammagard® S/D	PIDD	Baxter Healthcare Corp., USA	Frozen dried preparation	Detergent/solvent	[28]
	Iveegam EN	PIDD	Baxter Healthcare Corp., USA	Contains 50 ± 5 mg IgG, 50 mg glucose as a stabilizer, and 3 mg sodium chloride	Trypsin, PEG	[29]
	Cytogam®	Cytomegalovirus	Saol Therapeutics Inc., USA	5% sucrose and 1% albumin (Human)	Solvent/detergent	[30, 31]
	Nabi-HB®	Hepatitis B	Nabi Biopharmaceuticals, USA	0.042–0.108 M sodium chloride, 0.10–0.20 M glycine, and 0.005–0.050% polysorbate 80, at pH 5.8–6.5	Solvent/detergent (tri-n-butyl phosphate/Triton X-100) and virus filtration by Planova 35 nm	[32]
	BayTet®	Tetanus	Bayer Corp., USA	Solution containing 15–18% protein, 0.21–0.32 M glycine, pH 6.4–7.2	Solvent/detergent	[33]
Recombinant antibody	CroFab™	Anti-venom	BTG International Inc., USA	Freeze-dried powder, dibasic sodium phosphate, sodium chloride, and thimerosal	Pasteurization by heat treatment for 10 h at 60 °C	[34]
	DigiFab®	Digoxin toxicity	BTG International Inc., USA	40 mg digoxin immune Fab, 0.5 mg digoxin, 75 mg	Pasteurization	[35]

(continued)

Table 5.1 (continued)

Type	Product trade name	Therapeutic application	Manufacturing/ distributor company	Formulation	Viral inactivation/ reduction method	Reference
				mannitol USP, and 2 mg sodium acetate USP as a buffering agent		
	Anavip	Crotalidae venom	Instituto Bioclon, Mexico		Pasteurization	[36]
	Anascorp	Centruroides venom	Rare Disease Therapeutics Inc., USA	45–80 mg NaCl, 4.3–38.3 mg sucrose, and 6.6–94.9 mg glycine as stabilizer	Ammonium sulfate/heat treatment precipitation, digestion with pepsin, and nanofiltration	[37]
	Botulism antitoxin heptavalent (HBAT)	Botulinum toxin	Cangene Corp., Canada		Pasteurization	[24]

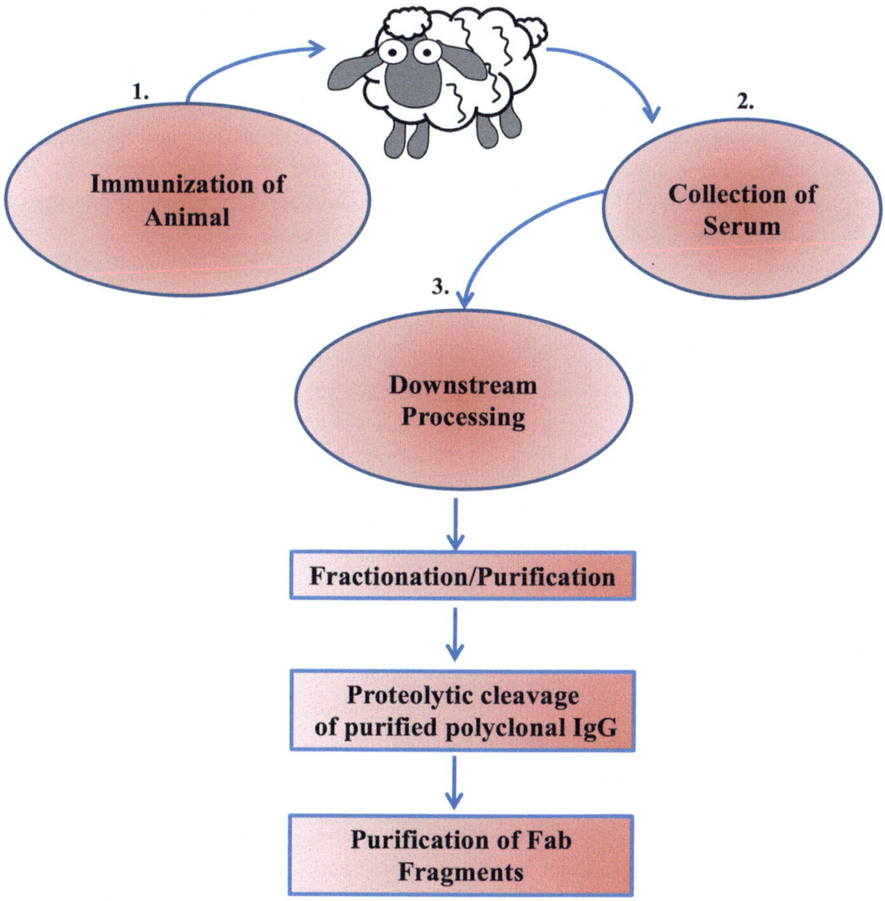

Fig. 5.2 Flow chart showing the production process of polyclonal antibodies (pAbs)

[39, 43]. Therefore, the downstream process additionally has a step to remove any contamination from the product (Fig. 5.2).

5.3.1.2 Intravenous Immunoglobulin G (IVIG)

Intravenous immunoglobulin G (IVIG) has been used regularly to treat individuals with autoimmune diseases, as replacement therapy for immunodeficiency, and during transplantation surgeries [44]. Earlier, Cohn fraction II was the only method for producing IgG products [45]. However, due to its intrinsic procedural limitations, the high levels of IgG aggregates were challenging to avoid. Moreover, the administration of IgG was accomplished only via the intramuscular route. Consequently, the serum levels of IgG were very low and inadequate to provide protection. In 1981, Imbach et al. demonstrated improvement in health conditions of patients suffering from thrombocytopenic purpura (ITP) upon intravenous administration of IgG

[46]. In recent years, the production of IVIG from plasma has been modified in order to prevent IgG aggregation and the inactivation of viruses. IVIG is produced from a large pool of blood serum donated by thousands of human donors [45, 47]. The initial steps involve fractionation of plasma by the modified Cohn-Oncley method, and the IgG fractions obtained are then subjected to filtration and purification [40, 45]. Due to the nature of the origin of serum and production processes, the composition and contaminants in the final IVIG product vary. To ensure the safety of IVIG products usage food drug administration (FDA) has put forth some regulations. For example, during cold ethanol fractionation of plasma, HIV is deactivated, but hepatitis C virus (HCV) is unaffected [39]. Therefore, the FDA has necessitated an additional step to remove or deactivate the virus [39]. Another safety concern is the tolerability of IVIG infusion by the patients. The tolerability of IVIG depends on the presence of IgA and preservatives used, such as sugars, salts, and pH [39, 43, 48]. IVIG infusions that contain sugars as preservatives pose health complications such as osmotic nephrosis and renal dysfunction. To prevent such difficulties, the infusion must be diluted to the minimum possible concentration and minimum infusion rate possible [49, 50].

5.3.1.3 Recombinant pAbs

With the advancement in recombinant DNA (rDNA) technology and high-throughput screening, pAbs are now available in different recombinant forms. The first report on producing recombinant antibodies goes back to 1994 by Sarantopoulos et al. [51]. The production of recombinant human pAbs at an industrial scale became possible with the development of an expression system by Symphogen, a Denmark-based company [52, 53]. The expression system developed by the company was a mammalian system called Symplex™. In this technology, the plasmid containing the gene for the antibody is integrated into the mammalian genome in a site-specific manner in a mammalian cell [53]. The use of an identical integration site assures the same level of antibody expression. The transformed cells expressing antibodies against different epitopes of the same antigen are combined to develop a polyclonal product [53].

5.3.1.4 Transgenic pAbs

Transgenic pAbs could be the future of antibody-based therapeutics. Transgenic pAbs are produced by constitutive expression of human antibodies in animals [54]. Complete human antibody gene loci are cloned into the artificial human chromosome (HAC) and transferred into animal embryonic fibroblast cells [55]. After immunizing animals with the antigen, they start producing functional human antibodies. The American biopharmaceutical company Hematech LLC has made transgenic calves using HAC technology, producing human pAbs [38, 52]. However, transgenic calves produce both human and bovine antibodies and leads to suppressed human pAbs expression. Also, the expression of chimeric antibodies consisting of both bovine and human antibody regions is another undesirable outcome [38, 52]. These limitations in transgenic pAbs are required to be overcome. Another two companies, namely Therapeutic Human Polyclonal Inc. and

Origen Therapeutics Inc. are developing rabbits and avian embryonic cells, respectively, to produce transgenic pAbs [38]. The production of transgenic pAbs is still in its initial stages but, once established, would revolutionize the production of pAbs.

5.3.2 Monoclonal Antibodies (mAbs)

The ingenious work of Georges Köhler and César Milstein in 1975 to develop the hybridoma technique for producing monoclonal antibodies (mAbs) revolutionized the use of antibodies in medicine [56]. Because of homogeneity with respect to their specificity towards a single epitope, mAbs were termed "magic bullets" with their binding ability to bind to targets without affecting healthy tissues [21, 57]. Initially, during the 1970s and 1980s, mAbs were produced successfully in murines. However, due to a lack of appropriate manufacturing standards, only a few mAbs were approved. Also, severe allergic reactions were observed in hosts when murine mAbs were administered [58]. With the advancement in mammalian cell culture techniques and standardized industrial manufacturing process, up to 5 g/L active mAb production can now be achieved [59]. Currently, more than 30 FDA-approved mAbs are on the market, and more than 250 are at various stages of clinical trials [58, 60].

5.3.2.1 Production of mAbs

Monoclonal antibodies are produced from single B-cell clones and are specific to a single epitope on an antigen. For industrial use, mAbs are produced by hybridoma technology. To prepare for hybridoma technology, a specific antigen is injected into animals, followed by screening for antibody titer every fortnight. Once the antibody titer reaches a desirable level, the B-lymphocytes are harvested. Myeloma cells that lack a salvage pathway due to the absence of hypoxanthine-guanine-phosphoribosyl transferase (HGPRT) are fused with B-lymphocytes by using either polyethylene glycol or Sendai virus [61]. The fused cells are called hybridoma cells, which are selected on a hypoxanthine-aminopterin-thymidine (HAT) medium. Unfused B-lymphocytes do not survive in the medium due to a limited life span, and *de novo* synthesis in unfused myeloma cells is inhibited in the presence of aminopterin in the HAT medium. Therefore, only fused hybridoma cells survive in the HAT medium [62–64]. Each hybridoma cell is produced by the fusion of different B-cell clones with myeloma cells, and consequently, the antibodies produced in the medium are polyclonal. The hybridoma cell suspension is then diluted to retain only a single cell in each culture plate well. Antibodies produced in each well are from a single hybridoma clone that is mAb. The mAbs produced in each well are further screened for specificity and affinity towards the antigen. The positive hybridoma cells are grown, subcloned, and further tested for their activity [59].

5.3.3 Modified mAbs

5.3.3.1 Chimeric Antibodies

Major drawbacks of using murine mAbs for therapeutic purposes in humans led to the quest for developing antibodies with higher immune tolerance in humans. It was perceived that the constant domains were responsible for the immunogenicity of the antibody [65]. Consequently, chimeric antibodies were constructed by Morrison et al. in 1984 by rDNA technology [66] that consisted of murine-derived variable regions, while constant regions were derived from humans. Subsequently, a series of chimeric mAbs were constructed and approved for therapeutic use (Table 5.2).

5.3.3.2 Humanized mAbs

Generating chimeric antibodies was undoubtedly a step forward in reducing murine mAb immunogenicity in humans. However, the immunogenicity of chimeric mAbs still existed. To reduce the immunogenicity of therapeutic mAbs, a new concept called humanized mAbs emerged [78]. In humanized antibodies, all the regions of the antibody are derived from humans except for the CDR sequences, which are replaced by murine CDR sequences [79]. CDR sequence governs the affinity of an antibody towards the antigen. As a result, a humanized antibody has its antigen-binding property derived from murine [62, 80]. Alemtuzumab was the first humanized antibody [81]. It consisted of CDRs derived from rat IgG2a, and the rest of the framework regions were derived from human myeloma IgG NEW and Bence-Jones protein REI [80, 82]. However, the replacement of murine CDRs alone reduced the antigen-binding affinity of the humanized mAbs. Later, studies suggested that the interaction between CDRs and framework residues around the CDRs is also essential for antigen binding. Therefore, to retain the binding affinity, either the interacting residues along with CDRs were grafted or corresponding interacting residues in the framework were mutated [80, 83]. For example, the humanization of pertuzumab [84] and bevacizumab [85] was done by grafting murine framework amino acids and CDRs. Also, humanized antibodies such as atezolizumab, elotuzumab, pembrolizumab, and obinutuzumab, have been generated by replacing non-human CDRs along with mutations in human CDR-interacting residues to obtain original mouse residues that would stabilize the CDR conformation [80].

5.3.4 Antibody Fragments

5.3.4.1 Single-Chain Variable Fragments (scVF)

It has been long known that antibodies bind to the epitope on an antigen with high avidity through CDRs located in the antigen-binding domain (Fab region) of the antibody. The amino-terminal region of light and heavy chain folds to form the Fab region, as evident from various biochemical studies and crystal structures [6]. The specificity of the antigen-binding region towards different antigens or epitopes on the same antigen is a consequence of sequence diversity generated by recombination

Table 5.2 Commercial monoclonal antibodies and their different formats

Type	Product trade name	Target	Therapeutic use	Manufacturing/ distributor company	Year of approval	Reference
Fully human antibodies	Aducanumab (Aduhelm)	Amyloid beta (Aβ)	Alzheimer's disease	Biogen Inc., USA	2021	[26]
	Adalimumab (Humira)	Tumor necrosis factor alpha (TNFα)	Rheumatoid arthritis disease	*AbbVie* pharmaceuticals, USA	2002	[27]
	Denosumab (Prolia)	RANK (receptor activator of nuclear factor kappa-β ligand)	Osteoporosis	Amgen, USA	2009	[28]
	Raxibacumab	Protective antigen of anthrax toxin	Anthrax	Emergent BioSolutions, USA	2012	[29]
	Bevacizumab (Avastin)	VEGF-A	Cancer	Roche, Switzerland	2004	[30]
	Ofatumumab (Arzerra)	CD20	Chronic lymphocytic leukemia	Novartis, Switzerland	2009	[31]
	Erenumab (Aimovig)	Calcitonin gene-related peptide receptor (CGRPR)	Prevention of migraine	Novartis, Switzerland and Amgen, USA	2018	[32]
Humanized antibodies	Eptinezumab (Vyepti)	CALCA and CALCB	Treatment of migraine	Lundbeck Seattle BioPharmaceuticals, USA	2020	[33]
	Romosozumab	Sclerostin	Treatment of osteoporosis	Amgen, USA	2019	[34]
	Inebilizumab (Uplizna)	CD19	Neuromyelitis Optica spectrum disorder	Viela Bio, USA	2020	[35]
	Ocrelizumab (Ocrevus)	CD20	Multiple sclerosis	Hoffmann-La Roche, USA	2017	[36]
		CD52		Sanofi, France	2001	[37]

(continued)

Table 5.2 (continued)

Type	Product trade name	Target	Therapeutic use	Manufacturing/ distributor company	Year of approval	Reference
	Alemtuzumab (Campath)		Chronic lymphocytic leukemia and multiple sclerosis			
	Ixekizumab (Taltz)	Interleukin 17A	Autoimmune diseases	Eli Lilly and Company, USA	2016	[67]
	Eculizumab	Complement protein C5	Paroxysmal nocturnal hemoglobinuria	Alexion Pharmaceuticals, USA	2007	[68]
Chimeric antibodies	Rituximab (Rituxan)	CD20	Autoimmune disease and type of cancers	Biogen, USA	1997	[69]
	Infliximab (Remicade)	TNF-α	Crohn's disease, ulcerative colitis, rheumatoid arthritis, ankylosing spondylitis, psoriasis, psoriatic arthritis, and Bechet's disease	Pfizer Inc., USA	1998	[70]
	Brentuximabvedotin (Adcetris)	CD30	Hodgkin lymphoma, anaplastic large cell lymphoma	Seattle Genetics Inc., USA	2011	[71]
	Ublituximab	CD20	Multiple sclerosis	TG Therapeutics, USA	Phase 3 clinical trials	[72]
	Dinutuximab (Unituxin)	GD2	Neuroblastoma	United Therapeutics Corporation, USA	2015	[73]
	Indatuximab ravtansine	SDC1	Multiple myeloma	Not known	Under clinical trial	[74]
Fab fragments/ Single-chain	Ranibizumab (Lucentis)	VEGF-A	Macular degeneration	Genentech, USA	2006	[75]

variable fragments (scFV)/VHH	Caplacizumab (Cablivi)	Von Willebrand factor (VWF)	Thrombocytopenic purpura and thrombosis	Ablynx, Belgium	2019	[76]
	Cimzia/ Certolizumab	TNF-α	Crohn's disease, rheumatoid arthritis, psoriatic arthritis, and ankylosing spondylitis	UCB, Belgium	2008	[77]

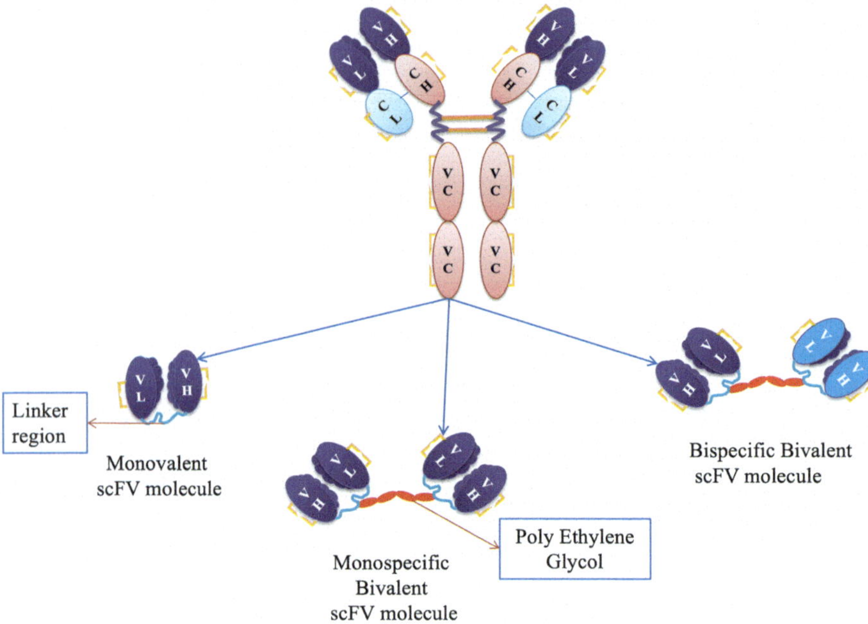

Fig. 5.3 Different therapeutic formats of a single-chain variable fragment (scFV)

of the genetic element in the B-cell germ line [2, 6]. Amino acid variability manifests in a specific folding pattern that governs the specific binding to an epitope. Even a single amino acid substitution in CDR has been demonstrated to perturb the antigen-binding specificity of an antibody [1].

At present, one can easily conceive the idea of using Fab fragment for a therapeutic purpose; however, initially, the fragment was used for cancer imaging [86]. Antibodies were the primary choice of researchers for imaging myeloma cells, but undesirable binding of heavy-chain constant region to non-myeloma cells led to exploration for alternatives [87]. The variable fragment (Fv), which lacks a constant region, served as a good alternative [88, 89]. But it was not easy and reliable to obtain Fv by proteolytic digestion of an mAb [90]. Also, the expression of V_L and V_H regions as distinct polypeptides in a bacterial system and then allowing both V_L and V_H folded domains to form a complete variable fragment was unsuccessful [91–93]. These difficulties laid the foundation for designing and synthesizing single-chain variable fragments (scFv) (Fig. 5.3).

Design and Synthesis of scFV

In scFV, the V_L and V_H domains are connected through a polypeptide linker region. Initially, the term single-chain binding-protein instead of scFV was commonly used in literature. Single-chain binding protein against bovine growth hormone (BGH) was the first successfully expressed protein of this kind in *Escherichia coli* [94]. The V_L and V_H regions were derived from a monoclonal anti-BGH antibody and were linked through a polypeptide linker [94]. The linker region mainly consists of alternating glycine and serine residues that provide flexibility to the linker region and thus allow the interaction of V_L and V_H domains to form an intact antigen-binding region [95]. The linkers with alternating glycine and serine are still the most commonly used linkers. It has been demonstrated that an increase in the length of polypeptide linker in anti-daidzin scFV (DZ-scFV) causes reduced reactivity of the scFV [95]. Both very short and long linkers would not allow the correct orientation of V_L and V_H domains to interact and form a complete scFV molecule. The most frequently used peptide linker length varies from 15 to 20 residues [95].

Applications of scFV

The size of an scFV molecule is approximately one-sixth of the complete antibody molecule [96]. A smaller size allows faster clearance of the molecule from the serum compared to a full-length antibody or Fab [97]. Furthermore, due to its small size, scFV can penetrate the tumors and thus provides a means to treat tumors specifically and effectively. The Fc region is highly glycosylated and, therefore, highly immunogenic [5]. Moreover, cells possess receptors for binding to Fc regions of an antibody and cause undesirable Fc-mediated immunogenic reactions when mAb is used as a therapeutic molecule [98]. On the contrary, the scFV molecule lacks an Fc region and, therefore, to an extent, mitigates the risk of immunogenic reaction from the body [96].

5.3.4.2 Heavy-Chain Antibodies (hcAbs) and Single-Domain Antibodies (sdAbs)

Most organisms produce antibodies with antigen-binding sites formed by both heavy and light chains, and their structural and functional characterization has been done extensively [7]. However, unusual homodimeric antibodies with only a heavy chain forming its antigen-binding site as a single domain have been found in sharks and camelids in addition to heterotetrameric antibodies [99]. The N-terminal region of the heavy chain folds in a specific manner to form a variable domain called single domain shark variable domain of new antigen receptor (VNAR) in sharks and variable heavy chain of heavy-chain antibodies (VHH) in camelids [100, 101]. This N-terminal variable domain of hcAbs, like the Fab region of conventional antibodies, binds to the specific antigen [102]. In camelids, three classes of IgG are naturally found: IgG1 comprises two light and two heavy chains, whereas IgG2 and IgG3 comprise only a heavy chain. Also, in IgG2 and IgG3, the CH1 domain is replaced by the hinge region [103, 104]. The hinge region directly connects the antigen-binding domain of (VHH) to the Fc fragment, which, in turn, is formed by CH2 and CH3 domains [100, 105]. The hinge region of IgG3 is comparatively shorter than the IgG2 type. While conventional antibody has a molecular weight of around 150 kDa, the hcAbs are approximately 1.5 times smaller, with a molecular weight of about 90 kDa [105]. The small size of hcAbs is believed to bind to sequestered antigens; however, the complete functions of hcAbs are not yet very clear. Nevertheless, hcAbs have been shown to activate immune cells by binding to the cell surface of monocytes and macrophages [106]. Certainly, the discovery of hcAbs has opened up new avenues for antibody-based therapeutics [107].

VHH Domain

In addition to scFv, a substantial effort has also moved to the construction and expression of single-domain antibodies (sdAbs). sdAbs are also known as nanobodies [107, 108]. Nanobodies are composed of only the VHH domain of the hcAbs. A VHH domain has a molecular weight of around 12–15 kDa (half of the molecular weight of the scFv molecule) and is only around 120 amino acids long. The variable characteristic of VHH essentially lies in three hypervariable (HV) regions, supported by a scaffold of conserved amino acids, which is designated as framework (FR) regions [109, 110]. Essentially, the native VHH domain is formed of nine antiparallel β-strands, arranged in two sheets, one with four β-strands and another with five β-strands connected through loops and a conserved disulfide bond [111], as shown in Fig. 5.4. The HV regions, located in the three loops, form the CDRs [100, 112]. When compared to the V_H of conventional antibodies, VHH loops are longer. The longer loops of VHH compensate for fewer loops and a lack of light chain in forming CDRs [113]. Longer loops of VHH aid in increased interface area for interacting with the antigen. However, longer loops are entropically unfavorable for interaction with an antigen [109, 113]. To an extent, the flexibility of the loops is constrained by interloop

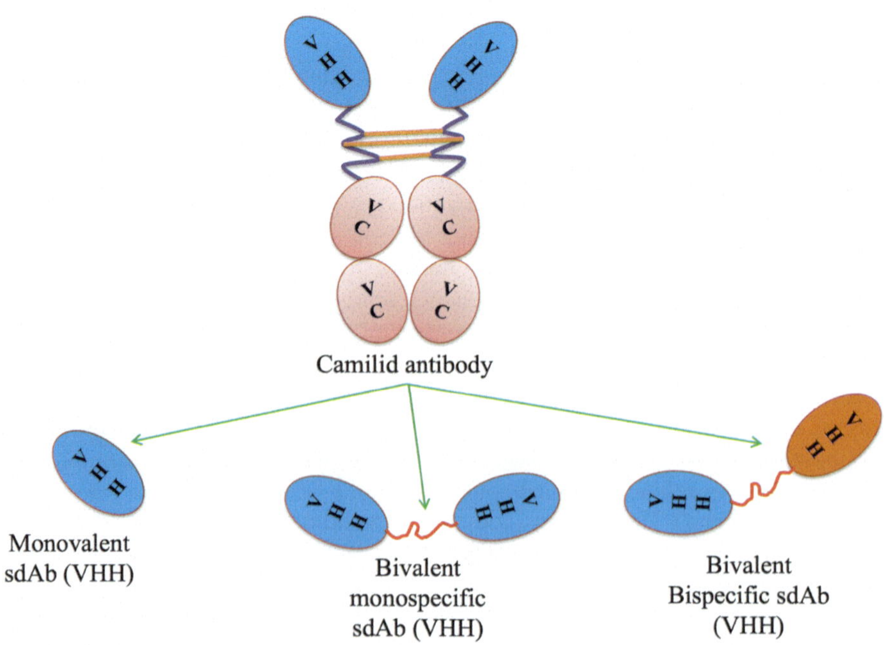

Fig. 5.4 Camelid heavy-chain antibody and different therapeutic formats of the variable heavy chain of heavy-chain antibodies (VHH)

disulfide bonds (Fig. 5.4) [113]. Moreover, llama VHH has been shown to have shorter loops with a less frequent extra interloop disulfide bond [114, 115].

Designing and Expression of sdAbs (Nanobodies)

Single-domain antibodies are selected in a multi-step process that involves initial immunization of the animal with antigen followed by cloning of cDNA for VHH domain of hcAbs obtained from lymph nodes, blood lymphocytes, and spleen into phagemid vector and selected based on their affinity for the antigen [112]. The method for the selection of high-affinity sdAbs is called biopanning. In biopanning, the immobilized antigen on cell culture plates, immunotubes, or ELISA plates are incubated with the VHH domain displayed on the surface of the phage [110, 112]. The filamentous M13 bacteriophage is the most commonly used phage for the phage library display method. The unbound or non-specific bound phages are washed, and phages bound to the antigens are eluted. The elution procedure for the bound phage depends upon the bacteriophage used [112]. For example, for the elution of M13 bacteriophage, glycine buffer, or mild HCl solutions are used. In addition, alkaline solutions of triethylamine and competitive elution with high antigen concentrations have been in practice [116]. The amplification of the eluted bacteriophage is achieved by further infection of *E. coli* cells during their exponential phase of growth. After selecting the high-affinity fragment, it is cloned and expressed in a bacterial expression system [91].

Applications of sdAbs

Targeting Tumors and Inflammation

Cell Surface Proteins

Various ecto-enzymes are expressed on the cell surface and are responsible for regulating cellular processes like apoptosis, cell trafficking, inflammation, cell-cell adhesion, and cell proliferation [117, 118], for instance, 5′ nucleotidase or CD73 (in addition to hydrolysis of AMP to adenosine also controls cell-cell adhesion and T-cell activation) [117–120], CD38 and cell surface monoamine oxidase (involved in dendritic cell trafficking and regulation of T-cell dependent immune response) [121, 122], mono ADP-ribosyltransferase 2 (ART2, is involved in cell death) [123], and E-NPP3 (or CD203c, negatively regulates the allergic responses) [124]. CD13, CD156b, CD13, and CD26 are other known cell surface peptidase and protease enzymes that modify the adhesion and chemotactic molecules by proteolytic cleavage [125]. The modification of various cellular processes by ecto-enzymes also results in the modulation of the immune system. Therefore, designing specific molecules that inhibit ecto-enzymes would help beneficially alter the immune system. Moreover, the orientation of ecto-enzymes catalytic/binding site is towards the extracellular matrix and is therefore easily accessible to the inhibitory molecules [126]. Usually, small molecules are used to develop as inhibitors, but their main drawback is low specificity. These small molecule inhibitors tend to bind to other members of the same protein family. Concerning specificity, mAbs can be instantly considered; however, the accessibility of their CDRs into the active site cleft of the enzyme is a major concern. The longer CDR loops in sdAbs aid in binding to the deeper active site cleft of the enzyme [107, 127]. Koch et al. have developed and demonstrated that sdAbs from llama against ART2.2 specifically bind to the enzyme and suppress its enzymatic and cytotoxic activities [126, 128]. They showed that sdAbs possess rapid activity (with a short time of 15 mins following the injection) and a rapid clearance time of 6 h in mice. Therefore, sdAbs against ART2.2 can potentially treat acute inflammation [126]. A bivalent version of anti-ART2.2 sdAbs has been constructed by Felix et al. with a fused mouse Fc fragment. This bivalent version has been proven effective for a longer period and could be developed for treating chronic inflammation [128]. CD38 is another ecto-enzyme that is upregulated in tumor cells [129]. Recently, nanobodies constructed against CD38 have been demonstrated to inhibit CD38 enzymatic activity, comparable to mAbs [130, 131].

In addition to ecto-enzymes, the leukocyte cell surface expresses numerous non-enzymatic receptors that mediate signal transmission through the cell membrane, transporters, and ion channels [126]. One of the receptors present on the surface of natural killer (NK) cells is FcγRIII. This receptor activates cytokine release and antibody-dependent cell cytotoxicity (ADCC) and thereby causes tumor cell death [132, 133]. Researchers have been able to produce anti-FcγRIII bispecific antibodies from the VHH domain produced in immunized llama [134]. These anti-FcγRIII bispecific antibodies have been shown to activate the NK cells [134]. For the treatment of cancer, sdAbs conjugates specific to carcinoembryonic antigen (CEA) have also been developed by fusing β-lactamase with sdAb through the camel hcAb hinge region. The β-lactamase enzyme catalyzes the conversion of a prodrug into a cytotoxic compound. Tumor xenografts have been shown to cure when anticancer prodrug was administered along with sdAb-β-lactamase conjugate and was localized in the affected area [135].

Cytokines and Other Soluble Proteins

The human immune system responds to bacterial infection by secreting cytokines. Cytokines are small peptides, glycoproteins, or regulatory proteins secreted by the white blood cells (WBCs) of the immune system [136]. A variety of cytokines are secreted by cells of the immune system, namely interferons, interleukins, tumor necrosis factors, colony-stimulating factors, and growth factors, which can have similar or overlapping effects [137]. For example, TNF-α acts on vascular endothelial cells and macrophages to induce the secretion of colony-stimulating factors, resulting in a transient increase in WBCs needed to alleviate infection [138]. However, persistent higher concentrations of TNF-α have been associated with the pathogenesis of psoriasis, spondyloarthropathies, and inflammatory disorders like rheumatoid arthritis (RA) [139]. For the treatment of RA, therapeutic antibody molecules have already been developed and are available on the market. Now with the advent of sdAbs, the limitations of the mAbs, which have already been discussed in this chapter, can be overcome. In the mice model, anti-TNF-α sdAbs have proved effective [140]. The level of another pro-inflammatory cytokine, IL23, is raised in Crohn's disease patients indicating its role in the pathogenesis of Crohn's disease [137]. A high-affinity anti-hIL23 nanobody has been shown to bind to IL23 with excellent efficacy [141]. In addition, the multivalent format of anti-hIL23 has been shown to have high efficacy than its monovalent counterpart [141].

sdAbs Against Infectious Agents and Their Toxins

After entering the host body, the pathogens trigger the host immune system either through cell surface antigens on their cell surface or through the toxins they produce [142, 143]. In the case of bacterial pathogens, lipopolysaccharides, membrane proteins, pili, and flagella proteins act as antigens [144–146]. The endotoxin produced by Gram-Negative bacteria can cause a severe immunogenic reaction known as septic shock syndrome. An anti-LPS VHH has been reported, which was produced by screening the phage library [147]. The anti-LPS VHH was found to successfully bind to LPS and hinder the elicitation of hyperimmune reaction [147]. Various surface proteins of bacteria have also been targeted to produce sdAbs. Proteolytically stable sdAbs (K922) against F4 fimbriae with significantly high stability in gastric and jejuna fluid have been shown to mitigate the *E. coli*-induced diarrhea in piglets [148]. *Streptococcus mutans* adheres to the teeth and forms dental caries. The sdAbs against *S. mutans* cells were developed by immunizing llama with *S. mutans* cells [149]. The sdAbs were expressed in yeast as a fusion protein with glucose oxidase [149]. Glucose oxidase produces hydrogen peroxide (which is toxic to bacterial cells) in the presence of glucose and oxygen. Upon treating mixed biofilm culture with sdAbs fused with glucose oxidase, the *S. mutans* cells were specifically bound and lysed [149].

In the case of viruses, antigenic epitopes are primarily located in either capsids or envelopes [150, 151]. Some viral epitopes are sequestered in deep clefts and are not easily accessible. For instance, domain I of envelope glycoprotein E is exposed only at the time of membrane fusion transition [152]. To detect such sequestered epitopes, the use of sdAbs is an ideal choice. VHH against gp120 protein of HIV-1 has been produced from llamas and has been shown to neutralize HIV-1 efficiently [153]. Despite possessing high immunogenicity, an inner capsid protein of rotavirus, VP6 lacks neutralizing antibody induction in the host. Recently, sdAbs have been produced from immunized llamas, which were able to neutralize rotavirus in an *in vitro* experiment [154]. Also, *in vivo* study showed that 60% of the neonatal mice population was protected from diarrhea caused due to rotavirus upon administration of the sdAbs [154]. Oral administration of lactobacillus, which constitutively expressed fused sdAb with *Lactobacillus* protease, reduced rotavirus-induced diarrhea in the mice model [155].

5.4 Challenges and Prospects of Therapeutic Antibodies

Antibodies have been proven efficient therapeutic molecules due to their specificity and high affinity towards the antigen [156]. Indeed, there had been problems associated with antibodies right from the beginning when serum was used for treating disease. With the advancement of technology and a greater understanding of the immune system, researchers have been able to solve a few problems to a great extent. One serious drawback of using therapeutic antibodies is their immunogenicity which leads to anaphylaxis [127]. Various factors, such as genetic predisposition of the patient for allergic reactions, antibody-drug formulation, manufacturing, and previous exposure to the molecule, are associated with the immunological effects of the antibody-based drug [98]. For example, adalimumab has been shown to produce immunogenic responses in around 30% of patients. The pathway to cause immunogenicity may also differ in different cases. In the case of adalimumab, the cause of immunogenicity was speculated to occur through T-cell-dependent pathway [78]. To resolve this problem, the amino acid sequences which are recognized by major histocompatibility factors of antigen-presenting cells and cause an undesirable immune response through T-cells must be identified and replaced accordingly. In addition, of the four subclasses of IgG, the IgG4 subclass is ineffective in mounting either antibody-dependent or complement-dependent cell cytotoxicity [157]. Importantly, antibodies can be altered at the genetic level to modulate their effector functions [158, 159].

Antibodies are proteins and therefore are prone to degradation even under the slightest change in conditions like temperature, agitation, pH, shear stress, and metal ions [160]. Therefore, formulation of antibody-based therapeutics plays a crucial role in protein stabilization and thus assures the safety of drug administration. Every therapeutic antibody is intrinsically different, and therefore formulation strategies differ accordingly. Data generated by general pre-formulation evaluation for a new therapeutic antibody can be utilized to optimize the formulation [161]. During optimization of formulation, the effect of different factors such as aggregation, deamination, oxidation [162], and fragmentation on protein stability is measured. Glutamine and asparagine are highly prone to deamidation [163] and are undesirable when present in CDRs [164]. Cysteine and methionine residues have a high propensity for oxidation [162]. In addition, phenylalanine, histidine, tryptophan, and tyrosine also undergo oxidation [165] and reduce the efficacy of the antibody.

For the treatment of chronic diseases, a high concentration of antibodies in therapeutic preparation is needed. Antibody preparation may contain protein concentrations greater than 100 mg/mL [49]. High concentration leads to two events, the first is protein aggregation, and the second is the increase in viscosity of the preparation [160, 166]. Aggregated antibody preparation is not suitable for intravenous administration. Moreover, the protein aggregation results in a lower dose concentration that might not be sufficient to produce desirable results [166]. Thus, antibody therapeutic drugs are stabilized by adding stabilizers such as surfactants. In recent times, the alteration of the glycosylation pattern of the constant region of the antibody has been utilized for stabilizing the antibody in therapeutic formulations. The thermal stability of bevacizumab has been improved by N-glycosylation of the constant region of the Fab fragment [160].

Recently, antibody fragments such as scFV and nanobodies have been developed. Unlike full-length antibody production, scFV and nanobodies can be successfully expressed in bacterial expression systems [42, 108]. This significantly reduces the cost of antibody-based therapeutic production. Their significance lies in their low molecular weight and significantly reduced immunogenicity [86]. Moreover, nanobodies and scFVs can be modified into different formats, for instance, bispecific fragments and bivalent fragments [167]. A bispecific nanobody or scFV fragment can bind to two different epitopes on the same antigen and can, therefore, increase their binding affinity [167, 168]. A recent surge in research and commercial interest for mAb therapeutics, engineered antibodies, and their different formats anticipates a vast market for these therapeutic molecules.

5.5 Conclusions

The coming pharmaceutical revolution will be in the field of personalized medicine, where antibody therapy is one of the foremost candidates. Therapeutic mAbs targeting the discrete antigen are among the prime targets in molecular medicine for developing a new generation of therapeutic agents [169]. Monoclonal antibodies have been shown to successfully reduce COVID-19 hospitalization rates when administered in the early phase of infection [170]. Some mAbs act on the immune system to stimulate it to target specific cells, particularly in cancer therapy. The research of mAb-induced immunity in the form of vaccines against cancer should be looked upon diligently as it has enormous market potential. The antibodies could be engineered to incorporate immunostimulatory motifs to effectively target tumor cells. Antibody therapy has potential in the treatment of age-related diseases like Alzheimer's and Parkinson's. Aging diseases result from the aggregation of misfolded proteins, and antibodies are among the attractive vehicle to target aberrant proteins that could bind to them and process them for degradation. The applications of antibody therapy have enormous potential in several clinical conditions, and it is expected to be at its peak in the following decades.

Acknowledgment R.S.R thanks to ICMR New-Delhi for SRF fellowship (file no. 5/3/8/12/ITR-F/2018-ITR).

References

1. Rudikoff S, Giustit AM, Cookt WD, Scharfft MD (1982) Single amino acid substitution altering antigen-binding specificity (immunoglobulin/mutation/phosphocholine/antibody diversity)
2. Sela-Culang I, Kunik V, Ofran Y (2013) The structural basis of antibody-antigen recognition. Front Immunol 4. https://doi.org/10.3389/fimmu.2013.00302
3. Kholodenko RV, Kalinovsky DV, Doronin II, Ponomarev ED, Kholodenko IV (2019) Antibody fragments as potential biopharmaceuticals for cancer therapy: success and limitations. Curr Med Chem 26:396–426. https://doi.org/10.2174/0929867324666170817152554
4. Schroeder HW, Cavacini L (2010) Structure and function of immunoglobulins. J Allergy Clin Immunol 125. https://doi.org/10.1016/j.jaci.2009.09.046
5. Arnold JN, Wormald MR, Sim RB, Rudd PM, Dwek RA (2007) The impact of glycosylation on the biological function and structure of human immunoglobulins. Annu Rev Immunol 25: 21–50. https://doi.org/10.1146/annurev.immunol.25.022106.141702
6. Decanniere K, Muyldermans S, Wyns L (2000) Canonical antigen-binding loop structures in immunoglobulins: more structures, more canonical classes? J Mol Biol 300:83–91. https://doi.org/10.1006/jmbi.2000.3839

7. Stanfield RL, Wilson IA, Crowe JE (2014) Antibody structure. https://doi.org/10.1128/microbiolspec.AID
8. Spiegelberg HL (1989) Biological role of different antibody classes. Appl Immunol 90:22–27
9. Krapp S, Mimura Y, Jefferis R, Huber R, Sondermann P (2003) Structural analysis of human IgG-Fc glycoforms reveals a correlation between glycosylation and structural integrity. J Mol Biol 325:979–989. https://doi.org/10.1016/S0022-2836(02)01250-0
10. Vincent A (2002) Unravelling the pathogenesis of myasthenia gravis. Nat Rev Immunol 2: 797–804. https://doi.org/10.1038/nri916
11. McInnes IB, Schett G (2011) Mechanism of disease the pathogenesis of rheumatoid arthritis. N Engl J Med 365:2205–2219
12. Sospedra M, Martin R (2005) Immunology of multiple sclerosis. Annu Rev Immunol 23:683–747. https://doi.org/10.1146/annurev.immunol.23.021704.115707
13. Davidson A, Diamond B (2001) Autoimmune diseases. N Engl J Med 345:340–350. www.nejm.org
14. Forthal DN (2015) Functions of antibodies. Antibodies Infect Dis 25–48. https://doi.org/10.1128/9781555817411.ch2
15. Dimitrov JD, Lacroix-Desmazes S (2020) Noncanonical functions of antibodies. Trends Immunol 41:379–393. https://doi.org/10.1016/j.it.2020.03.006
16. Vidarsson G, Dekkers G, Rispens T (2014) IgG subclasses and allotypes: from structure to effector functions. Front Immunol 5:1–17. https://doi.org/10.3389/fimmu.2014.00520
17. Mimura Y, Sondermann P, Ghirlando R, Lund J, Young SP, Goodall M, Jefferis R (2001) Role of oligosaccharide residues of IgG1-Fc in FcγRIIb binding. J Biol Chem 276:45539–45547. https://doi.org/10.1074/jbc.M107478200
18. Mimura Y, Church S, Ghirlando R, Ashton PR, Dong S, Goodall M, Lund J, Jefferis R (2000) The influence of glycosylation on the thermal stability and effector function expression of human IgG1-Fc: properties of a series of truncated glycoforms. Mol Immunol 37:697–706. https://doi.org/10.1016/S0161-5890(00)00105-X
19. Nezlin R, Ghetie V (2004) Interactions of immunoglobulins outside the antigen-combining site. Adv Immunol 82:155–215. https://doi.org/10.1016/S0065-2776(04)82004-2
20. Deisenhofer J (1981) Crystallographic refinement and atomic models of a human Fc fragment and its complex with fragment B of protein A from Staphylococcus aureus at 2.9- and 2.8-Å resolution. Biochemistry 20:2361–2370. https://doi.org/10.1021/bi00512a001
21. Zeitlin L, Cone RA, Moench TR, Whaley KJ (2000) Preventing infectious disease with passive immunization. Microbes Infect 2:701–708. https://doi.org/10.1016/S1286-4579(00)00355-5
22. Chen RT, Markowitz LE, Albrecht P, Stewart JA, Mofenson LM, Orenstein WA, Orenstein WA (1990) Measles antibody: reevaluation of protective titers. J Infect Dis 162:1036–1042. https://doi.org/10.1093/infdis/162.5.1036
23. Turnbull PCB, Broster MG, Carman JA, Manchee RJ, Melling J (1986) Development of antibodies to protective antigen and lethal factor components of anthrax toxin in humans and guinea pigs and their relevance to protective immunity. Infect Immun 52:356–363. https://doi.org/10.1128/iai.52.2.356-363.1986
24. Yu PA, Lin NH, Mahon BE, Sobel J, Yu Y, Mody RK, Gu W, Clements J, Kim HJ, Rao AK (2017) Safety and improved clinical outcomes in patients treated with new equine-derived heptavalent botulinum antitoxin. Clin Infect Dis 66:S57–S64. https://doi.org/10.1093/cid/cix816
25. Finne J, Leinonen M, Mäkelä PH (1983) Antigenic similarities between brain components and bacteria causing meningitis. Implications for vaccine development and pathogenesis. Lancet 322:355–357. https://doi.org/10.1016/S0140-6736(83)90340-9
26. Sevigny J, Chiao P, Bussière T, Weinreb PH, Williams L, Maier M, Dunstan R, Salloway S, Chen T, Ling Y, O'Gorman J, Qian F, Arastu M, Li M, Chollate S, Brennan MS, Quintero-Monzon O, Scannevin RH, Arnold HM, Engber T, Rhodes K, Ferrero J, Hang Y, Mikulskis A,

Grimm J, Hock C, Nitsch RM, Sandrock A (2016) The antibody aducanumab reduces Aβ plaques in Alzheimer's disease. Nature 537:50–56. https://doi.org/10.1038/nature19323
27. Bang LM, Keating GM (2004) Adalimumab: a review of its use in rheumatoid arthritis. BioDrugs 18:121–139. https://doi.org/10.2165/00063030-200418020-00005
28. Farahmand P, Ringe JD (2012) Denosumab. Chir Prax 75:541–545. https://doi.org/10.29309/tpmj/2012.19.02.2027
29. Mazumdar S (2009) Raxibacumab. MAbs 1:531–538. https://doi.org/10.4161/mabs.1.6.10195
30. McCormack PL, Keam SJ (2008) Bevacizumab: a review of its use in metastatic colorectal cancer. Drugs 68:487–506. https://doi.org/10.2165/11205090-000000000-00000
31. Zhang B (2009) Ofatumumab. MAbs 1:326–331. https://doi.org/10.4161/mabs.1.4.8895
32. Goadsby PJ, Reuter U, Hallström Y, Broessner G, Bonner JH, Zhang F, Sapra S, Picard H, Mikol DD, Lenz RA (2017) A controlled trial of erenumab for episodic migraine. N Engl J Med 377:2123–2132. https://doi.org/10.1056/nejmoa1705848
33. Ashina M, Saper J, Cady R, Schaeffler BA, Biondi DM, Hirman J, Pederson S, Allan B, Smith J (2020) Eptinezumab in episodic migraine: a randomized, double-blind, placebo-controlled study (PROMISE-1). Cephalalgia 40:241–254. https://doi.org/10.1177/0333102420905132
34. Bandeira L, Lewiecki EM, Bilezikian JP (2017) Romosozumab for the treatment of osteoporosis. Expert Opin Biol Ther 17:255–263. https://doi.org/10.1080/14712598.2017.1280455
35. Frampton JE (2020) Inebilizumab: first approval. Drugs 80:1259–1264. https://doi.org/10.1007/s40265-020-01370-4
36. Mulero P, Midaglia L, Montalban X (2018) Ocrelizumab: a new milestone in multiple sclerosis therapy. Ther Adv Neurol Disord 11. https://doi.org/10.1177/1756286418773025
37. Evana JR, Bozkurta SB, Thomasa NC, Bagnatoa F (2018) Alemtuzumab for the treatment of multiple sclerosis. Expert Opin Biol Ther 18:323–334. https://doi.org/10.1080/14712598.2018.1425388
38. Newcombe C, Newcombe AR (2007) Antibody production: polyclonal-derived biotherapeutics. J Chromatogr B Anal Technol Biomed Life Sci 848:2–7. https://doi.org/10.1016/j.jchromb.2006.07.004
39. Siegel J (2006) Safety considerations in IGIV utilization. Int Immunopharmacol 6:523–527. https://doi.org/10.1016/j.intimp.2005.11.004
40. Buchacher A, Iberer G (2006) Purification of intravenous immunoglobulin G from human plasma – Aspects of yield and virus safety. Biotechnol J 1:148–163. https://doi.org/10.1002/biot.200500037
41. Holliger P, Hudson PJ (2005) Engineered antibody fragments and the rise of single domains. Nat Biotechnol 23:1126–1136. https://doi.org/10.1038/nbt1142
42. Sandhu JS (1992) Protein engineering of antibodies. Crit Rev Biotechnol 12:437–462. https://doi.org/10.3109/07388559209114235
43. Gelfand EW (2006) Differences between IGIV products: impact on clinical outcome. Int Immunopharmacol 6:592–599. https://doi.org/10.1016/j.intimp.2005.11.003
44. Looney RJ, Huggins J (2006) Use of intravenous immunoglobulin G (IVIG). Best Pract Res Clin Haematol 19:3–25. https://doi.org/10.1016/j.beha.2005.01.032
45. Martin TD (2006) IGIV: contents, properties, and methods of industrial production – Evolving closer to a more physiologic product. Int Immunopharmacol 6:517–522. https://doi.org/10.1016/j.intimp.2005.11.005
46. Imbach P, Barandun S, Baumgartner C, Hirt A, Hofer F, Wagner HP (1981) High-dose intravenous gammaglobulin therapy of refractory, in particular idiopathic thrombocytopenia in childhood. Helv Paediatr Acta 36:81–86
47. Lemieux R, Bazin R, Néron S (2005) Therapeutic intravenous immunoglobulins. Mol Immunol 42:839–848. https://doi.org/10.1016/j.molimm.2004.07.046
48. Ballow M (2007) Safety of IGIV therapy and infusion-related adverse events. Immunol Res 38:122–132. https://doi.org/10.1007/s12026-007-0003-5
49. Daugherty AL, Mrsny RJ (2006) Formulation and delivery issues for monoclonal antibody therapeutics. Adv Drug Deliv Rev 58:686–706. https://doi.org/10.1016/j.addr.2006.03.011

50. Ballow M (2005) Clinical and investigational considerations for the use of IGIV therapy. Am J Heal Pharm. https://doi.org/10.2146/ajhp050283
51. Sarantopoulos S, Kao CY, Den W, Sharon J (1994) A method for linking VL and VH region genes that allows bulk transfer between vectors for use in generating polyclonal IgG libraries. J Immunol 152:5344–5351
52. Haurum JS (2006) Recombinant polyclonal antibodies: the next generation of antibody therapeutics? Drug Discov Today 11:655–660. https://doi.org/10.1016/j.drudis.2006.05.009
53. Nielsen LS, Baer A, Müller C, Gregersen K, Mønster NT, Rasmussen SK, Weilguny D, Tolstrup AB (2010) Single-batch production of recombinant human polyclonal antibodies. Mol Biotechnol 45:257–266. https://doi.org/10.1007/s12033-010-9270-9
54. Jakobovits A (1995) Production of fully human antibodies by transgenic mice. Curr Opin Biotechnol 6:561–566. https://doi.org/10.1016/0958-1669(95)80093-X
55. Green LL (1999) Antibody engineering via genetic engineering of the mouse: XenoMouse strains are a vehicle for the facile generation of therapeutic human monoclonal antibodies. J Immunol Methods 231:11–23. https://doi.org/10.1016/S0022-1759(99)00137-4
56. Kohler H (2021) The impact of the hybridoma technology on the R&D of idiotypic antibodies. Monoclon Antib Immonodiagn Immunother 40:2–5. https://doi.org/10.1089/mab.2020.0044
57. Strebhardt K, Ullrich A (2008) Paul Ehrlich's magic bullet concept: 100 years of progress. Nat Rev Cancer 8:473–480. https://doi.org/10.1038/nrc2394
58. Ribatti D (2014) From the discovery of monoclonal antibodies to their therapeutic application: an historical reappraisal. Immunol Lett 161:96–99. https://doi.org/10.1016/j.imlet.2014.05.010
59. Li F, Vijayasankaran N, Shen A, Kiss R, Amanullah A (2010) Cell culture processes for monoclonal antibody production. MAbs 2:466–479. https://doi.org/10.4161/mabs.2.5.12720
60. Ziegelbauer K, Light DR (2008) Monoclonal antibody therapeutics: leading companies to maximise sales and market share. J Commer Biotechnol 14:65–72. https://doi.org/10.1057/palgrave.jcb.3050081
61. Liddell E (2013) Antibodies. Immunoass Handb:245–265. https://doi.org/10.1016/B978-0-08-097037-0.00017-8
62. Yamashita M, Katakura Y, Shirahata S (2007) Recent advances in the generation of human monoclonal antibody. Cytotechnology 55:55–60. https://doi.org/10.1007/s10616-007-9072-5
63. Eren R, Lubin I, Terkieltaub D, Ben-Moshe O, Zauberman A, Uhlmann R, Tzahor T, Moss S, Ilan E, Shouval D, Galun E, Daudi N, Marcus H, Reisner Y, Dagan S (1998) Human monoclonal antibodies specific to hepatitis B virus generated in a human/mouse radiation chimera: the Trimera system. Immunology 93:154–161. https://doi.org/10.1046/j.1365-2567.1998.00426.x
64. Kipriyanov SM, Le Gall F (2004) Generation and production of engineered antibodies. Appl Biochem Biotechnol Part B Mol Biotechnol 26:39–60. https://doi.org/10.1385/MB:26:1:39
65. Khazaeli MB, Conry RM, LoBuglio AF (1994) human immune response to monoclonal antibodies. J Immunother 15:42–52
66. Morrison SL, Johnson MJ, Herzenberg LA, Oi VT (1984) Chimeric human antibody molecules: mouse antigen-binding domains with human constant region domains. Proc Natl Acad Sci USA 81:6851–6855. https://doi.org/10.1073/pnas.81.21.6851
67. Mease PJ, Van Der Heijde D, Ritchlin CT, Okada M, Cuchacovich RS, Shuler CL, Lin CY, Braun DK, Lee CH, Gladman DD (2017) Ixekizumab, an interleukin-17A specific monoclonal antibody, for the treatment of biologic-naive patients with active psoriatic arthritis: results from the 24-week randomised, double-blind, placebocontrolled and active (adalimumab)-controlled period of the phase III trail SPIRIT-P1. Ann Rheum Dis 76:79–87. https://doi.org/10.1136/annrheumdis-2016-209709
68. Dubois EA, Cohen AF (2009) Eculizumab. Br J Clin Pharmacol 68:318–319. https://doi.org/10.1111/j.1365-2125.2009.03491.x

69. Pierpont TM, Limper CB, Richards KL (2018) Past, present, and future of Rituximab—the world's first oncology monoclonal antibody therapy. Front Oncol 8. https://doi.org/10.3389/fonc.2018.00163
70. Ni C, Reddy SP, Wu JJ (2016) Chapter 9 – Infliximab. https://doi.org/10.1016/j.chtm.2016.07.034
71. Vaklavas C, Forero-Torres A (2012) Safety and efficacy of brentuximab vedotin in patients with Hodgkin lymphoma or systemic anaplastic large cell lymphoma. Ther Adv Hematol 3: 209–225. https://doi.org/10.1177/2040620712443076
72. Fox E, Lovett-Racke AE, Gormley M, Liu Y, Petracca M, Cocozza S, Shubin R, Wray S, Weiss MS, Bosco JA, Power SA, Mok K, Inglese M (2021) A phase 2 multicenter study of ublituximab, a novel glycoengineered anti-CD20 monoclonal antibody, in patients with relapsing forms of multiple sclerosis. Mult Scler J 27:420–429. https://doi.org/10.1177/1352458520918375
73. Hoy SM (2016) Dinutuximab: a review in high-risk neuroblastoma. Target Oncol 11:247–253. https://doi.org/10.1007/s11523-016-0420-2
74. Schönfeld K, Zuber C, Pinkas J, Häder T, Bernöster K, Uherek C (2017) Indatuximab ravtansine (BT062) combination treatment in multiple myeloma: pre-clinical studies. J Hematol Oncol 10. https://doi.org/10.1186/s13045-016-0380-0
75. Blick SKA, Keating GM, Wagstaff AJ (2007) Ranibizumab. Drugs 67:1199–1206
76. Scully M, Cataland SR, Peyvandi F, Coppo P, Knöbl P, Kremer Hovinga JA, Metjian A, de la Rubia J, Pavenski K, Callewaert F, Biswas D, De Winter H, Zeldin RK (2019) Caplacizumab treatment for acquired thrombotic thrombocytopenic purpura. N Engl J Med 380:335–346. https://doi.org/10.1056/nejmoa1806311
77. Connock M, Tubeuf S, Malottki K, Uthman A, Round J, Bayliss S, Meads C, Moore D (2010) Certolizumab pegol (CIMZIA®) for the treatment of rheumatoid arthritis. Health Technol Assess 14:1–10. https://doi.org/10.3310/hta14suppl2/01
78. Hwang WYK, Foote J (2005) Immunogenicity of engineered antibodies. Methods 36:3–10. https://doi.org/10.1016/j.ymeth.2005.01.001
79. Oliphant T, Engle M, Nybakken GE, Doane C, Johnson S, Huang L, Gorlatov S, Mehlhop E, Marri A, Chung KM, Ebel GD, Kramer LD, Fremont DH, Diamond MS (2005) Development of a humanized monoclonal antibody with therapeutic potential against West Nile virus. Nat Med 11:522–530. https://doi.org/10.1038/nm1240
80. Almagro JC, Daniels-Wells TR, Perez-Tapia SM, Penichet ML (2018) Progress and challenges in the design and clinical development of antibodies for cancer therapy. Front Immunol 8. https://doi.org/10.3389/fimmu.2017.01751
81. Riechmann L, Clark M, Waldmann H, Winter G (1988) Reshaping human antibodies for therapy. Nature 332:323–327. https://doi.org/10.1038/332323a0
82. Epp O, Lattman EE, Schiffer M, Huber R, Palm W (1975) Molecular structure of a dimer composed of the variable portions of the Bence-Jones protein REI refined at 2.0-Å resolution. Biochemistry 14:4943–4952. https://doi.org/10.1021/bi00693a025
83. Kim SJ, Park Y, Hong HJ (2005) Antibody engineering for the development of therapeutic antibodies. Mol Cells 20:17–29
84. Adams CW, Allison DE, Flagella K, Presta L, Clarke J, Dybdal N, McKeever K, Sliwkowski MX (2006) Humanization of a recombinant monoclonal antibody to produce a therapeutic HER dimerization inhibitor, pertuzumab. Cancer Immunol Immunother 55:717–727. https://doi.org/10.1007/s00262-005-0058-x
85. Presta LG, Chen H, O'Connor SJ, Chisholm V, Meng YG, Krummen L, Winkler M, Ferrara N (1997) Humanization of an anti-vascular endothelial growth factor monoclonal antibody for the therapy of solid tumors and other disorders. Cancer Res 57:4593–4599
86. Ahmad ZA, Yeap SK, Ali AM, Ho WY, Alitheen NBM, Hamid M (2012) ScFv antibody: principles and clinical application. Clin Dev Immunol 2012. https://doi.org/10.1155/2012/980250

87. Kaur S, Venktaraman G, Jain M, Senapati S, Garg PK, Batra SK (2012) Recent trends in antibody-based oncologic imaging Sukhwinder. Cancer Lett 315:97–111. https://doi.org/10.1016/j.canlet.2011.10.017.Recent
88. Hust M, Jostock T, Menzel C, Voedisch B, Mohr A, Brenneis M, Kirsch MI, Meier D, Dübel S (2007) Single chain Fab (scFab) fragment. BMC Biotechnol 7:1–15. https://doi.org/10.1186/1472-6750-7-14
89. Mazzocco C, Fracasso G, Germain-Genevois C, Dugot-Senant N, Figini M, Colombatti M, Grenier N, Couillaud F (2016) In vivo imaging of prostate cancer using an anti-PSMA scFv fragment as a probe. Sci Rep 6:1–10. https://doi.org/10.1038/srep23314
90. Akita EM, Nakai S (1993) Production and purification of Fab' fragments from chicken egg yolk immunoglobulin Y (IgY). J Immunol Methods 162:155–164. https://doi.org/10.1016/0022-1759(93)90380-P
91. Bruyns AM, De Jaeger G, De Neve M, De Wilde C, Van Montagu M, Depicker A (1996) Bacterial and plant-produced scFv proteins have similar antigen-binding properties. FEBS Lett 386:5–10. https://doi.org/10.1016/0014-5793(96)00372-9
92. Skerra A (1993) Bacterial expression of immunoglobulin fragments. Curr Opin Immunol 5: 256–262. https://doi.org/10.1016/0952-7915(93)90014-J
93. Franconi R, Roggero P, Pirazzi P, Arias FJ, Desiderio A, Bitti O, Pashkoulov D, Mattei B, Bracci L, Masenga V, Milne RG, Benvenuto E (1999) Functional expression in bacteria and plants of an scFv antibody fragment against tospoviruses. Immunotechnology 4:189–201. https://doi.org/10.1016/S1380-2933(98)00020-7
94. Bird RE, Hardman KD, Jacobson JW, Johnson S, Kaufman BM, Lee S-M, Lee T, Pope SH, Riordan GS, Whitlow M (n.d.) Single-chain antigen-binding proteins. www.sciencemag.org
95. Yusakul G, Sakamoto S, Pongkitwitoon B, Tanaka H, Morimoto S (2016) Effect of linker length between variable domains of single chain variable fragment antibody against daidzin on its reactivity. Biosci Biotechnol Biochem 80:1306–1312. https://doi.org/10.1080/09168451.2016.1156482
96. Miller KD, Weaver-Feldhaus J, Gray SA, Siegel RW, Feldhaus MJ (2005) Production, purification, and characterization of human scFv antibodies expressed in Saccharomyces cerevisiae, Pichia pastoris, and Escherichia coli. Protein Expr Purif 42:255–267. https://doi.org/10.1016/j.pep.2005.04.015
97. Weisser NE, Hall JC (2009) Applications of single-chain variable fragment antibodies in therapeutics and diagnostics. Biotechnol Adv 27:502–520. https://doi.org/10.1016/j.biotechadv.2009.04.004
98. Jefferis R (2016) Posttranslational modifications and the immunogenicity of biotherapeutics. J Immunol Res 2016. https://doi.org/10.1155/2016/5358272
99. Harmsen MM, De Haard HJ (2007) Properties, production, and applications of camelid single-domain antibody fragments. Appl Microbiol Biotechnol 77:13–22. https://doi.org/10.1007/s00253-007-1142-2
100. Muyldermans S (2001) Single domain camel antibodies: current status. Rev Mol Biotechnol 74:277–302. https://doi.org/10.1016/S1389-0352(01)00021-6
101. English H, Hong J, Ho M (2020) Ancient species offers contemporary therapeutics: an update on shark VNAR single domain antibody sequences, phage libraries and potential clinical applications. Antib Ther 3:1–9. https://doi.org/10.1093/ABT/TBAA001
102. Wu Y, Jiang S, Ying T (2017) Single-domain antibodies as therapeutics against human viral diseases. Front Immunol 8. https://doi.org/10.3389/fimmu.2017.01802
103. Arbabi Ghahroudi M, Desmyter A, Wyns L, Hamers R, Muyldermans S (1997) Selection and identification of single domain antibody fragments from camel heavy-chain antibodies. FEBS Lett 414:521–526. https://doi.org/10.1016/S0014-5793(97)01062-4
104. Leow CH, Cheng Q, Fischer K, McCarthy J (2018) The development of single domain antibodies for diagnostic and therapeutic applications. Antib Eng. https://doi.org/10.5772/intechopen.73324

105. Arbabi-Ghahroudi M (2017) Camelid single-domain antibodies: historical perspective and future outlook. Front Immunol 8:1–8. https://doi.org/10.3389/fimmu.2017.01589
106. Saccodossi N, de Simone EA, Leoni J (2012) Structural analysis of effector functions related motifs, complement activation and hemagglutinating activities in Lama glama heavy chain antibodies. Vet Immunol Immunopathol 145:323–331. https://doi.org/10.1016/j.vetimm.2011.12.001
107. Wesolowski J, Alzogaray V, Reyelt J, Unger M, Juarez K, Urrutia M, Cauerhff A, Danquah W, Rissiek B, Scheuplein F, Schwarz N, Adriouch S, Boyer O, Seman M, Licea A, Serreze DV, Goldbaum FA, Haag F, Koch-Nolte F (2009) Single domain antibodies: promising experimental and therapeutic tools in infection and immunity. Med Microbiol Immunol 198:157–174. https://doi.org/10.1007/s00430-009-0116-7
108. Muyldermans S (2013) Nanobodies: natural single-domain antibodies. Annu Rev Biochem 82:775–797. https://doi.org/10.1146/annurev-biochem-063011-092449
109. Hoey RJ, Eom H, Horn JR (2019) Structure and development of single domain antibodies as modules for therapeutics and diagnostics. Exp Biol Med 244:1568–1576. https://doi.org/10.1177/1535370219881129
110. Henry KA, Hussack G, Collins C, Zwaagstra JC, Tanha J, Mackenzie CR (2016) Isolation of TGF-β-neutralizing single-domain antibodies of predetermined epitope specificity using next-generation DNA sequencing. Protein Eng Des Sel 29:439–443. https://doi.org/10.1093/protein/gzw043
111. Bond CJ, Marsters JC, Sidhu SS (2003) Contributions of CDR3 to VHH domain stability and the design of monobody scaffolds for naive antibody libraries. J Mol Biol 332:643–655. https://doi.org/10.1016/S0022-2836(03)00967-7
112. Doyle PJ, Arbabi-Ghahroudi M, Gaudette N, Furzer G, Savard ME, Gleddie S, McLean MD, Mackenzie CR, Hall JC (2008) Cloning, expression, and characterization of a single-domain antibody fragment with affinity for 15-acetyl-deoxynivalenol. Mol Immunol 45:3703–3713. https://doi.org/10.1016/j.molimm.2008.06.005
113. Ewert S, Cambillau C, Conrath K, Plückthun A (2002) Biophysical properties of camelid VHH domains compared to those of human VH3 domains. Biochemistry 41:3628–3636. https://doi.org/10.1021/bi011239a
114. Harmsen MM, Ruuls RC, Nijman IJ, Niewold TA, Frenken LGJ, De Geus B (2000) Llama heavy-chain V regions consist of at least four distinct subfamilies revealing novel sequence features. Mol Immunol 37:579–590. https://doi.org/10.1016/S0161-5890(00)00081-X
115. Muyldermans S, Baral TN, Retamozzo VC, De Baetselier P, De Genst E, Kinne J, Leonhardt H, Magez S, Nguyen VK, Revets H, Rothbauer U, Stijlemans B, Tillib S, Wernery U, Wyns L, Hassanzadeh-Ghassabeh G, Saerens D (2009) Camelid immunoglobulins and nanobody technology. Vet Immunol Immunopathol 128:178–183. https://doi.org/10.1016/j.vetimm.2008.10.299
116. Bratkovič T (2010) Progress in phage display: evolution of the technique and its applications. Cell Mol Life Sci 67:749–767. https://doi.org/10.1007/s00018-009-0192-2
117. Koch-Nolte F, Adriouch S, Bannas P, Krebs C, Scheuplein F, Seman M, Haag F (2006) ADP-ribosylation of membrane proteins: unveiling the secrets of a crucial regulatory mechanism in mammalian cells. Ann Med 38:188–199. https://doi.org/10.1080/07853890600655499
118. Zimmermann H, Zebisch M, Sträter N (2012) Cellular function and molecular structure of ecto-nucleotidases. Purinergic Signal 8:437–502. https://doi.org/10.1007/s11302-012-9309-4
119. Resta R, Thompson LF (1997) T cell signalling through CD73. Cell Signal 9:131–139. https://doi.org/10.1016/S0898-6568(96)00132-5
120. Sträter N (2006) Ecto-5′-nucleotidase: structure function relationships. Purinergic Signal 2:343–350. https://doi.org/10.1007/s11302-006-9000-8
121. Partidá-Sánchez S, Rivero-Nava L, Shi G, Lund FE (2007) CD38: an ecto-enzyme at the crossroads of innate and adaptive immune responses. Adv Exp Med Biol 590:171–183

122. Jalkanen S, Salmi M (2001) Cell surface monoamine oxidases enzymes in.pdf. EMBO J 20: 3893–3901
123. Adriouch S, Hubert S, Pechberty S, Koch-Nolte F, Haag F, Seman M (2007) NAD + released during inflammation participates in T cell homeostasis by inducing ART2-mediated death of Naive T cells in vivo. J Immunol 179:186–194. https://doi.org/10.4049/jimmunol.179.1.186
124. Tsai SH, Kinoshita M, Kusu T, Kayama H, Okumura R, Ikeda K, Shimada Y, Takeda A, Yoshikawa S, Obata-Ninomiya K, Kurashima Y, Sato S, Umemoto E, Kiyono H, Karasuyama H, Takeda K (2015) The ectoenzyme E-NPP3 negatively regulates ATP-dependent chronic allergic responses by Basophils and mast cells. Immunity 42:279–293. https://doi.org/10.1016/j.immuni.2015.01.015
125. Salmi M, Jalkanen S (2005) Cell-surface enzymes in control of leukocyte trafficking. Nat Rev Immunol 5:760–771. https://doi.org/10.1038/nri1705
126. Koch-Nolte F, Reyelt J, Schößow B, Schwarz N, Scheuplein F, Rothenburg S, Haag F, Alzogaray V, Cauerhff A, Goldbaum FA (2007) Single domain antibodies from llama effectively and specifically block T cell ecto-ADP-ribosyltransferase ART2.2 in vivo. FASEB J 21:3490–3498. https://doi.org/10.1096/fj.07-8661com
127. Rossotti MA, Bélanger K, Henry KA, Tanha J (2021) Immunogenicity and humanization of single-domain antibodies. FEBS J. https://doi.org/10.1111/febs.15809
128. Scheuplein F, Rissiek B, Driver JP, Chen YG, Koch-Nolte F, Serreze DV (2010) A recombinant heavy chain antibody approach blocks ART2 mediated deletion of an iNKT cell population that upon activation inhibits autoimmune diabetes. J Autoimmun 34:145–154. https://doi.org/10.1016/j.jaut.2009.08.012
129. Van De Donk NWCJ, Richardson PG, Malavasi F (2018) CD38 antibodies in multiple myeloma: back to the future. Blood 131:13–29. https://doi.org/10.1182/blood-2017-06-740944
130. Li T, Qi S, Unger M, Hou YN, Deng QW, Liu J, Lam CMC, Wang XW, Xin D, Zhang P, Koch-Nolte F, Hao Q, Zhang H, Lee HC, Zhao YJ (2016) Immuno-targeting the multifunctional CD38 using nanobody. Sci Rep 6:1–11. https://doi.org/10.1038/srep27055
131. Fumey W, Koenigsdorf J, Kunick V, Menzel S, Schütze K, Unger M, Schriewer L, Haag F, Adam G, Oberle A, Binder M, Fliegert R, Guse A, Zhao YJ, Lee HC, Malavasi F, Goldbaum F, Van Hegelsom R, Stortelers C, Bannas P, Koch-Nolte F (2017) Nanobodies effectively modulate the enzymatic activity of CD38 and allow specific imaging of CD38+ tumors in mouse models in vivo. Sci Rep 7:1–13. https://doi.org/10.1038/s41598-017-14112-6
132. Kubota T, Niwa R, Satoh M, Akinaga S, Shitara K, Hanai N (2009) Engineered therapeutic antibodies with improved effector functions. Cancer Sci 100:1566–1572. https://doi.org/10.1111/j.1349-7006.2009.01222.x
133. Natsume A, Niwa R, Satoh M (2009) Improving effector functions of antibodies for cancer treatment: enhancing ADCC and CDC. Drug Des Develop Ther:7–16. https://doi.org/10.2147/dddt.s4378
134. Behar G, Sibéril S, Groulet A, Chames P, Pugnière M, Boix C, Sautès-Fridman C, Teillaud JL, Baty D (2008) Isolation and characterization of anti-FcγRIII (CD16) llama single-domain antibodies that activate natural killer cells. Protein Eng Des Sel 21:1–10. https://doi.org/10.1093/protein/gzm064
135. Cortez-Retamozo V, Backmann N, Senter PD, Wernery U, De Baetselier P, Muyldermans S, Revets H (2004) Efficient cancer therapy with a nanobody-based conjugate. Cancer Res 64: 2853–2857. https://doi.org/10.1158/0008-5472.CAN-03-3935
136. Dinarello CA (2000) Proinflammatory cytokines. Chest 118:503–508. https://doi.org/10.1378/chest.118.2.503
137. Schmitt H, Neurath MF, Atreya R (2021) Role of the IL23/IL17 pathway in Crohn's disease. Front Immunol 12. https://doi.org/10.3389/fimmu.2021.622934

138. Koeffler HP, Gasson J, Ranyard J, Souza L, Shepard M, Munker R (1987) Recombinant human TNF alpha stimulates production of granulocyte colony-stimulating factor. Blood 70: 55–59
139. Charles KA, Kulbe H, Soper R, Escorcio-Correia M, Lawrence T, Schultheis A, Chakravarty P, Thompson RG, Kollias G, Smyth JF, Balkwill FR, Hagemann T (2009) The tumor-promoting actions of TNF-α involve TNFR1 and IL-17 in ovarian cancer in mice and humans. J Clin Invest 119:3011–3023. https://doi.org/10.1172/JCI39065
140. Coppieters K, Dreier T, Silence K, De Haard H, Lauwereys M, Casteels P, Beirnaert E, Jonckheere H, Van De Wiele C, Staelens L, Hostens J, Revets H, Remaut E, Elewaut D, Rottiers P (2006) Formatted anti-tumor necrosis factor α VHH proteins derived from camelids show superior potency and targeting to inflamed joints in a murine model of collagen-induced arthritis. Arthritis Rheum 54:1856–1866. https://doi.org/10.1002/art.21827
141. Desmyter A, Spinelli S, Boutton C, Saunders M, Blachetot C, de Haard H, Denecker G, Van Roy M, Cambillau C, Rommelaere H (2017) Neutralization of human interleukin 23 by multivalent nanobodies explained by the structure of cytokine–nanobody complex. Front Immunol 8. https://doi.org/10.3389/fimmu.2017.00884
142. Wucherpfennig KW (2001) Mechanisms for the induction of autoimmunity by infectious agents. J Clin Invest 108:1097–1104. https://doi.org/10.1172/JCI200114235
143. Stijlemans B, Conrath K, Cortez-Retamozo V, Van Xong H, Wyns L, Senter P, Revets H, De Baetselier P, Muyldermans S, Magez S (2004) Efficient targeting of conserved cryptic epitopes of infectious agents by single domain antibodies: African trypanosomes as paradigm. J Biol Chem 279:1256–1261. https://doi.org/10.1074/jbc.M307341200
144. Smith PA, Contreras JR, Larenas JJ, Aguillon JC, Garces LH, Perez B, Fryer JL (1997) Immunization with bacterial antigens: piscirickettsiosis. Dev Biol Stand 90:161–166. http://europepmc.org/abstract/MED/9270845
145. Schoepfer AM, Schaffer T, Seibold-Schmid B, Müller S, Seibold F (2008) Antibodies to flagellin indicate reactivity to bacterial antigens in IBS patients. Neurogastroenterol Motil 20: 1110–1118. https://doi.org/10.1111/j.1365-2982.2008.01166.x
146. Farache J, Koren I, Milo I, Gurevich I, Kim KW, Zigmond E, Furtado GC, Lira SA, Shakhar G (2013) Luminal bacteria recruit CD103+ dendritic cells into the intestinal epithelium to sample bacterial antigens for presentation. Immunity 38:581–595. https://doi.org/10.1016/j.immuni.2013.01.009
147. El Khattabi M, Adams H, Heezius E, Hermans P, Detmers F, Maassen B, Van Der Ley P, Tommassen J, Verrips T, Stam J (2006) Llama single-chain antibody that blocks lipopolysaccharide binding and signaling: prospects for therapeutic applications. Clin Vaccine Immunol 13:1079–1086. https://doi.org/10.1128/CVI.00107-06
148. Harmsen MM, Van Solt CB, Van Zijderveld-Van Bemmel AM, Niewold TA, Van Zijderveld FG (2006) Selection and optimization of proteolytically stable llama single-domain antibody fragments for oral immunotherapy. Appl Microbiol Biotechnol 72:544–551. https://doi.org/10.1007/s00253-005-0300-7
149. Szynol A, De Soet JJ, Sieben-Van Tuyl E, Bos JW, Frenken LG (2004) Bactericidal effects of a fusion protein of llama heavy-chain antibodies coupled to glucose oxidase on oral bacteria. Antimicrob Agents Chemother 48:3390–3395. https://doi.org/10.1128/AAC.48.9.3390-3395.2004
150. Nyambi PN, Mbah HA, Burda S, Williams C, Gorny MK, Nádas A, Zolla-Pazner S (2000) Conserved and exposed epitopes on intact, native, primary human immunodeficiency virus type 1 virions of group M. J Virol 74:7096–7107. https://doi.org/10.1128/jvi.74.15.7096-7107.2000
151. Modis Y, Ogata S, Clements D, Harrison SC (2005) Variable surface epitopes in the crystal structure of dengue virus type 3 envelope glycoprotein. J Virol 79:1223–1231. https://doi.org/10.1128/jvi.79.2.1223-1231.2005

152. Kanai R, Kar K, Anthony K, Gould LH, Ledizet M, Fikrig E, Marasco WA, Koski RA, Modis Y (2006) Crystal structure of West Nile virus envelope glycoprotein reveals viral surface epitopes. J Virol 80:11000–11008. https://doi.org/10.1128/jvi.01735-06
153. Forsman A, Beirnaert E, Aasa-Chapman MMI, Hoorelbeke B, Hijazi K, Koh W, Tack V, Szynol A, Kelly C, McKnight Á, Verrips T, de Haard H, Weiss RA (2008) Llama antibody fragments with cross-subtype human immunodeficiency virus type 1 (HIV-1)-neutralizing properties and high affinity for HIV-1 gp120. J Virol 82:12069–12081. https://doi.org/10.1128/jvi.01379-08
154. Garaicoechea L, Olichon A, Marcoppido G, Wigdorovitz A, Mozgovoj M, Saif L, Surrey T, Parreño V (2008) Llama-derived single-chain antibody fragments directed to rotavirus VP6 protein possess broad neutralizing activity in vitro and confer protection against diarrhea in mice. J Virol 82:9753–9764. https://doi.org/10.1128/jvi.00436-08
155. Pant N, Hultberg A, Zhao Y, Svensson L, Pan-Hammarström Q, Johansen K, Pouwels PH, Ruggeri FM, Hermans P, Frenken L, Borén T, Marcotte H, Hammarström L (2006) Lactobacilli expressing variable domain of llama heavy-chain antibody fragments (lactobodies) confer protection against rotavirus-induced diarrhea. J Infect Dis 194:1580–1588. https://doi.org/10.1086/508747
156. Weiner GJ (2015) Building better monoclonal antibody-based therapeutics. Nat Rev Cancer 15:361–370. https://doi.org/10.1038/nrc3930
157. Grainger DW (2004) Controlled-release and local delivery of therapeutic antibodies. Expert Opin Biol Ther 4:1029–1044. https://doi.org/10.1517/14712598.4.7.1029
158. Idusogie EE, Wong PY, Presta LG, Gazzano-Santoro H, Totpal K, Ultsch M, Mulkerrin MG (2001) Engineered antibodies with increased activity to recruit complement. J Immunol 166: 2571–2575. https://doi.org/10.4049/jimmunol.166.4.2571
159. Shields RL, Namenuk AK, Hong K, Meng YG, Rae J, Briggs J, Xie D, Lai J, Stadlen A, Li B, Fox JA, Presta LG (2001) High resolution mapping of the binding site on human IgG1 for FcγRI, FcγRII, FcγRIII, and FcRn and design of IgG1 variants with improved binding to the FcγR. J Biol Chem 276:6591–6604. https://doi.org/10.1074/jbc.M009483200
160. Courtois F, Agrawal NJ, Lauer TM, Trout BL (2016) Rational design of therapeutic mAbs against aggregation through protein engineering and incorporation of glycosylation motifs applied to bevacizumab. MAbs 8:99–112. https://doi.org/10.1080/19420862.2015.1112477
161. Bazin R, Boucher G, Monier G, Chevrier MC, Verrette S, Broly H, Lemieux R (1994) Use of hu-IgG-SCID mice to evaluate the in vivo stability of human monoclonal IgG antibodies. J Immunol Methods 172:209–217. https://doi.org/10.1016/0022-1759(94)90108-2
162. Kroon DJ, Baldwin-Ferro A, Lalan P (1992) Identification of sites of degradation in a therapeutic monoclonal antibody by peptide mapping. Pharm Res An Off J Am Assoc Pharm Sci 9:1386–1393. https://doi.org/10.1023/A:1015894409623
163. Kamen MD (1970) controlled deamidation of peptides and proteins: an experimental hazard and a possible biological timer. PNAS 66:753–757. https://doi.org/10.1016/S0047-6374(01)00363-3
164. Huang L, Lu J, Wroblewski VJ, Beals JM, Riggin RM (2005) In vivo deamidation characterization of monoclonal antibody by LC/MS/MS. Anal Chem 77:1432–1439. https://doi.org/10.1021/ac0494174
165. Griffiths HR (2000) Antioxidants and protein oxidation. Free Radic Res 33:1–25
166. Lowe D, Dudgeon K, Rouet R, Schofield P, Jermutus L, Christ D (2011) Aggregation, stability, and formulation of human antibody therapeutics. Adv Protein Chem Struct Biol:41–61. https://doi.org/10.1016/B978-0-12-386483-3.00004-5

167. Kontermann RE, Brinkmann U (2015) Bispecific antibodies. Drug Discov Today 20:838–847. https://doi.org/10.1016/j.drudis.2015.02.008
168. Bathula NV, Bommadevara H, Hayes JM (2021) Nanobodies: the future of antibody-based immune therapeutics. Cancer Biother Radiopharm 36:109–122. https://doi.org/10.1089/cbr.2020.3941
169. Singh DB, Tripathi T (2020) Frontiers in protein structure, function, and dynamics. Springer Nature, Singapore
170. A neutralizing monoclonal antibody for hospitalized patients with covid-19. N Engl J Med (2021) 384:905–914. https://doi.org/10.1056/nejmoa2033130

Strategies for Formulation and Systemic Delivery of Therapeutic Proteins

6

Pallavi Shah and Basant

Abstract

Technological advancements in pharmaceutical biotechnology and bioengineering have made possible the use of a myriad of proteins and peptides as efficient and ideal drug candidates, such as growth factors, hormones, antibodies, and vaccines. Therapeutic proteins possess several advantages over their synthetic counterparts, especially concerning high specificity and activity, and thus have seen a significant rise in clinical applications in recent times. Despite being significant clinically, physiochemical and enzymatic destabilization of therapeutic proteins limits their successful use. The delivery systems used for therapeutic proteins often suffer from several limitations. The invasive routes like the parenteral route show poor protein stability in the systemic circulation. In contrast, the non-invasive routes like oral, transdermal, nasal, pulmonary, ocular, and rectal delivery systems face numerous physicochemical and biological barriers like enzymatic degradation, harsh pH environment, difficult transport across membranes, and poor bioavailability. This chapter highlights the different strategies for developing formulations to overcome these drawbacks for successful delivery and prolong the shelf-life of proteins in systemic circulation delivered via different delivery routes. It provides a comprehensive update on recent advances in nanotechnology and different methodologies adopted for improving both the delivery and efficacy of therapeutic proteins.

P. Shah (✉) · Basant
Division of Veterinary Biotechnology, ICAR-Indian Veterinary Research Institute, Izatnagar, Bareilly, Uttar Pradesh, India

D. B. Singh, T. Tripathi (eds.), *Protein-based Therapeutics*,
https://doi.org/10.1007/978-981-19-8249-1_6

Keywords

Therapeutic proteins · Invasive delivery · Non-invasive delivery · Nanotechnology

6.1 Introduction

Current advancements in biotechnology have made it possible to produce therapeutic proteins and peptides on a large scale to use in disease conditions. Therapeutic proteins and peptides have emerged as a growing class of drugs, constituting the majority of clinical biopharmaceuticals released recently [1]. A better understanding of disease pathologies at the molecular level and advancements in the field of biotechnology have made it possible to design and produce more specific, potent, long-lasting, and non-toxic therapeutic proteins [2]. These attributes make them better than the available low molecular weight chemical drugs, often showing non-specific targeting and producing harmful metabolites.

The therapeutic proteins and peptides can be classified into various categories based on their function, such as protein therapeutics with regulatory or enzymatic activity, protein therapeutics with high specificity for their targets, protein vaccines, and diagnostics based on therapeutic proteins [3]. These proteins have been used for the treatment of a wide range of diseases, including cancer, diabetes, immunoregulation, hepatitis, and genetic disorders [3–5]. However, even with these pros, the therapeutic proteins suffer from certain drawbacks, including poor permeability across biological barriers due to high molecular weight, instability, short half-life, and immunogenicity [6]. Several strategies have been designed to overcome these drawbacks, and the majority of them focus on developing suitable delivery systems for the administration of therapeutic proteins in their intact form without any kind of degradation accompanied by controlled release for improved efficacy and safety [7]. The therapeutic proteins can be delivered by invasive and non-invasive delivery systems, both having their set of advantages and limitations. Many methods involve modification of the protein structure, either by attachment of moieties covalently or by mutation, for example, generation of fusion proteins with albumin, Fc portion [8, 9]. Conjugation with polymers like polyethylene glycol (PEG) and other alternative molecules offers decent delivery and safety assurance. The recent use of liposomes and different polymer-based nanocarriers have further opened up new avenues for overcoming the existing limitations. This chapter sheds light on the current strategies for formulating the systematic delivery of therapeutic proteins. We discuss the different methods currently used to modify proteins and highlight the different modes of delivery, emphasizing how they influence the specificity, immunogenicity, and stability of the therapeutic proteins.

6.2 General Challenges in the Delivery of Therapeutic Proteins

The major drawbacks of therapeutic protein delivery include their short half-life, fragile structures, and distinct physicochemical properties. The invasive delivery systems involve parenteral administration through intravenous (IV), intramuscular (IM), and/or subcutaneous (SC) routes, which often require trained personnel or hospitalization [10]. Thus, recent times have focused on non-invasive delivery systems, which are more convenient and often guarantee better and sustained release of the therapeutics. Delivery systems are associated with challenges that range from affecting the therapeutic efficacy of the protein to targeted site-specific delivery for the desired effect. These challenges have made the controlled and sustained release-based delivery systems an urgent need. The steady-state delivery systems ensure efficient site-specific delivery of therapeutic proteins in their intact biological forms and have gained considerable attention in recent years (Fig. 6.1).

6.2.1 Chemical and Physical Instability

The biological activity of therapeutic proteins is often hampered by their chemical and physical instabilities. Peptides and protein drugs have to undergo a series of degradative steps based on the route of administration. Chemical degradation involves covalent modification of the primary protein structure through either bond formation or cleavage as in oxidation and deamidation reactions. On the other hand, physical degradation involves changes in higher three-dimensional (3D) conformational integrity by denaturation, aggregation, or precipitation. Both forms of degradation do not occur independently, but one may lead to the other [11]. Non-invasive methods like oral administration often lead to enzymatic degradation of proteins in the gastrointestinal (GI) tract. In contrast, when given parenterally, they get rapidly eliminated from the circulation by renal filtration, enzymatic degradation, or are taken up by the reticuloendothelial system [12]. Thus, efficient sustained release delivery systems ensure a longer half-life and stability of protein drugs.

6.2.2 Biopharmaceutical Challenges

The 3D structural integrity of proteins highly influences their therapeutic activity, which in turn, is often affected by external variables like temperature, pH, and excipient impurities. Determining structural changes in therapeutic protein formulations poses a major challenge for clinical use. Further, cases of protein drugs being immunogenic have been reported, and some generations of neutralizing antibodies have led to a loss of therapeutic effect. Parenteral administration through injections (i.e., IV, IM, or SC routes) often suffers from enzymatic degradation and rapid elimination, whereas non-invasive administration like oral administration has even lower bioavailability due to the acidic pH of the GI tract as well as the digestion

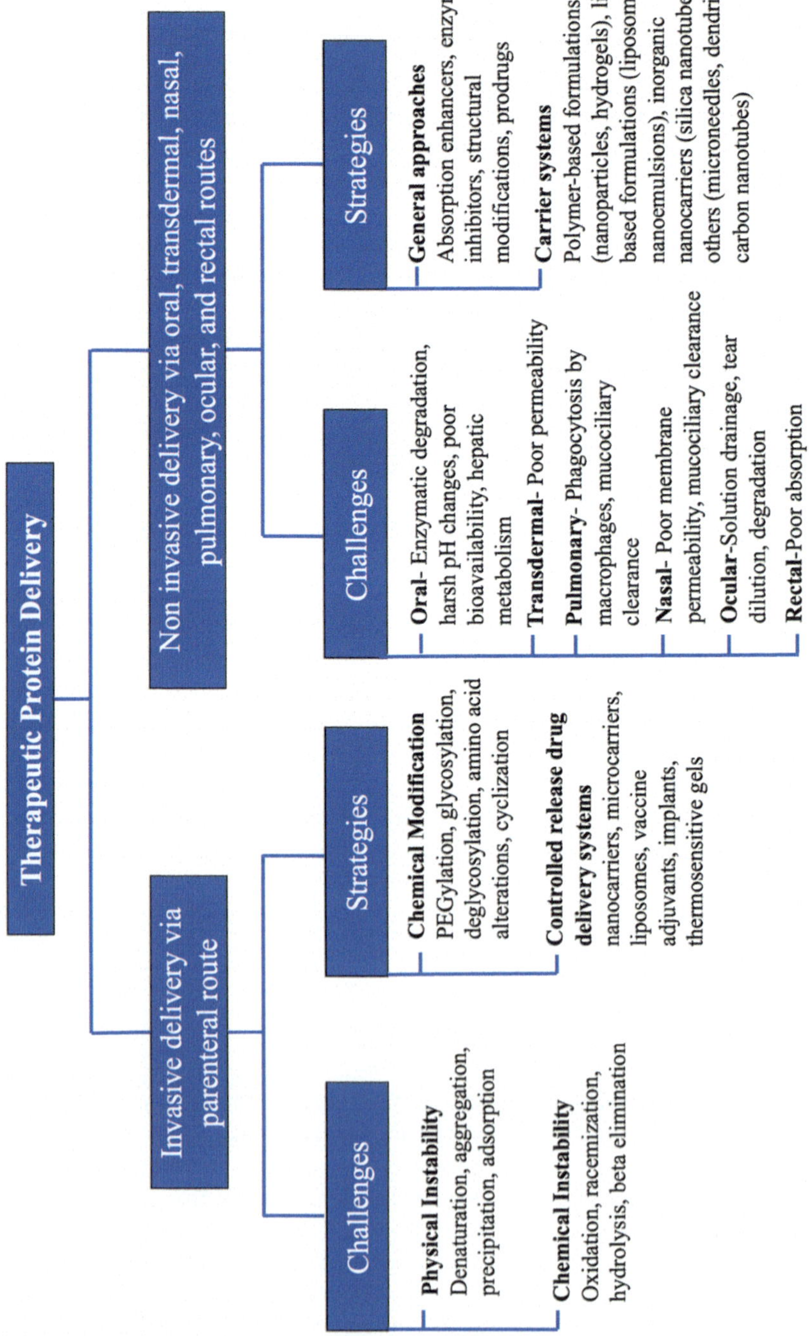

Fig. 6.1 Challenges and strategies to overcome the limitations for the delivery of therapeutic proteins via different routes of administration

by proteolytic enzymes in the intestinal lumen. The large molecular weight and hydrophilic nature of the protein drugs further make their transport across the intestinal cell lining difficult [13].

6.3 Invasive Delivery Systems and Delivery Challenges

Direct delivery of a drug into the bloodstream through parenteral routes like injections (i.e., IV, IM, or SC routes) are often considered invasive delivery systems. Once administered via these routes, the protein drug must reach its target after crossing several barriers. Invasive delivery systems often suffer from certain limitations like pain at the site of injection, undesirable pharmacokinetics, the requirement of trained personnel, and sometimes visits to hospitals. It is time-consuming and is further affected by the site of administration, dosage, muscular activity, and certain pathological conditions. Hence, while designing proteins for invasive delivery, it is essential to consider the effect of enzymatic degradation, protein solubility, reticuloendothelial system, targeted delivery, and cellular uptake. Presently, a number of formulations are available for invasive delivery, as mentioned in Table 6.1.

Successful formulation of a therapeutic protein for parenteral delivery involves considering both the pharmacokinetic and pharmacodynamic aspects of the drug. The pharmacokinetic aspect deals with the bioavailability, absorption, distribution, metabolism, and excretion of the drug. When therapeutic proteins are injected subcutaneously, the bioavailability of the drug is reduced due to proteolytic degradation at the site of injection. Also, bioavailability is not 100% for the intravenous route as proteolytic digestion starts in the blood itself. Apart from this, a series of factors like administration site, delivery system, injection depth, drug concentration for injection, injected volume, formulation-related factors, and patient-specific factors also affect bioavailability. Pharmacodynamics of therapeutic proteins include observing effects, and the time it takes to bring about that effect at a particular dose. Plasma proteins often interact with the injected therapeutic proteins and affect their pharmacodynamics. Several pharmacokinetic and pharmacodynamic models have been proposed to evaluate the concentration and response curves, protein binding, and immunogenicity of therapeutic proteins [14].

6.4 Non-Invasive Delivery Systems and Delivery Challenges

Presently, non-invasive routes for therapeutic protein delivery have gained much attention because of their decreased medical cost and better patient compliance than invasive routes. The past few decades have seen several non-invasive routes of protein drug administration, including oral, nasal, transdermal, pulmonary, rectal, and vaginal routes. These alternate routes can overcome the pain, needle phobia, and risk of infection associated with the parenteral route. There are significant biological and physicochemical barriers to be overcome while formulating protein-based drugs

Table 6.1 List of FDA-approved and commercially available therapeutic proteins and peptides delivered via parenteral route [7, 14, 15]

Brand name	Component	Route of administration	Treatment for/used as
Asparlas™	Calaspargase pegol	IV	Acute lymphoblastic leukemia
Crysvita™	Burosumab	IV	X-linked dominant hypophosphatemic rickets
Revcovi™	Elapegademase	IM	Adenosine deaminase severe combined immunodeficiency
Libtayo™	Cemiplimab	IV	Cutaneous squamous cell carcinoma
Aimovig™	Erenumab	SC	Migraine prevention
Gamifant™	Emapalumab-lzsg	IV	Hemophagocytic lymphohistiocytosis
Ajovy™	Fremanezumab-vfrm	SC	Migraine prevention
Trogarzo™	Ibalizumab-uiyk	IV	Multidrug-resistant HIV-1
Emgality™	Galcanezumab-gnlm	SC	Migraine prevention
Takhzyro™	Lanadelumab	SC	Hereditary angioedema attacks
Lumoxiti™	Moxetumomab pasudotox	IV	Relapsed or refractory hairy cell leukemia
Poteligeo™	Mogamulizumab-kpkc	IV	Mycosis fungoides and Sézary syndrome
Palynziq™	Pegvaliase-pqpz	SC	Pegylated enzyme Phenylketonuria
Ilumya™	Tildrakizumab	SC	Moderate-to-severe plaque psoriasis
Elzonris™	Tagraxofusp-erzs	IV	Blastic plasmacytoid dendritic cell neoplasm
Ultomiris™	Ravulizumab	IV	Paroxysmal nocturnal hemoglobinuria and atypical hemolytic uremic syndrome
Kineret®	IL-1Ra	SC	T2DM, inflammatory syndromes, RA
Lantus	Insulin glargine	SC	T1DM, T2DM
Enbrel	Etanercept	SC	RA, psoriatic arthritis, plaque psoriasis, ankylosing spondylitis
Sandimmune	Cyclosporine	Oral, IV	Prophylaxis, solid organ rejection
Neulasta	Pegfilgrastim	SC	Neutropenia
DDAVP®	Desmopressin	IV, IM, SC, Intranasal	Nocturnal enuresis
Angiomax®	Bivalirudin	IV	Anticoagulant
Victoza®	Liraglutide	SC	T2DM
Sandostatin®	Octreotide acetate	IV, SC, IM	Acromegaly, gigantism

to be delivered by these routes. Physicochemical properties like large molecular weight, poor membrane permeability, loss by enzymatic degradation, and physicochemical instability of therapeutic proteins often pose a challenge for this route of delivery [16]. Further, the surface charge of therapeutic protein is another physicochemical drug property that originates from the protein's amino acid sequence and the pH of its surroundings. It often arises due to isomerization, deamination, or post-translational modifications (PTMs) of amino acids leading to a change in the net charge and forming weak conjugate acids and bases [17]. The charge on the protein surface influences the interaction of protein drug with cell surface molecules or tissue components and subsequently affect their absorption, distribution, and elimination. Exposure to enzymatic degradation and physicochemical destabilization can occur at any point from the preparation of the formulation to systemic exposure after intake.

The biological barriers to non-invasive delivery include pH of the biological environments, enzymatic barriers, and mucosal barriers. The protein drugs are highly destabilized in the extreme acidic pH of the stomach (pH 1–3), whereas the alkaline pH in the intestines is less damaging [18]. Enzymatic barriers often degrade the protein drugs and lead to their poor bioavailability. Protease activity is highest in the small intestine, whereas it is comparatively lower in the colon when taken orally. This led to the development of colon-targeted drugs that last longer and have more time for absorption [19]. Besides these barriers, mucosal barriers like mucus and epithelial cell layers are the major absorption barrier against non-invasive therapeutic protein administration. Therefore, a better understanding of barriers may help design more efficient formulations for non-invasive delivery. Some commonly used formulations delivered via non-invasive routes are listed in Table 6.2.

6.5 Processing Strategies to Increase the Stability of Therapeutic Proteins

Even after having immense potential as therapeutics, protein drugs often suffer from many complications related to the preparation of their formulations. Contrary to their synthetic small drug counterparts, protein drugs possess complex secondary, tertiary, or quaternary structures along with various side chains. This structural integrity, when lost, can lead to loss of drug action as well. Thus, methods must be devised to maintain the 3D complexity of the protein drug for their effective activity.

6.5.1 Lyophilization

Lyophilization is among the most common methods for producing dry powder forms of protein drugs at low temperatures. Lyophilization consists of freeze-drying, where the sample is first frozen, followed by removal of frozen water by sublimation (also known as primary drying). Once the moisture content has been reduced to around 10%, secondary drying ensures removing the rest of the moisture content. This method generates solid-dried protein drugs with longer shelf lives. Despite its broad applicability and ease of use, this process often induces a certain variety of stresses, including pH changes, removal of protein hydration shell, and formation of dendritic ice crystals that may have derogatory effects on protein's structural integrity [20]. Cryoprotectants such as sugars, polymers, surfactants, and amino acids are

Table 6.2 List of commercially available protein and peptides delivered via non-invasive routes [7, 16]

Brand name	Component	Route of administration	Treatment for/used as
Lucents®	Ranibizumab	Ocular	Macular edema, myopic choroidal neovascularization (mCNV)
FluMist® Quadrivalent	Vaccine	Nasal	Influenza
Antepan®	Protirelin	Nasal	Hypothyroidism and acromegaly
Sandimmune®	Cyclosporine A	Oral	Immunosuppressant
Minirin®	Desmopressin	Oral, nasal	Diabetes insipidus and nocturia
Colomycin®	Colistin	Oral	Antimicrobial for Gram-negative bacteria-driven gut infection
Cachexon®	Glutathione	Oral	AIDS-related cachexia
Cytorest®	Cytochrome C	Oral	Leucopenia
Anginovag®	Tyrothricin	Oral	Pharyngitis
Ceredist® OD	Taltirelin	Oral	Spinocerebellar ataxia
Vancocin®	Vancomycin	Oral	Antimicrobial for oral infections, diarrhea
Synarel®	Nafarelin	Nasal	Endometriosis
Suprecur®	Buserelin	Nasal	Prostate cancer, endometriosis
Suprifact™	Buserelin	Nasal	Prostate cancer, endometriosis
Oral-Lyn™	Insulin	Buccal	Diabetes mellitus
Syntocinin™	Oxytocin	Nasal	Inducing uterine contractions
Desmospray®	Desmopressin	Nasal	Diabetes insipidus and nocturia
Fortical®	Salmon calcitonin	Nasal	Osteoporosis
Miacalcin®	Salmon calcitonin	Nasal	Osteoporosis
Kryptocur®	LHRH	Nasal	Cryptorchism
Eylea®	Aflibercept	Ocular	Macular edema

added to the preparation to overcome these stresses. Cycle conditions are also varied, and modifications can be made to drying rate, temperature, and thermal treatment conditions for better stability of freeze-dried protein drugs [21].

6.5.2 Spray Freeze-Drying

Water or moisture in any form is detrimental to the stability of several pharmaceutical products, and hence spray-freeze drying method has been devised to remove water. This method involves the same three basic steps of lyophilization, i.e., (i) freezing of the liquid sample containing the macromolecule, (ii) primary drying by sublimation of ice formed, and (iii) secondary drying, whereby the rest of the moisture is removed by desorption. The first step involves spraying the solution containing the protein drug into a vessel containing cryogenic liquid such as nitrogen, oxygen, or argon, which results in the freezing of the liquid due to the low boiling point of the cryogenic liquid. Spray freeze drying significantly improves the sample's shelf life and is often the method of choice for preparing protein drug formulations [22].

6.5.3 Other Common Methods

Supercritical fluids (SCFs), jet milling, and fluid bed spray coating methods are other less frequently used methods for processing therapeutic proteins. SCFs are often used in various analytical and extraction methods for the powder preparation of pharmaceuticals meant for inhalation. SCFs have properties of both gases and liquids above their critical point at specific temperatures and pressures. SCFs possess several advantages: high dissolving power, lower viscosity levels compared to other liquids, and higher diffusivity allowing a high mass transfer. Jet milling is used for particle size reduction using interparticle collisions and abrasion to produce small particles (1 to 20 μm) and for processing coarse peptides. Advances in fluidized-bed spray coating systems help in also coating particles of very small size. Fluidized-bed spray coating was applied to recombinant human deoxyribonuclease (rhDNase), and the coating was examined via scanning electron microscopy. It was found that the protein showed some aggregation, but with the use of calcium ions as stabilizers, this drawback was significantly reduced. This suggests that with proper stabilizers, fluid bed spray coating can be successfully used to coat therapeutic proteins to make them more stable [13].

6.6 Formulation Strategies for Overcoming Protein Drug Delivery Challenges

6.6.1 Strategies for Enhanced Delivery and Stability of Therapeutic Proteins Via Invasive Routes

Different strategies are being adapted to overcome the rapid clearance and drug instability often encountered during invasive drug delivery of therapeutic proteins. The two most common approaches to overcome these challenges include chemical modifications and colloidal carrier systems [23].

6.6.1.1 Chemical Modifications

The therapeutic potential of proteins can be enhanced by chemical modifications, including protein PEGylation, glycosylation, deglycosylation, amino acid alterations, and cyclization [14]. The most common chemical modification is the formulation of protein polymer conjugates that involves the chemical covalent conjugation of polymers like PEG and PEG analogues. Several natural and synthetic, biodegradable, and non-biodegradable polymers are used to conjugate and

Table 6.3 Polymers used for formulating therapeutic drugs for different delivery routes [16]

Biodegradable synthetic polymers	Biodegradable natural polymers	Non-biodegradable synthetic and semisynthetic polymers
Polyesters—Poly(glycolic acid) (PGA), poly(lactic acid) (PLA), poly(hydroxy butyrate) (PHB), poly (ε-caprolactone) (PCL), poly (α-malic acid) (PMA), poly (dioxanones), poly(lactic-co-glycolic acid) (PLGA)	Polysaccharides—Chitosan agarose, alginate, cyclodextrins, dextran, hyaluronic acid, pectin	Silica derivatives—Colloidal silica, polydimethylsiloxane (PDMS), polymethacrylates (PMA), poly (methyl methacrylate) (PMMA), polyhydro(ethylmethacrylate) (PEMA)
Polyamides—Poly(imino carbonates) (PIC), polyamino acids (PAAs)	Proteins—Gelatin, albumin, collagen	Cellulose derivatives—Carboxymethyl cellulose (CMC), cellulose acetate (CA), cellulose acetate propionate (CAP), ethyl cellulose (EC), hydroxypropyl methylcellulose (HMC)
Polyanhydrides—Poly (sebacic acid) (PSA), poly (adipic acid) (PAA), poly (terephthalic acid) (PTA)		Others—Polyvinyl pyrrolidone (PVP), ethyl vinyl acetate poloxamers, poloxamines
Polymers based on phosphorous—Polyphosphates, polyphosphonates polyphosphazenes		
Others—Polyurethane, polyacetals, poly (cyanoacrylates), polydihydropyrans, polyortho esters		

synthesize nanocarriers for therapeutic proteins (Table 6.3). Protein polymer conjugates impart essential properties to protein drugs like prolonged protected retention in circulation, low immunogenicity, and enhanced stability in different physiological conditions [24]. These properties arise as a result of altered solubility, molecular weight, size, and steric hinderance of the PEGylated proteins [7]. PEGylation avoids rapid degradation of protein drugs as reported for recombinant human thyroid-stimulating hormone, recombinant human growth hormone, and recombinant human tissue inhibitor of metalloproteinases-1 (TIMP-1) [25–27]. The half-life of these proteins increased significantly in systemic circulation after PEGylation. The pharmacokinetic and pharmacodynamic properties of the enzyme glucose oxidase have also been studied after glycosylation and deglycosylation. Glycosylation of glucose oxidase enhanced the receptor-mediated endocytosis of the enzyme by hepatic glucose receptors, whereas deglycosylation improved the half-life of the enzyme by ten-fold [28]. Proteins can further be modified by adding mannose and galactose to increase their receptor-mediated endocytosis.

6.6.1.2 Controlled Release Drug Delivery Systems

Several controlled-release injectable particulate carrier systems have been developed for the invasive delivery of therapeutic proteins, which include nanocarriers, microcarriers, liposomes, vaccine adjuvants, implants, thermosensitive gels, and customized particulate systems for pulsatile protein delivery (Fig. 6.2) [14, 29, 30]. These systems ensure the release of the required concentration of the therapeutic protein at a specific site, at the desired time, and the desired rate with minimum exposure to the systemic bioenvironment, reducing unwanted side effects like developing immunogenicity (Table 6.4).

Microspheres

Biodegradable polymer-based microspheres offer the most common and extensively used controlled drug release systems that can encapsulate therapeutic proteins and are easily administered parenterally. Ipsen Biotech, France, was the first company to release the microsphere formulation of luteinizing hormone-releasing hormone (LHRH) agonist [D-Trp6]-LHRH for the treatment of prostate cancer [54]. Currently, several microsphere-peptide formulations are available for the treatment of different diseases (Table 6.4). Most microsphere preparations are available in sterile powder form that must be reconstituted with a diluent containing a suspending agent and used immediately for injections. PLGA, PEG, and PEG-PLGA biodegradable polymers are often the materials of choice for synthesizing microspheres for protein and peptide drug release.

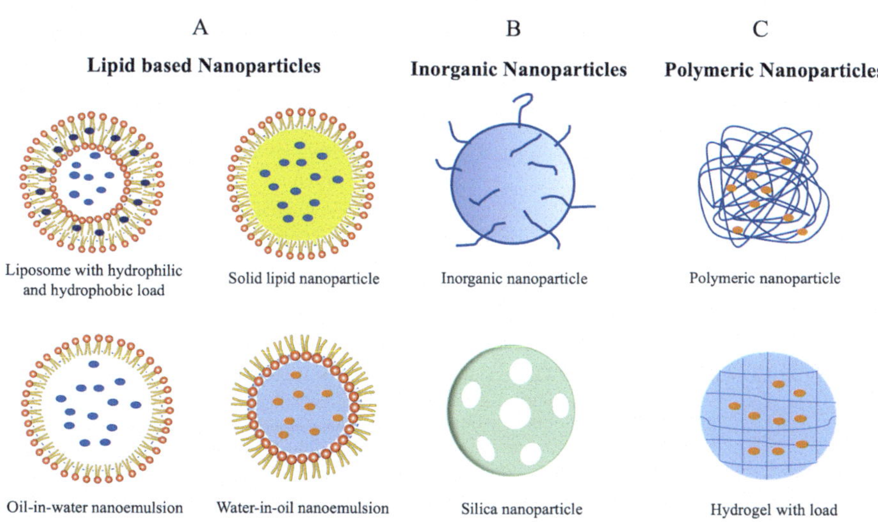

Fig. 6.2 Types of nanoparticles devised for controlled release of therapeutic proteins or peptides for invasive and non-invasive delivery routes. (**a**) Lipid-based nanoparticles (**b**) Inorganic nanoparticles (**c**) Polymeric nanoparticles

Table 6.4 Examples of controlled release colloidal carrier systems for enhanced delivery and stability of therapeutic proteins delivered via invasive routes

S. No.	Protein/peptide	Carrier system	Properties imparted	References
1.	Tetanus toxoid (Antigen)	Chitosan microspheres	Successful in maintaining antibody titer for a prolonged period	[31]
2.	Diphtheria toxoid (Antigen)	Poly-(ε-caprolactone) nanoparticles	Mucosal vaccine delivery system	[32]
3.	Hepatitis B core antigen (Antigen)	PLGA microspheres	Long-acting vaccine; gelatin used as a stabilizer	[33]
4.	Calcitonin (Hormone)	*In situ* biodegradable implant	PLA polymer	[34]
5.	Trypsin (Enzyme)	Heparin/Protamine-based approach	Delivery of enzyme drugs without associated toxic effects	[35]
6.	Protein C (Enzyme)	Monomethoxypoly (ethylene-oxide)-PLA nanoparticles	Improved loading efficiency	[36]
7.	Parathyroid (Hormone)	PLGA microspheres	Improved release profile and local delivery to bone	[37]
8.	Insulin-like growth factor (Cytokine)	PLGA microparticles	More stable, osteogenesis promoter	[38]
9.	Tumor necrosis factor (Cytokine)	PEGylated polycyanoacrylate nanoparticles	Controlled release	[39]
10.	Vascular endothelial growth factor/VEGF (Cytokine)	PLGA microsphere	Efficient local angiogenesis	[40]
11.	Interleukin-2; interferon-γ	Liposomal-conjugated	Prolonged biological effect with retained structural activity	[41, 42]
12.	Vasopressin (Hormone)	PEG conjugated to the phospholipid distearoyl phosphatidyl ethanolamine (PEG-DSPE)	Prolonged circulation as compared to phosphatidyl serine	[43, 44]
13.	Zoladex®-goserelin acetate	PLGA copolymeric implant	For treatment of prostatic carcinoma, preloaded in a special single-use syringe, administered every 12 weeks	[45]
14.	Vasopressin (Hormone)	Polypropylene/ collodion device	Showed reduction in urine production for at least 50 days on SC	[46]

(continued)

Table 6.4 (continued)

S. No.	Protein/peptide	Carrier system	Properties imparted	References
			administration in vasopressin-deficient Brattleboro rats	
15.	Nafarelin (GnRH analogue)	PLGA implant	Controlled release	[47]
16.	BSA	Encapsulated in PEG-PLGA nanoparticles	Improved BSA half-life and distribution	[48]
17.	Insulin, Thymocartin, Somatostatin	Encapsulated in SLNs	SLN prepared by solvent emulsification-evaporation method	[49, 50]
18.	Cyclosporin A	Encapsulated in SLNs	Microemulsion technique, prolonged drug release	[51]
19.	Influenza virus antigen	Polymeric particulate system as an adjuvant in vaccine delivery	Very effective vaccine with good antibody response reported in mice	[52]
20.	IgH, LH, Prostaglandin F_2	Polyacrylamide and polyvinylpyrrolidone gels	Used for controlled delivery	[53]

Implants

Another class of sustained-release systems are implants. Implantable polymeric depots offer site-specific and prolonged sustained release of protein drugs. Implants are often placed subcutaneously by a wide bore needle or minor surgery. A protein is released from an implant in three distinct phases: the initial burst phase, diffusion-controlled release, and erosion-controlled release. Various polymers are being used to synthesize implants, and often biodegradable ones such as PLGA are preferred as they do not need surgical procedures for removal. A more recent method for an injectable implant is the development of *in situ* forming parenteral drug delivery systems (ISFPD) [55]. ISFPD systems are liquid systems that form solid or semisolid matrices as soon as they come in contact with aqueous body fluids or release media when injected subcutaneously or intramuscularly.

Liposomes

Biodegradable lipid-based carriers, often called liposomes, offer another controlled-release system for therapeutic proteins. They can be targeted to specific sites and have also been reported to act as an adjuvant for vaccine delivery. Liposomes offer several advantages when used for protein and peptide drug delivery, which include the ease of altering the physical and chemical properties of liposomes by changing lipid composition, low toxicity, an aqueous core that maintains the protein in its natural state avoiding damage by dehydration, and that the liposomes act as circulating drug reservoirs for prolonged release while providing protection and reduced immunogenicity. Antibody-targeted liposomes, often termed immunoliposomes, have been studied for the site-specific release of protein drugs for cancer therapy. Hybrid liposomes like silica nanoparticle-coated liposomes are being used for the delivery of insulin. Diphtheria toxin-containing liposomes have been studied for the treatment of ovarian cancer [1, 56]. When administered via an intravenous route, the liposomes are often taken up by the reticuloendothelial system (RES); as a result, sterically shielded liposomes are designed using specialized phospholipids esterified with hydrophilic groups such as ganglioside GM1 and polyethylene derivatives [14].

Hydrogels

Hydrogels are hydrophilic polymer-based 3D networks that swell up in contact with water. They comprise significant amounts of water (70–99%) and a chemically or physically cross-linked polymer network. The high-water content makes them highly biocompatible and almost tissue-like and imparts them the property to easily encapsulate hydrophilic therapeutic proteins [57]. Physical gels are held in a network either by ionic forces, hydrophobic interactions, or physical entanglements. Biodegradable hydrogels are either made up of proteins and polysaccharides that undergo enzymatic degradation or degrade simply because they get dissolved by absorbing water. Several sustained-release systems for proteins and peptides have been developed based on the biodegradable hydrogels for tissue engineering. Recently, several systems have been developed to control the release of drugs both spatially and temporally. Here, the protein drug is first encapsulated in microparticles and then embedded into the hydrogel scaffold. Smart gels that respond to changes in pH, temperature, electrical stimuli, ultrasound, chemical or enzymatic stimuli have been prepared for the controlled and site-specific release of the therapeutic proteins [58, 59].

Nanoparticles

Various nanocomposites such as nanoparticles, nanoconjugates, nanomicelles, and nanocapsules have been formulated to deliver therapeutic proteins invasively. Nanoparticles offer several advantages over microparticles as their small nanometer size ($\leq$100 nm) helps in deeper penetration through fine capillaries to target organs like the lung, spleen, liver, lymph, and spinal cord. Their properties like polymer composition, pH, temperature, and ion sensitivity, and biodegradability can be modified to regulate site- and time-specific release of loaded drugs. They further reduce toxic side effects and prolong the drug life in the systemic circulation. Different methods like the catheter-based approach or surface modification with biospecific ligands like antibodies are used for the site-specific delivery of nanoparticles [60, 61]. Several procedures utilized for the synthesis of nanoparticles often involve the use of harsh organic chemicals and high shear forces that are often detrimental to proteins and peptides, and hence solvent displacement method is the current method of choice. Water-oil-water double emulsion method is also a widely used method for nanoencapsulation. Further, surface adsorption on nanoparticles is also used for drug loading, especially for proteins and peptides, as it is carried out in low temperatures and aqueous phases [62].

Polymeric nanoparticles are either synthesized by synthetic polymers such as PLA, PLGA, PCL, PMMA, PBCA (poly(n-butylcyanoacrylate)) or natural polymers such as chitosan, alginate, gelatin, and albumin. Nanoparticles have also been PEGylated to prolong their life in circulation and for controlled and site-specific delivery [63]. Smart polymers like PMAA (poly (methyl acrylic acid)) have been utilized for pH and temperature-specific release of hormones [64].

Solid Lipid Nanoparticles (SLNs) and Nanostructured Lipid Carriers (NLCs)

Solid lipid nanoparticles (SLNs) are colloidal particles made up of a solid lipid matrix stabilized by surfactants in an aqueous medium and range between 10 and – 1000 nm in size. SLNs overcome the drawbacks of other colloidal carrier systems like liposomes, polymeric nanoparticles, and emulsions meanwhile combining all their advantages. SLNs show advantages like higher encapsulation efficiency, stability, loading capacity, and target-specific release properties compared to other nanosystems. Therapeutic proteins and antigens may be incorporated or adsorbed onto SLN and are further administered by either parenteral routes or alternative routes such as oral, nasal, and pulmonary. SLNs have a higher bioavailability, prolonged-release profile, flexible application, and simple, scalable, and inexpensive production than phospholipids and biodegradable polymers used to produce other nanocarriers. Thus, they can be made at large scales and have industrial applicability [65]. Formulation of proteins and peptides in SLNs confers improved stability, circumvents proteolytic degradation, as well as the sustained release of the incorporated molecules. Peptides such as insulin [66] and calcitonin [67] have been incorporated into SLNs and are currently under investigation. Changing 30% lipid mass by liquid lipids can produce another form of nanoparticles known as nanostructured lipid carriers (NLCs). NLCs have additional advantages over SLNs, which include avoiding drug leakage by lipid crystallization during storage, possessing higher encapsulation efficiency, and lower water content, thereby showing better drug entrapment [68].

Vaccine Adjuvants

The development of new particulate adjuvants offers numerous advantages over the traditional adjuvant systems. Particulate systems can act both as delivery systems and adjuvants that induce cellular and humoral immune responses. These particulate systems confer a range of advantages, which include passive targeting to the antigen-presenting cells (APCs) present in the immune system, thermal stability, protection against degradation, antigenicity, stabilization of antigen, and sustained antigen release by providing chances of modification. Recent advances in pulsatile release systems can further aid in developing single-dose vaccines that can offer both primary and booster immunizations. Particulate systems like microspheres [69], liposomes [70], and nanoparticles have been used for peptide or protein antigen and DNA vaccine delivery. When introduced *in vivo,* the particles are taken up by APCs, and the released antigen is processed further for presentation and generates an adaptive immune response. The particles that are not taken up by APCs remain at the site of administration and act as a depot for the pulsatile release of the antigen for a prolonged immune response [71]. The most widely used polymers for encapsulating vaccine antigens are PLGA and its copolymers, PEO, and chitosan.

Others

Other nanocarriers and advances in parenteral delivery include microemulsions, self-microemulsifying drug delivery systems (SMEEDS), aquasomes, and microneedles. Microemulsions and SMEEDS are oil-in-water or water-in-oil dispersed systems with a surfactant/cosurfactant along with the incorporated therapeutic molecule. These systems act as thermodynamically stable drug delivery vehicles and offer high degrees of solubilization for several therapeutic proteins. Microemulsions are dispersed systems consisting of water, oil, and surfactant forming complex nanostructures like droplets or solutions, whereas SMEEDS, unlike microemulsions, form spontaneous microemulsion droplets only when introduced into the body. These systems offer great potential as protein and peptide delivery vehicles as hydrophilic molecules of different sizes can be incorporated, and their release can be regulated by adjusting parameters like droplet size, osmotic gradient, and volume fraction of the disperse phase [72]. Aquasomes, on the other hand, are nanoparticles made from ceramics, used mainly to increase protein and peptide stability and enhance the life of the drug in circulation. They possess a ceramic core often coated with a polymer like polyhydroxy oligomer for protein and peptide drug adsorption [73]. Microneedles, ranging in length between 150–1500 μm, 50–250 μm in width, and 1–25 μm in tip thickness, form micron-sized pores in the skin and enhance transdermal delivery [74].

6.6.2 Strategies for Enhanced Delivery of Therapeutic Proteins Via Non-invasive Routes

6.6.2.1 Oral Delivery

Oral delivery is one of the most favored routes of drug administration due to its high patient compliance. This route is cheaper and easier when compared to the parenteral route of drug administration. However, this route suffers from many drawbacks for the delivery of peptide and protein drugs, which include enzymatic proteolytic degradation in the GI tract and exposure to extreme pH changes in the GI tract that may lead to drug deactivation by hydrolysis, oxidation, or deamination. Epithelial barriers often reduce drug absorption, followed by active efflux pumps that can pump the drug back into the gut lumen. The first-pass mechanism in the liver further reduces the amount of drug that reaches the systemic circulation [73]. Several strategies have thus been designed to overcome these drawbacks for the successful delivery of therapeutic proteins via the oral route (Fig. 6.3).

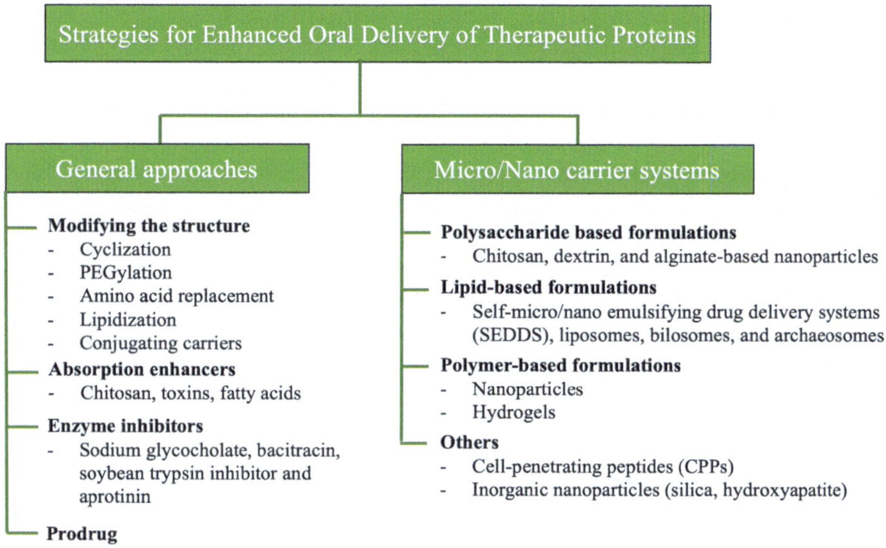

Fig. 6.3 Strategies for enhanced oral delivery of therapeutic proteins

Modifying the Structure

Modifications in the structure of protein and peptide drug significantly decrease their degradation in the GI tract and often aids in increasing their membrane permeability. Structural changes in therapeutic proteins and peptides can be induced by several methods such as cyclization, PEGylation, amino acid substitution, protein lipidation, and conjugation with carrier molecules [7, 73].

Cyclization of a peptide often involves the formation of a cyclic structure by linking one end to another or linking side chains via chemical bonds such as disulfide bonds and bridge formation. The increase in the bioavailability of therapeutic proteins after cyclization arises mainly due to increased stability against enzymatic degradation, flexibility reduction, and loss of intermolecular hydrogen bonds. Several naturally and synthetically derived peptides have been reported to possess cyclic structures imparting their biological activity. Vancomycin, cyclosporin A, and gramicidin are examples of cyclic peptides presently in clinical use [75]. Since not all proteins and peptides can undergo cyclization, they are PEGylated. PEGylation is the covalent conjugation of PEG with the protein, and as mentioned previously, it provides prolonged protected retention in circulation, low immunogenicity, and enhanced stability in different physiological conditions. PEGylated insulin has been combined with pH-responsive hydrogels for oral delivery. PEGylation helped improve the loading of hydrogels and maintain the sustained release of insulin [76]. PEGylation also enhances the resistance against enzymatic degradation and systemic clearance of salmon calcitonin [77].

Replacing the naturally present L-amino acids with D-amino acids in the protein or peptide offers another structural modification, which provides benefits like resistance to proteolysis by enzymes such as trypsin, chymotrypsin, pepsin, papain, and carboxypeptidases [78]. Replacing residues in MUC2 protein with D amino acids made the protein resistant to enzymatic proteolysis in lysosomal preparations [79]. Longer half-life, improved passage across biological membranes, and increased stability can be ensured by another structural modification known as protein lipidation. Lipidation involves the conjugation of therapeutic proteins with fatty acids. Salmon calcitonin, when lipidized using the reversible aqueous lipidation method, showed enhanced intestinal absorption, increased stability to intestinal enzymatic proteolysis, and prolonged life in systemic circulation [80]. Intestinal absorption of orally delivered therapeutic proteins can also be enhanced by covalently conjugating them with transport carrier molecules. Endogenous membrane receptors or transporters may recognize these carrier molecules and thus help increase the uptake of therapeutic proteins. Various transport carrier molecules have been associated with therapeutic proteins, such as vitamin, phosphate, and carbohydrate transporters [81].

Absorption Enhancers

Absorption enhancers are substances that, when used at a particular concentration, provide an absorption-enhancing effect across the intestinal wall. Orally delivered therapeutic proteins often suffer from poor bioavailability across the intestinal epithelium because of their large molecular weight and hydrophilic nature. Absorption enhancers are co-administered with the therapeutic protein and enhance the bioavailability by opening tight junctions, altering mucus viscosity, temporarily changing the membrane integrity, or enhancing membrane fluidity. Various absorption enhancers such as chitosan, fatty acids, and toxins have been used to enhance the bioavailability of therapeutic proteins [7]. A polymeric derivative of chitin, chitosan acts as an excellent absorption enhancer as it is non-toxic, biocompatible, and enhances the absorption of orally delivered therapeutic proteins across paracellular membranes [82]. Previous work has shown that chitosan improves the bioavailability of drugs like insulin, 8-R vasopressin, and atenolol [83]. Medium-chain fatty acids like laureate, caprylate, and caprate are also used as absorption enhancers [84]. Chitosan acts by binding to the intestinal epithelium by positive charges in its protonated form at a pH below 6.5 and redistributing the tight junctions and cytoskeletal actin, whereas fatty acids act by inducing dilation of the tight junctions. Toxins have also been employed as absorption enhancers at safe doses. One such example is zonula occludens toxin obtained from *Vibrio cholerae*, which enhanced the permeability of intestinal mucus membrane by modulating the tight junctions [85].

Enzyme Inhibitors

Oral delivery severely suffers from the drawback of enzymatic degradation in the GI tract. Various enzymes such as pepsin, trypsin, chymotrypsin, elastase, and carboxypeptidase cleave the therapeutic protein at different amino acid sites in different regions of the GI tract. This can be overcome by the co-administration of the enzymatic inhibitors with the therapeutic protein. Sodium glycocholate, bacitracin, soybean trypsin inhibitor, and aprotinin are common enzymatic inhibitors used to prevent protein degradation [86]. Chemical inhibitors and those derived from amino acids or modified amino acids often suffer from drawbacks like toxicity and systemic side effects. Hence inhibitors derived from peptides and modified peptides are being extensively studied [87]. A common example of a peptide-derived enzymatic inhibitor is aprotinin, which can inhibit both trypsin and chymotrypsin. Aprotinin increases the bioavailability of oral insulin and reduces blood glucose levels by 30% compared to insulin alone [88]. Soybean trypsin inhibitor is another peptide-based inhibitor that, when orally administered with therapeutic protein like insulin in rats and dogs, showed a significant rise in its bioavailability [89]. Recently discovered chicken and duck ovomucoids can inhibit trypsin and α-chymotrypsin and have been reported to provide 100% protection to oral insulin against these enzymes [90].

Micro/Nanocarrier Systems

A number of carrier systems have been developed to encapsulate therapeutic proteins to increase their bioavailability, protection against degradation, and targeted delivery. These include carrier systems based on polymers, polysaccharides, lipids, cell-penetrating peptides, and inorganic materials.

Formulations Based on Polysaccharides

Polysaccharides are natural polymers that are used to prepare carriers that are biocompatible, biodegradable, and safe. Most polysaccharides are protonated in acidic pH and thus bind tightly with the negatively charged mucus membrane of the intestinal epithelium, thereby influencing the membrane permeability by modifying the tight junctions and increasing the bioavailability of the encapsulated therapeutic protein [91]. Chitosan-based nanoparticles are thus being used for the oral delivery of therapeutic proteins. Chitosan, as mentioned earlier, possesses mucoadhesion properties and can bind to sialic acid residues of the mucus membrane and enhances the permeability of the drug by modifying the tight junctions. Chitosan loses its positive charge in the basic medium and hence is less effective in enhancing absorption after the duodenum. To overcome this drawback, several modified chitosan polymers, such as trimethyl chitosan, N-methylene phosphonic chitosan, O- and N-carboxymethyl chitosan, branched chitosan, and alkylated chitosan has been developed [92]. Dextran [93], alginate [94], and cellulose [95] are some other polysaccharides that have shown promise for the delivery of therapeutic proteins.

Formulations Based on Lipids

Lipid-based carriers possess the advantage of biocompatibility and the property to cross the intestinal barrier. Self-micro/nanoemulsifying drug delivery systems (SEDDS), liposomes, bilosomes, and archaeosomes are common lipid-based carriers being used for the oral delivery of therapeutic proteins [16, 73]. Bilosomes are vesicles made of phospholipids stabilized by bile salts such as deoxycholate and are being studied to deliver therapeutic proteins like insulin and calcitonin [96]. SEDDS offers a promising oral delivery option due to its intense physical strength and ease of production. They consist of oil, surfactants, cosurfactants, and drug. They are administered orally in gelatin capsules and form a spontaneous emulsion with micro/nanodroplets when they come in contact with GI fluids. Sakloetsakon et al. [97] prepared the thiolated chitosan-based SNEDDS for oral delivery of insulin, which ensured better insulin bioavailability. Liposomes have been used to encapsulate therapeutic proteins in their aqueous core, and their capability to fuse with biological membranes has been exploited for delivery. However, surface modifications are required to increase their stability against enzymatic degradation in the GI tract. Plant-based glycoproteins, known as lectins, have been used to coat liposomes for the delivery of calcitonin. Lectins can strongly bind to N-acetyl glucosamine (NAG) and sialic acid residues of the intestinal mucosa and increase liposomal absorption [98]. Archaeosomes are another class of liposomes made up of polar lipids from archaeobacteria. Their most significant feature is their high stability against harsh conditions like high temperature, low or high pHs, and enzymatic degradation, making them ideal candidates to withstand the same in the GI tract. Several studies have explored their potential as oral delivery agents for therapeutic proteins like insulin and antigens for oral vaccine formulations [99, 100].

Formulations Based on Polymers

Nanoparticles and hydrogels are two significant polymer-based carriers being explored for the oral delivery of therapeutic proteins. Nanoparticles offer increased physicochemical stability to therapeutic proteins in the GI tract by encapsulating/entrapping them in their polymeric matrix. Their nanometer size (10–1000 nm) influences their uptake, distribution, elimination, and activity in the GI tract. Various biocompatible, non-toxic, and often biodegradable natural and synthetic polymers are being used to synthesize nanoparticles. The most commonly used polymers include PLA, PLGA, chitosan, gelatin, polymethylmethacrylates, and poly-alkyl-cyanoacrylate [101]. Nanoparticles fabricated from chitosan and chitosan derivatives like N-trimethyl chitosan chloride (TMC) have been utilized to study the delivery of oral insulin in different *in vivo* and *in vitro* models [102, 103]. PLGA-based synthetic nanoparticles also offer enhanced therapeutic protein delivery via oral delivery. They suffer from drawbacks like burst release at acidic pH and poor mucosal penetration, but these have been overcome by surface modifications. PLGA nanoparticles coated with chitosan have shown enhanced insulin bioavailability [104]. Concanavalin A lectin-coated PLGA nanoparticles were targeted to M cells in the Peyer's patches and showed improved stability in acidic pH and steady, sustained release of encapsulated insulin [105]. In another study, targeted delivery of nanoparticles loaded with an oral protein was achieved by conjugating it with the Fc portion, which specifically binds to the FcRn (neonatal Fc receptor) found in the apical region of the small intestine and colon epithelial cells. PEG-PLGA-based nanoparticles loaded with exenatide, when conjugated with Fc, were reported to enhance its bioavailability and hypoglycemic activity [106]. Nanoparticles with surfaces modified with specific targeting ligands can help overcome the drawback of poor bioavailability of orally delivered therapeutic proteins and can be explored to treat numerous diseases.

As mentioned previously, hydrogels are 3D polymeric networks capable of swelling in an aqueous medium. Hydrogels can be designed to respond to physiological stimuli like pH, ionic strength, temperature, ultrasound, chemicals, and certain enzymes. When given their specific stimuli, the hydrogels can undergo changes in network structure, permeability, swelling, and mechanical strength. pH-responsive hydrogels are most studied for the oral delivery of therapeutic proteins for protection against the harsh environment in the GI tract. Poly (methacrylic acid-grafted-ethylene glycol) (Poly (MAA-g-EG))-based hydrogels are being used as carriers for the oral delivery of therapeutic proteins, antibodies, and vaccines [107]. BSA-loaded alginate and xanthan gum/poly (N-vinyl imidazole)-based hydrogels exhibited pH-dependent swelling and increased physiological activity [108, 109].

Others

Cell-penetrating peptides (CPPs) and inorganic nanoparticles are two other categories of carriers being explored for oral drug delivery. The CPPs are short peptides made up of positively charged amino acids like lysine and arginine and possess membrane-penetrating properties. These increase the membrane permeability of orally delivered therapeutic proteins when co-administered. Hydrophobic amino acids like tryptophan can be included in the CPP sequence that favor their translocation across the lipid bilayer. Covalent conjugation of CPPs or simple co-administration with the therapeutic proteins has been reported to enhance their bioavailability [110].

Inorganic nanoparticles are also being explored for oral delivery carriers as they possess excellent stability both in harsh acidic and enzymatic environments of the GI tract. Silica [111], zirconium phosphate [112], hydroxyapatite [113], and titanium dioxide [114] are some inorganic nanocarriers being explored for the delivery of oral proteins.

6.6.2.2 Transdermal Delivery

Therapeutic proteins can be delivered by the transdermal route and thus don't have to go through the harsh environment in the GI tract and first-pass effect in the liver. The ease of administration, better patient compliance, and prolonged sustained drug release are advantages of this mode of delivery. Skin acts as a major barrier to this mode of delivery and often limits the entry of large molecules (>500 Da). The skin consists of three stratiform tissues, i.e., epidermis, dermis, and subcutaneous tissue. The epidermis is made up of stratum corneum and viable epidermis. The stratum corneum is the hydrophobic fibrous structure responsible for blocking the entry of any exogenous molecule, especially large hydrophilic molecules [115]. The tight junctions present in the epidermis, aside from adhesion proteins, pose more resistance to entry of any molecule. These drawbacks in transdermal delivery are mitigated by using biochemical enhancers such as glycols and terpenes and with the help of physical penetration methods like electroporation, sonophoresis, iontophoresis, and microneedles [116]. Recent advances in both active and passive transport methods have made transdermal delivery viable for therapeutic proteins.

Formulations Based on Lipids

Emulsions and liposomes are two important lipid-based carrier systems often used for the transdermal delivery of therapeutic proteins. Nanoemulsions have been formulated, which are isotropic and low viscosity dispersed systems containing two immiscible liquids prepared by methods like microfluidization and high-pressure homogenization [117]. These nanoemulsion preparations often suffer from physical instability and may flocculate during storage. This, however, is overcome by optimizing the particle size and the type of surfactant used. Nanoemulsions have been studied for transcutaneous immunization as vaccine delivery vehicles. The superior T cell response was reported using nanoemulsions for transcutaneous immunization for viral infections [118]. Transdermal delivery of insulin was reported using microemulsions [119]. The encapsulation of therapeutic proteins inside liposomes for the transdermal delivery has been successfully used in topical applications for the skin. The similarity between the lipid composition of the skin and liposomes makes their absorption in the epidermis easier with deeper penetration. Conventional liposomes have been modified for more efficient transdermal delivery, especially to overcome the delivery barrier across the stratum corneum. Several flexible liposomes such as Transferosomes®, invasomes, niosomes, and ethosomes have been developed for enhanced and deeper delivery across the skin [120]. Transferosomes® are elastic liposomes with flexibility derived from an incorporated edge activator, i.e., a single-chain surfactant. Several studies have demonstrated the transdermal delivery efficiency of Transferosomes®, such as enhanced insulin permeation when made into a transferosomal gel and prolonged hypoglycemic effect in rats with alloxan-induced diabetes [121].

Formulations Based on Polymers

Nanoparticles can alter the extracellular lipids of the stratum corneum and can thus enhance skin permeability. Charge, size, and material used for synthesis are the major factors that influence skin penetration by nanoparticles [122]. Apart from these properties, the dose of nanoparticles, their morphology [123], bio-adhesiveness [124], and dissociation in *in vivo* system also play a role in skin penetration [125]. Transdermal patches consisting of chitosan nanoparticles have been developed for insulin delivery [126]. Microneedle-mediated intradermal delivery of ovalbumin (OVA)-loaded PLGA nanoparticles elicited protective T cell-mediated immunity to *Listeria monocytogenes* [127]. Self-dissolving hyaluronan-based microneedles were used for intradermal delivery of OVA-loaded PLGA nanoparticles, and their immunogenicity was assessed [128].

Others

Other methods that ensure effective transdermal delivery include electroporation, sonophoresis, iontophoresis, and microneedles [116]. Microneedles are either solid or hollow needles (50–900 μm) that can pass through the stratum corneum to create microchannels for drug delivery. Iontophoresis involves transdermal drug delivery by using a mild electric current to the skin ranging <0.5 mA/cm^2 of the skin. Sonophoresis is another physical method that uses ultrasound waves to enhance the skin penetration of drugs, whereas electroporation is an electrical technique where high voltage electronic pulses for short durations of time are applied to increase the skin permeability by opening the aqueous pores reversibly. Iontophoresis was used for the delivery of interferon alpha-2b (IFNalpha2b) in hairless rats [129], and electroporation was used for the transdermal delivery of insulin [130].

6.6.2.3 Intranasal Delivery

Intranasal administration of therapeutic proteins has gained loads of attention for local and systemic effects in recent years because of several advantages over other invasive and non-invasive methods of delivery. These advantages include non-invasiveness, ease of application, enhanced accessibility due to a thin and porous epithelial barrier, and highly vascularization. Furthermore, this route bypasses hepatic first-pass metabolism and has a low activity of proteolytic enzymes, both of which aid in the rapid onset of activity of the delivered therapeutic protein. Apart from these, it is a highly patient-friendly route and bypasses the blood–brain barrier by allowing drugs to be delivered directly to the brain tissue or cerebrospinal fluid through olfactory neurons [131]. These attributes make nasal delivery ideal for both local and systemic delivery.

However, nasal delivery suffers from certain drawbacks like the short residential period of the applied drug, mucus layer, and mucociliary clearance, limiting the permeability of the drug across the nasal mucosa. Therapeutic substances are transported across the nasal mucosa via transcellular or paracellular transport. Passive or facilitated diffusion of molecules occurs during transcellular transport, whereas the molecules diffuse only passively through aqueous channels during paracellular transport. Hydrophilic proteins follow paracellular transport, but this route favors the transport of only small molecular weight molecules. Membrane-bound proteolytic enzymes and other enzymes like monooxygenase, reductase, and transferase, can degrade the proteins in the nasal cavity [132]. Various strategies have been developed to overcome these challenges, which include the use of enzyme inhibitors to overcome proteolytic degradation, absorption enhancers to increase permeability across nasal mucosa, mucoadhesive formulations to increase the nasal residential time of drug, and the use of a prodrug approach to impart important physicochemical attributes [133].

Formulations Based on Polymers

Polymer-based nanoparticles and nanogels are the most sought-after carriers for nasal delivery these days. Mucoadhesive nanoparticles have been designed to increase the residential time of the protein drug in the nasal cavity. These include chitosan-based preparations and have been widely used for the delivery of insulin. Positively charged chitosan interacts with the mucosal membrane and prolongs the contact between insulin and the nasal membrane, thereby enhancing its bioavailability. The nasal-associated lymphoid tissue (NALT) is mainly targeted during intranasal vaccination, and polymeric nanoparticles of chitosan, PLGA, and polystyrene have been found effective with respect to antigen uptake by NALT [134]. Nanogels are cross-linked polymeric nanoparticles that swell in contact with water and have a size range of 10–100 nm. They can be tailored by modifying parameters like size, charge, density, and functional groups [135]. Intranasal administration of covalently attached insulin with poly(N-vinyl pyrrolidone)-based nanogel overcame the blood–brain barrier and provided better neuroprotection against amyloid β-induced dysfunction compared to free-insulin [136]. Research in this regard thus shows the effectiveness of intranasal nanopreparations for the treatment of neurogenerative disorders. Intranasal vaccine delivery systems have also been developed and show great potential and broad applications for treating other diseases [137].

Formulations Based on Lipids

Drug-loaded liposomes are being used for the nasal administration of therapeutics for the treatment of central nervous system (CNS) disorders and have been found effective as it allows direct nose-to-brain drug delivery through lipid nanoparticles. Liposomal vaccines have been developed for intranasal administration to treat lung cancer. Surface modifications effectively targeted the vaccine and showed efficient anti-lung tumor activity [138]. Spray-dried polymer-coated liposomes have been used to improve the penetration of proteins across the nasal mucosa. TMC-coated liposomes not only show the high entrapment efficiency of bovine serum albumin (BSA)but also retain the structural integrity of the entrapped protein [139].

6.6.2.4 Pulmonary Delivery

The pulmonary route is efficient for delivering therapeutic proteins due to advantages such as large surface area, thin alveolar epithelium, and high vascularization. This route surpasses the first-pass effect and is non-invasive with high efficacy even at low doses. It is applicable for both local and systemic delivery [140]. Pulmonary tissue has lower enzymatic activity than the GI tract, with the pulmonary epithelium rich in immunological responses. This route also suffers from certain limitations, which include respiratory mucus, mucociliary clearance, alveolar epithelium with tight junctions, pulmonary enzymes, and macrophages that secrete peroxidases and proteases. Even though the pulmonary epithelium has good permeability for many lipophilic therapeutic compounds, the large size often puts a constraint on this route to deliver larger protein molecules [141]. Absorption enhancers and enzyme inhibitors have been used to increase the bioavailability of proteins delivered via this route. Apart from these, numerous nanotechnology-based approaches are being used to deliver therapeutic proteins through this route.

Formulations Based on Lipids

Liposomes are used as effective pulmonary carriers for protein drugs, offering effective and controlled drug release with biodegradable, biocompatible, and non-immunogenic attributes [142]. Liposomes impart a hydrophobic nature to the encapsulated drug, thereby enhancing its membrane permeability across the alveolar epithelium. They further reduce the mucociliary clearance of the drug as a result of their surface viscosity. Apart from liposomes, solid lipid nanoparticles are also being used to deliver proteins via the pulmonary route [143].

Formulations Based on Polymers

Polymer-based nanoparticles are the carriers of choice for pulmonary delivery as they offer attributes like biodegradability, biocompatibility, and targeted delivery using surface modifications. Chitosan, gelatin, and alginate are the most used natural polymers, whereas PLGA and PEG are the most widely used synthetic polymers for nanoparticle synthesis for pulmonary delivery. *In vivo* studies in rats have revealed the prolonged alveolar deposition of gelatin and PLGA nanoparticles after nebulization by avoiding phagocytosis by alveolar macrophages as a result of their small particle size (<200 nm) [144]. In another study, BSA-loaded biodegradable polymer poly(glycerol adipate-co-ω-pentadecalactone) (PGA-co-PDL)-based nanoparticles were developed for the pulmonary delivery and were found to effectively target the dendritic cells after inhalation [145].

6.6.2.5 Rectal Delivery

Rectal drug administration is an efficient alternative to oral and parenteral routes of delivery for local and systemic effects of therapeutic molecules for different diseases. It offers benefits like a stable physicochemical and enzymatic environment of the colorectal mucosa and bypasses the hepatic first-pass effect. Large-scale rectal vasculature, extensive lymphatic system, and lower enzymatic activity play a key role in enhanced drug absorption in the rectal route of administration. These attributes make it an efficient route for delivering therapeutic proteins that often suffer from enzymatic degradation and poor bioavailability [146]. However, this route does have a few disadvantages like a limiting absorbing surface area, poor or erratic absorption across the rectal mucosa, dissolution problems due to the small fluid content of the rectum, and drug metabolism in microorganisms and rectal mucosa [147]. A series of methods and approaches have been developed to overcome these drawbacks, like the use of enzyme inhibitors, absorption enhancers, prodrugs, and nanoformulations. All these general approaches aim to increase the stability and bioavailability of the therapeutic proteins. The nanoformulation approach has been less explored for this route of delivery but can play a significant role in adapting this route for the delivery of proteins and peptides.

Formulations Based on Lipids

Liposomes of nanometer dimensions are being studied as a mode of delivery for therapeutic proteins via the rectal route. A recent study modified nanosized liposomes loaded with hepatitis B surface antigen for immunization by intracolonic administration in rats. They were composed of 1,2-dipalmitoyl-sn-glycero-3-phosphocholine bilayer encapsulating a solid glyceryl tripalmitate core and used lipid A as an adjuvant [148]. These hybrid nanoliposomes could elicit prolonged, stable, and significant immunological responses in rats. Solid lipid nanoparticles are also being studied for rectal delivery of therapeutic proteins but with less success than the available methods [149].

Formulations Based on Polymers

Hydrogels and nanoparticles are the two chiefly studied polymeric formulations for the delivery of proteins via the rectal route. Previous work has shown the effectiveness of these formulations in rectal delivery. In a study, insulin-loaded hydroxypropyl methyl cellulose-co-polyacrylamide-co-methacrylic acid (HPMC-co-PAM-co-PMAA) hydrogel was used as a rectal suppository to treat diabetic rats. After application, a significant hypoglycemic effect was observed in the treatment groups [150]. PLGA, PLA, chitosan and its derivatives, and methacrylic acid copolymers are commonly used for formulating polymeric nanoparticles for rectal drug delivery. Surface modifications of these nanoparticles can target them to specific sites for efficient uptake and longer circulation time. Recently, it has been shown that the non-covalent dense surface modification of PLGA nanoparticles with short-chain PEG enhanced drug distribution and retention after rectal delivery [151]. Advances in nanotechnology, such as the recent use of carbon nanotubes (CNTs), have opened new avenues in the field of efficient drug delivery and are being explored for the rectal delivery of proteins.

6.6.2.6 Ocular Delivery

Recent advances in therapeutics for ophthalmology have shown promise for treating chronic ocular diseases. Protein-based therapeutics have been successfully studied to cure several ocular disorders with advantages like low toxicity, high potency, less drug–drug interaction, and broader spectra of diverse formulations [152]. However, delivery of therapeutic proteins via the ocular route does suffer from certain drawbacks, which include degradation, poor permeability, short half-life, and in some cases, immunogenic response [153]. As a result, there is an urgent need to develop novel formulations for efficient action and targeted delivery to specific ocular tissue. Conventional eye drops have poor bioavailability and, subsequently, low therapeutic efficacy. Protein aggregation is a major concern for ocular therapeutics, and hence recently, protein aggregation inhibitors (chemical chaperons) are being used with protein therapeutics to avoid misfolding and inhibit the self-assembly of sequences prone to aggregation in native protein structures [154]. The co-administration of proteins with recombinant hyaluronidases has been utilized to enhance protein delivery across the ocular tissue [155]. Several nanocarriers are being extensively explored for controlled and site-specific delivery of therapeutic proteins.

Formulations Based on Polymers

Nanoparticles, dendrimers, polymeric micelles, nanowafers, and hydrogels are some of the most recent polymer-based formulations devised for the ocular delivery of therapeutic proteins. Topical, intravitreal, periocular, and suprachoroidal administration of nanoparticles are being studied for their efficiency and shortcomings. Cho et al. developed a Chitosan-based thermogelling system [156] as a new drug carrier for topical drug delivery to the eye. Brimonidine-loaded hexanoyl glycol chitosan (HGC), a thermosensitive system, showed a prolonged duration of action and better bioavailability than the Alphagan P formulation available in the market.

Polymeric micelles are self-assembling nanocarriers made up of amphiphilic block copolymers with hydrophilic chains forming a shell and hydrophobic chains forming a core ranging in size from 10 to 100 nm. Polymeric micelles have been widely studied for ocular delivery and can be modified for better bioavailability, controlled release, targeted delivery, and low toxicity [157]. Previous works demonstrate the efficacy of polymeric micelles in ocular delivery, such as anti-Flt1 peptide (an antagonistic peptide for vascular endothelial growth factor receptor 1 (VEGFR1 or Flt1)) when amidated to tetra-n-butyl ammonium modified hyaluronate (HATBA) showed the formation of self-assembling micelles in aqueous solution and were used for the treatment of retinal neovascularization and diabetic retinopathy [158].

Dendrimers are radially symmetric, immensely branched, well-organized nanostructures that can be utilized as nanocarriers [159]. They are based on different polymers such as polyamines, polyamides, polyamidoamines, and carbohydrates. Their molecular weight and charge determine their carrying and release capacity. Dendrimers can either encapsulate or conjugate the protein for delivery. A large number of functional groups on their surface make them ideal for modifications for target-specific delivery and ligand-receptor binding and can be used for more efficient stimuli response. Nanowafers constitute another class of nanocarriers being utilized for the ocular delivery of therapeutic proteins. These are small transparent membranes made up of nanoreservoirs loaded with protein/drug for their controlled and prolonged release [153]. These are made up of PVP, PVA, or CMC and can be placed directly on the patient's eye. They can adhere strongly and avoid removal due to constant blinking and release the drug slowly for a long time which improves the drug efficacy and availability. *In situ* hydrogels are another polymer-based formulation available for ocular delivery. Recently, stimuli-responsive gels have been developed, for example, the ion-activated Timoptic-XE® and pH-activated Virgan® for treating ocular hypertension and herpes simplex virus infection in the eye, respectively [160].

Formulations Based on Lipids

Liposomes, SLNs, and niosomes are some lipid-based formulations being utilized for ocular delivery. Immuno nanoliposomes (INLs) loaded with pigment epithelium-derived factor (PEDF) have been developed by Li et al. [161] for the treatment of choroidal neovascularization. SLNs also offer enhanced corneal delivery of both hydrophilic and hydrophobic proteins [162]. Niosomes are self-assembling nanovesicles made up of non-ionic surfactants that are recently being explored for ocular delivery due to several advantages like high stability, low immunogenicity, biodegradability, low toxicity, and high biocompatibility [163].

6.7 Computer-Aided Preparation of Formulations

Computational methods are being employed for the preparation of formulations. They aid in determining the components and excipients to be used and help predict the absorption, distribution, metabolism, excretion, and toxicity (ADMET) of the encapsulated drug. They further help determine the toxicity, stability, and activity of different delivery systems. Several simulation models and tools are available, which include molecular modelling, discrete and finite element modelling, fluid dynamics, molecular mechanics, and ADMET prediction models [164]. Computer simulations can help predict self-assembly, structural dynamics of aggregation, loading capacity, mechanism and rate of drug release, interactions between drug and delivery systems, and design of targeted drug delivery systems [165]. Computational approaches thus aid in understanding the mechanism of action and interactions of different delivery systems with drugs, other biomolecules, and biomembranes. It can hence expedite the design and development of efficient delivery systems.

6.8 Conclusion and Future Prospects

Protein-based therapeutics play a crucial role in current medical practice. However, they suffer from physicochemical and enzymatic destabilization that limits their successful use. The delivery systems used for therapeutic proteins often suffer from several limitations. The invasive routes like the parenteral route show poor protein stability in the systemic circulation. In contrast, the non-invasive routes face numerous physicochemical and biological barriers like enzymatic degradation, harsh pH environment, difficult transport across membranes, and poor bioavailability. Therefore, several formulation approaches have been devised to overcome the obstacles associated with different delivery systems. Nanotechnology offers an excellent opportunity to improve protein delivery via both invasive and non-invasive routes. Various polymer-based, lipid-based, and other nanomaterials have helped improve the delivery and specificity of therapeutic proteins. Advances in biotechnology, nanotechnology, and computational methods will further help device efficient delivery options for successful, safe, and effective systemic delivery of therapeutic proteins [166–168].

References

1. Moncalvo F, Espinoza Martinez MI, Cellesi F et al (2020) Nanosized delivery systems for therapeutic proteins: clinically validated technologies and advanced development strategies. Front Bioeng Biotechnol. https://doi.org/10.3389/fbioe.2020.00089
2. Yin L, Chen X, Vicini P, Rup B, Hickling TP et al (2015) Therapeutic outcomes, assessments, risk factors and mitigation efforts of immunogenicity of therapeutic protein products. Cell Immunol 295:118–126. https://doi.org/10.1016/j.cellimm.2015.03.002
3. Leader B, Baca QJ, Golan DE et al (2008) Protein therapeutics: a summary and pharmacological classification. Nat Rev Drug Discov 7(1):21–39. https://doi.org/10.1038/nrd2399
4. Serna L, Sánchez-García U, Unzueta R, Díaz E, Vázquez R, Mangues A, Villaverde et al (2018) Protein-based therapeutic killing for cancer therapies. Trends Biotechnol 36(3): 318–335. https://doi.org/10.1016/j.tibtech.2017.11.007
5. Rekha MR, Sharma CP (2013) Oral delivery of therapeutic protein/peptide for diabetes—future perspectives. Int J Pharm 440(1):48–62. https://doi.org/10.1016/j.ijpharm.2012.03.056

6. Kintzing JR, Filsinger Interrante MV, Cochran JR et al (2016) Emerging strategies for developing next-generation protein therapeutics for cancer treatment. Trends Pharmacol Sci 37:993–1008. https://doi.org/10.1016/j.tips.2016.10.005
7. Rehman K, Akash MSH, Akhtar B, Tariq M, Mahmood A, Ibrahim M et al (2016) Delivery of therapeutic proteins: challenges and strategies. Curr Drug Targets 17(10):1172–1188. https://doi.org/10.2174/1389450117666151209120139
8. Lagassé H, Alexaki A, Simhadri V, Katagiri N, Jankowski W, Sauna Z et al (2017) Recent advances in (therapeutic protein) drug development. F1000Res 6:113. https://doi.org/10.12688/f1000research.9970.1
9. Levin D, Golding B, Strome SE, Sauna ZE et al (2015) Fc fusion as a platform technology: potential for modulating immunogenicity. Trends Biotechnol 33:27–34. https://doi.org/10.1016/j.tibtech.2014.11.001
10. Rhee YS, Park CW, DeLuca PP, Mansour HM et al (2010) Sustained-release injectable drug delivery. Pharm Tech 11:16–22
11. Werle M, Bernkop-Schnürch A (2006) Strategies to improve plasma half life time of peptide and protein drugs. Amino Acids 30(4):351–367
12. Torchilin VP, Lukyanov AN (2003) Peptide and protein drug delivery to and into tumors: challenges and solutions. Drug Discov Today 8(6):259–266
13. Feridooni T, Hotchkiss A, Agu RU (2016) Noninvasive strategies for systemic delivery of therapeutic proteins—prospects and challenges. Smart Drug Deliv Syst. https://doi.org/10.5772/61266
14. Agrawal H, Thacker N, Misra A (2011) Parenteral delivery of peptides and proteins. In: Challenges in delivery of therapeutic genomics and proteomics (1st ed). https://doi.org/10.1016/B978-0-12-384964-9.00011-6
15. Mitchell MJ, Billingsley MM, Haley RM, Wechsler ME, Peppas NA, Langer R (2021) Engineering precision nanoparticles for drug delivery. Nat Rev Drug Discov 20(2):101–124. https://doi.org/10.1038/s41573-020-0090-8
16. Bajracharya R, Song JG, Back SY, Han HK (2019) Recent advancements in non-invasive formulations for protein drug delivery. Comput Struct Biotechnol J 17:1290–1308. https://doi.org/10.1016/j.csbj.2019.09.004
17. Tibbitts J, Canter D, Graff R, Smith A, Khawli LA et al (2016) Key factors influencing ADME properties of therapeutic proteins: a need for ADME characterization in drug discovery and development. MAbs 8:229–245
18. Muheem A, Shakeel F, Jahangir MA, Anwar M, Mallick N et al (2016) A review on the strategies for oral delivery of proteins and peptides and their clinical perspectives. Saudi Pharm J 24:413–428
19. Thwala LN, Santander-Ortega MJ, Lozano MV, Csaba NS et al (2018) Functionalized polymeric nanostructures for mucosal drug delivery. In: Sarmento B, das Neves J (eds) Biomedical applications of functionalized nanomaterials. Elsevier, Amsterdam, pp 449–487
20. Luthra S, Obert JP, Kalonia DS, Pikal MJ et al (2007) Investigation of drying stresses on proteins during lyophilization: differentiation between primary and secondary drying stresses on lactate dehydrogenase using a humidity controlled mini freeze dryer. J Pharm Sci 96(1):61–70
21. Wang W (2000) Lyophilization and development of solid protein pharmaceuticals. Int J Pharm 203:1–60
22. Shoyele SA (2008) Engineering protein particles for pulmonary drug delivery. Methods Mol Biol 437:149–160
23. Pisal DS, Kosloski MP, Balu-Iyer SV et al (2010) Delivery of therapeutic proteins. J Pharm Sci 99:2557–2575
24. Suthiwangcharoen N, Nagarajan R (2017) Nanoarmoring of proteins by conjugation to block copolymer micelles. In: Methods in enzymology (1st ed). https://doi.org/10.1016/bs.mie.2017.01.013

25. Batra J, Robinson J, Mehner C et al (2012) PEGylation extends circulation half-life while preserving in vitro and in vivo activity of tissue inhibitor of metalloproteinases-1 (TIMP-1). PLoS One 7:e50028
26. da Silva Freitas D, Mero A, Pasut G et al (2013) Chemical and enzymatic site specific PEGylation of hGH. Bioconjug Chem 24:456–463
27. Qiu H, Boudanova E, Park A et al (2013) Site-specific PEGylation of human thyroid stimulating hormone to prolong duration of action. Bioconjug Chem 24:408–418
28. Hashida M, Nishikawa M, Yamashita F, Takakura Y et al (1994) Targeting delivery of protein drugs by chemical modification. Drug Dev Ind Pharm 20:581–590
29. Tan ML, Choong PF, Dass CR et al (2010) Recent developments in liposomes, microparticles and nanoparticles for protein and peptide drug delivery. Peptides 31:184–193
30. Beg S, Samad A, Nazish I et al (2013) Colloidal drug delivery systems in vaccine delivery. Curr Drug Targets 14:123–137
31. Dineshkumar B, Dhanaraj SA, Santhi K, Vijayan P, Chandrasekhar R (2010) Single dose vaccine delivery system of tetanus toxoid formulation based on chitosan microspheres. Int J Adv Pharmaceut Sci 1(1)
32. Singh J, Pandit S, Bramwell VW, Alpar HO (2006) Diphtheria toxoid loaded poly-(ε-caprolactone) nanoparticles as mucosal vaccine delivery systems. Methods 38(2):96–105
33. Uchida T, Shiosaki K, Nakada Y, Fukada K, Eda Y, Tokiyoshi S, Matsuyama K (1998) Microencapsulation of hepatitis B core antigen for vaccine preparation. Pharm Res 15(11): 1708–1713
34. Prabhu S, Tran LP, Betageri GV (2005) Effect of co-solvents on the controlled release of calcitonin polypeptide from in situ biodegradable polymer implants. Drug Deliv 12(6): 393–398
35. Liang JF, Li YT, Yang VC (2000) A novel approach for delivery of enzyme drugs: preliminary demonstration of feasibility and utility in vitro. Int J Pharm 202(1–2):11–20
36. Zambaux MF, Bonneaux F, Gref R, Dellacherie E, Vigneron C (2001) Protein C-loaded monomethoxypoly (ethylene oxide)–poly (lactic acid) nanoparticles. Int J Pharm 212(1):1–9
37. Wei G, Pettway GJ, McCauley LK, Ma PX (2004) The release profiles and bioactivity of parathyroid hormone from poly (lactic-co-glycolic acid) microspheres. Biomaterials 25(2): 345–352
38. Meinel L, Illi OE, Zapf J, Malfanti M, Merkle HP, Gander B (2001) Stabilizing insulin-like growth factor-I in poly (D,L-lactide-co-glycolide) microspheres. J Control Release 70(1–2): 193–202
39. Li YP, Pei YY, Zhou ZH, Zhang XY, Gu ZH, Ding J, Gao XJ (2001) PEGylated polycyanoacrylate nanoparticles as tumor necrosis factor-α carriers. J Control Release 71(3): 287–296
40. Cleland JL, Duenas ET, Park A, Daugherty A, Kahn J, Kowalski J, Cuthbertson A (2001) Development of poly-(D,L-lactide–coglycolide) microsphere formulations containing recombinant human vascular endothelial growth factor to promote local angiogenesis. J Control Release 72(1–3):13–24
41. Kanaoka E, Takahashi K, Yoshikawa T, Jizomoto H, Nishihara Y, Hirano K (2003) Continuous release of interleukin-2 from liposomal IL-2 (mixture of interleukin-2 and liposomes) after subcutaneous administration to mice. Drug Dev Ind Pharm 29(10):1149–1153
42. Van Slooten ML, Visser AJWG, Van Hoek A, Storm G, Crommelin DJA, Jiskoot W (2000) Conformational stability of human interferon-gamma on association with and dissociation from liposomes. J Pharm Sci 89(12):1605–1619
43. Woodle MC, Storm G, Newman MS, Jekot JJ, Collins LR, Martin FJ, Szoka FC Jr (1992) Prolonged systemic delivery of peptide drugs by long-circulating liposomes: illustration with vasopressin in the Brattleboro rat. Pharm Res 9(2):260–265
44. Torchilin VP (2005) Recent advances with liposomes as pharmaceutical carriers. Nat Rev Drug Discov 4(2):145–160

45. Chaubal M (2002) Polylactides/glycolides-excipients for injectable drug delivery and beyond. Drug Deliv Technol 2:34–36
46. Kruisbrink J, Boer GJ (1986) The use of [3H] vasopressin for in-vivo studies of controlled delivery from an Accurel/collodion device in the Brattleboro rat. J Pharm Pharmacol 38(12): 893–897
47. Sanders LM, Kell BA, McRae GI, Whitehead GW (1986) Prolonged controlled-release of nafarelin, a luteinizing hormone-releasing hormone analogue, from biodegradable polymeric implants: influence of composition and molecular weight of polymer. J Pharm Sci 75(4): 356–360
48. Li YP, Pei YY, Zhang XY, Gu ZH, Zhou ZH, Yuan WF, Gao XJ (2001) PEGylated PLGA nanoparticles as protein carriers: synthesis, preparation and biodistribution in rats. J Control Release 71(2):203–211
49. Trotta M, Cavalli R, Carlotti ME, Battaglia L, Debernardi F (2005) Solid lipid micro-particles carrying insulin formed by solvent-in-water emulsion–diffusion technique. Int J Pharm 288(2): 281–288
50. Almeida AJ, Souto E (2007) Solid lipid nanoparticles as a drug delivery system for peptides and proteins. Adv Drug Deliv Rev 59(6):478–490
51. Ugazio E, Cavalli R, Gasco MR (2002) Incorporation of cyclosporin A in solid lipid nanoparticles (SLN). Int J Pharm 241(2):341–344
52. Kreuter J, Liehl E (1981) Long-term studies of microencapsulated and adsorbed influenza vaccine nanoparticles. J Pharm Sci 70(4):367–371
53. Davis BK (1974) Diffusion in polymer gel implants. Proc Natl Acad Sci 71(8):3120–3123
54. Horvath JE, Bajo AM, Schally AV, Kovacs M, Herbert F, Groot K et al (2002) Effects of long-term treatment with the luteinizing hormone-releasing hormone (LHRH) agonist Decapeptyl and the LHRH antagonist Cetrorelix on the levels of pituitary LHRH receptors and their mRNA expression in rats. Proc Natl Acad Sci USA 99:15048–15053
55. Chen S, Singh J (2008) Controlled release of growth hormone from thermosensitive triblock copolymer systems: in vitro and in vivo evaluation. Int J Pharm 352:58–65
56. Nissander UK, Steerenberg PA, Poppe H, Storm G, Jap PHK, Poels LG et al (1992) In vivo targeting of OV-TL3 immunoliposomes to ascitic ovarian carcinoma cells (OVCAR-3) in athymic nude mice. Cancer Res 52:646–653
57. Li J, Mooney D (2016) Designing hydrogels for controlled drug delivery. Nat Rev Mater 1: 16071. https://doi.org/10.1038/natrevmats.2016.71
58. Kasiński A, Zielińska-Pisklak M, Oledzka E, Sobczak M et al (2020) Smart hydrogels – synthetic stimuli-responsive antitumor drug release systems. Int J Nanomedicine 15:4541–4572. https://doi.org/10.2147/IJN.S248987
59. Xia LW, Xie R, Ju XJ et al (2013) Nano-structured smart hydrogels with rapid response and high elasticity. Nat Commun 4:22–26. https://doi.org/10.1038/ncomms3226
60. Moghimi SM, Hunter AC, Murray JC et al (2001) Long-circulating and target specific nanoparticles: theory to practice. Pharmacol Rev 53:283–318
61. Solaro R, Chiellini F, Battisti A et al (2010) Targeted delivery of protein drugs by nanocarriers. Materials (Basel) 3(3). https://doi.org/10.3390/ma3031928
62. Jung T, Kamm W, Breitenbach A, Klebe G, Kissel T et al (2002) Loading of tetanus toxoid to biodegradable nanoparticles from branched poly (sulfobutyl-polyvinyl alcohol)-γ-(lactide-co-glycolide) nanoparticles by protein adsorption: a mechanistic study. Pharm Res 19:1105–1113
63. Suk JS, Xu Q, Kim N, Hanes J, Ensign LM et al (2016) PEGylation as a strategy for improving nanoparticle-based drug and gene delivery. Adv Drug Deliv Rev 99:28–51. https://doi.org/10.1016/j.addr.2015.09.012
64. Sawant RM, Hurley JP, Salmaso S, Kale A, Tolcheva E, Levchenko TS et al (2006) "SMART" drug delivery systems: double-targeted pH-responsive pharmaceutical nanocarriers. Bioconjug Chem 17:943–949
65. Chadha S, Jogaiah S, Singh HB, Fraceto LF, de Lima R, et al (2021) Recent advances in nano-encapsulation technologies for controlled release of biostimulants and antimicrobial agents. In:

Advances in nano-fertilizers and nano-pesticides in agriculture, 29–55. https://doi.org/10.1016/B978-0-12-820092-6.00002-1
66. Zhang N, Ping Q, Huang G, Xu W, Cheng Y, Han X et al (2006) Lectin-modified solid lipid nanoparticles as carriers for oral administration of insulin. Int J Pharm 327:153–159
67. Wang L, Wang CY, Zhang Y, Fu HJ, Gao Y, Zhang KR et al (2019) Preparation and characterization of solid lipid nanoparticles loaded with salmon calcitonin phospholipid complex. J Drug Deliv Sci Technol 52:838–845. https://doi.org/10.1016/j.jddst.2019.05.045
68. Elmowafy M, Al-Sanea MM (2021) Nanostructured lipid carriers (NLCs) as drug delivery platform: Advances in formulation and delivery strategies. Saudi Pharmaceut J 29(9): 999–1012. https://doi.org/10.1016/j.jsps.2021.07.015
69. Jia J, Liu Q, Yang T, Wang L, Ma G et al (2017) Facile fabrication of varisized calcium carbonate microspheres as vaccine adjuvants. J Mater Chem B 5(8):1611–1623. https://doi.org/10.1039/c6tb02845d
70. Vyas SP, Goyal AK, Khatri K et al (2010) Mannosylated liposomes for targeted vaccines delivery. Methods Mol Biol 605:177–188
71. Bharali DJ, Mousa SA, Thanavala Y et al (2007) Micro- and nanoparticle-based vaccines for hepatitis B. In: Michael RS, Yuri SS (eds) Immune mediated diseases (Part B). Springer, New York, NY, pp 415–421
72. Rao SV, Yajurvedi K, Shao J et al (2008) Self-nanoemulsifying drug delivery system (SNEDDS) for oral delivery of protein drugs: III. In vivo oral absorption study. Int J Pharm 362(1–2):16–19. https://doi.org/10.1016/j.ijpharm.2008.05.015
73. Asfour MH (2021) Advanced trends in protein and peptide drug delivery: a special emphasis on aquasomes and microneedles techniques. Drug Deliv Transl Res 11(1):1–23. https://doi.org/10.1007/s13346-020-00746-z. PMID: 32337668
74. Waghule T, Singhvi G, Dubey SK, Pandey MM, Gupta G, Singh M, Dua K et al (2019) Microneedles: a smart approach and increasing potential for transdermal drug delivery system. Biomed Pharmacother 109:1249–1258. https://doi.org/10.1016/j.biopha.2018.10.078
75. Joo SH (2012) Cyclic peptides as therapeutic agents and biochemical tools. Biomol Ther (Seoul) 20(1):19–26. https://doi.org/10.4062/biomolther.2012.20.1.019
76. Tuesca AD, Reiff C, Joseph JI, Lowman AM et al (2009) Synthesis, characterization and in vivo efficacy of pegylated insulin for oral delivery with complexation hydrogels. Pharm Res 26(3):727–739. https://doi.org/10.1007/s11095-008-9816-8
77. Youn YS, Jung JY, Oh SH, Yoo SD, Lee KC et al (2006) Improved intestinal delivery of salmon calcitonin by Lys18-amine specific PEGylation: stability, permeability, pharmacokinetic behavior and in vivo hypocalcemic efficacy. J Control Release 114(3):334–342
78. Qvit N, Rubin SJS, Urban TJ, Mochly-Rosen D, Gross ER et al (2017) Peptidomimetic therapeutics: scientific approaches and opportunities. Drug Discov Today 22(2):454–462. https://doi.org/10.1016/j.drudis.2016.11.003
79. Tugyi R, Uray K, Ivan D, Fellinger E, Perkins A, Hudecz F et al (2005) Partial D-amino acid substitution: improved enzymatic stability and preserved Ab recognition of a MUC2 epitope peptide. Proc Natl Acad Sci USA 102:413–418
80. Wang J, Chow D, Heiati H, Shen WC et al (2003) Reversible lipidization for the oral delivery of salmon calcitonin. J Control Release 88(3):369–380
81. Russell-Jones GJ (2004) Use of targeting agents to increase uptake and localization of drugs to the intestinal epithelium. J Drug Target 12:113–123
82. Thanou M, Verhoef J, Junginger H et al (2001) Chitosan and its derivatives as intestinal absorption enhancers. Adv Drug Deliv Rev 50:S91–S101
83. Cano-Cebrian MJ, Zornoza T, Granero L, Polache A et al (2005) Intestinal absorption enhancement via the paracellular route by fatty acids, chitosans and others: a target for drug delivery. Curr Drug Deliv 2(1):9–22
84. Lindmark T, Nikkila T, Artursson P et al (1995) Mechanisms of absorption enhancement by medium chain fatty acids in intestinal epithelial Caco-2 cell monolayers. J Pharmacol Exp Ther 275(2):958–964

85. Fasano A, Uzzau S (1997) Modulation of intestinal tight junctions by Zonula occludens toxin permits enteral administration of insulin and other macromolecules in an animal model. J Clin Invest 99:1158–1164
86. Mahato RI, Narang AS, Thoma L, Miller DD et al (2003) Emerging trends in oral delivery of peptide and protein drugs. Crit Rev Ther Drug Carrier Syst 20:153–214
87. Zhu Q, Chen Z, Paul PK, Lu Y, Wu W, Qi J et al (2021) Oral delivery of proteins and peptides: challenges, status quo and future perspectives. Acta Pharm Sin B 11(8):2416–2448. https://doi.org/10.1016/j.apsb.2021.04.001
88. Ziv E, Lior O, Kidron M et al (1987) Absorption of protein via the intestinal wall. A quantitative model. Biochem Pharmacol 36(7):1035–1039
89. Fujii S, Yokoyama T, Ikegaya K, Sato F, Yokoo N et al (1985) Promoting effect of the new chymotrypsin inhibitor FK-448 on the intestinal absorption of insulin in rats and dogs. J Pharm Pharmacol 37(8):545–549
90. Agrawal V, Reddy IK, Khan MA et al (2000) Oral delivery of proteins: effect of chicken and duck ovomucoid on the stability of insulin in the presence of a-chymotrypsin and trypsin. Pharm Pharmcol Commun 6:223–227
91. Liu Z, Jiao Y, Wang Y, Zhou C, Zhang Z et al (2008) Polysaccharides-based nanoparticles as drug delivery systems. Adv Drug Deliv Rev 60(15):1650–1662
92. Garg U, Chauhan S, Nagaich U, Jain N et al (2019) Current advances in chitosan nanoparticles based drug delivery and targeting. Adv Pharmaceut Bullet 9(2):195–204. https://doi.org/10.15171/apb.2019.023
93. Soudry-Kochavi L, Naraykin N, Nassar T, Benita S et al (2015) Improved oral absorption of exenatide using an original nanoencapsulation and microencapsulation approach. J Control Release 217:202–210
94. Lee S, Kim YC, Park JH et al (2016) Zein-alginate based oral drug delivery systems: protection and release of therapeutic proteins. Int J Pharm 515(1–2):300–306
95. Song Y, Gan W, Li Q, Guo Y et al (2011) Alkaline hydrolysis and flocculation properties of acrylamide modified cellulose polyelectrolytes. Carbohydr Polym 86(1):171–176
96. Ahmad J, Singhal M, Amin S, Rizwanullah M et al (2017) Bile salt stabilized vesicles (Bilosomes): a novel nano-pharmaceutical design for oral delivery of proteins and peptides. Curr Pharm Des 23(11):1575–1588
97. Sakloetsakun D, Dünnhaupt S, Barthelmes J, Perera G, Bernkop-Schnürch A et al (2013) Combining two technologies: multifunctional polymers and self-nanoemulsifying drug delivery system (SNEDDS) for oral insulin administration. Int J Biol Macromol A 61:363–372
98. Makhlof A, Fujimoto S, Tozuka Y, Takeuchi H et al (2011) In vitro and in vivo evaluation of WGA–carbopol modified liposomes as carriers for oral peptide delivery. Eur J Pharm Biopharm 77:216–224
99. Li Z, Chen J, Sun W, Xu Y et al (2010) Investigation of archaeosomes as carriers for oral delivery of peptides. Biochem Biophys Res Commun 394:412–417
100. Li Z, Zhang L, Sun W, Ding Q, Hou Y et al (2011) Archaeosomes with encapsulated antigens for oral vaccine delivery. Vaccine 29:5260–5266
101. Ma X, Williams RO (2018) Polymeric nanomedicines for poorly soluble drugs in oral delivery systems: an update. J Pharm Investig 48:61–75
102. Mukhopadhyay P, Sarkar K, Chakraborty M, Bhattacharya S, Mishra R et al (2013) Oral insulin delivery by self-assembled chitosan nanoparticles: in vitro and in vivo studies in diabetic animal model. Mater Sci Eng C 33:376–382
103. Sandri G, Bonferoni MC, Rossi S, Ferrari F, Boselli C et al (2010) Insulin-loaded nanoparticles based on N-trimethyl chitosan: in vitro (Caco-2 model) and ex vivo (excised rat jejunum, duodenum, and ileum) evaluation of penetration enhancement properties. AAPS Pharm Sci Tech 11:362–371
104. Zhang X, Sun M, Zheng A, Cao D, Bi Y et al (2012) Preparation and characterization of insulin-loaded bioadhesive PLGA nanoparticles for oral administration. Eur J Pharm Sci 45: 632–638

105. Hurkat P, Jain A, Jain A, Shilpi S, Gulbake A et al (2012) Concanavalin A conjugated biodegradable nanoparticles for oral insulin delivery. J Nanopart Res 14:1219
106. Shi Y, Sun X, Zhang L, Sun K, Li K et al (2018) Fc-modified exenatide-loaded nanoparticles for oral delivery to improve hypoglycemic effects in mice. Sci Rep 8:726
107. Yoshida M, Kamei N, Muto K, Kunisawa J, Takayama K et al (2017) Complexation hydrogels as potential carriers in oral vaccine delivery systems. Eur J Pharm Biopharm 112:138–142
108. Lima DS, Tenório-Neto ET, Lima-Tenório MK, Guilherme MR, Scariot DB et al (2018) pH-responsive alginate-based hydrogels for protein delivery. J Mol Liq 262:29–36
109. Sabaa MW, Hanna DH, Abu Elella MH, Mohamed RR et al (2019) Encapsulation of bovine serum albumin within novel xanthan gum-based hydrogel for protein delivery. Mater Sci Eng C 94:1044–1055
110. Kristensen M, Nielsen HM (2016) Cell-penetrating peptides as carriers for oral delivery of biopharmaceuticals. Basic Clin Pharmacol Toxicol 118(2):99–106
111. Sreeharsha N, Philip M, Krishna SS, Viswanad V, Sahu RK, Shiroorkar PN, Venugopala KN et al (2022) Multifunctional mesoporous silica nanoparticles for oral drug delivery. Coatings 12(3). https://doi.org/10.3390/coatings12030358
112. Díaz A, David A, Pérez R, González ML, Báez A, Wark SE et al (2010) Nanoencapsulation of insulin into zirconium phosphate for or a l delivery application. Biomacromolecules 11(9): 2465–2470
113. Zhang Y, Zhang L, Ban Q, Li J, Li CH, Guan YQ et al (2018) Preparation and characterization of hydroxyapatite nanoparticles carrying insulin and gallic acid for insulin oral delivery. Nanomedicine 14(2):353–364
114. Kamari Y, Ghiaci P, Ghiaci M et al (2017) Study on montmorillonite/insulin/TiO2 hybrid nanocomposite as a new oral drug-delivery system. Mater Sci Eng C 75:822–828
115. Zeb A, Arif ST, Malik M, Shah FA, Din FU et al (2018) Potential of nanoparticulate carriers for improved drug delivery via skin. J Pharm Investig 48:1–33
116. Jain S, Patel N, Shah MK, Khatri P, Vora N et al (2017) Recent advances in lipid-based vesicles and particulate carriers for topical and transdermal application. J Pharm Sci 106:423–445
117. Escobar-Chávez JJ, Rodríguez-Cruz IM, Domínguez-Delgado CL, Díaz-Torres R, Revilla-Vázquez AL et al (2012) Nanocarrier systems for transdermal drug delivery. In: Recent advances in novel drug carrier systems. InTech, pp 201–240
118. Lopez PA, Denny M, Hartmann AK, Alflen A, Probst HC et al (2017) Transcutaneous immunization with a novel imiquimod nanoemulsion induces superior T cell responses and virus protection. J Dermatol Sci 87:252–259
119. Malakar J, Sen SO, Nayak AK, Sen KK et al (2011) Development and evaluation of microemulsions for transdermal delivery of insulin. ISRN Pharmaceutics
120. Ashtikar M, Nagarsekar K, Fahr A et al (2016) Transdermal delivery from liposomal formulations – evolution of the technology over the last three decades. J Control Release 242:126–140
121. Malakar J, Sen SO, Nayak AK, Sen KK et al (2012) Formulation, optimization and evaluation of transferosomal gel for transdermal insulin delivery. Saudi Pharm J 20:355–363
122. Palmer BC, DeLouise LA (2016) Nanoparticle-enabled transdermal drug delivery systems for enhanced dose control and tissue targeting. Molecules (Basel, Switzerland) 21(12):1719. https://doi.org/10.3390/molecules21121719
123. Plascencia-Villa G, Bahena D, Rodríguez AR, Ponce A, José-Yacamán M et al (2013) Advanced microscopy of star-shaped gold nanoparticles and their adsorption-uptake by macrophages. Metall Integr Biometal Sci 5:242–250. https://doi.org/10.1039/c3mt20202j
124. Deng Y, Ediriwickrema A, Yang F, Lewis J, Girardi M, Saltzman WM et al (2015) A sunblock based on bioadhesive nanoparticles. Nat Mater 14:1278–1285. https://doi.org/10.1038/nmat4422
125. Larese Filon F, Mauro M, Adami G, Bovenzi M, Crosera M et al (2015) Nanoparticles skin absorption, new aspects for a safety profile evaluation. Regulat Toxicol Pharm 72:310–322. https://doi.org/10.1016/j.yrtph.2015.05.005

126. Sadhasivam L, Dey N, Francis AP, Devasena T et al (2015) Transdermal patches of chitosan nanoparticles for insulin delivery. Int J Pharm Pharm Sci 7:84–88
127. de Groot AM, Du G, Monkare J, Platteel ACM, Broere F et al (2017) Hollow microneedle-mediated intradermal delivery of model vaccine antigen-loaded PLGA nanoparticles elicits protective T cell-mediated immunity to an intracellular bacterium. J Control Release 266:27–35
128. Monkare J, Pontier M, van Kampen EEM, Du G, Leone M et al (2018) Development of PLGA nanoparticle loaded dissolving microneedles and comparison with hollow microneedles in intradermal vaccine delivery. Eur J Pharm Biopharm 129:111–121
129. Badkar AV, Smith AM, Eppstein JA, Banga AK (2007) Transdermal delivery of interferon alpha-2B using microporation and iontophoresis in hairless rats. Pharm Res 24:1389–1395
130. Mohammad EA, Elshemey WM, Elsayed AA, Abd-Elghany AA et al (2016) Electroporation parameters for successful transdermal delivery of insulin. Am J Ther 23:1560–1567
131. Duan X, S. (2010) Mao New strategies to improve the intranasal absorption of insulin. Drug Discov Today 15:416–427
132. Jadhav KR, Gambhire MN, Shaikh IM, Kadam VJ, Pisal SS et al (2007) Nasal drug delivery system-factors affecting and applications. Current Drug Therapy 2:27–38
133. Yuba E, Kono K (2014) Nasal delivery of biopharmaceuticals. In: das Neves J, Sarmento B (eds) Mucosal delivery of biopharmaceuticals: biology, challenges and strategies. Springer, Boston, MA, pp 197–220
134. Dombu CY, Betbeder D (2013) Airway delivery of peptides and proteins using nanoparticles. Biomaterials 34:516–525
135. Aderibigbe B, T. (2018) Naki Design and efficacy of nanogels formulations for intranasal administration. Molecules 23:1241
136. Picone P, Ditta LA, Sabatino MA, Militello V, San Biagio PL et al (2016) Ionizing radiation-engineered nanogels as insulin nanocarriers for the development of a new strategy for the treatment of Alzheimer's disease. Biomaterials 80:179–194
137. Nochi T, Yuki Y, Takahashi H, Sawada S, Mejima M et al (2010) Nanogel antigenic protein-delivery system for adjuvant-free intranasal vaccines. Nat Mater 9:685
138. Kakhi Z, Frisch B, Heurtault B, Pons F et al (2016) Liposomal constructs for antitumoral vaccination by the nasal route. Biochimie 130:14–22
139. Chen KH, Di Sabatino M, Albertini B, Passerini N, Kett VL et al (2013) The effect of polymer coatings on physicochemical properties of spray-dried liposomes for nasal delivery of BSA. Eur J Pharm Sci 50:312–322
140. Emami F, Mostafavi Yazdi SJ, Na DH et al (2019) Poly (lactic acid)/poly (lactic-co-glycolic acid) particulate carriers for pulmonary drug delivery. J Pharm Investig 49:1–16
141. Patton JS, Fishburn CS, Weers JG et al (2004) The lungs as a portal of entry for systemic drug delivery. Proc Am Thorac Soc 1:338–344
142. Sharma PKJ, Bansal S, Banik A et al (2011) Noninvasive routes of proteins and peptides drug delivery. Indian J Pharm Sci 73:367–375
143. Gaspar DP, Serra C, Lino PR, Gonçalves L, Taboada P et al (2017) Microencapsulated SLN: an innovative strategy for pulmonary protein delivery. Int J Pharm 516:231–246
144. Menon JU, Ravikumar P, Pise A, Gyawali D, Hsia CCW et al (2014) Polymeric nanoparticles for pulmonary protein and DNA delivery. Acta Biomater 10:2643–2652
145. Alfagih I, Kunda N, Alanazi F, Dennison SR, Somavarapu S et al (2015) Pulmonary delivery of proteins using nanocomposite microcarriers. J Pharm Sci 104:4386–4398
146. Shah B, Surti N, Misra A et al (2011) Other routes of protein and peptide delivery: transdermal, topical, uterine, and rectal, 1st edn. Elsevier. https://doi.org/10.1016/B978-0-12-384964-9.00012-8
147. Jannin V, Lemagnen G, Gueroult P, Larrouture D, Tuleu C et al (2014) Rectal route in the 21st century to treat children. Adv Drug Deliv Rev 73:34–49. https://doi.org/10.1016/j.addr.2014.05.012

148. Sahu KK, Pandey RS (2016) Immunological evaluation of colonic delivered Hepatitis B surface antigen loaded TLR-4 agonist modified solid fat nanoparticles. Int Immunopharmacol 39:343–352
149. Melo M, Nunes R, Sarmento B, das Neves J et al (2018) Rectal administration of nanosystems: from drug delivery to diagnostics. Mater Today Chem 10:128–141
150. Shi Y, Xue J, Sang Y, Xu X, Shang Q et al (2019) Insulin-loaded hydroxypropyl methyl cellulose-co-polyacrylamide-co-methacrylic acid hydrogels used as rectal suppositories to regulate the blood glucose of diabetic rats. Int J Biol Macromol 121:1346–1353
151. Nunes R, Araújo F, Tavares J, Sarmento B, das Neves J et al (2018) Surface modification with polyethylene glycol enhances colorectal distribution and retention of nanoparticles. Eur J Pharm Biopharm 130:200–206
152. Vyas SP, Paliwal R, Paliwal SR et al (2011) Ocular delivery of peptides and proteins. In: Van Der Walle C (ed) Peptide and protein delivery. Academic Press, Boston, pp 87–103
153. Mandal A, Pal D, Agrahari V, Trinh HM, Joseph M et al (2018) Ocular delivery of proteins and peptides: challenges and novel formulation approaches. Adv Drug Deliv Rev 126:67–95
154. Schymkowitz J, Rousseau F (2016) Protein aggregation: a rescue by chaperones. Nat Chem Biol 12:58
155. Stern R, Jedrzejas MJ (2006) Hyaluronidases: their genomics, structures, and mechanisms of action. Chem Rev 106:818–839
156. Cho IS, Park CG, Huh BK, Cho MO, Khatun Z et al (2016) Thermosensitive hexanoyl glycol chitosan-based ocular delivery system for glaucoma therapy. Acta Biomater 39:124–132
157. Mandal A, Bisht R, Rupenthal ID, Mitra AK (2017) Polymeric micelles for ocular drug delivery: from structural frameworks to recent preclinical studies. J Control Release 248:96–116
158. Oh EJ, Choi JS, Kim H, Joo CK, Hahn SK (2011) Anti-Flt1peptide–hyaluronate conjugate for the treatment of retinal neovascularization and diabetic retinopathy. Biomaterials 32(3): 115–123
159. Yavuz B, Bozdağ Pehlivan S, Ünlü N (2013) Dendrimeric systems and their applications in ocular drug delivery. Sci World J 2013:732340. https://doi.org/10.1155/2013/732340
160. Weng Y, Liu J, Jin S, Guo W, Liang X, Hu Z (2017) Nanotechnology-based strategies for treatment of ocular disease. Acta Pharm Sin B 7(3):281–291. https://doi.org/10.1016/j.apsb.2016.09.001
161. Li T, Zhang M, Han Y, Zhang H, Xu L et al (2010) Targeting therapy of choroidal neovascularization by use of polypeptide- and PEDF-loaded immunoliposomes under ultrasound exposure. J Huazhong Univ Sci Technol Med Sci 30:798–803
162. Alvarez-Trabado J, Diebold Y, Sanchez A (2017) Designing lipid nanoparticles for topical ocular drug delivery. Int J Pharm 532:204–217
163. Sahoo SK, Dilnawaz F, Krishnakumar S (2008) Nanotechnology in ocular drug delivery. Drug Discov Today 13:144–151
164. Huynh L, Neale C, Pomès R, Allen C (2012) Computational approaches to the rational design of nanoemulsions, polymeric micelles, and dendrimers for drug delivery. Nanomedicine 8:20–36
165. Ramezanpour M, Leung SSW, Delgado-Magnero KH, Bashe BYM, Thewalt J, Tieleman DP (2016) Computational and experimental approaches for investigating nanoparticle-based drug delivery systems. Biochim Biophys Acta Biomembr 1858:1688–1709. https://doi.org/10.1016/j.bbamem.2016.02.028
166. Saudagar P, Tripathi T (2023) Advanced spectroscopic methods to study biomolecular structure and dynamics, 1st edn. Academic Press, Cambridge, MA
167. Singh DB, Tripathi T (2020) Frontiers in protein structure, function, and dynamics. Springer Nature, Singapore
168. Tripathi T, Dubey VK (2022) Advances in protein molecular and structural biology methods, 1st edn. Academic Press, Cambridge, MA

Approved Protein Therapeutics and Their Biochemical Targets

7

Rajendran Ramesh and Ramakrishnan Ravichandran

Abstract

Therapeutic proteins have shown unprecedented success and enjoy a significant share of the biotechnology market. These include diverse molecules such as monoclonal antibodies (mAbs), peptide hormones, growth factors, plasma proteins, enzymes, and hemolytic factors. The biochemical aspects of these molecules demand considerable attention as the exact roles or mechanism of action remains elusive in some cases. Nevertheless, these drug molecules have shown notable success in clinical trials and have been successfully approved by the regulatory bodies. With a brief discussion on the drug approval process and the therapeutic proteins classification, we cover the biochemistry and rationale behind the design aspects of some of the recently approved protein therapeutics. A few examples from various classes of protein therapeutics and the biochemistry underlying design and target(s) selection are also discussed.

Keywords

Therapeutic proteins · Drug targets · Antibodies · Enzymes · Hormones

Rajendran Ramesh and Ramakrishnan Ravichandran contributed equally with all other contributors.

R. Ramesh (✉)
Centre for BioSeparation Technology, Vellore Institute of Technology, Vellore, Tamil Nadu, India

R. Ravichandran
Biologics Development Centre, Dr. Reddy's Laboratories, Hyderabad, Telangana, India

D. B. Singh, T. Tripathi (eds.), *Protein-based Therapeutics*,
https://doi.org/10.1007/978-981-19-8249-1_7

7.1 Overview of Protein Therapeutics Approval Process

The discovery and development of a new drug usually take several years [1]. The newly developed therapeutic molecule has to go through the four main phases of the drug discovery process, i.e., discovery phase, development phase, preclinical and clinical studies, and data review and approval by the regulatory bodies [2]. As an example, we will discuss the development and approval process in the United States, which is regulated by the US Food and Drug Administration (FDA). The discovery phase involves discovering and validating the potential drug targets, hit discovery, and identifying the lead molecules through high throughput screening and optimization. Also, if required, further research is carried out to enhance the drug's activity and reduce the potential adverse effects. In the development phase, the biological drug is tested on volunteers through phased clinical trials (I–III) [3]. Before starting clinical trials, the investigational new drug (IND) application must be submitted to the FDA. The biological drug will be granted approval for the trials provided the regulatory body's safety, efficacy, and quality standards are met. Finally, with the availability of the best drug molecule with proven quality, safety and efficacy, and with the data from clinical trials, the application for the review and approval of the biological drug, referred to as therapeutic biologics application (BLA), is submitted to the regulatory body, i.e., US FDA. The biological drug is granted approval after evaluating several factors (quality, safety, and efficacy) and if the beneficial effects are proven to outweigh the risks involved [4–7].

The list of currently available FDA-approved therapeutic biological drugs, their substitutable drug products, and reference drugs are available online in "the purple book" at https://www.fda.gov/home. Other countries or region-specific regulatory bodies that oversee the regulatory processes include the European Medicines Agency (Europe), Pharmaceutical and Medical Devices Agency (Japan), Medicines and Healthcare Products Regulatory Agency (UK), Therapeutic Goods Administration (Australia), Health Canada (Canada), and Central Drug Standard Control Organization (India) [8, 9].

7.2 Classification of Protein Therapeutics

As per the FDA, any impact or activity of a single or combination of drug products intended to detect, cure, alleviate, treat, or prevent disease, or influence the structure or function of the body, is referred to as "therapeutic" action. Although several classes of biological molecules find application as therapeutics, protein and peptide therapeutics are the most widely used and will be the focus of the chapter. A therapeutic biological product can be one of the following classes of molecules: monoclonal antibodies (mAbs), cytokines, enzymes, growth factors, thrombolytics, and other non-vaccine protein therapeutics. These protein products are either produced by recombinant DNA (rDNA) technology using bacteria, mammalian cell lines, insect cell lines, yeast, and transgenic systems or purified directly from natural sources, such as plants and animals or microorganisms. However, in recent decades,

Table 7.1 Classification of protein therapeutics proposed by Leader et al. [13]

Group I	Group II	Group III	Group IV
Protein therapeutics with enzyme activity or regulatory function i. substituting a missing or faulty protein (e.g., recombinant insulin, coagulation factors factor VIII) ii. Enhancing an existing pathway (granulocyte-monocyte colony-stimulating factor, lipase, and proteases) iii. Deliver a new function (e.g., collagenase and papain)	Protein therapeutics with a high level of target specificity. i. Interference with the molecule or organism ii. Targeted delivery of compounds or proteins (e.g., recombinant therapeutic monoclonal antibodies, antibody-drug conjugates and fusion proteins)	Recombinant protein vaccines (e.g., Hepatitis B vaccine and human papillomavirus vaccine)	Diagnostics (e.g., human immunodeficiency virus diagnosis and cancer detection)

recombinant protein therapeutics have been widely preferred over natural sources due to benefits such as efficient and low-cost production, higher activity of the protein product due to expression of the exact gene product, reduced chances of contamination (viruses and bacteria) from the source, and the ability to engineer the recombinant product further to improve its effectiveness or functionality [10–12].

Different classification for therapeutic proteins has been proposed. Leader et al., for instance, suggested a classification strategy for protein therapies based on their pharmacological activity [13]. Accordingly, the existing protein drugs were categorized into four primary groups: (i) protein therapeutics with enzyme activity or regulatory function, (ii) protein therapeutics with defined targets, (iii) protein vaccines, and (iv) protein diagnostics. Examples of protein drugs under each of these four categories are listed in Table 7.1. Protein therapeutics can also be classified according to the nature of the molecule (therapeutic mAbs, coagulation factors, anticoagulants, blood factors, bone morphogenetic proteins, Fc fusion proteins, enzymes, growth factors, hormones, cytokines, and thrombolytics) or its biological mechanisms (non-covalent binding for targeting as in mAbs, modify/affect covalent bonds as in enzymes and activity through non-specific interactions as seen in serum albumin) [14, 15].

7.3 Protein Therapeutics: An Update

In the latter half of the nineteenth century, protein therapies were approved for use in humans for the first time. Human insulin, developed using rDNA technology, was first approved in 1982 (Humulin®) and is still widely used in clinical practice [16]. As evident from Fig. 7.1, the number of approved protein therapeutics has steadily risen over the past years. Until 2015, FDA approved 239 therapeutic

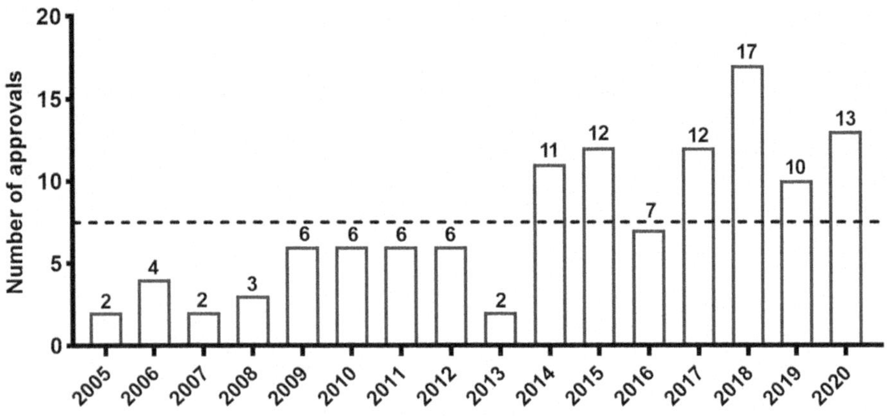

Fig. 7.1 Biologics license applications approved by the Food and Drug Administration (FDA). The number of approvals has gradually increased over the past 15 years, with a mean approval rate of 7.4 therapeutic protein drugs per year (dotted line) [19]

proteins for clinical use [17]. The recent trend among the approved protein therapeutics shows that mAbs continue to dominate the approved biological drugs (~50%), followed by coagulation factors (19%), enzymes (11%), and other classes like hormones and fusion proteins. These approved protein drugs find wide therapeutic applications in treating cancer (26%), hematological diseases (29%), and the remaining distributed among immunological, genetic, cardiological, and pulmonary diseases. This heterogeneous distribution of therapeutic proteins shows the importance of this class of molecules in treating diverse diseases in different patient populations [18]. This chapter focuses on some recently approved protein therapeutics and the biochemistry underlying their design aspects.

7.4 Approved Protein Therapeutics and Their Biochemical Action

Protein therapeutics are considered impactful depending on multiple factors like the number of patients who benefitted from the treatment, minimal adverse reactions, and the projected market value of the drug. Some of the recently approved protein drugs which are predicted to be impactful based on the above factors are discussed in this section. The focus will be on the protein drug's biological targets and the underlying biochemical mechanism through which the drug molecule exerts its therapeutic effect(s). Table 7.2 lists some recently approved therapeutic protein drugs whose biochemical mechanisms are discussed in detail.

Table 7.2 List of protein therapeutics discussed in this chapter

Monoclonal antibodies (mAbs)	Antibody-drug conjugates	Bispecific mAbs	Fusion proteins	Peptides	Enzymes
Teprotumumab Risankizumab	Sacituzumab govitecan Belantamab mafodotin	Blinatumomab	Luspatercept Efmoroctocog alfa	Afamelanotide Macimorelin Pegcetacoplan Lonapegsomatropin Setmelanotide Galium-68-DOTATOC	Pegvaliase Avalglucosidase alfa

7.4.1 Teprotumumab

Teprotumumab (Tepezza®; Horizon Therapeutics) is the first FDA-approved therapeutic IgG1 mAb-based medication for the treatment of Graves' orbitopathy, an ophthalmological manifestation of Grave's disease [20]. The autoimmune condition is characterized by the infiltration of fibrocytes and lymphocytes into the orbit (the crater-like structure that accommodates the eye). As illustrated in Fig. 7.2, post-infiltration, lymphocytes secrete inflammatory cytokines such as IL-1ß, IL-6, IL-16, and tumor necrosis factor-α (TNF-α), and the fibrocytes differentiate into fibroblasts. Adipogenesis and production of glycosaminoglycans, primarily hyaluronan, occur due to the interaction between these cells, resulting in enhanced extraocular muscle growth and orbital fat storage [21]. In principle, Grave's disease is characterized by the increased secretion of thyroid hormone by the thyroid gland (hyperthyroidism). Autoantibody-mediated secretion of thyroid hormone is one of the proven mechanisms contributing to the pathology of Grave's disease. Autoantibodies target the thyroid-stimulating hormone receptor (TSHR), and this binding triggers the secretion of thyroid hormone [22]. In addition, a second class of receptors called insulin-like growth factor-1 receptors (IGF-1R) are also implicated in the pathology of Grave's orbitopathy. IGF-1R is frequently overexpressed in fibroblasts and lymphocytes (T and B cells) isolated from patients with Grave's disease [23]. IGF-1 receptor activation is associated with hyaluronan production and release

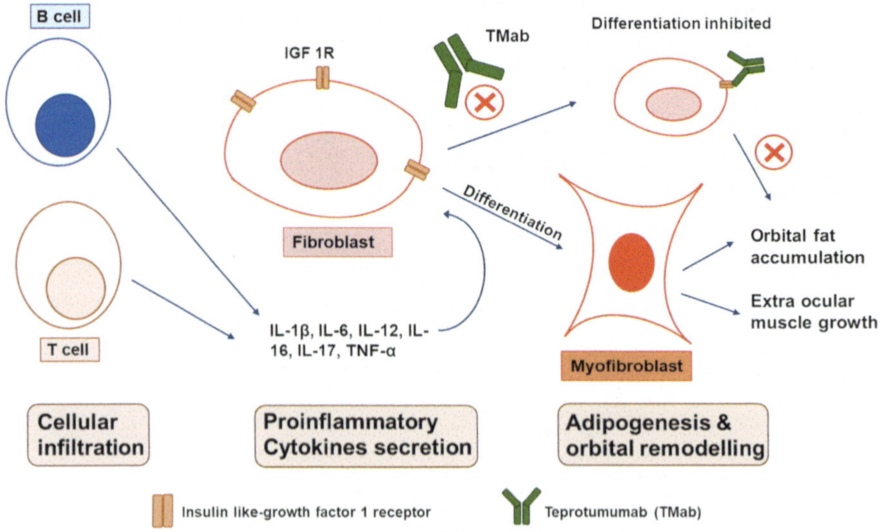

Fig. 7.2 The pathological mechanism in Grave's orbitopathy. The pathology is initiated by the infiltration of lymphocytes and fibrocytes in orbit, followed by cytokine secretion. The interaction between these cells through IGF-1 signalling leads to the differentiation and activation of fibroblasts resulting in orbital fat accumulation and ocular growth [24]. The IGF-1R-mediated signalling in the fibroblasts can be successfully blocked using Teprotumumab (TMab)-based therapy

of cytokines and chemokines [IL-16 and Regulated on Activation, Normal T-cell Expressed and Secreted (RANTES)]. It was also shown that IGF-1 receptor forms complexes with TSHR and contributes to the release of glycosaminoglycans [24, 25]. These discoveries formed the rationale for targeting the IGF-1R using mAbs, which could potentially ameliorate the disease symptoms and improve the outcomes in patients [26].

The IGF-1R is a transmembrane receptor belonging to the class of tyrosine kinase receptors. IGF-1R is a tetrameric protein consisting of two alpha and two beta subunits. The beta subunits extend into the cytoplasm, while the alpha subunits stretch from the membrane to the extracellular area. Disulfide bridges link the two subunits together [27]. The receptor gets activated upon binding its natural ligands, insulin-like growth factor-1 and insulin-like growth factor-2 (IGF-2). The alpha chains phosphorylate the beta chains in response to ligand binding or activation by autoantibodies (as seen in Grave's disease), which sets off a chain of intracellular processes that leads to the synthesis of glycosaminoglycans [23]. Teprotumumab (TmAb) is shown to effectively bind to the extracellular alpha subunit and block the IGF-1R mediated signalling in the fibroblasts, thus inhibiting the synthesis of glycosaminoglycans [26]. Teprotumumab is commercially available as a lyophilized powder (500 mg) for intravenous administration.

7.4.2 Risankizumab

Risankizumab (Skyrizi®; Abbvie) is a humanized mAb of the IgG1 class that was approved by the FDA for treating moderate to severe plaque psoriasis (a form of psoriasis) in adults [28]. Psoriasis is a chronic inflammatory skin disease marked by elevated red plaques and is often encased in silver scales formed from dead cells. Although multiple subtypes of psoriasis are reported, plaque psoriasis is predominant in the general population [29]. Psoriasis is a persisting inflammatory condition with elevated pro-inflammatory cytokines such as tumor necrosis factor- α (TNF-α), interleukin-1 (IL-1), IL-12, IL-17, and IL-23 [30]. Among the cytokines, the IL-17 and IL-23 mediated immune pathways are primarily implicated in the pathogenesis of psoriasis [31]. IL-17 is produced by a class of T-helper cells called Th17. These cells, along with regulatory T-cells and other T-helper cells such as Th1 cells, are usually present in the skin and believed to take part in the protective function. External factors such as infection, injury, smoking, stress, and medication can trigger a disbalance between Th17 and Th1 cells, resulting in chronic inflammation. The proliferation and levels of Th17 cells are regulated by the cytokine IL-23 produced by the myeloid cells [32]. The potential trigger factors mentioned above can activate the skin keratinocytes resulting in the secretion of cytokines IL-1β, TNF-α, IL-6, and IL-36 (Fig. 7.3). These cytokines, in turn, can activate the myeloid cells, such as dendritic cells and macrophages, producing IL-23, which triggers the differentiation (and maintenance) of the Th17 cells [33]. IL-17 secreted by these cells is shown to be responsible for the extravasation of neutrophils, T-cell infiltration, and neutrophil and platelet recruitment in the skin leading to plaque formation. Thus, blocking the

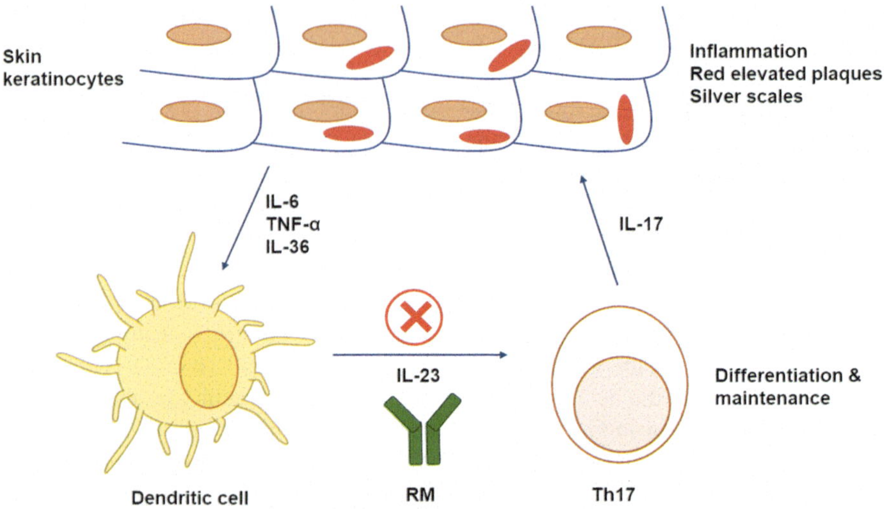

Fig. 7.3 Mechanism of action of Risankizumab (RM). The Mab binds to the cytokine IL-23 and prevents the differentiation of Th17 cells and the subsequent secretion of IL-17, contributing to inflammation and tissue damage in psoriasis patients

IL-17/IL-23 axis could potentially prevent the proliferation and differentiation of Th17 cells, reduce inflammation, and contribute to favorable outcomes in psoriasis patients [34, 35].

The biological target of Risankizumab, IL-23, in its active forms, consists of two subunits: p19 and p40, and their interaction is considered crucial for the biological function [36]. While the p19 subunit is specific for IL-23, the p40 subunit is shared with the cytokine IL-12, which, unlike IL-23, is shown to exert beneficial anti-inflammatory effects [37]. Risankizumab (RM) acts through the binding to the p19 subunit of IL-23 and thus blocking the differentiation of Th17 cells, the release of pro-inflammatory cytokines, and downstream inflammatory cascades. Other approved mAbs such as Guselkumab (Janssen Pharmaceuticals) and Tildrakizumab (Almirall S.A) target the p40 subunit of IL-23 and are currently used for the treatment of psoriasis. These Mabs can also bind to IL-12 (due to the shared subunits between IL-12 and IL23) and potentially hinder the anti-inflammatory effects of IL-23 [38]. Risankizumab was designed to overcome these complications and render therapeutic effects by explicitly recognizing and binding the p40 subunit of IL-23. Risankizumab is commercially available in solution form (150 mg/mL) for intravenous administration.

7.4.3 Sacituzumab Govitecan

Targeted therapy using Antibody-Drug Conjugates (ADC) is considered one of the significant advances in cancer therapy. The primary goal of ADCs is to selectively target the tumor cells and deliver the cytotoxic drug load to improve the therapeutic efficiency and reduce the detrimental effect of the drug on the system. In the United States, presently, there are nine FDA-approved and commercially available ADCs, demonstrating their strength as therapeutic medicines in oncology [39, 40]. Sacituzumab govitecan (Trodelvy®; Immunomedics) is one such ADC that was recently granted fast-track approval by the FDA for the treatment of metastatic triple-negative breast cancer [41]. Sacituzumab govitecan (SG) is humanized IgG1 mAb designed to target the trophoblast cell-surface antigen-2 (TROP-2), which is abundantly expressed in the surface of triple-negative breast cancer cells (compared to the normal cells). TROP-2 is a transmembrane protein shown to regulate signalling pathways associated with the migration and proliferation of tumor cells. TROP-2 expression is elevated in approximately 90% of triple-negative breast cancer patients and correlates with the disease severity [42–44]. Sacituzumab govitecan (SG) is a humanized mAb of Ig1 class, conjugated with the cytotoxic topoisomerase-1 inhibitor drug SN-38. The typical drug-to-antibody ratio (DAR) for SG is estimated to be 7–8. As shown in Fig. 7.4, SG binds the TROP-2 antigen on the tumor cell surface. The antigen-antibody complex gets internalized, resulting in the delivery of the cytotoxic drug through the hydrolysis of the CL2A linker (sensitive to pH change) [45, 46]. The cytotoxic drug SN-38 induces cell death through DNA damage by binding to topoisomerase I. Other than that, SN-38 is also shown to block the expression of oncogenes by preventing the binding of far upstream binding protein-1 (FUBP1) to Far upstream elements [47]. SG is commercially available in powder form (180 mg) and administered intravenously.

7.4.4 Belantamab Mafodotin

Belantamab mafodotin (Blenrep®; GlaxoSmithKline) is an FDA-approved ADC used in the treatment of relapsed and refractory multiple myeloma (RRMM) [19]. Belantamab mafodotin (BM) is an Ig1 class mAb conjugated to a cytotoxic drug, monomethyl auristatin-F (MMAF). The mAb targets the B-cell maturation antigen (BCMA), which is expressed in high levels in CD138+ myeloma cells [48]. BCMA is a type-III transmembrane protein belonging to the tumor necrosis factor (TNF) receptor superfamily and plays a vital role in the differentiation of B cells into plasma cells. BCMA is expressed by normal plasma cells; however, their levels are significantly higher in the myeloma cells [49, 50]. As shown in Fig. 7.5, the binding of the natural BCMA receptor ligands such as B-cell-activating factor (BAFF) or a proliferation-inducing ligand (APRIL) can activate multiple downstream signalling pathways like nuclear factor kappa B (NFκB) signalling, mitogen-activated protein (MAP) kinase pathway, Elk-1-mediated signalling, c-Jun N-terminal kinases (JNK), and p38 kinase which are crucial in the survival and proliferation of myeloma cells [51, 52]. *In vivo* studies on myeloma cells have shown that elevated levels of BCMA or binding of APRIL to BCMA can promote the expression of anti-apoptotic proteins, rendering protection to the tumor cells from

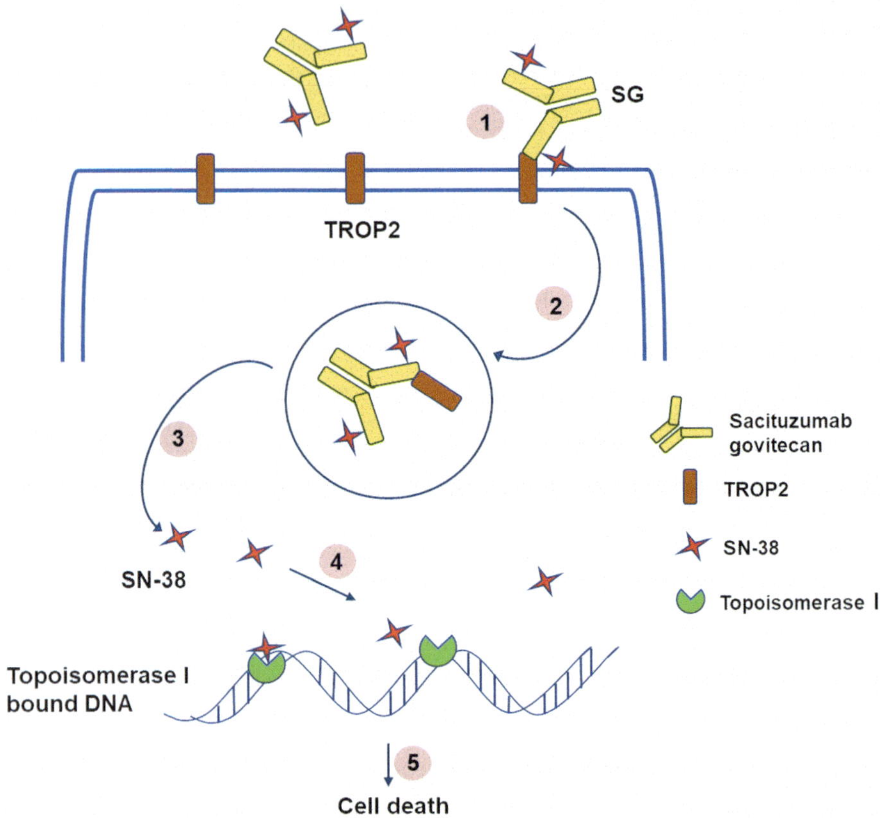

Fig. 7.4 Stepwise illustration of the action of sacituzumab govitecan (SG) on tumor cells. (1) The binding of SG to the target receptor, trophoblast cell-surface antigen 2 (TROP2), on the surface of tumor cells, (2) Internalization of the SG-TROP2 complex, (3) Release of SN-38 by hydrolysis of the linker and (4) and (5) indicates the binding of SN-38 to topoisomerase 1 and induction of cell death [46]

apoptosis [53]. Thus, blocking the BCMA receptor function using therapeutic mAbs could be an effective treatment for relapsed and refractory multiple myeloma.

Belantamab mafodotin is designed and works similar to other ADCs except for a few notable improvements related to the glycosylation and the cytotoxic payload, as claimed by the manufacturers. Once bound to the BCMA receptor on the target myeloma cells, the mAb, apart from blocking the receptor function, can deliver the cytotoxic payload to the tumor cells. MMAF is a microtubule disrupting agent that binds to tubulin and arrests the cell cycle at the DNA damage checkpoints, leading to apoptosis [54]. The drug has an estimated DAR of 4. The mAb is afucosylated at the Fc region to promote efficient antibody-dependent cellular cytotoxicity (ADCC). The drug MMAF is covalently attached to the mAbs through a highly protease-resistant maleimidocaproyl linker. BM effectively eliminates myeloma cells through

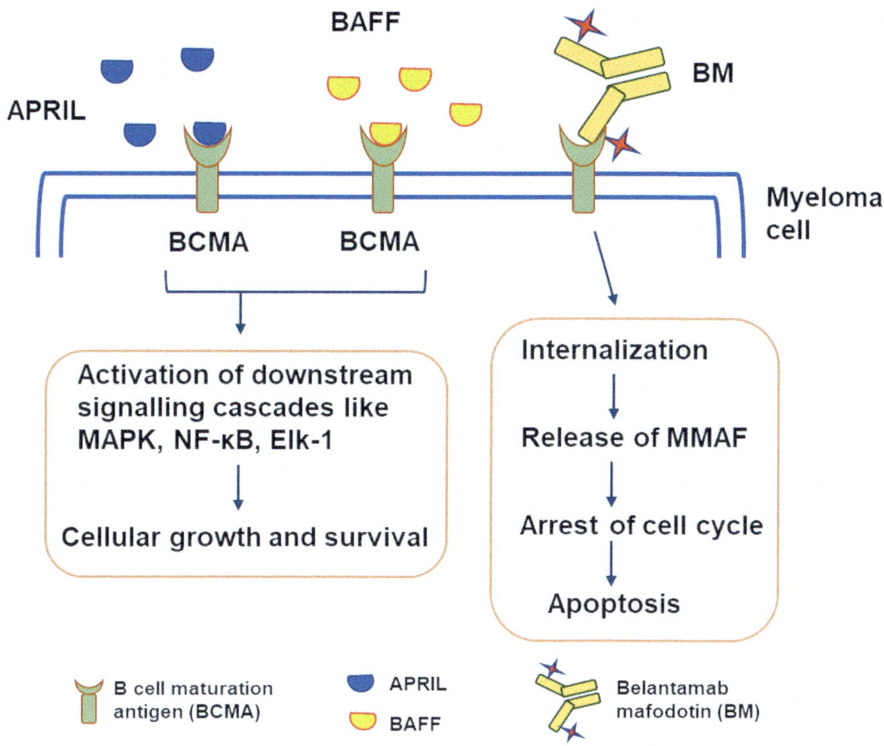

Fig. 7.5 Mechanism of action of Belantamab mafodotin (BM). B-cell maturation antigen (BCMA) receptor expressed on the surface of myeloma cells can be activated by either B cell-activating factor (BAFF) or a proliferation-inducing ligand (APRIL). The activation can lead to further activation of downstream signalling molecules, which trigger the expression of proteins that supports the growth and survival of myeloma cells. However, blocking the BCMA by Belantamab mafodotin inhibits cellular growth and destroys myeloma cells through the action of cytotoxic monomethyl auristatin-F (MMAF)

multiple mechanisms like (i) inhibitory action of cytotoxic MMAF, (ii) ADCC, and (iii) antibody-dependent cellular phagocytosis (ADCP) [48, 55, 56]. The ADC is commercially available as a lyophilized powder (50 and 100 mg) and administered intravenously.

7.4.5 Blinatumomab

One of the recent advances in targeted therapy is the development of bispecific antibodies that can target two antigens simultaneously. These antibodies exhibit bivalent specificity and have found successful applications in cancer immunotherapy [57, 58]. The conventional mAbs used in the therapy exert their action through ADCC or ADCP involving immune cells like monocytes, neutrophils, and

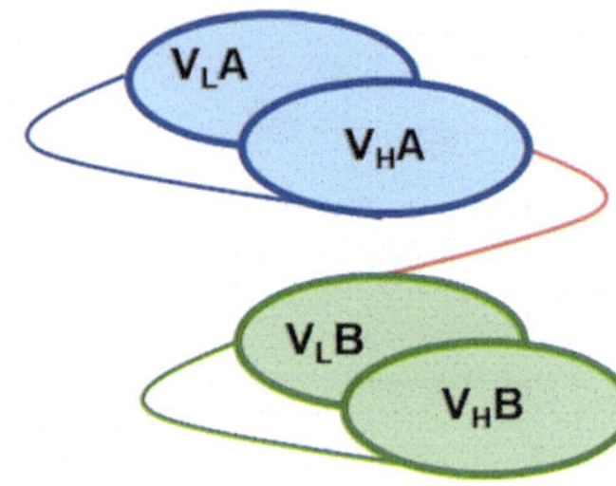

Fig. 7.6 Schematic representation of a bispecific antibody

macrophages [55]. As evidenced in tumor cells, the primary limitation of employing the mAbs is their inability to directly activate the T-cell-mediated immune response, thus providing a channel for tumor cells to escape [59]. Bispecific antibodies can address these difficulties through their dualistic binding abilities, as they are comparatively more flexible regarding their functional abilities. The three distinct and contrasting abilities of bispecific antibodies are (i) the ability to act as a bridge between target (tumor) cells and the cells of the immune system, (ii) targeting the immune checkpoints to restore or activate the immune system, and (iii) simultaneously block two signalling pathways [60]. The production of bispecific antibodies begins with the screening and recombinant expression of appropriate single-chain variable fragments (ScFvs) of two different antibodies. A biologically active ScFv (single polypeptide) can be produced by combining the heavy chain ScFv (V_H) with the light chain ScFv (V_L) of an antibody. Bispecific antibodies are produced by connecting two ScFvs using a short amino acid linker sequence. These linkers usually comprise glycine residues and prevent the non-covalent self-association of ScFvs, which can potentially impair the antigen binding. As shown in Fig. 7.6, bispecific antibodies are produced in the following arrangement: the variable heavy and light chains of two different antibodies (A and B) linked through glycine linkers to generate two polypeptide chains, V_HA-V_LA and V_HB-V_LB, capable of recognizing two different antigens.

Blinatumomab (Blincyto®; Amgen) is a bispecific antibody approved for the treatment of relapsed or refractory acute lymphoblastic leukemia (RR-ALL) [61]. The two arms of Blinatumomab (BtMab) are made up of ScFvs capable of binding to CD19 and CD3 ligands of the B-cell lymphoblasts and the T-cells, respectively. Blinatumomab is designed in the following configuration: V_LCD19–V_HCD19 linked to V_LCD3–V_HCD3. A small non-immunogenic peptide linker of serine and glycine residues connects the CD19 and CD3 ScFvs and renders considerable flexibility [62]. Blinatumomab is approximately 55 kDa, which is significantly smaller than a typical mAb (~150 kDa). The smaller size is beneficial, as it enhances the tumor cell permeation, brings about effective cell–cell interactions due to the proximity, and reduces potential immunogenicity [63]. Blinatumomab belongs to a class of bispecific antibodies called bispecific T-cell engagers (BiTEs) that facilitates the action of T-cells (through CD3) on tumor cells via the tumor antigen, i.e., CD19. Blinatumomab induces the apoptosis of tumor cells through the following steps: The higher affinity of the BtMab ScFv arm ($K_d \sim 10^{-9}$ M) favors the

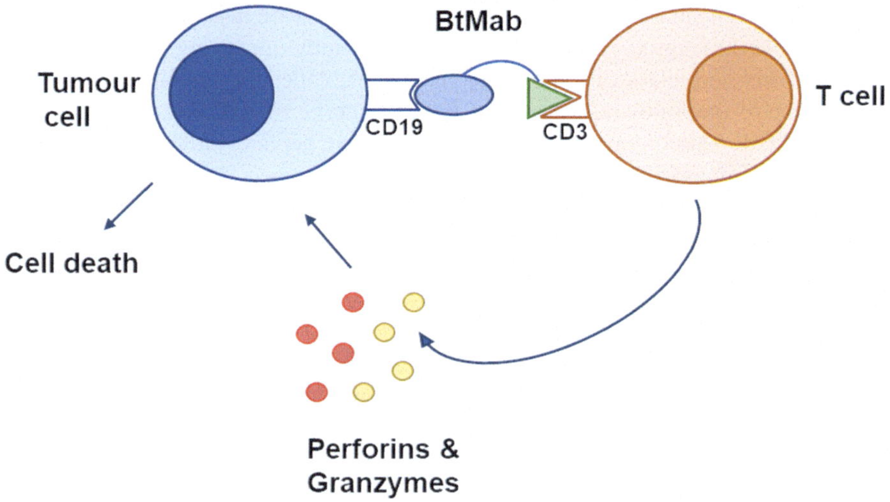

Fig. 7.7 Mechanism of action of Blinatumomab (BtMab). Cell bridging by the bispecific T-cell engagers (BiTE) class BtMab brings tumor cells (B-cell lymphoblast) and T-cells into proximity, facilitating direct interaction between CD3 and CD19. Cell bridging by BtMab initiates T-cell-mediated cytotoxicity on tumor cells by releasing perforins and granzymes [66]

binding of BtMab with CD19 ligand, leads to the build-up of BtMab-coated tumor cells. In the subsequent steps, T-cells bind to the other arm of Blinatumomab, forming the immunologic synapse. Blinatumomab allows the activation of the immunologic synapse without the typical interactions involving the T-cell receptor, other signalling complexes, and costimulatory signals with the antigen presentation (by tumor cells) through the major histocompatibility complex (Fig. 7.7) [64]. In principle, the antigen presentation through MHC complexes is hindered in the tumor cells due to their evolved protective mechanisms that downregulate the MHC expression [65]. Upon binding to T-cells (via CD3), Blinatumomab effectively reduces the proximity between the tumor cell and T-cell, initiating the interaction between CD3 and CD19, resulting in the formation of the immunological synapse. These events result in T-cell activation and elicit T-cell-mediated cytotoxicity by releasing perforin and granzyme. Perforins are lytic proteins that form channels in the membrane of tumor cells, allowing for the entry of calcium ions and granzymes with subsequent activation of caspases and apoptosis, leading to the death of tumor cells [66].

7.4.6 Luspatercept

Peptide and protein therapeutics are often rapidly cleared from circulation. This results in an increased frequency of drug administration or an increase in the effective drug dose to maintain the optimal drug concentration for therapeutic effect

[67]. To overcome these drawbacks, therapeutic proteins are fused with the fragment crystallizable (Fc) region of IgG antibodies, significantly improving their circulation half-life. This new class of therapeutic molecules, called Fc fusion proteins, can contain one of the following molecules fused to the Fc region, viz., an engineered extracellular domain of a receptor, biologically active peptide, biological trap, or an enzyme [68–70].

Luspatercept (Reblozyl®; Celgene) is an FDA-approved Fc fusion protein used to treat anemia associated with β-thalassemia [71]. Ineffective erythropoiesis (red blood cell generation in the bone marrow) is a symptom of this condition caused by a faulty or decreased synthesis of hemoglobulin beta subunits. This defect leads to subsequent accumulation of heme and precipitation of free alpha globins, causing severe anemia in β-thalassemia patients. Patients develop hypergenesis of the bone marrow and spleen (splenomegaly), iron homeostasis disturbances, and an accumulation of reactive oxygen species in the erythroid cells, all contributing to anemia [72, 73]. It is well recognized that the late-stage differentiation of erythroid cells is impeded in β-thalassemia; therapeutics promoting late-stage erythropoiesis could potentially minimize the hemolysis and improve the disease outcomes.

The transforming growth factor-beta (TGF-β) signalling regulates the growth and differentiation of hematopoietic stem cells [74]. TGF-β superfamily comprises more than 30 soluble, secreted factors, including but not limited to TGF-β, bone morphogenetic proteins (BMPs), growth differentiation factors (GDFs), and activins. The binding of these ligands triggers the assembly of a ternary complex that controls erythroblast differentiation through the activation of downstream effector proteins [75]. The ternary receptor complex consists of heteromeric type I and type II receptors. The type I and type II receptors have been categorized into seven and five subtypes, respectively. The type of ligand binding to the receptors dictates the combination of receptor complex formed and the subsequent activation of downstream signalling events [76]. Activin receptor type IIB (ACTRIIB) belongs to the TGF- β superfamily receptors and is activated by binding its natural ligands (activin A, GDF8, GDF11, BMP11). The binding of these ligands is shown to activate effector proteins of the Smad family, particularly Smad 2/3. These proteins translocate into the nucleus and trigger the expression of a specific set of genes that suppress the differentiation of erythroblasts [77, 78]. Interestingly, among the Smad proteins, the levels of Smad 2/3 are found to be elevated in the bone marrow of β-thalassemia patients and other myelodysplastic syndromes, indicating the ACTRIIB-mediated suppression of erythroblast differentiation [79].

Luspatercept is a recombinant fusion protein containing a modified ACTRIIB domain (TGF-β receptor) fused with the Fc region of human IgG1. The therapeutic fusion protein effectively inhibits the TGF-β signalling by trapping the ligands and subsequently blocking the activation of the receptors on the cell surface [80]. The recombinant fusion protein reduced hemolysis and successfully promoted late-stage erythropoiesis in β-thalassemia patients. In the early phase studies, the drug was shown to increase hemoglobin levels in a dose-dependent manner [81]. Further, phase III clinical trials data showed that treatment with Luspatercept significantly reduced the blood transfusion dependence in the patient's group compared to the placebo group with minimal adverse effects [82]. Luspatercept differs from approved treatment options like Aranesp™ and Erythropoietin™, which promote early-stage erythroblast differentiation [83]. Luspatercept is commercially available as Reblozyl® and administered subcutaneously.

7.4.7 Efmoroctocog Alfa

Efmoroctocog alfa (Eloctate®; Swedish Orphan Biovitrum AB) is a recombinant fusion protein comprising recombinant coagulation Factor VIII fused with polyethylene glycol is covalently linked to the dimeric IgG1 Fc domain. Efmoroctocog alfa was approved as an enzyme replacement therapy by the FDA to treat the rare genetic disorder Haemophilia A [84]. Patients diagnosed with Haemophilia A lack the coagulation Factor VIII and suffer from continuous bleeding and excessive blood loss in the event of an injury or trauma [85].

Human Factor VIII (FVIII) is a globular protein consisting of heavy (200 kDa) and light chains (80 kDa) connected through covalent bonds. FVIII is complexed with von Willebrand Factor and stabilized by calcium and copper ions in an inactive form. In case of an injury, FVIII separates from the complex and participates in the blood coagulation cascade. The active FVIII serves as a cofactor for coagulation Factor IXa, which in complex with calcium ions and phospholipids, is responsible for the activation of Factor X. This results in a series of downstream events, with the formation of thrombin and eventually leading to the conversion of Fibrin from Fibrinogen and the formation of blood clot [86, 87]. The lack of active FVIII in Haemophilia A patients leads to impaired coagulation pathways and the failure to produce blood clots. Thus, enzyme replacement therapy using recombinant FVIII could be an effective treatment option for Haemophilia A. Although effective, the previous generation of recombinant FVIII (Moroctocog alfa) was rapidly cleared and required frequent injections. Efmoroctocog alfa is a new generation drug designed to extend the half-life of the recombinant protein drug [88]. Efmoroctocog alfa, used as a prophylactic treatment, effectively reduces the bleeding in case of trauma or surgery. It also reduces the inhibitory action exerted by neutralizing antibodies [84]. Efmoroctocog alfa is available in the injection form for intravenous administration.

7.4.8 Afamelanotide

Afamelanotide (Scenesse®; Clinuvel) is a synthetic peptide analog of α-melanocyte-stimulating hormone (α-MSH) approved for the treatment of skin damage in patients suffering from erythropoietic protoporphyria (EP) [28]. The disease is characterized by the deficiency of a key enzyme involved in the heme biosynthetic pathway called ferrochelatase. The enzyme catalyzes the addition of iron into the porphyrin ring to produce a biologically active heme. Thus, the lack of ferrochelatase in EP patients results in the accumulation of a chemical intermediate called protoporphyrin IX in

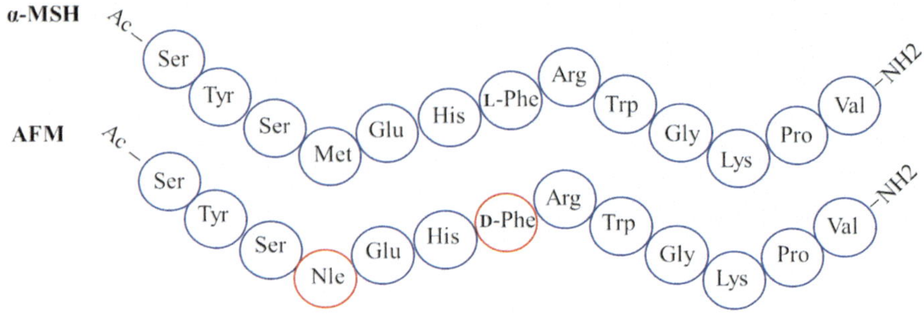

Fig. 7.8 Schematic representation of the peptide hormone, alpha-melanocyte stimulating hormone (α-MSH), and its chemically synthesized analog afamelanotide (AFM). The peptide analog is modified at positions four and seven (shown in red) with norleucine (Nle) and D-phenylalanine (D-Phe), respectively. MSH and AFM can bind to its natural receptor melanocortin-1 receptor and increase the expression of the enzyme tyrosinase, which is involved in the synthesis of eumelanin from tyrosine in a multi-step process [92]

the liver and the external surface of the skin. The exposure of the EP patient's skin to sunlight results in the production of reactive oxygen species from protoporphyrin IX that causes skin damage [89, 90].

Alpha-melanocyte stimulating hormone (α-MSH) is a peptide hormone (13 amino acids) belonging to the melanocortin family. The hormone is produced endogenously by the keratinocytes and pituitary gland [91]. The binding of α-MSH to the melanocortin-1 receptor leads to the G-protein coupled receptor (GPCR)-mediated activation of the melanogenesis pathway. The binding of α-MSH leads to the activation of multiple kinases and transcriptional factors, resulting in the expression of tyrosinase, the enzyme responsible for synthesizing eumelanin from tyrosine [92]. The final product, eumelanin, is the skin's dark pigment that renders protection against UV-induced skin damage. Apart from its photoprotective role, eumelanin is also shown to exhibit anti-inflammatory effects [93]. Thus, increasing the expression of eumelanin could have protective effects in EP patients who frequently experience skin damage due to porphyrin accumulation and UV exposure.

Afamelanotide (AFM) is a chemically synthesized peptide analog of the hormone α-MSH. As illustrated in Fig. 7.8, AFM differs from the natural hormone at two positions, i.e., methionine and phenylalanine at the fourth and seventh positions are replaced with norleucine and D-phenylalanine, respectively. Such a modification improves the affinity of the peptide to the receptor, biological activity, and half-life. The binding of AFM to the melanocortin-1 receptor activates the GPCR-mediated signalling pathway with the subsequent increase in the expression of tyrosinase and eumelanin synthesis [94, 95]. Data from the phase 3 trials of AFM showed that the peptide drug exerted photoprotective effects and significantly improved the "pain-free time" in EP patients after 6 months. AFM is commercially available as implants (16 mg).

7.4.9 Macimorelin

Macimorelin (Macrilen®; Æterna Zentaris) is a novel tripeptide growth hormone secretagogue (GHS) approved for the diagnosis and treatment of adult growth hormone deficiency [96]. Growth hormone secretagogues are synthetic compounds capable of binding to growth hormone secretagogue receptors (GHSR) present in the hypothalamus and pituitary gland. The stimulation of GHSR by its natural ligand, ghrelin, results in the secretion of growth hormones and subsequent promotion of growth and development. Ghrelin is a 28 amino acid polypeptide produced by the endocrine cells of the stomach. Macimorelin is a chemically synthesized analog of ghrelin that mimics the natural ligand's structure and biological activity triggering the production of growth hormones [97].

Growth hormone deficiency is characterized by the inability of the body to produce the required amount of growth hormones. The condition is observed in childhood and adulthood and poses challenges for diagnosis, particularly in adults [98]. Deficiency of growth hormones in adults manifests with the following symptoms: accumulation of body fat around the waist, weakness, reduced sexuality, depression, lean body mass, reduced bone density, dyslipidemia, and can gradually lead to glucose intolerance. Adult growth hormone deficiency (AGHD) can arise due to damage to the pituitary or hypothalamus arising from a head injury, radiation, surgery, presence of a tumor, or hormone deficiency arising from genetic factors [99, 100]. Further, growth hormones in the circulation tend to get absorbed rapidly by the body. Due to their rapid absorption and clearance, blood tests are usually ineffective in diagnosing growth hormone deficiency. To overcome this, clinicians stimulate and check the pituitary gland's ability to secrete the hormone instead of directly measuring the growth hormones. Also, with the symptoms of AGHD being non-specific, at least one growth hormone stimulation test is mandatory to confirm the growth hormone deficiency [101, 102].

Macimorelin is shown to effectively stimulate the GHSR and trigger the secretion of growth hormones into the circulation. Macimorelin exhibits a similar affinity to GHSR as ghrelin (the natural ligand), with superior stability and oral bioavailability. The drug showed dose-dependent release of growth hormones in healthy volunteers, with maximum levels achieved in just 1 h [103]. Validation of diagnostic efficiency and safety of Macimorelin on AGHD patients revealed 82% sensitivity, 92% specificity, and 13% misclassification rate [104]. Apart from diagnosis, Macimorelin is also used to treat growth deficiency in children and cachexia in adults. Cachexia is the characteristic weight loss, muscle wasting, and loss of body fat experienced by patients with cancer, cystic fibrosis, human immune deficiency virus (HIV), chronic kidney disease, and autoimmune diseases [105]. Macimorelin is available as sachets containing granules of Macimorelin acetate (60 mg) and administered orally.

7.4.10 Pegcetacoplan

Pegcetacoplan (Empaveli®; Apellis Pharmaceuticals) is a PEGylated peptide (covalently conjugated with polyethylene glycol, PEG) approved for the treatment of the rare hematological condition, paroxysmal nocturnal hemoglobinuria (PNH), [106]. The disease is characterized by the mutation of the phosphatidylinositol N-acetyl glucosaminyltransferase subunit A (*PIGA*) gene required to synthesize the N-acetylglucosaminyl phosphatidylinositol (GPI) anchor. This defect results in the inability of cells to anchor proteins at the membrane surface, leading to a loss of function. In PNH, hemopoietic cells with defective *PIGA* gene are incapable of anchoring two key membrane proteins, CD55 and CD59, due to which the cells undergo complement-mediated lysis [107]. CD55 and CD59 are critical regulators of the complement system and protect the host cells from complement-mediated clearing by inhibiting select complement factors [108]. Thus, a lack of CD55 and CD59 in PNH cells increases their sensitivity to complement factors and uncontrolled complement activation, resulting in hemolysis, thrombosis, and bone marrow abnormalities [109]. Thus, targeting the complement factors could serve as a potential therapy and ameliorate disease symptoms in PNH.

The alternative complement pathway is initiated by the spontaneous activation of C3 convertase (C3bBb), producing C3a and C3b from C3 (Fig. 7.9). Next, C3b bound Factor B is cleaved by Factor D into C3bBa and C3bBb fragments. The larger C3bBb fragment can initiate a vicious cycle of further complement activation and production of C3b. The complement factor, C3b, can directly attach to target cells, acting as an opsonin and promoting phagocytosis [110, 111]. CD55 or complement decay-accelerating factor (DAF) is a GPI-anchored type I surface protein capable of protecting the host cells from complement-mediated attack. CD55 is a glycoprotein expressed on the surface of all hemopoietic (erythrocytes) and non-hemopoietic cells. CD55 recognizes and binds complement proteins, C3b (alternative pathway), and C4b (Classical/Lectin pathway). CD55, which is anchored to the cell membrane via a GPI anchor, binds the C3 convertase (C3bBb) of the alternative pathway and dislodges the Bb component from the C3 convertases rendering the C3 convertases ineffective. CD55 accelerates the decay of C3 convertases and is often referred to as complement decay-accelerating factor (DAF) [108]. As the alternative pathway proceeds (Fig. 7.9), C5 is cleaved into C5a and C5b by a complex called C5 convertase (consists of C3bBbC3b). The C5b and complement factors C6, C7, C8, and C9 form the membrane attack complex (MAC) on the cell surface, causing cell death. CD59 or Protectin is a GPI-anchored inhibitory protein that prevents the formation of MAC. CD59 binds to C8 of the C5bC6C7C8 complex and inhibits the binding of C9 and the formation of MAC and subsequent cell lysis [110, 112]. To summarize, complement factor C3 can be considered as a critical target to prevent hemolysis for the following reasons: (i) capacity to spontaneously activate alternative complement pathway, (ii) act through a vicious positive feedback loop generating more C3b complement factor, (iii) promote phagocytosis through opsonization, and (iv) renders some of the existing therapies (discussed below) ineffective due to the C3's involvement in the upstream of the MAC formation.

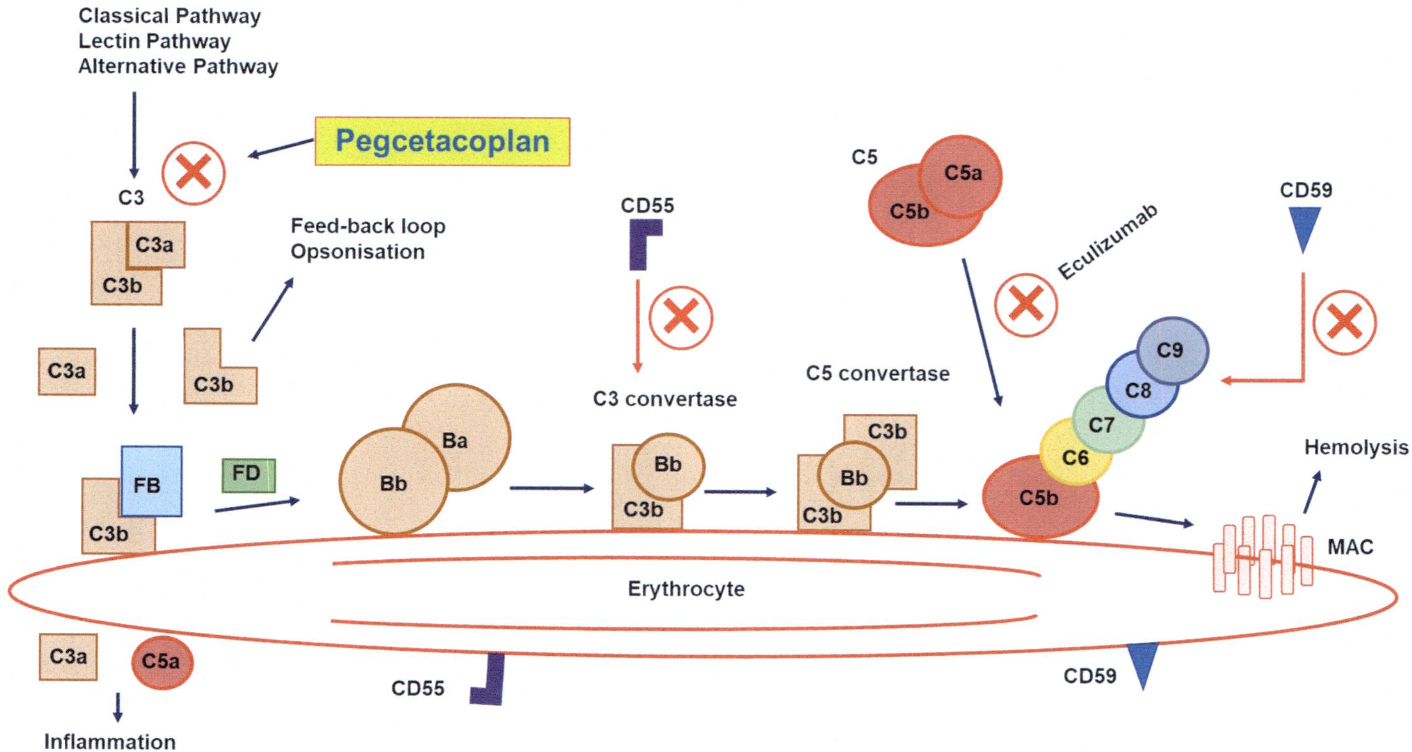

Fig. 7.9 Illustration of complement pathway and the role of each complement factor. Also shown are the target and inhibitory action of membrane-anchored natural inhibitors of the complement pathway, CD55 and CD59, and the commercially approved complement factor inhibitors, Pegcetacoplan and Eculizumab

Intravascular hemolysis, which occurs in blood vessels, and extravascular haemolysis in the liver and spleen, are two types of hemolysis observed in PNH patients [106]. C5 inhibitors were primarily used to treat PNH patients before the recent approval of Pegcetacoplan (PCN). For instance, Eculizumab (Soliris®; Alexion Pharmaceuticals) is a therapeutic mAb approved for the treatment of PNH. Eculizumab is shown to inhibit the conversion of C5 into C5a and C5b, thus preventing the formation of MAC. Eculizumab is shown to be particularly effective in treating intravascular hemolysis primarily caused by MAC formation [113]. However, extravascular hemolysis is observed in the liver and spleen of PNH patients as C3 fragments opsonize erythrocytes and reduce their circulating half-life. Thus, despite treatment with EM, PNH patients experienced persistent hemolytic anemia and remained blood transfusion dependent [114]. Pegcetacoplan is a PEGylated peptide inhibitor designed to inhibit the C3 complement factors and overcome the drawbacks associated with Eculizumab. Pegcetacoplan is a cyclic peptide comprising 15 amino acids and binds to C3 and C3b complement factors. Pegcetacoplan was effective in treating PNH patients, controlling both intravascular and extravascular hemolysis [115]. A randomized Phase III trial showed that PNH patients treated with Pegcetacoplan showed improved hemoglobulin levels and favorable outcomes in controlled extravascular and intravascular hemolysis [106]. Pegcetacoplan is available in solution form (1080 mg/20 mL) and administered subcutaneously.

7.4.11 Lonapegsomatropin

Lonapegsomatropin (Skytrofa®; Ascendis Pharma) is a carrier-conjugated growth hormone approved by the FDA for treating pediatric growth deficiency for patients with the following characteristics: Age > one year and weight > 11.5 kg [116]. The hormone conjugate comprises somatotropin attached to the carrier molecule methoxypolyethylene glycol. Upon reaching the bloodstream, the prodrug undergoes self-cleavage releasing somatotropin from the carrier molecule. Somatotropin binds to the growth hormone receptor (GHR) and promotes cell division and growth. The activation of GHR is necessary for multiple physiological processes like long bone epiphyses, chondrocyte production, hydrolysis of glycogen in the liver, and lipolysis [117]. Unlike the previous generation drugs like Genotropin (Pfizer), which required daily administration, Lonapegsomatropin can be used as a weekly administered drug due to its design that imparts considerable stability and improves the half-life of the hormone prodrug [118].

Growth hormone receptors (GHR) are homodimeric proteins belonging to the family of class I cytokine receptors [119]. The binding of somatotropin induces structural changes in the dimeric GHR leading to the activation of Janus kinase 2 (JAK2), which phosphorylates the receptor at multiple tyrosine residues. The SM binding-induced conformational changes of the GHR allow the binding of transcriptional factors called a signal transducer and activator of transcription (STAT). Subsequently, Janus kinase (JAK) phosphorylates different members of the STAT family transcriptional factors like STAT1, STAT3, STAT5a, and STAT5b and initiates the STAT signalling. The phosphorylated transcriptional factors translocate into the nucleus and initiate the transcription of specific genes [120]. The stimulation of GHR can also activate other signalling pathways like Ras/extracellular signal-regulated kinase (ERK) and PI 3-kinase/Akt. Thus, the activation of GHR by growth hormones can activate multiple signalling cascades and initiate the transcription of genes responsible for several physiological processes, including growth and development [121, 122].

The secondary effect of the stimulation of GHR by growth hormones is the production of a major growth factor, IGF1, in the liver. IGF1 acts through the IGF1 receptors present on all the cells in the body. IGF1 can either act together with other growth hormones or independently promote the growth of cells; however, the production of IGF1 requires the stimulation of GHR by growth hormones [123]. The physiological process associated with the growth hormone (GH)/Insulin-like growth factor 1 (IGF1) axis helps us understand their role in growth hormone deficiency-related diseases and develop potential therapies. The stimulation of GHR is shown to be essential for the differentiation of mesenchymal stem cells into chondrocytes (cartilages) and osteoblast (bones), thus taking part in physiological processes of cartilage and bone formation, respectively [124, 125]. The GH/IGF1 axis also plays a significant role in the growth and development of skeletal muscles. It was shown that IGF1 promotes skeletal muscle cell hypertrophy by regulating the calcium-calmodulin-dependent phosphatase called calcineurin [126]. Besides, GH has been proven to increase skeletal muscle cell development by stimulating the integration of myoblasts with newly formed myotubes [127]. Thus, stimulation of GHR with growth hormones can serve as an effective treatment option for deficiencies associated with growth hormones. Lonapegsomatropin is available in lyophilized form for subcutaneous administration.

7.4.12 Setmelanotide

The FDA has approved Setmelanotide (Imcivree®, Rhythm Pharmaceuticals) for treating obesity arising from the deficiency of proopiomelanocortin (POMC), proprotein convertase subtilisin/kexin type 1 (PCSK1), or leptin receptor (LEPR) [19]. Setmelanotide is a synthetic peptide with a cyclic structure designed using proopiomelanocortin as a template and acts as a melanocortin 4 receptor (MC4R) agonist. Unlike α-melanocyte-stimulating hormone (the endogenous ligand of MC4R), Setmelanotide is shown to be approximately 20-fold more potent with half-maximal effective concentration (EC_{50}) in the nanomolar concentration [128]. Further, Setmelanotide is shown to simultaneously occupy both allosteric and orthosteric binding sites of the MC4R and exhibit high affinity and specificity [129]. Setmelanotide is commercially available in solution form (10 mg/mL) for subcutaneous administration.

Neuronal signalling pathways regulate hunger and body weight through the action of peripherally derived molecules. In line with that, the MC4R, a class of GPCR, expressed throughout the CNS, and the activation events, both upstream and downstream of this receptor (discussed below), can impact food intake and obesity [130]. Proopiomelanocortin (POMC) is a product of the *POMC* gene, synthesized as a precursor peptide (241 amino acids) in the anterior pituitary (by the corticotrope cells), arcuate nucleus of the hypothalamus, and skin melanocytes. POMC undergoes cleavage by a series of enzymatic action and post-translational modifications depending on the tissue and the enzyme availability, generating multiple biologically active peptides. One of the enzymes involved in POMC cleavage, proprotein convertase subtilisin/Kexin type 1, is shown to cleave multiple substrates such as POMC (pituitary), proinsulin, and proglucagon (pancreas). The cleavage of POMC in the neurons generates a biologically active tetrapeptide called α-melanocyte-stimulating hormone (α-MSH), which acts as a natural ligand of MC4R [131, 132].

Neurons expressing POMC and Agouti-related protein (AgRP) are two different classes of neurons in the hypothalamus that regulate hunger and feeding behavior [133]. Ghrelin is a growth hormone that acts on AgRP class neurons triggering the production of Agouti-related protein (AgRP) and neuropeptide YY (NPY), which activates the orexigenic pathway (Fig. 7.10). Similarly, leptin, the hormone secreted by the adipose tissue, acts on POMC neurons triggering the production of α-MSH, which binds to MC4R and activates the anorexigenic pathway [134]. Besides their stimulatory effects, the two hormones, ghrelin and leptin, are also capable of regulatory functions. The activation of AgRP neurons by ghrelin leads to the production of AgRP, inhibiting the anorexigenic pathway by binding to MC4R (Fig. 7.10). On the other hand, leptin can also bind to receptors on the AgRP neurons and inhibit the production of AgRP and NPY. This shows the tight regulation of orexigenic and anorexigenic pathways by peripherally derived signals [131, 135]. Overall, activation of MC4R and the downstream signalling events are critical for satiety and weight loss.

Setmelanotide has shown promising results in treating obesity arising from the deficiency of POMC, PCSK1, and LEPR. Phase III clinical trials evaluating the efficiency of ST showed that 45% of patients with leptin receptor deficiency and 80% of patients with proopiomelanocortin deficiency showed a 10% weight reduction with minimal adverse effects [136]. Setmelanotide is currently tested on other rare obesity-associated genetic disorders such as Alström syndrome, Bardet–Biedl syndrome, and other diseases associated with MC4R pathway deficiency [137].

7.4.13 Galium-68-DOTATOC

Dotatate gallium (Ga-68) (Gallium-68 Dotatate®; Advanced Accelerator Applications) is a radioactive agent approved for use in positron emission tomography (PET)-based diagnosis of neuroendocrine tumors (NET) in both adults and pediatric patients [138]. Ga-68 dotatate is an eight amino acid cyclic peptide conjugated to the chelator dodecane tetraacetic acid (DOTA). Ga-68 dotatate has the following sequence, H-D-Phe-Cys-Tyr-D-Trp-Lys-Thr-Cys-Thr-OH, with a disulfide bond linking the two cysteines. The radioactive peptide is an analog of

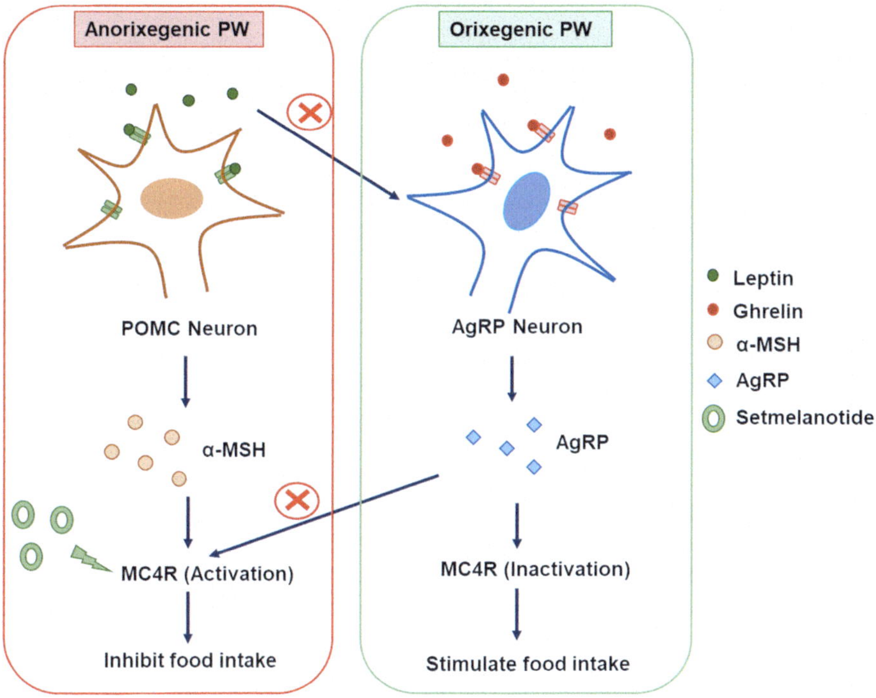

Fig. 7.10 Schematic representation of the regulation of food intake and hunger by anorexigenic and orexigenic pathways. Also shown are the different hormones (leptin and ghrelin) and peptides/proteins (Agouti-related protein/AgRP and α-melanocyte-stimulating hormone/α-MSH) that regulate the anorexigenic and orexigenic pathways. Setmelanotide is the approved peptide drug that effectively binds to Melanocortin 4 receptor (MC4R) and activates signalling cascades, subsequently reducing food intake and resulting in weight loss

somatostatin subtype 2 receptors (ST-2R), frequently overexpressed in NETs [139]. Ga-68 dotatoc binds to ST-2R with high affinity making it an efficient PET tracer (a molecule that binds to specific targets and emits positron). Upon intravenous injection, the radioactive peptide binds to ST-2R. The radioactive nuclide in the peptide tracer decays and results in an annihilation event recorded using PET systems. Ga-68 dotatoc can be used for diagnosis, quantitative imaging, and localization of NETs [140, 141]. Ga-68 DOTATOC injection contains 3.6 μg/mL peptide prepared in a sterile buffered solution.

7.4.14 Pegvaliase

Pegvaliase (Palynziq®; BioMarin Pharmaceutical) was approved as an enzyme replacement therapy to treat phenylketonuria by the FDA in 2018 [138]. The rare autosomal recessive disorder is characterized by the deficiency of the enzyme phenylalanine hydroxylase, which catalyzes the conversion of phenylalanine into tyrosine. Phenylalanine hydroxylase deficiency leads to the excessive accumulation of phenylalanine which causes brain dysfunction affecting the patient's intellectual abilities and behavior. Further, patients also experience neurological and psychiatric symptoms [142]. As per the established guidelines [143], it is mandatory for phenylketonuria patients to maintain blood phenylalanine levels between 120 and 360 μmol/L. In addition to the accumulation of phenylalanine from enzyme deficiency, phenylalanine levels can also increase from the intake of certain high-protein food products like meat, fish, milk, and milk products [144].

Pegvaliase is a recombinant form of the naturally occurring enzyme phenylalanine ammonia lyase sourced from *Anabaena variabilis*. The recombinant enzyme is PEGylated to promote pharmacological stability and minimize adverse immunogenic responses that could stem from the species variability. Pegvaliase converts phenylalanine into *trans*-cinnamic acid and ammonia (Fig. 7.11), reducing phenylalanine levels in the blood of phenylketonuria patients. The degraded products are further broken down in the liver and can be detected in the excretion [145, 146]. Prior to the approval of PG, Saproterin (Kuvan®; BioMarin Pharmaceutical Inc.) was used for the treatment of patients with phenylketonuria [147]. The drug is a chemically synthesized form of tetrahydrobiopterin, which is an essential cofactor required for the action of phenylalanine hydroxylase. However, reports indicate that a significant number of patients show a lack of responsiveness to Saproterin and continue to experience elevated levels of phenylalanine in the circulation. Also, the patients experience diet-associated difficulties and nutritional deficiency in the long term, as it is tough to follow a restricted diet. Pegvaliase can help address some of these drawbacks and serve as an effective alternative treatment option [145, 148]. In the phase III clinical trial comprising 261 phenylketonuria patients, the Phe levels in blood were restored in 60.7% of patients at the end of 2 years, demonstrating the drug's therapeutic efficiency [149]. Pegvaliase is available in solution form (20 mg/mL) and is administered subcutaneously.

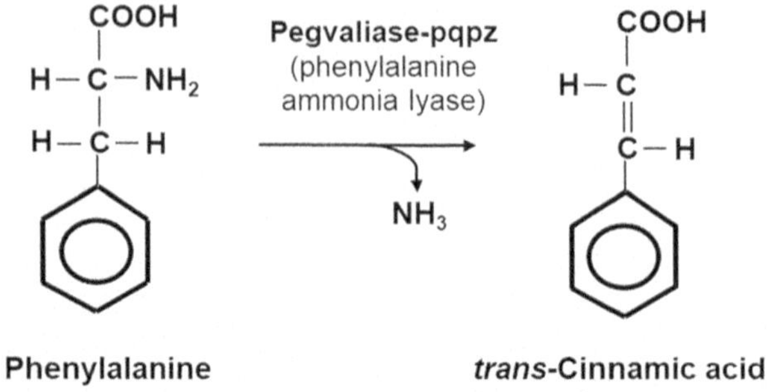

Fig. 7.11 The conversion of phenylalanine to *trans*-cinnamic acid by the action of Pegvaliase-pqpz

7.4.15 Avalglucosidase Alfa

Avalglucosidase alfa (Nexviazyme; Genzyme Corporation) was approved by the FDA as an enzyme replacement therapy for the treatment of late-onset Pompe disease (PD) [116]. The rare hereditary neuromuscular disorder is characterized by the deficiency of acid α-glucosidase (GAA), an enzyme that catalyzes the breakdown of glycogen into glucose in the lysosome [150]. The body cells will subsequently utilize the glucose produced as the breakdown product as an energy source. The deficiency of GAA in patients results in the inability to breakdown glycogen leading to its accumulation in the muscle cells and disrupting normal cellular function. The accumulation of lysosomal glycogen is frequently observed in the skeletal, cardiac, and smooth muscle. Patients suffering from Pompe disease experience characteristic disease symptoms such as muscle weakness and motor function impairment. There are two forms of the disease: infantile-onset and late-onset Pompe disease. Patients suffering from the common form of the disease, i.e., infantile-onset Pompe disease IOPD die before they reach the age of one year, mostly due to cardiovascular failure from hypertrophy. The disease that sets in the late stage is referred to as late-onset Pompe disease in both adults and children who suffer from progressive myopathy with lesser cardiac involvement [151–153].

The only approved treatment for Pompe disease is enzyme replacement therapy [154, 155]. Although enzyme replacement therapy has made remarkable progress in terms of clinical outcomes, it is widely accepted that the treatment only reduces the pace of disease progression rather than reversing the condition. Besides, the clinical data remains heterogeneous among patients for factors that remain to be explored [156]. Also, the enzyme used for replacement therapy frequently triggers an immunogenic response generating neutralizing antibodies in the patients [157]. However, to date, no statistical data link the antibody titers and the performance of the therapeutic enzyme in Pompe disease. One of the other issues to be addressed in Pompe disease is the poor response by the skeletal muscles to the therapy. The uptake of the therapeutic enzyme is shown to be dependent on the cell surface receptor called cation-independent mannose-6-phosphate receptor (CIMPR) [158]. Unlike other cell surface receptors, CIMPR is shown to be expressed at relatively lower levels. Further, the currently used therapeutic enzyme, Alglucosidase alfa (Genzyme Corporation), demonstrates poor affinity to CIMPR due to the limited mannose-6-phosphate content. To address these drawbacks associated with sugar-phosphate content, the external glycan structures of the recombinant enzyme were modified to increase the mannose-6-phosphate content [159]. The improved form of Alglucosidase alfa is the Avalglucosidase alfa, which contains higher bis-phosphorylated mannose-6-phosphate content and thereby exhibits higher affinity to CIMPR [156]. Upon binding to the surface CIMPRs, the enzyme is internalized, undergoes proteolytic cleavage in the lysosomes, and catalyzes the conversion of glycogen to glucose. Data from clinical trials showed that patients treated with Avalglucosidase alfa showed significant improvement in symptoms and safety compared to those treated with alglucosidase alfa [160]. The enzyme (Nexviazyme; Genzyme Corporation) is available in lyophilized powder form (100 mg/10 mL) and administered through intravenous injection.

7.5 Conclusion and Future Perspectives

We have reached a new age where protein therapeutics are developed and approved at a record pace. Protein therapeutics ranging from mAbs, cytokines, hormones, and enzymes dominate the biotechnology market and find applications in treating multiple diseases ranging from autoimmune disorders, cancers, hormone deficiencies, infection, and other conditions. The recently developed ADCs and bispecific antibody proteins have extended the application of existing mAbs and proved their utility, particularly in oncology. Further, advances in rDNA technology and protein engineering have continued to improve existing therapeutic drugs and paved the way for discovering novel molecules [161–163]. This chapter provides an overview of the biochemical aspects of some of the recently approved protein therapeutics. The future of biological therapeutics, particularly proteins and peptides, appears promising, with many drugs in the pipeline and some already in clinical trials. With the new-age technologies and the availability of affordable protein drugs, the future potential of protein therapeutics could become immeasurable, improving lives worldwide.

References

1. DiMasi JA, Feldman L, Seckler A, Wilson A (2010) Trends in risks associated with new drug development: success rates for investigational drugs. Clin Pharmacol Ther 87:272–277
2. Drews J (2000) Drug discovery: a historical perspective. Science (New York, NY) 287:1960–1964
3. Darrow JJ, Avorn J, Kesselheim AS (2020) FDA approval and regulation of pharmaceuticals, 1983–2018. JAMA 323:164–176
4. Andrews L, Ralston S, Blomme E, Barnhart K (2015) A snapshot of biologic drug development: challenges and opportunities. Hum Exp Toxicol 34:1279–1285
5. Brown DG, Wobst HJ (2021) A decade of FDA-approved drugs (2010–2019): trends and future directions. J Med Chem 64:2312–2338
6. Scannell JW, Bosley J (2016) When quality beats quantity: decision theory, drug discovery, and the reproducibility crisis. PLoS One 11:e0147215
7. Umscheid CA, Margolis DJ, Grossman CE (2011) Key concepts of clinical trials: a narrative review. Postgrad Med 123:194–204
8. Bighelli I, Barbui C (2012) What is the European medicines agency? Epidemiol Psychiatr Sci 21:245–247
9. Hojo T (2017) [Regulatory science in practice (Pharmaceuticals and medical devices agency)], Yakugaku zasshi. J Pharmaceut Soc Jpn 137:439–442
10. Sinha R, Shukla P (2019) Current trends in protein engineering: updates and progress. Curr Protein Pept Sci 20:398–407
11. Brannigan JA, Wilkinson AJ (2002) Protein engineering 20 years on. Nat Rev Mol Cell Biol 3: 964–970

12. Bertolini LR, Meade H, Lazzarotto CR, Martins LT, Tavares KC, Bertolini M, Murray JD (2016) The transgenic animal platform for biopharmaceutical production. Transgenic Res 25: 329–343
13. Leader B, Baca QJ, Golan DE (2008) Protein therapeutics: a summary and pharmacological classification. Nat Rev Drug Discov 7:21–39
14. Carter PJ (2011) Introduction to current and future protein therapeutics: a protein engineering perspective. Exp Cell Res 317:1261–1269
15. Dimitrov DS (2012) Therapeutic proteins. Methods Mol Biol (Clifton, NJ) 899:1–26
16. Kinch MS (2015) An overview of FDA-approved biologics medicines. Drug Discov Today 20:393–398
17. Usmani SS, Bedi G, Samuel JS, Singh S, Kalra S, Kumar P, Ahuja AA, Sharma M, Gautam A, Raghava GPS (2017) THPdb: database of FDA-approved peptide and protein therapeutics. PLoS One 12:e0181748
18. Lagassé HA, Alexaki A, Simhadri VL, Katagiri NH, Jankowski W, Sauna ZE, Kimchi-Sarfaty C (2017) Recent advances in (therapeutic protein) drug development. F1000Research 6:113
19. Mullard A (2020) FDA drug approvals. Nat Rev Drug Discov 20(2021):85–90
20. Markham A (2020) Teprotumumab: first approval. Drugs 80:509–512
21. Bahn RS (2010) Graves' ophthalmopathy. N Engl J Med 362:726–738
22. Kahaly GJ, Diana T, Olivo PD (2020) TSH receptor antibodies: relevance & utility. Endocr Pract 26:97–106
23. Davies TF, Andersen S, Latif R, Nagayama Y, Barbesino G, Brito M, Eckstein AK, Stagnaro-Green A, Kahaly GJ (2020) Graves' disease. Nat Rev Dis Primers 6:52
24. Chen H, Mester T, Raychaudhuri N, Kauh CY, Gupta S, Smith TJ, Douglas RS (2014) Teprotumumab, an IGF-1R blocking monoclonal antibody inhibits TSH and IGF-1 action in fibrocytes. J Clin Endocrinol Metab 99:E1635–E1640
25. Paik JS, Kim SE, Kim JH, Lee JY, Yang SW, Lee SB (2020) Insulin-like growth factor-1 enhances the expression of functional TSH receptor in orbital fibroblasts from thyroid-associated ophthalmopathy. Immunobiology 225:151902
26. Douglas RS, Kahaly GJ, Patel A, Sile S, Thompson EHZ, Perdok R, Fleming JC, Fowler BT, Marcocci C, Marinò M, Antonelli A, Dailey R, Harris GJ, Eckstein A, Schiffman J, Tang R, Nelson C, Salvi M, Wester S, Sherman JW, Vescio T, Holt RJ, Smith TJ (2020) Teprotumumab for the treatment of active thyroid eye disease. N Engl J Med 382:341–352
27. Hakuno F, Takahashi SI (2018) IGF1 receptor signaling pathways. J Mol Endocrinol 61:T69–t86
28. Mullard A (2019) FDA drug approvals. Nat Rev Drug Discov 19(2020):79–84
29. Griffiths CE, Barker JN (2007) Pathogenesis and clinical features of psoriasis. Lancet (London, England) 370:263–271
30. Armstrong AW, Read C (2020) Pathophysiology, clinical presentation, and treatment of psoriasis: a review. JAMA 323:1945–1960
31. Hawkes JE, Yan BY, Chan TC, Krueger JG (1950) Discovery of the IL-23/IL-17 signaling pathway and the treatment of psoriasis. J Immunol (Baltimore, MD) 201(2018):1605–1613
32. Di Cesare A, Di Meglio P, Nestle FO (2009) The IL-23/Th17 axis in the immunopathogenesis of psoriasis. J Invest Dermatol 129:1339–1350
33. Zhou F, Zhu Z, Gao J, Yang C, Wen L, Liu L, Zuo X, Zheng X, Shi Y, Zhu C, Liang B, Yin X, Wang W, Cheng H, Shen S, Tang X, Tang H, Sun L, Zhang A, Yang S, Zhang X, Sheng Y (2018) NFKB1 mediates Th1/Th17 activation in the pathogenesis of psoriasis. Cell Immunol 331:16–21
34. Gu C, Yang J (2019) Risankizumab for the treatment of psoriasis. Expert Rev Clin Pharmacol 12:851–857
35. Gordon KB, Strober B, Lebwohl M, Augustin M, Blauvelt A, Poulin Y, Papp KA, Sofen H, Puig L, Foley P, Ohtsuki M, Flack M, Geng Z, Gu Y, Valdes JM, Thompson EHZ, Bachelez H (2018) Efficacy and safety of risankizumab in moderate-to-severe plaque psoriasis (UltIMMa-

1 and UltIMMa-2): results from two double-blind, randomized, placebo-controlled and ustekinumab-controlled phase 3 trials. Lancet (London, England) 392:650–661
36. Haugh IM, Preston AK, Kivelevitch DN, Menter AM (2018) Risankizumab: an anti-IL-23 antibody for the treatment of psoriasis. Drug Des Devel Ther 12:3879–3883
37. Kulig P, Musiol S, Freiberger SN, Schreiner B, Gyülveszi G, Russo G, Pantelyushin S, Kishihara K, Alessandrini F, Kündig T, Sallusto F, Hofbauer GF, Haak S, Becher B (2016) IL-12 protects from psoriasiform skin inflammation. Nat Commun 7:13466
38. Bai F, Li GG, Liu Q, Niu X, Li R, Ma H (2019) Short-term efficacy and safety of IL-17, IL-12/23, and IL-23 inhibitors Brodalumab, Secukinumab, Ixekizumab, Ustekinumab, Guselkumab, Tildrakizumab, and Risankizumab for the treatment of moderate to severe plaque psoriasis: a systematic review and network meta-analysis of randomized controlled trials. J Immunol Res 2019:2546161
39. Thomas A, Teicher BA, Hassan R (2016) Antibody-drug conjugates for cancer therapy. Lancet Oncol 17:e254–e262
40. Chau CH, Steeg PS, Figg WD (2019) Antibody-drug conjugates for cancer. Lancet (London, England) 394:793–804
41. Syed YY (2020) Sacituzumab Govitecan: first approval. Drugs 80:1019–1025
42. Goldenberg DM, Cardillo TM, Govindan SV, Rossi EA, Sharkey RM (2015) Trop-2 is a novel target for solid cancer therapy with sacituzumab govitecan (IMMU-132), an antibody-drug conjugate (ADC). Oncotarget 6:22496–22512
43. Goldenberg DM, Stein R, Sharkey RM (2018) The emergence of trophoblast cell-surface antigen 2 (TROP-2) as a novel cancer target. Oncotarget 9:28989–29006
44. Lenárt S, Lenárt P, Šmarda J, Remšík J, Souček K, Beneš P (2020) Trop2: Jack of all trades, master of none. Cancer 12
45. Spring LM, Nakajima E, Hutchinson J, Viscosi E, Blouin G, Weekes C, Rugo H, Moy B, Bardia A (2021) Sacituzumab Govitecan for metastatic triple-negative breast cancer: clinical overview and management of potential toxicities. Oncologist 26:827–834
46. Weiss J, Glode A, Messersmith WA, Diamond J (2019) Sacituzumab govitecan: breakthrough targeted therapy for triple-negative breast cancer. Expert Rev Anticancer Ther 19:673–679
47. Khageh Hosseini S, Kolterer S, Steiner M, von Manstein V, Gerlach K, Trojan J, Waidmann O, Zeuzem S, Schulze JO, Hahn S, Steinhilber D, Gatterdam V, Tampé R, Biondi RM, Proschak E, Zörnig M (2017) Camptothecin and its analog SN-38, the active metabolite of irinotecan, inhibit binding of the transcriptional regulator and oncoprotein FUBP1 to its DNA target sequence FUSE. Biochem Pharmacol 146:53–62
48. Becnel MR, Lee HC (2020) The role of belantamab mafodotin for patients with relapsed and/or refractory multiple myeloma. Therapeut Adv Hematol 11:2040620720979813
49. Pieper K, Grimbacher B, Eibel H (2013) B-cell biology and development. J Allergy Clin Immunol 131:959–971
50. Shah N, Chari A, Scott E, Mezzi K, Usmani SZ (2020) B-cell maturation antigen (BCMA) in multiple myeloma: rationale for targeting and current therapeutic approaches. Leukemia 34: 985–1005
51. Mackay F, Schneider P, Rennert P, Browning J (2003) BAFF and APRIL: a tutorial on B cell survival. Annu Rev Immunol 21:231–264
52. Ryan MC, Grewal IS (2009) Targeting of BAFF and APRIL for autoimmunity and oncology. Adv Exp Med Biol 647:52–63
53. Moreaux J, Legouffe E, Jourdan E, Quittet P, Rème T, Lugagne C, Moine P, Rossi JF, Klein B, Tarte K (2004) BAFF and APRIL protect myeloma cells from apoptosis induced by interleukin 6 deprivation and dexamethasone. Blood 103:3148–3157
54. Sokka IK, Ekholm FS, Johansson MP (2019) Increasing the potential of the Auristatin cancer-drug family by shifting the conformational equilibrium. Mol Pharm 16:3600–3608
55. Sheikh S, Lebel E, Trudel S (2020) Belantamab mafodotin in the treatment of relapsed or refractory multiple myeloma. Future Oncol (London, England) 16:2783–2798

56. Lonial S, Lee HC, Badros A, Trudel S, Nooka AK, Chari A, Abdallah AO, Callander N, Lendvai N, Sborov D, Suvannasankha A, Weisel K, Karlin L, Libby E, Arnulf B, Facon T, Hulin C, Kortüm KM, Rodríguez-Otero P, Usmani SZ, Hari P, Baz R, Quach H, Moreau P, Voorhees PM, Gupta I, Hoos A, Zhi E, Baron J, Piontek T, Lewis E, Jewell RC, Dettman EJ, Popat R, Esposti SD, Opalinska J, Richardson P, Cohen AD (2020) Belantamab mafodotin for relapsed or refractory multiple myeloma (DREAMM-2): a two-arm, randomized, open-label, phase 2 study. Lancet Oncol 21:207–221
57. Labrijn AF, Janmaat ML, Reichert JM, Parren P (2019) Bispecific antibodies: a mechanistic review of the pipeline. Nat Rev Drug Discov 18:585–608
58. Krishnamurthy A, Jimeno A (2018) Bispecific antibodies for cancer therapy: a review. Pharmacol Ther 185:122–134
59. Trabolsi A, Arumov A, Schatz JH (1950) T cell-activating bispecific antibodies in cancer therapy. J Immunol (Baltimore, MD) 203(2019):585–592
60. Thakur A, Huang M, Lum LG (2018) Bispecific antibody based therapeutics: strengths and challenges. Blood Rev 32:339–347
61. Sanford M (2015) Blinatumomab: first global approval. Drugs 75:321–327
62. Le Jeune C, Thomas X (2016) Potential for bispecific T-cell engagers: role of blinatumomab in acute lymphoblastic leukemia. Drug Des Devel Ther 10:757–765
63. Rogala B, Freyer CW, Ontiveros EP, Griffiths EA, Wang ES, Wetzler M (2015) Blinatumomab: enlisting serial killer T-cells in the war against hematologic malignancies. Expert Opin Biol Ther 15:895–908
64. Zimmerman Z, Maniar T, Nagorsen D (2015) Unleashing the clinical power of T cells: CD19/CD3 bi-specific T cell engager (BiTE®) antibody construct blinatumomab as a potential therapy. Int Immunol 27:31–37
65. Johnson AM, Bullock BL, Neuwelt AJ, Poczobutt JM, Kaspar RE, Li HY, Kwak JW, Hopp K, Weiser-Evans MCM, Heasley LE, Schenk EL, Clambey ET, Nemenoff RA (2020) Cancer cell–intrinsic expression of MHC class II regulates the immune microenvironment and response to anti-PD-1 therapy in lung adenocarcinoma. J Immunol 204:2295–2307
66. Bassan R (2012) Toward victory in adult ALL: blinatumomab joins in. Blood 120:5094–5095
67. Zaman R, Islam RA, Ibnat N, Othman I, Zaini A, Lee CY, Chowdhury EH (2019) Current strategies in extending half-lives of therapeutic proteins. J Control Release 301:176–189
68. Duivelshof BL, Murisier A, Camperi J, Fekete S, Beck A, Guillarme D, D'Atri V (2021) Therapeutic Fc-fusion proteins: current analytical strategies. J Sep Sci 44:35–62
69. Jafari R, Zolbanin NM, Rafatpanah H, Majidi J, Kazemi T (2017) Fc-fusion proteins in therapy: an updated view. Curr Med Chem 24:1228–1237
70. Yang C, Gao X, Gong R (2018) Engineering of Fc fragments with optimized physicochemical properties implying improvement of clinical potentials for Fc-based therapeutics. Front Immunol 8
71. Markham A (2020) Luspatercept: first approval. Drugs 80:85–90
72. Cao A, Galanello R (2010) Beta-thalassemia. Genet Med 12:61–76
73. Thein SL (2018) Molecular basis of β thalassemia and potential therapeutic targets. Blood Cells Mol Dis 70:54–65
74. Vander Ark A, Cao J, Li X (2018) TGF-β receptors: in and beyond TGF-β signaling. Cell Signal 52:112–120
75. Hata A, Chen YG (2016) TGF-β signaling from receptors to Smads. Cold Spring Harb Perspect Biol 8
76. Feng XH, Derynck R (2005) Specificity and versatility in TGF-beta signaling through Smads. Annu Rev Cell Dev Biol 21:659–693
77. Derynck R, Zhang YE (2003) Smad-dependent and Smad-independent pathways in TGF-beta family signalling. Nature 425:577–584
78. Han S (2011) Crystal structure of activin receptor type IIB kinase domain. Vitam Horm 85:29–38

79. Martinez PA, Suragani RN, Bhasin M, Li R, Pearsall RS, Kumar R (2015) Rap-536 (murine ACE-536/Luspatercept) inhibits Smad2/3 signaling and promotes erythroid differentiation by restoring GATA-1 function in murine b-Thalassemia. Blood 126:751–751
80. Cappellini MD, Taher AT (2021) The use of luspatercept for thalassemia in adults. Blood Adv 5:326–333
81. Piga A, Perrotta S, Gamberini MR, Voskaridou E, Melpignano A, Filosa A, Caruso V, Pietrangelo A, Longo F, Tartaglione I, Borgna-Pignatti C, Zhang X, Laadem A, Sherman ML, Attie KM (2019) Luspatercept improves hemoglobin levels and blood transfusion requirements in a study of patients with β-thalassemia. Blood 133:1279–1289
82. Cappellini MD, Viprakasit V, Taher AT, Georgiev P, Kuo KHM, Coates T, Voskaridou E, Liew HK, Pazgal-Kobrowski I, Forni GL, Perrotta S, Khelif A, Lal A, Kattamis A, Vlachaki E, Origa R, Aydinok Y, Bejaoui M, Ho PJ, Chew LP, Bee PC, Lim SM, Lu MY, Tantiworawit A, Ganeva P, Gercheva L, Shah F, Neufeld EJ, Thompson A, Laadem A, Shetty JK, Zou J, Zhang J, Miteva D, Zinger T, Linde PG, Sherman ML, Hermine O, Porter J, Piga A (2020) A phase 3 trial of luspatercept in patients with transfusion-dependent β-thalassemia. N Engl J Med 382:1219–1231
83. Suragani RN, Cadena SM, Cawley SM, Sako D, Mitchell D, Li R, Davies MV, Alexander MJ, Devine M, Loveday KS, Underwood KW, Grinberg AV, Quisel JD, Chopra R, Pearsall RS, Seehra J, Kumar R (2014) Transforming growth factor-β superfamily ligand trap ACE-536 corrects anemia by promoting late-stage erythropoiesis. Nat Med 20:408–414
84. Frampton JE (2021) Efmoroctocog Alfa: a review in Haemophilia A. Drugs 81:2035–2046
85. Samuelson Bannow B, Recht M, Négrier C, Hermans C, Berntorp E, Eichler H, Mancuso ME, Klamroth R, O'Hara J, Santagostino E, Matsushita T, Kessler C (2019) Factor VIII: long-established role in haemophilia A and emerging evidence beyond haemostasis. Blood Rev 35: 43–50
86. Fay PJ (2006) Factor VIII structure and function. Int J Hematol 83:103–108
87. Fay PJ, Jenkins PV (2005) Mutating factor VIII: lessons from structure to function. Blood Rev 19:15–27
88. Tiede A (2015) Half-life extended factor VIII for the treatment of hemophilia A. J Thrombosis Haemostasis JTH 13(Suppl 1):S176–S179
89. Phillips JD (2019) Heme biosynthesis and the porphyrias. Mol Genet Metab 128:164–177
90. Balwani M (2019) Erythropoietic protoporphyria and X-linked protoporphyria: pathophysiology, genetics, clinical manifestations, and management. Mol Genet Metab 128:298–303
91. Switonski M, Mankowska M, Salamon S (2013) Family of melanocortin receptor (MCR) genes in mammals-mutations, polymorphisms and phenotypic effects. J Appl Genet 54:461–472
92. Wolf Horrell EM, Boulanger MC, D'Orazio JA (2016) Melanocortin 1 receptor: structure, function, and regulation. Front Genet 7:95
93. Singh M, Mukhopadhyay K (2014) Alpha-melanocyte stimulating hormone: an emerging anti-inflammatory antimicrobial peptide. Biomed Res Int 2014:874610
94. Wensink D, Wagenmakers MAEM, Langendonk JG (2021) Afamelanotide for prevention of phototoxicity in erythropoietic protoporphyria. Expert Rev Clin Pharmacol 14:151–160
95. Yuan S, Yu B, Liu H-M (2020) New drug approvals for 2019: synthesis and clinical applications. Eur J Med Chem 205:112667
96. Mullard A (2017) FDA drug approvals. Nat Rev Drug Discov 17(2018):81–85
97. Bluet-Pajot MT, Tolle V, Zizzari P, Robert C, Hammond C, Mitchell V, Beauvillain JC, Viollet C, Epelbaum J, Kordon C (2001) Growth hormone secretagogues and hypothalamic networks. Endocrine 14:1–8
98. Chinoy A, Murray PG (2016) Diagnosis of growth hormone deficiency in the paediatric and transitional age. Best Practice Res Clin Endocrinol Metab 30:737–747
99. Jørgensen JO, Müller J, Møller J, Wolthers T, Vahl N, Juul A, Skakkebaek NE, Christiansen JS (1994) Adult growth hormone deficiency. Horm Res 42:235–241

100. Kato Y, Murakami Y, Sohmiya M, Nishiki M (2002) Regulation of human growth hormone secretion and its disorders. Intern Med (Tokyo, Japan) 41:7–13
101. Ghigo E, Aimaretti G, Corneli G (2008) Diagnosis of adult GH deficiency. Growth Horm IGF Res 18:1–16
102. Díez JJ, Sangiao-Alvarellos S, Cordido F (2018) Treatment with growth hormone for adults with growth hormone deficiency syndrome: benefits and risks. Int J Mol Sci 19
103. Klaus B, Sachse R, Ammer N, Kelepouris N, Ostrow V (2020) Safety, tolerability, pharmacokinetics, and pharmacodynamics of macimorelin in healthy adults: results of a single-dose, randomized controlled study. Growth Horm IGF Res 52:101321
104. Garcia JM, Swerdloff R, Wang C, Kyle M, Kipnes M, Biller BM, Cook D, Yuen KC, Bonert V, Dobs A, Molitch ME, Merriam GR (2013) Macimorelin (AEZS-130)-stimulated growth hormone (GH) test: validation of a novel oral stimulation test for the diagnosis of adult GH deficiency. J Clin Endocrinol Metab 98:2422–2429
105. Holubová M, Spolcová A, Demianová Z, Sýkora D, Fehrentz JA, Martinez J, Stofková A, Jurčovičová J, Drápalová J, Lacinová Z, Haluzík M, Zelezná B, Maletínská L (2013) Ghrelin agonist JMV 1843 increases food intake, body weight and expression of orexigenic neuropeptides in mice. Physiol Res 62:435–444
106. Hillmen P, Szer J, Weitz I, Röth A, Höchsmann B, Panse J, Usuki K, Griffin M, Kiladjian J-J, de Castro C, Nishimori H, Tan L, Hamdani M, Deschatelets P, Francois C, Grossi F, Ajayi T, Risitano A, de la Tour RP (2021) Pegcetacoplan versus eculizumab in paroxysmal nocturnal hemoglobinuria. N Engl J Med 384:1028–1037
107. Brodsky RA (2014) Paroxysmal nocturnal hemoglobinuria. Blood 124:2804–2811
108. Ruiz-Argüelles A, Llorente L (2007) The role of complement regulatory proteins (CD55 and CD59) in the pathogenesis of autoimmune hemocytopenias. Autoimmun Rev 6:155–161
109. Hill A, DeZern AE, Kinoshita T, Brodsky RA (2017) Paroxysmal nocturnal haemoglobinuria. Nat Rev Dis Primers 3:17028
110. Schröder-Braunstein J, Kirschfink M (2019) Complement deficiencies and dysregulation: pathophysiological consequences, modern analysis, and clinical management. Mol Immunol 114:299–311
111. Dobó J, Kocsis A, Gál P (2018) Be on target: strategies of targeting alternative and lectin pathway components in complement-mediated diseases. Front Immunol 9
112. Harboe M, Mollnes TE (2008) The alternative complement pathway revisited. J Cell Mol Med 12:1074–1084
113. Kulasekararaj AG, Hill A, Rottinghaus ST, Langemeijer S, Wells R, Gonzalez-Fernandez FA, Gaya A, Lee JW, Gutierrez EO, Piatek CI, Szer J, Risitano A, Nakao S, Bachman E, Shafner L, Damokosh AI, Ortiz S, Röth A, Peffault de Latour R (2019) Ravulizumab (ALXN1210) vs eculizumab in C5-inhibitor-experienced adult patients with PNH: the 302 study. Blood 133: 540–549
114. McKinley CE, Richards SJ, Munir T, Griffin M, Mitchell LD, Arnold L, Riley K, Copeland N, Newton DJ, Hill A, Hillmen P (2017) Extravascular hemolysis due to C3-loading in patients with PNH treated with eculizumab: defining the clinical syndrome. Blood 130:3471–3471
115. de Castro C, Grossi F, Weitz IC, Maciejewski J, Sharma V, Roman E, Brodsky RA, Tan L, Di Casoli C, El Mehdi D, Deschatelets P, Francois C (2020) C3 inhibition with pegcetacoplan in subjects with paroxysmal nocturnal hemoglobinuria treated with eculizumab. Am J Hematol 95:1334–1343
116. Urquhart L (2021) FDA new drug approvals in Q3 2021. Nat Rev Drug Disc
117. Brooks AJ, Waters MJ (2010) The growth hormone receptor: mechanism of activation and clinical implications. Nat Rev Endocrinol 6:515–525
118. Thornton PS, Maniatis AK, Aghajanova E, Chertok E, Vlachopapadopoulou E, Lin Z, Song W, Christoffersen ED, Breinholt VM, Kovalenko T, Giorgadze E, Korpal-Szczyrska M, Hofman PL, Karpf DB, Shu AD, Beckert M (2021) Weekly lonapegsomatropin in treatment-naïve children with growth hormone deficiency: the phase 3 heiGHt trial. J Clin Endocrinol Metab

119. Liongue C, Ward AC (2007) Evolution of Class I cytokine receptors. BMC Evol Biol 7:120
120. Reh CS, Geffner ME (2010) Somatotropin in the treatment of growth hormone deficiency and Turner syndrome in pediatric patients: a review. Clin Pharmacol Adv Appl 2:111–122
121. Dehkhoda F, Lee CMM, Medina J, Brooks AJ (2018) The growth hormone receptor: mechanism of receptor activation, cell signaling, and physiological aspects. Front Endocrinol 9
122. Carter-Su C, Schwartz J, Argetsinger LS (2016) Growth hormone signaling pathways. Growth Hormone IGF Res 28:11–15
123. Laron Z (2001) Insulin-like growth factor 1 (IGF-1): a growth hormone. Mol Pathol 54:311–316
124. Wang J, Zhou J, Cheng CM, Kopchick JJ, Bondy CA (2004) Evidence supporting dual, IGF-I-independent and IGF-I-dependent, roles for GH in promoting longitudinal bone growth. J Endocrinol 180:247–255
125. Wu S, Yang W, De Luca F (2015) Insulin-like growth factor-independent effects of growth hormone on growth plate chondrogenesis and longitudinal bone growth. Endocrinology 156: 2541–2551
126. Semsarian C, Wu MJ, Ju YK, Marciniec T, Yeoh T, Allen DG, Harvey RP, Graham RM (1999) Skeletal muscle hypertrophy is mediated by a Ca2+-dependent calcineurin signalling pathway. Nature 400:576–581
127. Sotiropoulos A, Ohanna M, Kedzia C, Menon RK, Kopchick JJ, Kelly PA, Pende M (2006) Growth hormone promotes skeletal muscle cell fusion independent of insulin-like growth factor 1 up-regulation. Proc Natl Acad Sci USA 103:7315–7320
128. Collet TH, Dubern B, Mokrosinski J, Connors H, Keogh JM, Mendes de Oliveira E, Henning E, Poitou-Bernert C, Oppert JM, Tounian P, Marchelli F, Alili R, Le Beyec J, Pépin D, Lacorte JM, Gottesdiener A, Bounds R, Sharma S, Folster C, Henderson B, O'Rahilly S, Stoner E, Gottesdiener K, Panaro BL, Cone RD, Clément K, Farooqi IS, Van der Ploeg LHT (2017) Evaluation of a melanocortin-4 receptor (MC4R) agonist (Setmelanotide) in MC4R deficiency. Mol Metab 6:1321–1329
129. Falls BA, Zhang Y (2019) Insights into the allosteric mechanism of setmelanotide (RM-493) as a potent and first-in-class melanocortin-4 receptor (MC4R) agonist to treat rare genetic disorders of obesity through an in silico approach. ACS Chem Neurosci 10:1055–1065
130. Adan RA, Tiesjema B, Hillebrand JJ, la Fleur SE, Kas MJ, de Krom M (2006) The MC4 receptor and control of appetite. Br J Pharmacol 149:815–827
131. Millington GWM (2007) The role of proopiomelanocortin (POMC) neurones in feeding behaviour. Nutr Metab 4:18
132. Patel CA, Acharya SR (2019) Energy homeostasis and obesity: the therapeutic role of anorexigenic and orexigenic peptide. Int J Pept Res Ther 25:919–932
133. Belgardt BF, Okamura T, Brüning JC (2009) Hormone and glucose signalling in POMC and AgRP neurons. J Physiol 587:5305–5314
134. Fani L, Bak S, Delhanty P, van Rossum EFC, van den Akker ELT (2014) The melanocortin-4 receptor as target for obesity treatment: a systematic review of emerging pharmacological therapeutic options. Int J Obes 38:163–169
135. Jéquier E (2002) Leptin signaling, adiposity, and energy balance. Ann NY Acad Sci 967:379–388
136. Clément K, van den Akker E, Argente J, Bahm A, Chung WK, Connors H, De Waele K, Farooqi IS, Gonneau-Lejeune J, Gordon G, Kohlsdorf K, Poitou C, Puder L, Swain J, Stewart M, Yuan G, Wabitsch M, Kühnen P (2020) Efficacy and safety of setmelanotide, an MC4R agonist, in individuals with severe obesity due to LEPR or POMC deficiency: single-arm, open-label, multicentre, phase 3 trials. Lancet Diab Endocrinol 8:960–970
137. Markham A (2021) Setmelanotide: first approval. Drugs 81:397–403
138. Mullard A (2018) FDA drug approvals. Nat Rev Drug Discov 18(2019):85–89
139. Dash A, Chakraborty S, Pillai MR, Knapp FF Jr (2015) Peptide receptor radionuclide therapy: an overview. Cancer Biother Radiopharm 30:47–71

140. Shastry M, Kayani I, Wild D, Caplin M, Visvikis D, Gacinovic S, Reubi JC, Bomanji JB (2010) Distribution pattern of 68Ga-DOTATATE in disease-free patients. Nucl Med Commun 31:1025–1032
141. Hennrich U, Benešová M (2020) [(68)Ga]Ga-DOTA-TOC: the first FDA-approved (68)Ga-radiopharmaceutical for PET imaging. Pharmaceuticals (Basel, Switzerland) 13
142. van Spronsen FJ, Blau N, Harding C, Burlina A, Longo N, Bosch AM (2021) Phenylketonuria. Nat Rev Dis Primers 7:36
143. Vockley J, Andersson HC, Antshel KM, Braverman NE, Burton BK, Frazier DM, Mitchell J, Smith WE, Thompson BH, Berry SA (2014) Phenylalanine hydroxylase deficiency: diagnosis and management guideline. Genet Med 16:188–200
144. Evans S, Adam S, Adams S, Allen H, Ashmore C, Bailey S, Banks J, Churchill H, Cochrane B, Cook J, Dale C, Daly A, Dixon M, Dunlop C, Ellerton C, Emm A, Firman S, Ford S, French M, Gribben J, Grimsley A, Herlihy I, Hill M, Judd S, Lang K, Males J, McDonald J, McStravick N, Millington C, Newby C, Noble C, Pereira R, Pinto A, Robertson L, Robotham A, Ross K, Singleton K, Skeath R, Terry A, Van Wyk K, White F, White L, Wildgoose J, Woodall A, MacDonald A (2020) Uniformity of food protein interpretation amongst dietitians for patients with phenylketonuria (PKU): 2020 UK National Consensus Statements. Nutrients 12
145. Trefz FK, Burton BK, Longo N, Casanova MM, Gruskin DJ, Dorenbaum A, Kakkis ED, Crombez EA, Grange DK, Harmatz P, Lipson MH, Milanowski A, Randolph LM, Vockley J, Whitley CB, Wolff JA, Bebchuk J, Christ-Schmidt H, Hennermann JB (2009) Efficacy of sapropterin dihydrochloride in increasing phenylalanine tolerance in children with phenylketonuria: a phase III, randomized, double-blind, placebo-controlled study. J Pediatr 154:700–707
146. Mahan KC, Gandhi MA, Anand S (2019) Pegvaliase: a novel treatment option for adults with phenylketonuria. Curr Med Res Opin 35:647–651
147. Blau N (2013) Sapropterin dihydrochloride for the treatment of hyperphenylalaninemias. Expert Opin Drug Metab Toxicol 9:1207–1218
148. Burton BK, Longo N, Vockley J, Grange DK, Harding CO, Decker C, Li M, Lau K, Rosen O, Larimore K, Thomas J (2020) Pegvaliase for the treatment of phenylketonuria: results of the phase 2 dose-finding studies with long-term follow-up. Mol Genet Metab 130:239–246
149. Thomas J, Levy H, Amato S, Vockley J, Zori R, Dimmock D, Harding CO, Bilder DA, Weng HH, Olbertz J, Merilainen M, Jiang J, Larimore K, Gupta S, Gu Z, Northrup H (2018) Pegvaliase for the treatment of phenylketonuria: results of a long-term phase 3 clinical trial program (PRISM). Mol Genet Metab 124:27–38
150. Parenti G, Andria G (2011) Pompe disease: from new views on pathophysiology to innovative therapeutic strategies. Curr Pharm Biotechnol 12:902–915
151. van den Hout HM, Hop W, van Diggelen OP, Smeitink JA, Smit GP, Poll-The BT, Bakker HD, Loonen MC, de Klerk JB, Reuser AJ, van der Ploeg AT (2003) The natural course of infantile Pompe's disease: 20 original cases compared with 133 cases from the literature. Pediatrics 112:332–340
152. Cupler EJ, Berger KI, Leshner RT, Wolfe GI, Han JJ, Barohn RJ, Kissel JT (2012) Consensus treatment recommendations for late-onset Pompe disease. Muscle Nerve 45:319–333
153. van der Ploeg AT, Reuser AJ (2008) Pompe's disease. Lancet (London, England) 372:1342–1353
154. Rossi M, Parenti G, Della Casa R, Romano A, Mansi G, Agovino T, Rosapepe F, Vosa C, Del Giudice E, Andria G (2007) Long-term enzyme replacement therapy for Pompe disease with recombinant human alpha-glucosidase derived from Chinese hamster ovary cells. J Child Neurol 22:565–573
155. Van den Hout JM, Kamphoven JH, Winkel LP, Arts WF, De Klerk JB, Loonen MC, Vulto AG, Cromme-Dijkhuis A, Weisglas-Kuperus N, Hop W, Van Hirtum H, Van Diggelen OP, Boer M, Kroos MA, Van Doorn PA, Van der Voort E, Sibbles B, Van Corven EJ, Brakenhoff JP, Van Hove J, Smeitink JA, de Jong G, Reuser AJ, Van der Ploeg AT (2004) Long-term

intravenous treatment of Pompe disease with recombinant human alpha-glucosidase from milk. Pediatrics 113:e448–e457
156. Xu S, Lun Y, Frascella M, Garcia A, Soska R, Nair A, Ponery AS, Schilling A, Feng J, Tuske S, Valle MCD, Martina JA, Ralston E, Gotschall R, Valenzano KJ, Puertollano R, Do HV, Raben N, Khanna R (2019) Improved efficacy of a next-generation ERT in murine Pompe disease. JCI Insight 4
157. de Vries JM, Kuperus E, Hoogeveen-Westerveld M, Kroos MA, Wens SC, Stok M, van der Beek NA, Kruijshaar ME, Rizopoulos D, van Doorn PA, van der Ploeg AT, Pijnappel WW (2017) Pompe disease in adulthood: effects of antibody formation on enzyme replacement therapy. Genet Med 19:90–97
158. Wang Y, MacDonald RG, Thinakaran G, Kar S (2017) Insulin-like growth factor-II/cation-independent mannose 6-phosphate receptor in neurodegenerative diseases. Mol Neurobiol 54: 2636–2658
159. Zhu Y, Li X, McVie-Wylie A, Jiang C, Thurberg BL, Raben N, Mattaliano RJ, Cheng SH (2005) Carbohydrate-remodelled acid alpha-glucosidase with higher affinity for the cation-independent mannose 6-phosphate receptor demonstrates improved delivery to muscles of Pompe mice. Biochem J 389:619–628
160. Kushlaf H, Attarian S, Borges JL, Bouhour F, Chien Y-H, Choi Y-C, Clemens P, Day J, Diaz-Manera J, Erdem-Ozdamar S, Goker-Alpan O, Illarioshkin S, Kishnani P, Kostera-Pruszczyk-A, Ladha S, Mozaffar T, Roberts M, Straub V, Toscano A, van der Ploeg A, Haack KA, Hug C, Huynh-Ba O, Zhou T, Johnson J, Dimachkie M, Schoser B (2021) Efficacy and safety results of the Avalglucosidase alfa phase 3 COMET trial in late-onset Pompe Disease patients (4195). Neurology 96:4195
161. Singh DB, Tripathi T (2020) Frontiers in protein structure, function, and dynamics. Springer Nature, Singapore
162. Saudagar P, Tripathi T (2023) Advanced spectroscopic methods to study biomolecular structure and dynamics, 1st edn. Academic Press
163. Tripathi T, Dubey VK (2022) Advances in protein molecular and structural biology methods, 1st edn. Academic Press, Cambridge, MA

Pharmacogenetic Biomarkers of Protein Therapeutics

8

Priya Swaminathan

Abstract

Genes implicated in mechanisms that alter the pharmacokinetics of small-molecule therapeutics have historically been the focus of pharmacogenetic research. The FDA allows new protein therapeutics in the market based on pharmacogenetics for clinical applications. It is a time-old knowledge that genetic polymorphisms affect the efficacy and toxicity of an approved drug. The complexity of biological organisms allows protein therapeutics development to proceed cautiously. Methodologies using pharmacogenetic information about the protein therapeutic candidates and their protein receptors can yield successful protein-based drugs. Immunogenicity of protein therapies is becoming a major issue in the development and approval of biologics, indicating that a pharmacogenetic strategy is required. Hence, the hallmarks of therapeutic reactions should be listed and analyzed in order to improve biologics and their success.

Keywords

Biologics · Pharmacogenetics · Immunogenicity · Protein therapeutics

P. Swaminathan (✉)
Department of Biotechnology, SRM Institute of Science and Technology, Kattankulathur, India
e-mail: priyas@srmist.edu.in

D. B. Singh, T. Tripathi (eds.), *Protein-based Therapeutics*,
https://doi.org/10.1007/978-981-19-8249-1_8

8.1 Introduction

Pharmacogenetics, developed in the early 1990s, is the science where the adverse effect of a drug on a specific individual, rather than the entire population, is focused. A set of polymorphisms could affect the way a therapeutic drug acts in a body [1]. This discovery became part of functional genomics, which resulted in biomarkers for therapeutic drug efficiency for an individual. The Food and Drug Administration (FDA) provides guidelines and rules for developing a biomarker. In generality, a clinically used biomarker is validated by testing its effectiveness as a prognosis for a disease. Pharmacogenetic biomarkers can monitor clinical responses to therapeutic intervention. The different types of biomarkers are (i) known biomarkers, which can be tested consistently and have well-established clinical consequences correlations, (ii) probable biomarkers, whose endpoints have an accepted scientific foundation but, due to some technicality, not widely replicated, and (iii) exploratory biomarkers, which have been studied in clinical trials, but the therapeutic decisions are not made based on them [2].

Recombinant protein treatments are the latest, less toxic options and more specific for diseases. They are being developed as a one-of-a-kind treatment option for autoimmune illnesses, cancer, and metabolic disorders. There are approved protein drugs in the market, such as growth factors, cytokines, enzymes, soluble receptors, antibodies, and enzyme inhibitors [4]. The ever-growing science of immunomodulatory proteins is complemented by protein therapeutics, which are more specific and targeted on a disease pathway (Fig. 8.1).

There are various protein targets for protein-based therapeutics (Fig. 8.2). Although protein therapeutics are specific, drug success depends on identifying patients who will benefit the most from the treatment [5]. For example, Herceptin

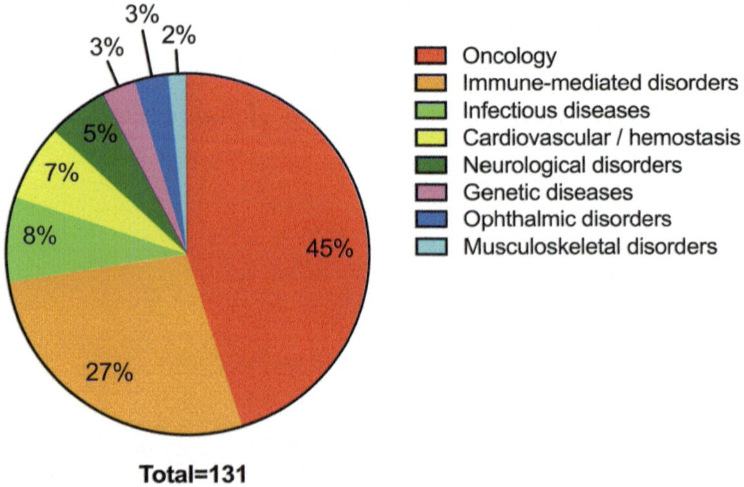

Fig. 8.1 Different diseases where protein therapeutics are usually administered [3]

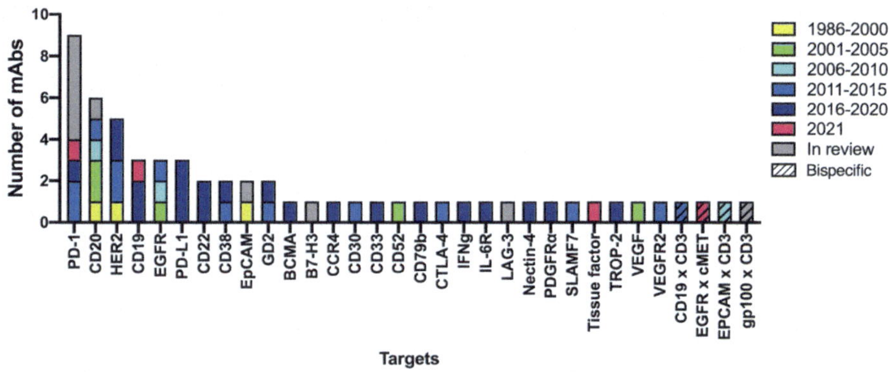

Fig. 8.2 Different targets of protein therapeutics [3]

(Trastuzumab) is a drug developed by Genentech and Roche that is exclusively effective in people with HER2-positive breast cancer [6]. Hence, genotyping the patients and decreasing the risk of drug therapy toxicity is a new approach biotechnological and pharmaceutical companies are adopting. Before delivering a protein drug to the market, a detailed study of pharmacogenetic biomarkers will provide valuable indicators for predicting a patient's likely reaction to the treatment. Pharmacogenetic studies associate polymorphisms and microsatellite repeats with characteristic drug response endpoints. Protein therapeutics cause repeatable changes in phenotype, which is easier to link to genotype than other complex traits [7]. Also, the biology of protein therapeutics is less complex than the conventional drugs that react with multiple targets. Protein therapeutics focuses on pharmacogenetic studies of pharmacodynamic targets, unlike small molecule pharmacogenetic studies, which focus on drug transporters and metabolizers [8].

8.2 Pharmacogenetics Based on Pharmacodynamics of Protein Therapeutics

The availability of gene sequencing technologies and population data related to polymorphism allows protein therapeutics to focus on drug target pathways and their polymorphisms as a starting point for pharmacodynamics studies [9]. The challenge is to design unique solutions for every polymorphism in the disease target [10], which is quite large; hence, there are fewer examples of pharmacodynamics-centered trials for protein treatments [11].

8.2.1 Protein Therapeutics and Their Effect on Upstream and Downstream Pathways

Varied reactions to protein therapeutics might be linked to the variable drug target or variable upstream or downstream signaling pathways. Candidate genes are essential to pharmacodynamics-based pharmacogenetic investigations (Fig. 8.3). Functional polymorphisms could also include exon-intron changes that affect mRNA manifestations, maturity, and degradation. In theory, a mechanistic underpinning for employing a genetic polymorphism as a biomarker will provide a prognosis of a variable phenotypic response when a protein-based drug is administered to a patient [12]. Pharmacodynamics-based assay interpretations take time and effort and are specific for every subset of polymorphisms in the patient.

Sequence analysis can uncover intron/exon splice variants, polymorphism sites in the protein target, and upstream and downstream gene products that allow drug action, which is a powerful tool. Sequence analysis is becoming an essential part of the early discovery steps for most protein therapeutics. The kinetics of the drug and knowledge of the most prevalent variations in its target will guide the genotype-to-phenotype association that follows. Often, genetic variances are responsible for the variation in disease response to protein drugs [14]. The variability of the response could be due to polymorphisms in the proteins upstream or downstream of the drug target in the pathway (Table 8.1) [15].

8.2.1.1 Interferon-Alpha (IFNα)

In chronic Hepatitis C virus (HCV) infection, the interferon-alpha (IFNα) and its variable responses are an example of host factors in disease response to protein

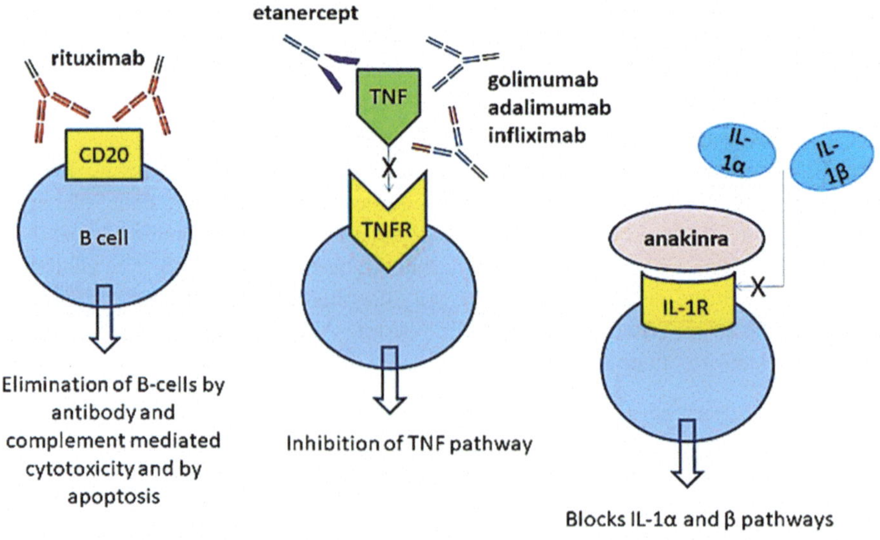

Fig. 8.3 Different modes of action of protein therapeutics in the market [13]

Table 8.1 Upstream and downstream factors that affect protein therapeutics

Disease in which pharmacogenetic studies were done	Protein therapeutics commercial name	Genotype/allele marker	Effect	Ref.
Acute myocardial infarction	Fibrinolytic therapy (recombinant human enzyme-like tPA)	Fibrin-stabilizing factor (FXIII)—Leu34	Clots are less permeable.	[29]
Cerebral ischemia	Fibrinolytic therapy (recombinant human tPA)	Matrix metallopeptidase 9 (MMP9)—C1562T	Hemorrhagic risk increases.	[30]
Colorectal cancer	Cetuximab (humanized anti-EGFR mAb)	High levels of cyclooxygenase-2 (COX2)/Epidermal growth factor receptor (EGFR)/Interleukin 8 (IL-8)	Angiogenesis and VEGF activation of the tumor cells.	[31]
Chronic hepatitis C	Interferon-α (recombinant human cytokine; IFN α -2a, IFN α -2b)	Suppressor of cytokine signaling 1 (SOCS1) expression decreases	SOCS1 cytokine signaling and transcription of IFN-α-responsive genes. High SOCS1 expression is directly proportional to a high viral load of HCV.	[32]
Rheumatoid arthritis	Anakinra (recombinant human IL-1 receptor antagonist; IL-1RA)	Interleukin 1A (IL1 α) +4845 bps and Interleukin 1 Beta (IL1β) +3954 bps	Increases IL-1 α and IL-1 β concentration in the system and endogenous IL-1R α expression.	[33]
Waldenstrom's macroglobulinemia	Rituximab (CD20 against monoclonal antibody)	Immunoglobulin gamma Fc region receptor III-A (FCGR3A)-Val158	FCGR3A Val158 had a higher response leading to target engagement.	[34]
Systemic lupus erythematosus	Rituximab (anti-CD20 monoclonal antibody)	Immunoglobulin gamma Fc region receptor III-A (FCGR3A)—Val158	Increases efficiency of B-cell clearance.	[35]

therapeutics. IFNα signaling cascades and receptor subunits have polymorphisms along with genes controlling inflammatory cytokines and antiviral responses. This, along with viral load, host ancestry, and gender, contributes to a unique model of IFNα immunotherapy in HCV patients. When IFNα was used for multiple sclerosis patients, exploratory biomarkers like IL-1b and IL-8 showed decreased expression, indicating that the protein therapy was working in the patient [16, 17]. Although pharmacodynamic models have been constructed for various genes that are altered by IFNα therapy, the efficiency of responses is susceptible to interindividual genetic heterogeneity [18].

8.2.1.2 Rituximab

Monoclonal antibodies (mAbs) against cellular targets work by activating phagocytic and cytotoxic actions within the cell. Polymorphisms in the gene targets that mediate this contact may influence protein drug efficacy. Several studies have examined the impacts of polymorphisms in the immunoglobulin Fc receptor (FcR) for mAb treatment. For example, the Val158 variation of the receptor for the Fc portion of immunoglobulin G3A (FcGR3A) has been demonstrated in in vitro to have a greater affinity for Fc and to activate natural killer cells more effectively [19, 20]. IgG1 mAb rituximab (Rituxan; Genentech) gives a better response in Val158 carriers for follicular lymphoma where it targets the CD20 antigen on B cells [21–24]. Homozygotes for FcGR2A His131 had greater responses in a multivariate assessment of prognostic variables, and each genotype was significant. Val158 carriers were more successful with B-cell reduction in the Rituximab dose-dependent lupus study [25]. FcGR2A -Arg131 and FcGR3A -Val158 exhibited higher removal of Rh + ve erythrocytes from treated patients, according to a study on novel anti-RhD mAb being developed for Rh prophylaxis. A gene-dosage effect was seen in the erythrocyte elimination curves [26].

8.2.1.3 Infliximab

Effector cell mechanisms do not directly play a role in soluble cytokine removal. It also does not directly approve the obviousness of investigating FcR variants in the context of therapies targeting soluble cytokine factors. Variations in FcR, on the other hand, have been published in relation to tumor necrosis factor (TNF) inhibiting protein drugs. Val158 variation of FcGR3A was linked with decreasing C-reactive protein (CRP) in the blood of Crohn's disease patients treated with Infliximab (Remicade; Centocor), a TNF inhibitor. Val158 was used as a biomarker for disease activity efficacy. According to the researchers, this outcome could be due to the removal of membrane-bound TNF68-overexpressing cells. In a second investigation, the same group of researchers looked into the linkage disequilibrium [LD] between FcGR3A polymorphisms at position 158 in the CRP locus and found that CRP was not linked to FcGR3A and infliximab responsiveness [27]. The FcGR3A homozygous Phe158 genotype, which encodes the low-affinity FcR, was related to favorable treatment responses in a trial using TNF inhibitors in inflammatory arthritis [28].

8.2.1.4 Etanercept

The genetic polymorphisms influence both the circulation of the target and the milieu in which the therapeutic protein functions. Polymorphisms that modify TNF production may be considered upstream of the action of protein drugs like Infliximab or Etanercept. Variable TNF production control could affect disease levels and treatment-induced regulatory mechanisms. Multiple polymorphisms in the *TNF* α and lymphotoxin A (*LTA*) genes have been associated with TNF antagonists like Etanercept responses (Enbrel; Amgen). These alleles could influence increased transcriptional output or mRNA stability or may be related to other functional alleles [36, 37]. Increased TNF production was seen in the TNF α positions –308A and –857C alleles in reaction to stimulation of peripheral blood mononuclear cells (PBMCs) in vitro [37, 38]. Because TNF α and LTA are both found inside the HLA complex, which is linked to autoimmune diseases, they are probably associated with functional polymorphisms in related genes [39]. HLA-DRB1, a common epitope used as a predictive factor in inflammatory illnesses, also correlates to protein drug efficacy [40, 41]. Repeats in the interleukin-10 (IL10) promoter region, as well as polymorphisms in death receptor ligand-FAS ligand (FASL) and caspase-9 (CASP9) [42, 43], have been associated with varied TNF-blocking protein drug responses.

8.2.1.5 Anakinra

The cytokine interleukin-1 is regulated in inflammatory diseases by Anakinra (Kineret; Amgen), a recombinant IL-1 receptor antagonist (IL-1Ra), and has been approved to treat rheumatoid arthritis. In a randomized clinical trial, patients with minor alleles of IL1α and IL1beta had a higher reaction to Anakinra. *IL1α* and *IL1β* genes control the expression of IL-1, which competes with anakinra for receptor occupancy and the production of internal IL-1Ra [44]. Increased IL-1 expression could signal a disease type that responds favorably to treatment.

8.2.1.6 Tissue Plasminogen Activator (tPA)

Patients can risk treatment-related problems due to differences in upstream pathways. Tissue plasminogen activator (tPA) is used to prevent further thrombosis in artery clotting diseases such as heart attack or stroke. The efficacy of tPA therapy has been linked to an exon mutation in the fibrin-stabilizing factor FXIII46. Insufficient reperfusion and unfavorable outcomes, such as mortality, reinfarction, and percutaneous coronary intervention such as angioplasty, were more likely in Leu34 allele carriers. Nonsmokers have less plasma fibrinogen than smokers, and the Leu34 allele carrying FXIII46 generates taut clots at low fibrinogen concentrations and more leaky clots [45] at high fibrinogen concentrations, which is confined to nonsmokers.

Treatment with tPA reduces ischemia-related harm in stroke patients and increases the risk of hemorrhagic complications. In multiple investigations, increased risk after tPA therapy was linked to serum levels of matrix metalloproteinase-9 (MMP9), and tissue plasminogen activator enhanced MMP9 activity in animal studies [46–48]. According to these investigations, MMP9 activation was linked to increased fluid retention, inflammatory permeates, and internal bleeding following tPA reperfusion. However, no pharmacogenetic correlations with tPA treatment results were found when a TATA box-region variation in the *MMP9* gene was investigated; hence, MMP9 protein levels are unlikely to be influenced by this polymorphism [49].

8.2.1.7 Peptide Immunotherapy with Amyloid Beta

Experimental peptide immunotherapy using amyloid-beta 42 (Aβ42) was performed for the treatment of Alzheimer's disease (Fig. 8.4), but the treatment was terminated after 6% of the inoculated patients developed meningoencephalitis. Pretreatment gene expression profiling (GEP) analysis in PBMC revealed patterns linked to adverse drug reactions that were distinct from GEP patterns associated with therapeutic responses. At the start of the study, meningoencephalitis patients exhibited greater levels of apoptosis and inflammation-related mRNA transcripts than those who did not have meningoencephalitis. On the other hand, mRNA transcripts associated with therapeutic response were mostly connected to protein synthesis or cell cycle control. These patterns were determined using rules-based machine learning methods to distinguish between therapeutic and adverse drug response mRNA transcripts [50]. Another example is in chronic HCV infection, where Cetuximab can be continued for treatment if the mRNA expression profile includes EGFR, IL-8, and Ribavirin [52].

8.2.2 Protein Therapeutics and Their Effect on Genetic Variations in Drug Targets

Drug responses can be influenced by functional differences in the coding sequence or copy number of a drug target transcript. This eventually leads to changes in protein-based drug and protein-receptor interactions and following downstream signaling. Most trials have focused on identifying individuals who will respond to mAb targeting gene products of the ErbB family of epidermal growth factor receptor (EGFR/HER1) and HER2/neu. Both EGFR and HER2/neu are upregulated in solid tumors, and their activation promotes tumor cell proliferation and survival. Some strategies used to block ErbB mediated signals include inhibition of ligand interaction and prompting downregulation of the EGFR and HER2/neu receptors, reducing

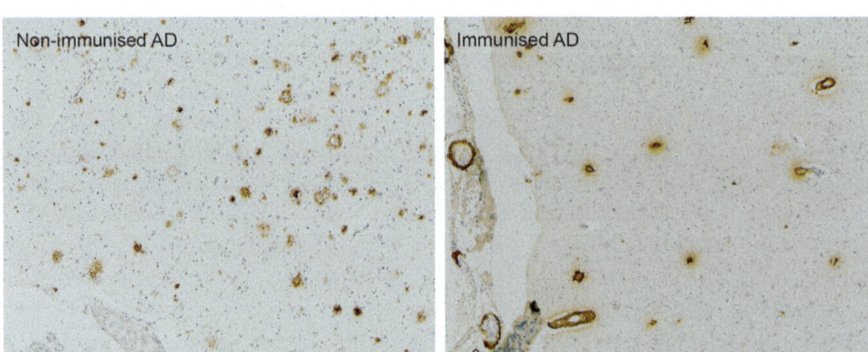

Fig. 8.4 Peptide immunotherapy with amyloid-beta untreated and treated brain for Alzheimer's disease [AD] treatment [51]

Table 8.2 Pharmacogenetic markers of protein therapeutics and their target variations

Disease in which Pharmacogenetic studies were done	Protein therapeutics commercial name	Genotype/allele marker	Ref.
Chronic hepatitis C	IFN α (recombinant human cytokine; IFN α -2a, IFN α -2b)	Interferon receptor-IFNAR1 should have GT repeats ranging from 5 to 14.	[64]
Non–growth hormone-deficient children with growth delay	Somatropin (recombinant human cytokine; rGH)	Growth hormone receptor-D3-GHR should lack exon 3.	[65]
Metastatic breast cancer	Trastuzumab (humanized anti-HER2/neu mAb)	HER2/*neu* detection in IHC should be 3+ level.	[66]
Colorectal cancer (CRC)	Cetuximab (humanized anti-EGFR mAb	IHC should be negative for EGFR.	[67]
Renal cell carcinoma (RCC)	Panitumumab	Score of EGFR should be 2+ or 3+ score in IHC.	[68]

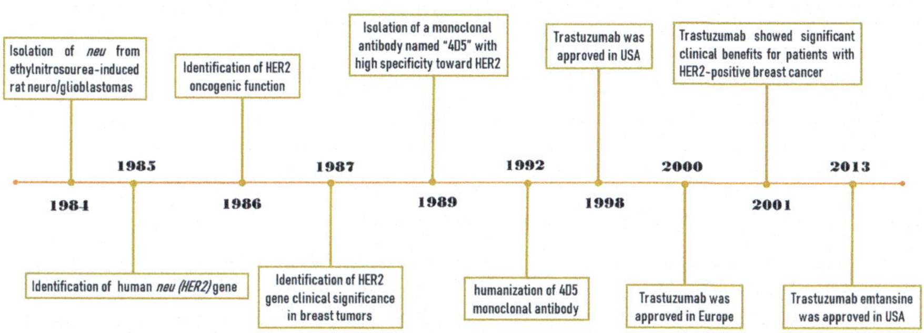

Fig. 8.5 Timeline of Trastuzumab development [58]

HER2/neu extracellular domain proteolysis, preventing downstream signaling to ErbB heterodomains, and stopping the tyrosine kinase activity in EGFR [53]. The genetic diversity of EGFR and HER2/neu and its direct link to tumor treatment like anti-ErbB therapy techniques have been thoroughly investigated. The vigorous processes that drive tumor formation do impact the identification of protein drug targets in oncology (Table 8.2). Numerous polymorphisms in the ErbB pathways may provide tumors an advantage in terms of growth, resulting in dozens of disease subtypes and genotypes and a wide range of therapeutic targets.

8.2.2.1 Trastuzumab

Trastuzumab (Erbitux) was the first specifically produced mAb to target HER2/neu (Fig. 8.5). It generates a response in tumors with high HER2/neu and EGFR gene copy numbers. Trastuzumab therapy for breast cancer is usually paired with an assay of HER2/neu overexpression. In breast cancers with HER2/neu overexpression, Trastuzumab has been highly successful [54–57]. HER2/neu overexpression assay identifies the subset of breast cancer patients who will favorably react to Trastuzumab.

When patient subgroups were assessed for gene copy numbers using fluorescence in situ hybridization (FISH) rather than immunohistochemistry (IHC), more correlations were discovered [59, 60]. The lack of correlation in IHC could be due to technical aspects of the assays or biological phenomena such as receptor degradation or endocytosis of the receptor or uneven distribution of gene copy number in the oncology sample, or due to progression to metastatic foci [61]. In contrast, the effect of Pertuzumab (Omnitarg; Genentech), a new HER2/neu dimerization inhibitor, can be administered in patients without HER2/neu overexpression and has been proven to be not dependent on HER2/neu levels in preclinical testing [62]. The advantages of genotyping, gene amplification, and overexpression of members of the ErbB family may depend on the cell environment and the factors that affect the anticancer protein drug due to the varied modes of action [63].

In Gefitinib (Iressa; AstraZeneca) and Erlotinib (Tarceva; Genentech/OS), which are antagonists to EGFR tyrosine kinase, the role of genetic diversity affects the clinical effectiveness [69, 70]. Inheritable mutations in the EGFR tyrosine kinase domain have been associated with clinical reactions to their inhibitors like Gefitinib and Erlotinib. HER2/neu and *EGFR* gene amplifications are generally unresponsive to kinase domain modifications.

8.3 Pharmacokinetics-Based Pharmacogenetics

Small-molecule drugs are administered and metabolized differently than protein-based drugs. The bulk of protein treatments is given by injection, and their half-life ranges from a few minutes to a few days in vivo. Drug metabolism comprises a variety of broad activities, such as proteolysis and degradation, as well as specialized activities such as receptor-mediated responses. Because protein drugs have a high affinity for their targets, lower dissociation off-rates indicate strong binding to the target and lower nonspecific binding. Their pharmacokinetic features are typically linked to the drug target's metabolic fate. Drug targets in circulation, lymphatic system, or exterior tissue can all affect drug metabolism. Because drug targets can bind a massive load of the therapeutic protein dose, clearance mechanisms need drug-target binding dynamics, an essential variation from most small-molecular therapies. Internalization and destruction of a protein drug-target complex within the cell, cleavage, release into circulation, or apoptosis or phagocytic cell consumption are options. As a result, the nature of protein drugs is closely linked to their protein target binding effects.

Nonspecific processes in the body remove most of the protein drug from Tolerx, an anti-CD4 antibody TRX1, while receptor-mediated endocytosis eliminates the rest of the drug when a single dose is administered [71]. Panitumumab, an anti-EGFR antibody, has nonlinear kinetics that could be explained by EGFR site saturation [68]. Furthermore, IFN research has shown that receptor-mediated ingestion is the primary mode of removal, while renal processing and hydrolysis of the protein drug act as minor mechanisms [72]. The pharmacodynamic response of protein therapies, like that of small-molecule pharmaceuticals, can alter the copy number of targets. Target modification can influence drug removal, while other changes can influence drug responsiveness without affecting target copy numbers. These alterations could be indicated by changes in the kinetic properties of the protein-based drugs, such as half-life or maximum plasma concentration, brought about by side-chain substitutions and torsion angle optimizations of the protein drug.

Another example is the overexpression of suppressors of cytokine signaling proteins (SOCS) after an IFN treatment. SOCS stops cell-to-cell signaling through the Janus kinase/signal transducers and JAK/STAT transcription pathway and hence reduces IFN-responsive gene transcription [73]. In chronic HCV infection, patients with high SOCS1 protein levels in the liver were less likely to produce durable antiviral responses to IFN therapy [74]. The importance of genetic variants and their influence on SOCS protein generation is unknown, even though *SOCS* genes are hypermethylated in the promoter and have somatic mutations in tumors [75].

8.4 Pharmacogenetics-Based Immunogenicity Prediction of Protein Therapeutics

One of the most critical risk factors in the progress of protein therapeutics is its immunogenicity. Antibody responses to nonnative proteins are often initiated by the formation of peptides, and an MHC II pathway–mediated T-cell response (Fig. 8.6). The similarity of a protein drug to certain native proteins could cause adverse effects on the patient [76]; hence, antigenic synthetic variants of endogenous proteins are a

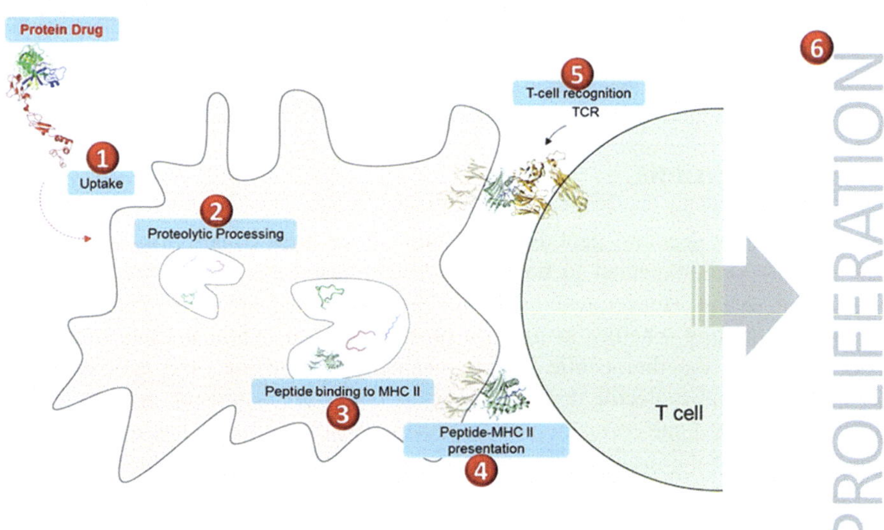

Fig. 8.6 Immunogenicity pathway for a protein-based drug [78]

major source of concern. Patients with greater sensitivity to antibody reactions to protein dugs have been reported in recent pharmacogenetic research. For example, patients with the major histocompatibility antigen HLA-B-5701 are more likely to be hypersensitive to the antiretroviral medication Abacavir. Antibodies to IFN have also been linked to particular HLA haplotypes in patients with multiple sclerosis [77]. The long-term efficacy of IFN therapy is reduced in patients who generate neutralizing antibodies.

It is conceivable to take a pharmacogenetic approach to predict the immunogenicity risks of a protein drug [79]. The steps where the pharmacogenetics approach can be applied are as follows: (i) When an endogenous component is injected with a therapeutic protein, the native protein of the patient is regarded as a self-protein, and immunogenic tolerance is granted. As a result, while sequence variations between endogenous and infused proteins (protein drugs) are necessary for generating an immune response, they are insufficient to cause further complications to the patient. (ii) The immunological response can now be split into two distinct types: Firstly, foreign peptides are synthesized based on differences in proteolytic enzymes in a population. Secondly, even if the peptides are synthesized, they will bind with varying degrees of affinity to the patient's MHC-II variation. (iii) MHC-II affinity and half-life predictions are used as predictors of antigenicity. MHC proteins are one of the polymorphic proteins in the human genome that has multiple ramifications. The MHC-II variations in distinct populations is crucial knowledge that needs to be meticulously considered throughout the pharmaceutical development of a peptide. For example, a foreign protein sequence used as a linker will concatenate two different domains during protein engineering experiments and facilitate binding to a few MHC-II proteins with a high affinity. Binding a foreign peptide to a rare MHC protein will have a lesser utility effect than attaching to a largely shared MHC allele, because MHC proteins do not evolve at the same rate. (iv) Finally, the number of MHC proteins in each population differs. This means that a protein drug tailored for one population may induce a higher immune response in another population.

8.5 Conclusions

For the past few decades, peptide therapeutics have been consistently authorized, and this trend is expected to continue. Protein-based treatments have a unique position in the pharmaceutical landscape outperforming small molecules. Early decisions regarding whether to move a protein lead into clinical trials should be made based on the therapeutic target, mechanism of action, and difficulties in generating a small-molecule drug for the same target. Because protein therapeutics are so concentrated, the correlation of pharmacogenetics can be studied in both Phase 1 and phase 2 of the drug development process. Generating genetic biomarkers for assessing treatment activity is time consuming but has been continuously on the rise. By addressing the impact of genotype on the phenotype of the effect of the protein-based drug early in the drug development cycle, researchers can generate a dynamic basis to use pharmacogenetic markers in clinical trials success analysis. These indicators could help identify response rates and side effects, conduct efficacy trials, and refer patients to the most successful treatments.

References

1. Gummadi AC, Guddati AK (2021) Genetic polymorphisms in pharmaceuticals and chemotherapy. World J Oncol 12(5):149–154. https://doi.org/10.14740/wjon1405
2. Sepulveda AR, Hamilton SR, Allegra CJ et al (2017) Molecular biomarkers for the evaluation of colorectal cancer: guideline from the American Society for Clinical Pathology, College of American Pathologists, Association for Molecular Pathology, and American Society of Clinical Oncology. J Mol Diagn 19(2):187–225.9. https://doi.org/10.1016/j.jmoldx.2016.11.001
3. Kaplon H, Chenoweth A, Crescioli S, Reichert JM (2022) Antibodies to watch in 2022. MAbs 14(1):2014296. https://doi.org/10.1080/19420862.2021.2014296
4. Shepard HM, Phillips GL, Thanos CD, Feldmann M (2017) Developments in therapy with monoclonal antibodies and related proteins. Clin Med (Lond) 17(3):220–232. https://doi.org/10.7861/clinmedicine.17-3-220
5. Dimitrov DS (2012) Therapeutic proteins. Methods Mol Biol 899:1–26. https://doi.org/10.1007/978-1-61779-921-1_1
6. Wilson FR, Coombes ME, Wylie Q et al (2017) Herceptin® (Trastuzumab) in HER2-positive early breast cancer: protocol for a systematic review and cumulative network meta-analysis. Syst Rev 6(1):196. https://doi.org/10.1186/s13643-017-0588-2
7. Roden DM, Wilke RA, Kroemer HK, Stein CM (2011) Pharmacogenomics: the genetics of variable drug responses. Circulation 123(15):1661–1670. https://doi.org/10.1161/CIRCULATIONAHA.109.914820
8. Nerenz RD (2018) Chapter 32 - pharmacogenomics and personalized medicine in the treatment of human diseases. In: Coleman WB, Tsongalis GJ (eds) Molecular pathology, 2nd edn. Academic Press, pp 731–743
9. Roses A (2008) Pharmacogenetics in drug discovery and development: a translational perspective. Nat Rev Drug Discov 7:807–817. https://doi.org/10.1038/nrd2593
10. Taylor D (2015) The pharmaceutical industry and the future of drug development. In: Pharmaceuticals in the environment, pp 1–33. https://doi.org/10.1039/9781782622345-00001
11. Lindpaintner K, Foot E, Caulfield M et al (2001) Pharmacogenetics: focus on pharmacodynamics. Int J Pharm Med 15:74–82. https://doi.org/10.1097/00124363-200104000-00009
12. Ahmed S, Zhou Z, Zhou J, Chen SQ (2018) Pharmacogenomics of drug metabolizing enzymes and transporters: relevance to precision medicine Genomics Proteomics Bioinf. 2016;14(5): 298–313. https://doi.org/10.1016/j.gpb.2016.03.008
13. Sakano H, Harris JP (2019) Emerging options in immune-mediated hearing loss. Laryng Investig Otolaryngol 4:102–108. https://doi.org/10.1002/lio2.205
14. Madian AG, Wheeler HE, Jones RB, Dolan ME (2012) Relating human genetic variation to variation in drug responses. Trends Genet 28(10):487–495. https://doi.org/10.1016/j.tig.2012.06.008
15. Lagassé HA, Alexaki A, Simhadri VL et al (2017) Recent advances in (therapeutic protein) drug development. F1000Res 6:113. Published 2017 Feb 7. https://doi.org/10.12688/f1000research.9970.1
16. Sturzebecher S et al (2003) Expression profiling identifies responder and non-responder phenotypes to interferon-beta in multiple sclerosis. Brain 126:1419–1429
17. Hong J, Zang YC, Hutton G, Rivera VM, Zhang JZ (2004) Gene expression profiling of relevant biomarkers for treatment evaluation in multiple sclerosis. J Neuroimmunol 152:126–139
18. Weinstock-Guttman B et al (2003) Genomic effects of IFN-a in multiple sclerosis patients. J Immunol 171:2694–2702

19. Dall'Ozzo S et al (2004) Rituximab-dependent cytotoxicity by natural killer cells: influence of FCGR3A polymorphism on the concentration-effect relationship. Cancer Res 64:4664–4669
20. Bowles JA, Weiner GJ (2005) CD16 polymorphisms and NK activation induced by monoclonal antibody-coated target cells. J Immunol Methods 304:88–99
21. Weng WK, Levy R (2003) Two immunoglobulin G fragment C receptor polymorphisms independently predict response to rituximab in patients with follicular lymphoma. J Clin Oncol 21:3940–3947
22. Cartron G et al (2002) Therapeutic activity of humanized anti-CD20 monoclonal antibody and polymorphism in IgG Fc receptor FcGRIIIa gene. Blood 99:754–758
23. Farag SS et al (2004) FcGRIIIa and FcG RIIa polymorphisms do not predict response to Rituximab in B-cell chronic lymphocytic leukemia. Blood 103:1472–1474
24. Treon SP et al (2005) Polymorphisms in FcGRIIIA (CD16) receptor expression are associated with clinical response to Rituximab in Waldenstrom's macroglobulinemia. J Clin Oncol 23: 474–481
25. Anolik JH et al (2003) The relationship of FcGRIIIa genotype to degree of B cell depletion by Rituximab in the treatment of systemic lupus erythematosus. Arthritis Rheum 48:455–459
26. Miescher S et al (2004) A single recombinant anti-RhD IgG prevents RhD immunization: association of RhD- positive red blood cell clearance rate with polymorphisms in the FcGRIIA and FcGIIIA genes. Blood 103:4028–4035
27. Willot S et al (2006) No association between C-reactive protein gene polymorphisms and decrease of C-reactive protein serum concentration after infliximab treatment in Crohn's disease. Pharmacogenet Genomics 16:37–42
28. Tutuncu Z et al (2005) FcG receptor type IIIA polymorphisms influence treatment outcomes in patients with inflammatory arthritis treated with tumor necrosis factor beta-blocking agents. Arthritis Rheum 52:2693–2696
29. Silvain J, Pena A, Vignalou JB et al (2011) FXIII-A Leu34 genetic variant in premature coronary artery disease: a genotype--phenotype case control study. Thromb Haemost 106(3): 511–520. https://doi.org/10.1160/TH11-01-0027
30. Halade GV, Jin YF, Lindsey ML (2013) Matrix metalloproteinase (MMP)-9: a proximal biomarker for cardiac remodeling and a distal biomarker for inflammation. Pharmacol Ther 139(1):32–40. https://doi.org/10.1016/j.pharmthera.2013.03.009
31. Vallböhmer D, Zhang W, Gordon M et al (2005) Molecular determinants of cetuximab efficacy. J Clin Oncol 23(15):3536–3544. https://doi.org/10.1200/JCO.2005.09.100
32. López-Rodríguez R, Hernández-Bartolomé Á, Borque MJ et al (2017) Interferon-related tic markers of necroinflammatory activity in chronic hepatitis C. PLoS One 12(7):e0180927. https://doi.org/10.1371/journal.pone.0180927
33. Cvetkovic RS, Keating G (2002) Anakinra. BioDrugs 16(4):303–314. https://doi.org/10.2165/00063030-200216040-00005
34. Zheng YH, Xu L, Cao C et al (2019) Rituximab-based combination therapy in patients with Waldenström macroglobulinemia: a systematic review and meta-analysis. Onco Targets Ther 12:2751–2766. https://doi.org/10.2147/OTT.S191179
35. Zhong M, van der Walt A, Campagna MP et al (2020) The pharmacogenetics of rituximab: potential implications for anti-CD20 therapies in multiple sclerosis. Neurothera 17:1768–1784. https://doi.org/10.1007/s13311-020-00950-2
36. Martinez A et al (2004) Association of the major histocompatibility complex with response to infliximab therapy in rheumatoid arthritis patients. Arthritis Rheum 50:1077–1082
37. Louis E et al (1998) Tumour necrosis factor (TNF) gene polymorphism influences TNF-a production in lipopolysaccharide (LPS)-stimulated whole blood cell culture in healthy humans. Clin Exp Immunol 113:401–406
38. van Heel DA et al (2002) Inflammatory bowel disease is associated with a TNF polymorphism that affects an interaction between the OCT1 and NFκB transcription factors. Hum Mol Genet 11:1281–1289

39. Shiina T, Inoko H, Kulski JK (2004) An update of the HLA genomic region, locus information and disease associations: 2004. Tissue Antigens 64:631–649
40. Kang CP, Lee KW, Yoo DH, Kang C, Bae SC (2005) The influence of a polymorphism at position-857 of the tumour necrosis factor α gene on clinical response to etanercept therapy in rheumatoid arthritis. Rheumatology (Oxford) 44:547–552
41. Criswell LA et al (2004) The influence of genetic variation in the HLA-DRB1 and LTA-TNF regions on the response to treatment of early rheumatoid arthritis with methotrexate or Etanercept. Arthritis Rheum 50:2750–2756
42. Schotte H et al (2005) Interleukin 10 promoter microsatellite polymorphisms are associated with response to long term treatment with Etanercept in patients with rheumatoid arthritis. Ann Rheum Dis 64:575–581
43. Hlavaty T et al (2005) Polymorphisms in apoptosis genes predict response to infliximab therapy in luminal and fistulizing Crohn's disease. Aliment Pharmacol Ther 22:613–626
44. Camp NJ et al (2005) Evidence of a pharmacogenomic response to interleukin-l receptor antagonist in rheumatoid arthritis. Genes Immun 6:467–471
45. Lim BC, Ariens RA, Carter AM, Weisel JW, Grant PJ (2003) Genetic regulation of fibrin structure and function: complex gene-environment interactions may modulate vascular risk. Lancet 361:1424–1431
46. Horstmann S, Kalb P, Koziol J, Gardner H, Wagner S (2003) Profiles of matrix metalloproteinases, their inhibitors, and laminin in stroke patients: influence of different therapies. Stroke 34:2165–2170
47. Montaner J et al (2003) Matrix metalloproteinase-9 pre-treatment level predicts intracranial hemorrhagic complications after thrombolysis in human stroke. Circulation 107:598–603
48. Tsuji K et al (2005) Tissue plasminogen activator promotes matrix metalloproteinase-9 upregulation after focal cerebral ischemia. Stroke 36:1954–1959
49. Montaner J et al (2003) Safety profile of tissue plasminogen activator treatment among stroke patients carrying a common polymorphism (C-1562T) in the promoter region of the matrix metalloproteinase-9 gene. Stroke 34:2851–2855
50. O'Toole M et al (2005) Risk factors associated with beta- amyloid(1–42) immunotherapy in preimmunization gene expression patterns of blood cells. Arch Neurol 62:1531–1536
51. Nimmo JT, Kelly L, Verma A, Carare RO, Nicoll JAR, Dodart JC (2021) Amyloid-β and α-synuclein immunotherapy: from experimental studies to clinical trials. Front Neurosci 15: 733857. https://doi.org/10.3389/fnins.2021.733857
52. Chen L et al (2005) Hepatic gene expression discriminates responders and nonresponders in treatment of chronic hepatitis C viral infection. Gastroenterology 128:1437–1444
53. Wieduwilt MJ, Moasser MM (2008) The epidermal growth factor receptor family: biology driving targeted therapeutics. Cell Mol Life Sci 65(10):1566–1584. https://doi.org/10.1007/s00018-008-7440-8
54. Saltz LB et al (2004) Phase II trial of cetuximab in patients with refractory colorectal cancer that expresses the epidermal growth factor receptor. J Clin Oncol 22:1201–1208
55. Cunningham D et al (2004) Cetuximab monotherapy and Cetuximab plus irinotecan in irinotecan-refractory metastatic colorectal cancer. N Engl J Med 351:337–345
56. Burtness B, Goldwasser MA, Flood W, Mattar B, Forastiere AA (2005) Phase III randomized trial of cisplatin plus placebo compared with cisplatin plus Cetuximab in metastatic/recurrent head and neck cancer: an eastern cooperative oncology group study. J Clin Oncol 23:8646–8654
57. Chung KY et al (2005) Cetuximab shows activity in colorectal cancer patients with tumors that do not express the epidermal growth factor receptor by immunohistochemistry. J Clin Oncol 23: 1803–1810
58. Maadi H, Soheilifar MH, Choi WS, Moshtaghian A, Wang Z (2021) Trastuzumab mechanism of action; 20 years of research to unravel a dilemma. Cancers (Basel) 13(14):3540. https://doi.org/10.3390/cancers13143540

59. Moroni M et al (2005) Gene copy number for epidermal growth factor receptor (EGFR) and clinical response to antiEGFR treatment in colorectal cancer: a cohort study. Lancet Oncol 6: 279–286
60. Mass RD et al (2005) Evaluation of clinical outcomes according to HER2 detection by fluorescence in situ hybridization in women with metastatic breast cancer treated with Trastuzumab. Clin Breast Cancer 6:240–246
61. Dai Z, Barbacioru C, Huang Y, Sadée W (2006) Prediction of anti-cancer drug potency from expression of genes involved in growth factor signaling. Pharm Res 23:336–349
62. Agus DB et al (2005) Phase I clinical study of pertuzumab, a novel HER dimerization inhibitor, in patients with advanced cancer. J Clin Oncol 23:2534–2543
63. Hynes NE, Lane HA (2005) ERBB receptors and cancer: the complexity of targeted inhibitors. Nature Rev Cancer 5:341–354
64. Karamitros T, Papatheodoridis G, Paraskevis D et al (2018) Impact of interferon-α Receptor-1 promoter polymorphisms on the transcriptome of the hepatitis B virus-associated hepatocellular carcinoma. Front Immunol 9:777. https://doi.org/10.3389/fimmu.2018.00777
65. van der Klaauw AA, van der Straaten T, Baak-Pablo R et al (2008) Influence of the d3-growth hormone (GH) receptor isoform on short-term and long-term treatment response to GH replacement in GH-deficient adults. J Clin Endocrinol Metab 93(7):2828–2834. https://doi.org/10.1210/jc.2007-2728
66. Won HS, Ahn J, Kim Y et al (2022) Clinical significance of HER2-low expression in early breast cancer: a nationwide study from the Korean Breast Cancer Society. Breast Cancer Res 24:22. https://doi.org/10.1186/s13058-022-01519-x
67. Zhang W, Gordon M, Schultheis AM et al (2007) FCGR2A and FCGR3A polymorphisms associated with clinical outcome of epidermal growth factor receptor expressing metastatic colorectal cancer patients treated with single-agent cetuximab. J Clin Oncol 25(24):3712–3718. https://doi.org/10.1200/JCO.2006.08.8021
68. Kast J, Dutta S, Upreti VV (2021) Panitumumab: a review of clinical pharmacokinetic and pharmacology properties after over a decade of experience in patients with solid tumors. Adv Ther 38:3712–3723. https://doi.org/10.1007/s12325-021-01809-4
69. Pao W, Miller VA (2005) Epidermal growth factor receptor mutations, small-molecule kinase inhibitors, and non-small-cell lung cancer: current knowledge and future directions. J Clin Oncol 23:2556–2568
70. Giaccone G (2005) Epidermal growth factor receptor inhibitors in the treatment of non-small-cell lung cancer. J Clin Oncol 23:3235–3242
71. Ng CM, Stefanich E, Anand BS, Fielder PJ, Vaickus L (2006) Pharmacokinetics/pharmacodynamics of nondepleting anti-CD4 monoclonal antibody (TRX1) in healthy human volunteers. Pharm Res 23:95–103
72. Mager DE, Jusko WJ (2001) General pharmacokinetic model for drugs exhibiting target-mediated drug disposition. J Pharmacokinet Pharmacodyn 28:507–532
73. Valentino L, Pierre J (2006) JAK/STAT signal transduction: regulators and implication in hematological malignancies. Biochem Pharmacol 71:713–721
74. Mafanda EK, Kandhi R, Bobbala D et al (2019) Essential role of suppressor of cytokine signaling 1 (SOCS1) in hepatocytes and macrophages in the regulation of liver fibrosis. Cytokine 124:154501
75. Sutherland K, Lindeman G, Choong D et al (2004) Differential hypermethylation of *SOCS* genes in ovarian and breast carcinomas. Oncogene 23:7726–7733

76. Koren E, Zuckerman LA, Mire-Sluis AR (2002) Immune responses to therapeutic proteins in humans--clinical significance, assessment and prediction. Curr Pharm Biotechnol 3:349–360
77. Barbosa MD, Vielmetter J, Chu S, Smith DD, Jacinto J (2005) Clinical link between MHC class II haplotype and interferon-g (IFN-g) immunogenicity. Clin Immunol 118:42–50
78. Rosenberg AS, Sauna ZE (2018) Immunogenicity assessment during the development of protein therapeutics. J Pharm Pharmacol 70:584–594
79. Pandey GS, Sauna ZE (2014) Pharmacogenetics and the immunogenicity of protein therapeutics. J Interf Cytokine Res 34(12):931–937

Immunogenicity of Therapeutic Proteins 9

Mohammad Yasir, Alok Shiomurti Tripathi, Prashant Shukla, and Rahul Kumar Maurya

Abstract

Therapeutic proteins are potent medications that have shown substantial promise in treating a wide range of illnesses and conditions. Immunogenicity is a unique obstacle that must be overcome when dealing with protein therapy. The immune response is triggered when there is a change in the structure of the protein, which may occur because of posttranslational changes such as the administration, storage, or manufacturing process of the product. The structure of the protein may affect immunotolerance; for example, low-abundance proteins have a lower overall tolerance. Antidrug antibodies (ADAs) may influence the pharmacokinetics, pharmacodynamics, and efficacy of therapeutic proteins. When it comes to the development of ADAs, one of the primary sources of concern is the interaction that therapeutic proteins have with endogenous proteins. In this chapter, we discuss the significant elements associated with the immunogenicity of therapeutic proteins.

Keywords

Aggregation · Antidrug antibodies · Deimmunization · Immunogenicity · Therapeutic proteins

M. Yasir · R. K. Maurya (✉)
Amity Institute of Pharmacy, Lucknow, Amity University Uttar Pradesh, Noida, India
e-mail: myasir@lko.amity.edu; rkmaurya@lko.amity.edu

A. S. Tripathi
Department of Pharmacology, S.K. Patel College of Pharmaceutical Education and Research, Ganpat University, Mehsana, India

P. Shukla
Department of Pharmaceutical Sciences, School of Health Sciences, University of Petroleum Studies, Dehradun, India

D. B. Singh, T. Tripathi (eds.), *Protein-based Therapeutics*,
https://doi.org/10.1007/978-981-19-8249-1_9

9.1 Introduction

Proteins are the most versatile and active biomacromolecules in the human body that are involved in several processes like biochemical activity, membrane receptors and channels, intracellular and extracellular scaffolding, molecule transportation, and other functions [1–4]. Therapeutic proteins are potent and fast-acting medicines that have shown significant promise in the treatment of a broad variety of diseases and ailments. As our understanding of the molecular mechanisms behind disease expands, the likelihood of protein-based therapeutics will increase [5, 6].

In today's pharmaceutical industry, protein-based therapeutics are major players. Drug target discovery, protein engineering and design, purification, drug delivery, and marketing are a few areas that need to be improved for these technologies to become more widely used in clinical practice. Even though there are a lot of potentially therapeutic proteins in the human genome, they did not evolve to be therapeutic, so they do not have the best stability, specificity affinity, or activity for treating disease. The stability and immunogenicity of the proteins are two of the biggest problems that make it challenging for protein therapies to work [7]. All medications must be proven safe for use before their approval. Protein therapies provide a distinct challenge in terms of immunogenicity. Pathogens are recognized by the human immune system by their proteins or protein-processed compounds. The immune system of individuals who receive protein therapies may also produce antidrug antibodies against the protein drug. With such immune reactions, the effectiveness of the therapeutic protein molecule may be compromised and, in rare instances, can result in life-threatening conditions. Hence, immunogenicity testing is vital for developing protein therapies [8–10].

9.2 Immunogenicity

Immunogenicity is the ability of antigenic substances to stimulate the immune response in an individual, which is considered physiologically an undesirable effect. Immunogenicity is an immune response that contributes to the production of anti-drug antibodies (ADAs), which induce adverse drug reactions and inactivate the therapeutic effect. It is challenging for biotherapy to predict the immunogenicity potential of novel therapeutic proteins [11] (Table 9.1).

9.2.1 Factors Responsible for Immunogenicity [24]

Several factors contribute to the production of antibodies against any therapeutic protein. It is broadly characterized in two sections, that is, patient-related factors and product-related factors.

Table 9.1 Strategies to overcome the immunogenic responses of therapeutic proteins

Therapeutic protein	Immunogenic response	Strategies to overcome the response	References
Clotting factor VIII	Develop neutralizing anti-FVIII antibodies inhibitors	• Longer-lasting or prolonged half-life clotting factors have recently been released for therapeutic and preventative use in hemophilia A. Fc and albumin fusion proteins, as well as PEGylated FVIII, are among them. • When FVIII inhibitors are used, they negatively impact coagulation; hence, an alternative method is the development of bypassing agents that either do not need FVIII or that imitates the function of coagulation. Emicizumab and Fitusiran are examples of new medicines.	[12–14]
Insulin	Increases cross-reactive antibodies as well as antiinsulin antibody production	• Choice of insulin should be based on the low differences between the amino acid sequence of human insulin and insulin analogue.	[15]
Urokinase	Anaphylaxis or angioedema	• Reduction of anaphylactoid symptoms may be achieved with histamine receptor blockers (H1 and H2). Urokinase formulated with 5% albumin prevents severe hypersensitivity reactions.	[16]
Abatacept (CTLA4-Fc fusion protein)	Development of resistance	• Abatacept is used for the management of autoimmune diseases like rheumatoid arthritis. However, experimental investigations revealed that it was ineffective against preclinical transplantation models due to alteration of T-cell activation by incomplete blockage of the B7 costimulatory receptor. Thus, another drug molecule, i.e., Belatacept, with similar characteristics, has a higher affinity for CD80/CD86, which is used for organ transplantation.	[17–20]
INF-beta	Produces antibodies due to exposure to the antigenic epitope	• Immunogenicity of it reduced by hyperglycosylation, which improves activity, solubility, and stability.	[21–23]

9.2.1.1 Patient-Related Factors

Age

Age is one of the major factors contributing to immune response; thus, it is not possible to project the immunogenicity data of one age group with another. Patient age affects the production of the immune response against therapeutic proteins. Maturation of the immune system among the pediatric population is observed to be different among the different ages. If a drug is intended for children, clinical investigations in the pediatric population are necessary. Maturation of the immune system is observed to differ among the ages.

Genetic Factors

Genetic variation is present among everyone, which is also there in T cell receptors and major histocompatibility complex (MHC) molecules that modify or alter the level of immunity. Patients with genetic anomalies may have autoimmune disorders, which are considered to be at high risk for the development of ADA [25, 26].

Diseases

Diseases are also an important factor that causes the development of immune responses in an individual. Patients who suffer from autoimmune diseases, allergies, and chronic infections have an activated immune system against the therapeutic proteins. Moreover, diseases such as HIV, malignant disease, and malnutrition are immunocompromised conditions with less immune response. These diseases affect the development and production of antibodies against therapeutic proteins, which depend on the different stages of the disease. A systematic review suggests that anti-TNF agents such as infliximab, adalimumab, and certolizumab show immunogenicity of 25.3%, 14.1%, and 6.9%, respectively, in patients suffering from rheumatoid arthritis [27].

Pre-Existing Antibodies

Pre-existing antibodies are endogenous antibodies with a similar epitope of existing protein that overlaps with the epitope of a therapeutic protein. The exact underlying cause is yet to be determined. For example, antibodies against gal-alpha-1,3-gal are found in most humans and can react with the corresponding antigen on mAbs generated in mice. As a result, patients treated with Erbitux have hyposensitive responses [28].

9.2.1.2 Product-Related Factors

Several product-related factors such as the nature and origin of therapeutic protein, modifications, impurities, degradation, and formulation excipients alter the immunogenicity of therapeutic proteins. These factors are discussed in detail below.

Protein Source

Heterologous proteins are reported to be immunogenic in nature, with proteins derived from animal sources being noticed first. A nonnative recombinant version of proteins causes the formation of antibodies like salmon calcitonin, bacterial staphylokinase, and streptokinase. The immunogenicity of a heterologous protein can be predicted with the analysis of protein sequence and can be reduced. However, native human proteins are less immunogenic than other therapeutic proteins [24].

Protein Structure

The structure of protein affects immunotolerance, as low-abundance proteins have weaker general tolerance than high-abundance proteins. Furthermore, healthy individuals have a low level of cytokines; thus, autoantibodies are not uncommon. Alteration in the structure of the therapeutic protein triggers the immune reaction. This alteration occurs due to posttranslational modifications that develop due to the administration, storage, and manufacturing process of the product. In therapeutic proteins, alteration in amino acid sequence modifies the T cell epitopes [29]. Moreover, European guidelines suggest that the biological and physicochemical properties of proteins can also be influenced by glycosylation (http://www.emea.europa.eu). The immunogenicity of a therapeutic protein is altered by both presence and absence of carbohydrate moieties. Notably, carbohydrate binding alters the protein conformation, which leads to a change in protein immunogenicity [30]. Shielding of immunogenic epitopes reduces immunogenicity due to glycosylation and PEGylation.

Formulation and Packaging

The preparation of formulations and their compositions are explicitly specifically so that no alteration occurs in the structure of therapeutic proteins. Excipients used in the formulation should not interact with the active therapeutic proteins and packaging materials, since the interaction may alter the protein structure and impart impurities. Furthermore, a suitable formulation depends on the physical and chemical nature of excipients and the packaging material used. For example, the IFN-α2a formulation is immunogenic in nature; however, it is oxidized at room temperature and becomes more immunogenic in an individual [31].

Treatment Duration and Frequency of Dose

The frequency of drug administration is directly proportional to the development of immunogenicity. Clinically intramuscular administration of interferon beta-1 (INF-beta-1) three times a week enhances the development of antibodies compared to the administration once a week. Moreover, chronic administration improves antibody formation more than short-term administration [32].

Route of Administration

Protein administered intradermally or subcutaneously has a larger likelihood of developing immunogenicity than protein administered orally, intraperitoneally, or intravenously. Furthermore, subcutaneous administration of INF-beta1 is reported to be greater than intramuscular administration in multiple sclerosis patients [33].

9.2.2 Aggregation

Protein aggregation is a process in which monomers interact or self-assemble in their natural quaternary structure. During aggregation, proteins assemble in diverse forms that differ in biophysical and biochemical characteristics and sizes, ranging from visible particles to dimers [34]. Protein aggregates can be soluble or insoluble, associated with covalent or noncovalent linkages, and be reversible or irreversible in their formation. Several neurodegenerative disorders, such as Huntington's disease, Parkinson's disease, amyloidosis, amyotrophic lateral sclerosis, and Alzheimer's disease, are associated with protein aggregation (Fig. 9.1) [35].

9.2.2.1 Aggregation and Immunogenicity

Molecules that induce immune reactions are discussed under the heading "immunogenic and therapeutic proteins". They are reported to augment the immune response due to the formation of ADAs through protein aggregation. Furthermore, aggregation is used to predict the immunogenicity of therapeutic proteins [36]. Thus, the development of therapeutic protein formulations ensures a lower concentration of aggregations; however, the hazard of unforeseen aggregations depends on the kind and size of the aggregation. The method of production used in biotherapeutics can also cause aggregation, and manufacturers take this into account to prevent aggregation-related immunogenicity [37].

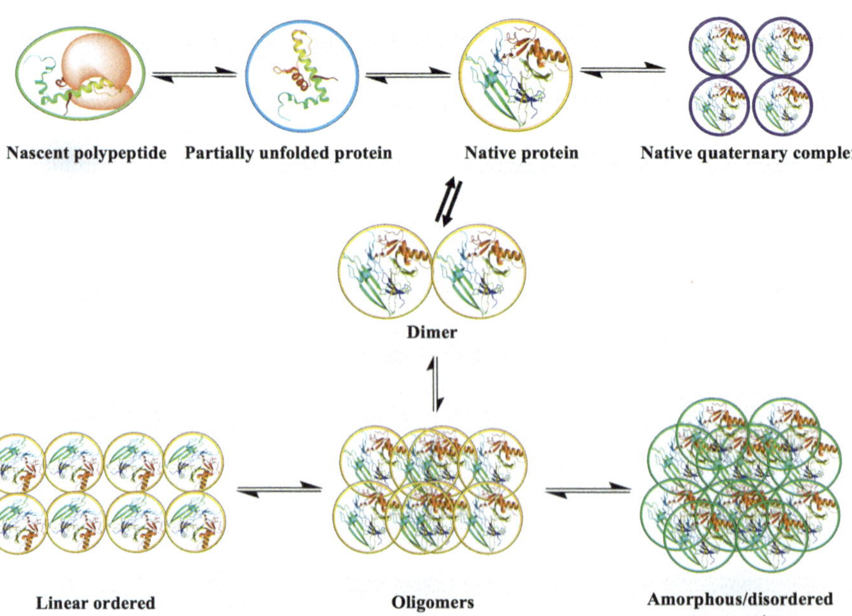

Fig. 9.1 Schematic representation of protein aggregation

9.2.2.2 Interaction of Protein Aggregate with Immunity

Adaptive immunity is an essential part of the immune system that provides responses to specific antigens by activating B and T cells. Adaptive immunity creates immunological memory, allowing cells to protect the body after repeated antigen exposure, while innate immunity, which is also an integral part of the human defense, generates a rapid response against antigens [38]. Both types of immunity defend the body against foreign objects or infections.

9.2.2.3 Innate Immunity Against Antigens

Protein aggregation is the first point of contact for innate immunity, as shown in the etiology of Alzheimer's disease, where the deposition of aggregated amyloid-beta (Aβ) plaques activates the immune system. TNF production from monocytes is induced by toll-like receptors 2 and 4 (TLR2 and TLR4). A study suggests that secretion of inflammatory cytokine is enhanced by the aggregation of three different monoclonal antibodies (mAbs) through the stimulation of TLR2 and TLR4 from the peripheral blood mononuclear cells, which is a part of innate immunity response [39]. Moreover, several other pathways are involved in innate immunity during protein aggregation.

9.2.2.4 Antidrug Antibody Formation

The T lymphocytes are involved in the production of antibodies. ADAs are antibodies formed against therapeutic protein molecules. The ADAs impact the potency, pharmacodynamics, and pharmacokinetics of therapeutic protein molecules. The interaction of therapeutic proteins with endogenous proteins is a prominent source of concern in the development of ADAs. The molecular characteristics of therapeutic protein aggregates influence their interaction with the immune system. The development of ADAs is the subject of various hypotheses. According to one hypothesis, the antigen-presenting cells (APCs) may recognize and trap aggregates, activating B cells for antibody formation via a linear interaction with the T cell epitope. However, according to another hypothesis, large therapeutic protein aggregates, particularly rigid and regular ones, stimulate ADA production by activating TLRs, which increases B-cell proliferation (Fig. 9.2) [40].

9.3 Therapeutic Proteins

Therapeutic proteins are proteins that have been genetically modified in the laboratory to be used as pharmaceuticals. Noncovalent binders, proteins that impact covalent bonding (nearly all enzymes), albumin, mAbs, blood products (clotting factor VIII), neuromuscular antagonists (botulinum toxin), hormones (glucagon), cellular growth factors (GM-CSF), and recombinant human cytokines (α and β interferon) are examples of therapeutic proteins, also known as biological therapeutics or macromolecular therapeutic proteins [10]. Therapeutic proteins have transformed disease management, because they are very efficacious in vivo. Therapeutic proteins are classified into five classes based on their pharmacological action:

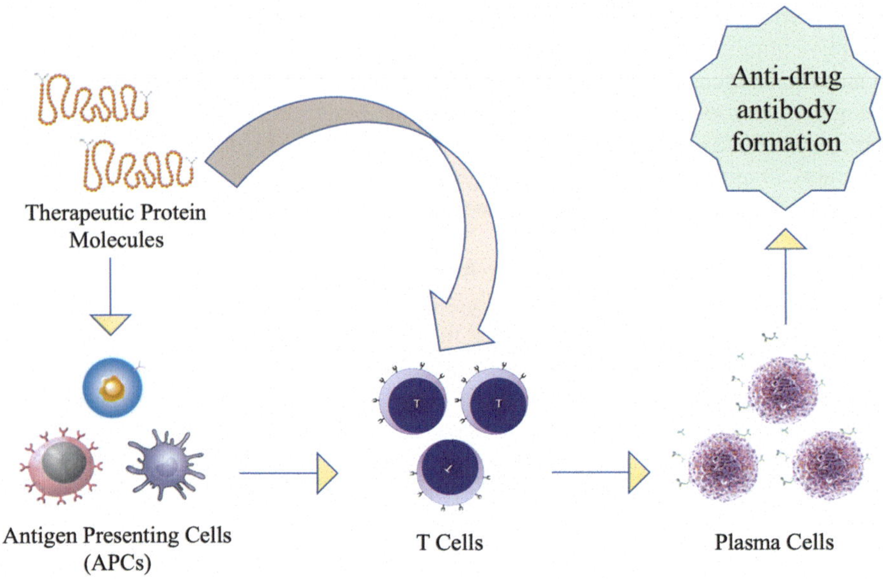

Fig. 9.2 Mechanism of antidrug antibody formation

(i) protein replacement that is abnormal or deficient, (ii) existing pathway augmentation, (iii) novel activity or function, (iv) interfering with a molecule or organism, and thereby blocking the function, causing destruction or activating different signaling pathways, and (v) delivering other proteins or compounds, like a cytotoxic drug, radionuclide, or effector proteins at specific sites [1].

9.3.1 Delivery of Therapeutic Proteins

Most therapeutic proteins are administered orally, which has several drawbacks, including enzymatic breakdown, low solubility, and nonlinear pharmacokinetics. Furthermore, mucosal, intranasal, intravaginal, intravenous, and transdermal administration routes have been employed to deliver therapeutic proteins (Fig. 9.3) [41]. Therapeutic proteins are delivered noninferiorly and suitably through the subcutaneous route than intravenous infusion. However, in a subgroup of individuals, there is the possibility of increased immunogenicity to subcutaneously injected proteins. Aside from the route of administration as an immunogenicity regulator, there are a variety of product-related adverse effects that are especially relevant to proteins administered subcutaneously [42]. Moreover, the pharmacokinetics and effectiveness of multiple biologics are influenced by undesired ADAs with repeated delivery of proteins or mAbs. The subcutaneous route of administration is associated with product-related risk factors like particle concentration,

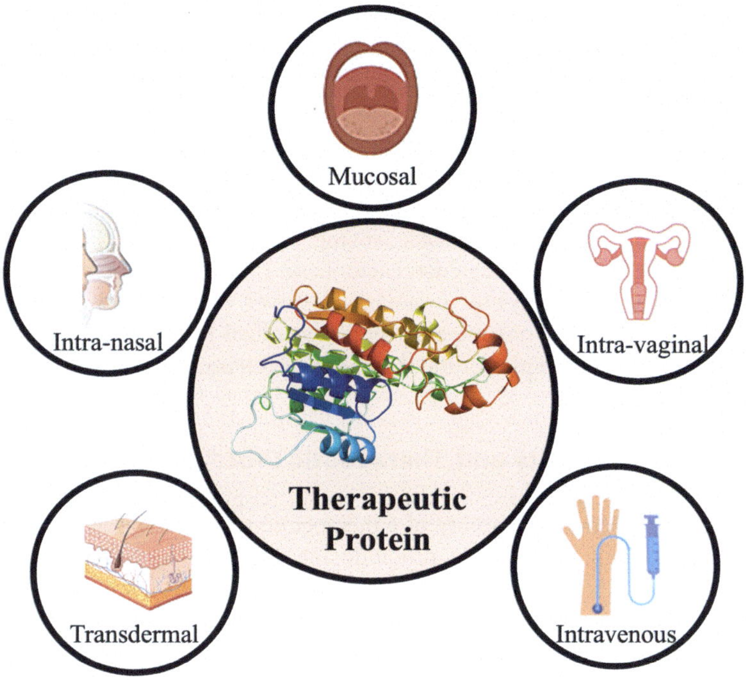

Fig. 9.3 Delivery of therapeutic proteins

aggregation, biophysical characteristics, and pollutants as a contributor to immunogenicity.

Because of the enzymatic breakdown, renal clearance, hepatic metabolism, and immunogenicity, protein therapies encounter the most serious issues during clinical uses [43]. This implies that many protein treatments require regular dosage, leading to problems that include causing an immunological response [10, 44, 45]. Furthermore, compared to other existing therapies, frequent dosing puts protein therapeutics at a serious disadvantage and increases the chance of patient noncompliance with the dosing schedule. Therapeutic protein molecules are susceptible to renal filtration and liver metabolism due to their size and hydrophobicity. Another disadvantage of several therapeutic proteins is that they may have dose-limiting solubility issues, which may preclude their use as a therapy. Extending the half-life of protein therapeutics artificially is considered a revolutionary approach that could be achieved by linking the protein to a molecule like PEG to increase the size or alter the amino acid chain to induce the formation of aggregates or receptor-mediated recycling [46, 47]. Glycosylation [47], protein fusion [48], and the formation of albumin conjugate [49] are some other methods for increasing the half-life of protein therapeutics in the blood.

The PEG moiety has several advantages in improving the stability and circulation half-life of a protein. The flexibility, hydrophilicity, changeable size, compatibility, and low toxicity of PEG make it ideal for a variety of applications [50]. Furthermore, the United State Food and Drug Administration (FDA) has declared PEG to be "generally regarded as safe [51]." One of the advantages of fusion proteins is that they may be engineered to interact with numerous targets and operate on two or more separate disease pathways simultaneously. The potential to build fusion proteins with bi- or multifunctional specificity is highly enticing, and it is a key focus of ongoing therapeutic fusion protein research and development. Serum albumin has a highly extended half-life in humans, owing to Fc receptor (FcRn)-mediated recycling, which protects it against catabolism [52]. Therefore, albumin is a promising carrier of therapeutic proteins. However, the absence of acceptable preclinical animal models has impeded pharmaceutical research and development of albumin-conjugate therapies of the therapeutic proteins.

9.3.2 Immunogenicity and Therapeutic Proteins

Immunogenicity refers to the ability of a therapeutic protein to elicit an unintended immune response against itself. Recurrent dosing and even single administration of therapeutic protein might promote/start ADA response. ADAs with neutralizing or binding capacities impact the effectiveness of therapeutic protein directly or indirectly [10]. ADAs are a problem associated with the therapeutic protein business and clinical formulations since they might produce adverse effects (such as endogenous protein neutralization) or diminish a biotherapeutic's effectiveness. Immune responses might be predicted early in the clinical development process using trustworthy nonclinical approaches. The subcutaneous administration of therapeutic proteins is favored for compliance and convenience. However, it might be difficult owing to immunogenic potential or undesired immune reactions [53]. Furthermore, it can potentially extend the therapeutic's half-life in circulation [54]. However, chronic administration's immunogenic potential and long-term consequences are frequently overlooked in clinical studies [55]. The subcutaneous route of administration does not enhance immunogenicity in some preclinical investigations [56] when compared to intravenous administration. For example, compared to intravenous administration, the relative immunogenicity of IFN B-1b is lower when given subcutaneously [57].

9.3.3 Safety and Efficacy

The link among dosage, exposure, effectiveness, and toxicity of therapeutic proteins must be understood before developing them. Despite progression in therapeutic proteins, there are still considerable obstacles to overcome the stability and efficacy during the manufacturing process and prolonging their storage. Furthermore, deamidation, misfolding, inappropriate glycosylation, aggregation and oxidation, and protease degradation are potential degradative processes that affect the efficacy of therapeutic proteins [58]. In theory, misfolded proteins undergo endosomal proteolysis, with the amino acids recovered in the unfolded protein solution and the endoplasmic reticulum overburden response [58, 59]. However, at significant protein concentrations, cells might become overwhelmed by misfolded recombinant proteins and cause misfolded and aggregated proteins to be released [60]. Multisubunit proteins like recombinant IgG and blood coagulation proteins like Factor VIII might be particularly problematic [61]. Misfolding of protein and aggregation induces abnormalities in the function of cells and tissues and may cause pathogenic diseases, similar to the behavior of infectious prions in prion diseases [60]. Furthermore, protein aggregation tendency is also influenced by its surroundings, such as pH, ionic strength, cosolute concentration, and exposure to various bulk liquid-fluid and liquid-solid interfaces. For several practical reasons, only a few protein therapeutics are produced, manufactured, and kept at neutral pH or in salt concentrations isotonic to blood plasma [62]. Because pH is one of the most significant modifiable parameters in therapeutic proteins [62], new attempts that emphasize the role of charged amino acids protein stability are also being made [63].

9.3.4 Quality

Glycosylation, aggregation, charge variations, and sequence variants are all quality features that impact the activity, efficiency, safety, pharmacokinetics, and pharmacodynamics of therapeutic proteins [64].

9.3.5 Biopharmaceutics of Therapeutic Proteins

Immunogenicity is a word used in the biopharmaceutical business to characterize immunological responses to protein or peptide medications that are unwanted. Immunological responses to therapeutic proteins can be triggered by one of two mechanisms: a traditional immune response or the breakdown of the tolerance [65]. Components inherent to the product, host cell proteins that migrate with the medicines during purification or protein therapeutics-related factors like excipients all have a role in immunogenicity. ADAs are frequently used to assess immunogenicity in clinical samples. T cells, which detect drug-derived sequences displayed on extremely varied Human Leukocyte Antigens (HLAs), play a vital role in defining an individual's immunological response [66]. Hypersensitivity-related events can also be linked to the immunogenicity of a therapeutic protein. Type 1 hypersensitivity is often associated with IgE isotype ADA [67]. IgE-mediated anaphylaxis is caused by ADA-IgE complexes binding and crosslinking the Fcε on basophils and mast cells. Furthermore, IgG ADA may form immunological complexes with the therapeutic protein, which can crosslink Fcγ receptors on neutrophils, releasing platelet-activating substances that are similar to histamine [68].

Protein therapies have complicated architectures that result in distinctive pharmacokinetic and pharmacodynamic (PK/PD) characteristics. Therapeutic proteins interact with pharmaceutical target structures and other endogenous proteins with a high affinity [69]. When produced in high amounts, these high-affinity ADAs have a greater possibility of altering and neutralizing the therapeutic effects of therapeutic proteins. Characterization of these ADA reactions faces bioanalytical challenges [70].

The immunogenicity of therapeutic proteins is altered by several variables [70] categorized as disease-related, patient-related, or product-related variables in the European Medicine Agency's guideline paper and other evaluations. The alteration of immune function in inflammatory diseases, inflammatory responses in response to a pathogenic infection, or an existing immunological reaction in a patient in a medical state are examples of disease variables. A patient-related factor might be a patient history with the MHC, which has been demonstrated to affect the immunogenicity of therapeutic protein treatments [71]. There have also been reports of product-related parameters that can impact immune response to a therapeutic protein [72]. Furthermore, it has been shown that administering an immunomodulatory biopharmaceutical causes an unanticipated inflammatory reaction in healthy patients [73].

9.4 Prediction of Immunogenicity of Therapeutic Proteins

Immunogenicity risk assessment is vital when developing proteins as successful therapeutic products. The immunogenicity of therapeutic proteins leads to loss of response (i.e., diminished efficacy on repeated administration) and hypersensitivity reactions. A prior potential immunogenicity risk assessment in preclinical stages ensures avoidance of negative surprises leading to economic and patient health-related implications. Regulatory guidelines of the European Medical Association (EMA) and FDA focus on developing standards for immunogenicity assays to clarify the requirement of tests before or following the drug approval process.

9.4.1 Analysis of ADAs

During clinical trials for assessing the safety and efficacy of therapeutic proteins, the regulatory bodies of different countries have indicated that the analysis of ADAs and neutralizing ADAs during the clinical trials can be very useful for predicting the immunogenicity of therapeutic proteins and their formulations. In addition, these ADA assays are also a critical part of the preclinical toxicology of therapeutic proteins as there are chances that ADA can mask the potential toxicity or side effect of therapeutic proteins during preclinical toxicity assessment. In most cases, it has been found that the immunogenicity assessment can be accurately predicted in humans during clinical trials. The experimental sequence to analyze the immunogenicity is shown in Fig. 9.4.

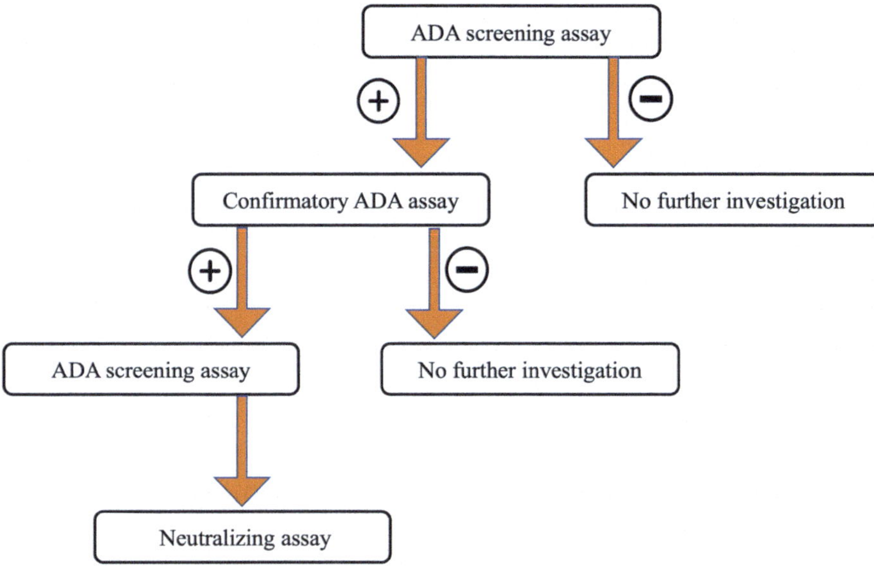

Fig. 9.4 Stepwise approach to accesses immunogenicity in therapeutic proteins

9.4.2 *In silico* Models for Predicting Immunogenicity

Immunogenic reactions to the foreign proteins usually depend on the T cell and B cell epitopes. T cell epitopes are present in proteins and produce interferon γ (IFN-γ) and cytokines (interleukin 2, 4, and 5) depending on the epitopes. Immunoinformatics is the field that deals with the in silico characterization of various biochemical phenomena involved in various immunological mechanisms. It has also been used to predict the immunogenicity of therapeutic proteins [74]. Immunoinformatics holds great promise as a technique for predicting antigenic epitopes as conventional techniques such as nuclear magnetic resonance (NMR) and X-ray crystallography are tedious and time consuming. There are several T cell epitope databases, such as Syfpeithi, IEDB (immune epitope databases) and IMGT (the international ImMunoGeneTics information system) (https://www.iedb.org/), that are frequently employed to predict the immunogenicity of proteins using immunoinformatics [75]. Currently, several servers and programs to predict the binding of peptides to any MHC II molecules are available. For example, an online server-based program, NetMHCIIpan-4.0, uses artificial neural networks to perform predictions. This program is trained on an extensive database (more than 500,000 measurements) of binding affinity and ligand mass spectroscopy. This database also covers three types of human MHCII molecules, namely, HLA-DQ, HLA-DR, and HLA-DP. Another program used for T cell epitope prediction is TE predict, which uses data from IEDB (http://tools.iedb.org/main/tcell/) and models constructed by partial least square linear regression for predictions. It can also predict the proteasomal processing of antigens [76]. Although in silico methods can predict immunogenicity, they are based on the linear sequence of peptides and often are overpredictive; hence, correlation with in vivo activity is required [77].

The primary shortcoming of the *in silico* approach is that the accuracy of the model depends upon the quality of large experimental datasets used to predict immunogenicity [56]. Most tools used for the prediction of immunogenicity focus on the interaction between peptide sequences and MHC molecules; however, in reality, an immunological response is a very complex phenomenon involving the uptake and processing of antigen by the APC, T cell receptor (TCR) mediated T cell activation, development of tolerance to the epitopes encountered by T cells, and the involvement of other immune cells. Therefore, *in silico* methods have very low reliability, and essentially in vivo studies are needed to confirm the outcomes predicted by *in silico* methods. So, *in silico* and other in vitro models may be used for screening, but final confirmation is always obtained by in vivo studies [78].

9.5 Exclusion and Avoidance of Immunogenicity of Therapeutic Proteins

As described in the previous section, the epitopes of nonhuman origin play a vital role in the induction of immunogenicity via activation of helper T cells leading to cytokine release, thereby initiating an immune response, which finally leads to the generation of B-cell-derived ADAs [9].

The deimmunization process, which involves the removal of T cell epitopes, is central to the mitigation of T-cell-dependent immunogenicity risk. In addition, regulatory T cell (Treg) epitopes reduce the immunogenicity of the protein containing humanized components. Thus, an ideal protein having minimal immunogenicity should have minimum T cell epitopes and maximum Treg epitopes [40]; for example, Adalimumab (Humaira) is less immunogenic clinically due to its properties mentioned above [79]. Humanizing proteins via grafting complementarity-determining regions (CDRs) onto antibody frameworks of human origin and removing T cell epitopes sequences found through the combination of epitope prediction logarithms are two mutually nonexclusive techniques for deimmunization [80].

Tolerization, along with deimmunization, also helps mitigate the immunogenicity of therapeutic proteins. This process involves the introduction of Treg sequences [81]. This is particularly useful in peptides used as replacement therapy (i.e., hormones, etc.). Furthermore, the Pompe disease (glycogen storage disease due to deficiency of acid α-glucosidase) is usually treated with recombinant human alglucosidase alfa (rhGAA); however, rapid tolerance to the therapy occurs due to the development of antidrug antibodies. De Groot et al. predicted the Treg peptides using an epimatrix program in recombinant human GAA and were enhanced in the model peptide and evaluated for immunogenicity potential by using Tetanus Toxoid Bystander Suppression Assay (TTBSA). The experiment resulted in the reduction of immunogenicity of the synthesized model protein [82].

9.6 Limitations and Future Perspectives

Therapeutic proteins are not optimal drug molecules due to production constraints, and their development is frequently impeded. The problem of spontaneous chemical degradation at the asparagine and aspartic acid residues significantly impacts the development of protein therapeutics [83, 84]. Chemical degradation concerns are typically identified in the later stages of the development, at which point repeated protein-engineering efforts are used to minimize these alterations. Many therapeutic proteins have properties that make drug development more complex, such as nonlinear elimination and distribution, inadequate absorption due to subcutaneous and intramuscular injection, and poor oral bioavailability [85–87].

9.6.1 Physicochemical Instability

Extensive formulation and process development screening investigations are used to generate potent biologic medicine. Nonetheless, even in the best formulation, a protein might degrade spontaneously during production, storage, and usage. Chemical alterations to amino acid residues, like oxidation of methionine or tryptophan, or change in charge, like carbonylation or deamidation, lead to conformational changes in the protein structure, which can result in alteration in the physical and chemical stability [88]. The safety and quality of therapeutic proteins are also affected due to nonenzymatic posttranslational modifications (PTMs) of isomerization of aspartic acid and deamidation of asparagine residues, which occurs during storage and production [89]. It is challenging to detect isomerization, since the molecular weight of aspartic acid and iso-aspartic acid is the same, with no change in charges. High-throughput bioluminescent tests have been used to calculate the amount of iso-aspartic acid formed [89]. Furthermore, mechanical and interfacial stresses can cause aggregation, resulting in physical instabilities during the manufacturing process.

The monoclonal antibody (mAb) abatacept (CTLA4-Ig) is used to treat rheumatoid arthritis. Monoclonal antibodies are stabilized using buffers and other stabilizing agents [90]. However, selecting a suitable buffer is a time-consuming procedure, since the ideal buffer is determined after testing various buffers and analyzing the results. Ligand binding is the most common way to improve the thermodynamic stability of proteins in buffers. In general, it has been shown that increasing thermodynamic stability reduces the rate of protein aggregation [91–93]. Many investigations involving proteins in buffer solutions have been undertaken, but little is known about the mechanism of protein-buffer interactions.

9.6.2 Pharmacokinetics of Therapeutic Proteins

Protein engineering efforts have recently become more complex to create protein therapies with desired features. Next-generation therapeutic proteins like engineered antibodies and receptors, antibody mimetic new scaffolds, bi- or multispecific proteins, and immunoconjugates are discovered by this process [94, 95]. These new protein therapies have distinct physicochemical features and function via different mechanisms of action, which contribute to their pharmacokinetics and pharmacodynamics. The molecular size of these proteins ranges from 10 kDa to >200 kDa, which leads to alteration in its pharmacokinetic features [96–100].

9.6.3 Manufacturing System

Recombinant production of therapeutic proteins has enormous potential for disease management. However, several fundamental challenges must be addressed to produce a therapeutic recombinant protein in a large amount [101]. These issues include protein size [102], folding and solubility [103, 104], PTMs [105], safety [106, 107], yield [108], genetic engineering [109], growth conditions and rate [110], etc. Protein-drug interactions must be thoroughly investigated while formulating a therapeutic protein to improve overall stability and reduce of risk of immunogenicity [111]. Liposomes and protein-polymer conjugates are established nanosystems with a long list of medically authorized products, and several types of protein-loaded therapeutic nanocarriers of different compositions, shapes, and sizes have been investigated [112–114]. These options are particularly appealing due to the utilization of various nanodelivery systems and the design of nanomaterials with customizable targeting strategies, releasing mechanisms, and physicochemical features [115].

9.7 Conclusions

Scientists and doctors are becoming increasingly conscious of the need to monitor the immunogenicity of new therapeutic proteins when they are introduced and the immunogenicity of existing therapeutics when they are modified or their production method is altered. Monitoring for antibodies during clinical trials and postmarketing surveillance is still a critical concern for all therapeutic proteins, despite the fact that numerous methods have been proposed to minimize the immunogenicity of therapeutic proteins.

References

1. Leader B, Baca QJ, Golan DE (2008) Protein therapeutics: a summary and pharmacological classification. Nat Rev Drug Discov 7:21–39. https://doi.org/10.1038/nrd2399
2. Singh DB, Tripathi T (2020) Frontiers in protein structure, function, and dynamics. Springer Nature, Singapore
3. Saudagar P, Tripathi T (2023) Advanced spectroscopic methods to study biomolecular structure and dynamics, 1st edn. Academic Press, Cambridge, MA

4. Tripathi T, Dubey VK (2022) Advances in protein molecular and structural biology methods, 1st edn. Academic Press, Cambridge, MA
5. Lagassé HAD, Alexaki A, Simhadri VL, Katagiri NH, Jankowski W, Sauna ZE, Kimchi-Sarfaty C (2017) Recent advances in (therapeutic protein) drug development. F1000Research 6:113. https://doi.org/10.12688/f1000research.9970.1
6. Kimchi-Sarfaty C, Schiller T, Hamasaki-Katagiri N, Khan MA, Yanover C, Sauna ZE (2013) Building better drugs: developing and regulating engineered therapeutic proteins. Trends Pharmacol Sci 34:534–548. https://doi.org/10.1016/j.tips.2013.08.005
7. Oprea TI, Bologa CG, Brunak S, Campbell A, Gan GN, Gaulton A, Gomez SM, Guha R, Hersey A, Holmes J, Jadhav A, Jensen LJ, Johnson GL, Karlson A, Leach AR, Ma'ayan A, Malovannaya A, Mani S, Mathias SL, McManus MT, Meehan TF, von Mering C, Muthas D, Nguyen D-T, Overington JP, Papadatos G, Qin J, Reich C, Roth BL, Schürer SC, Simeonov A, Sklar LA, Southall N, Tomita S, Tudose I, Ursu O, Vidović D, Waller A, Westergaard D, Yang JJ, Zahoránszky-Köhalmi G (2018) Unexplored therapeutic opportunities in the human genome. Nat Rev Drug Discov 17:317–332. https://doi.org/10.1038/nrd.2018.14
8. Barbosa MDFS, Celis E (2007) Immunogenicity of protein therapeutics and the interplay between tolerance and antibody responses. Drug Discov Today 12:674–681. https://doi.org/10.1016/j.drudis.2007.06.005
9. De Groot AS, Moise L (2007) Prediction of immunogenicity for therapeutic proteins: state of the art. Curr Opin Drug Discov Devel 10:332–340
10. De Groot AS, Scott DW (2007) Immunogenicity of protein therapeutics. Trends Immunol 28:482–490. https://doi.org/10.1016/j.it.2007.07.011
11. Schellekens H (2002) Immunogenicity of therapeutic proteins: clinical implications and future prospects. Clin Ther 24:1720–1740.; discussion 1719. https://doi.org/10.1016/s0149-2918(02)80075-3
12. Sherman A, Biswas M, Herzog RW (2017) Innovative approaches for immune tolerance to factor VIII in the treatment of hemophilia A. Front Immunol 8. https://doi.org/10.3389/fimmu.2017.01604
13. Scott DW, Pratt KP (2019) Factor VIII: perspectives on immunogenicity and tolerogenic strategies. Front Immunol 10:3078. https://doi.org/10.3389/fimmu.2019.03078
14. Pasi KJ, Rangarajan S, Georgiev P, Mant T, Creagh MD, Lissitchkov T, Bevan D, Austin S, Hay CR, Hegemann I, Kazmi R, Chowdary P, Gercheva-Kyuchukova L, Mamonov V, Timofeeva M, Soh C-H, Garg P, Vaishnaw A, Akinc A, Sørensen B, Ragni MV (2017) Targeting of antithrombin in hemophilia A or B with RNAi therapy. N Engl J Med 377:819–828. https://doi.org/10.1056/NEJMoa1616569
15. Fineberg SE, Kawabata TT, Finco-Kent D, Fountaine RJ, Finch GL, Krasner AS (2007) Immunological responses to exogenous insulin. Endocr Rev 28:625–652. https://doi.org/10.1210/er.2007-0002
16. Pechlaner C, Knapp E, Wiedermann CJ (2001) Hypersensitivity reactions associated with recombinant tissue-type plasminogen activator and urokinase. Blood Coagul Fibrinolysis 12:491–494. https://doi.org/10.1097/00001721-200109000-00010
17. Moreland L, Bate G, Kirkpatrick P (2006) Abatacept. Nat Rev Drug Discov 5:185–186. https://doi.org/10.1038/nrd1989
18. Kirk AD, Harlan DM, Armstrong NN, Davis TA, Dong Y, Gray GS, Hong X, Thomas D, Fechner JHJ, Knechtle SJ (1997) CTLA4-Ig and anti-CD40 ligand prevent renal allograft rejection in primates. Proc Natl Acad Sci U S A 94:8789–8794. https://doi.org/10.1073/pnas.94.16.8789
19. Larsen CP, Pearson TC, Adams AB, Tso P, Shirasugi N, Strobert E, Anderson D, Cowan S, Price K, Naemura J, Emswiler J, Greene J, Turk LA, Bajorath J, Townsend R, Hagerty D, Linsley PS, Peach RJ (2005) Rational development of LEA29Y (belatacept), a high-affinity variant of CTLA4-Ig with potent immunosuppressive properties. Am J Transplant Off J Am Soc Transplant Am Soc Transpl Surg 5:443–453. https://doi.org/10.1111/j.1600-6143.2005.00749.x

20. Adams AB, Ford ML, Larsen CP (2016) Costimulation blockade in autoimmunity and transplantation: the CD28 pathway. J Immunol 197:2045–2050. https://doi.org/10.4049/jimmunol.1601135
21. Ramsden L, Rider CC (1992) Selective and differential binding of interleukin (IL)-1 alpha, IL-1 beta, IL-2 and IL-6 to glycosaminoglycans. Eur J Immunol 22:3027–3031. https://doi.org/10.1002/eji.1830221139
22. Kaneko N, Kurata M, Yamamoto T, Morikawa S, Masumoto J (2019) The role of interleukin-1 in general pathology. Inflamm Regen 39:12. https://doi.org/10.1186/s41232-019-0101-5
23. Macdougall IC (2000) Novel erythropoiesis stimulating protein. Semin Nephrol 20:375–381
24. Schellekens H (2005) Factors influencing the immunogenicity of therapeutic proteins. Nephrol Dial Transplant Off Publ Eur Dial Transpl Assoc - Eur Ren Assoc 20 Suppl 6:vi3-9. https://doi.org/10.1093/ndt/gfh1092
25. Scepanovic P, Alanio C, Hammer C, Hodel F, Bergstedt J, Patin E, Thorball CW, Chaturvedi N, Charbit B, Abel L, Quintana-Murci L, Duffy D, Albert ML, Fellay J (2018) Human genetic variants and age are the strongest predictors of humoral immune responses to common pathogens and vaccines. Genome Med 10:59. https://doi.org/10.1186/s13073-018-0568-8
26. Fakhir FZ, Lkhider M, Badre W, Alaoui R, Meurs EF, Pineau P, Ezzikouri S, Benjelloun S (2018) Genetic variations in toll-like receptors 7 and 8 modulate natural hepatitis C outcomes and liver disease progression. Liver Int 38:432–442. https://doi.org/10.1111/liv.13533
27. Thomas SS, Borazan N, Barroso N, Duan L, Taroumian S, Kretzmann B, Bardales R, Elashoff D, Vangala S, Furst DE (2015) Comparative immunogenicity of TNF inhibitors: impact on clinical efficacy and tolerability in the management of autoimmune diseases. A systematic review and meta-analysis. BioDrugs 29:241–258. https://doi.org/10.1007/s40259-015-0134-5
28. Kelly J (2008) Pre-existing antibodies explain severe cetuximab reactions. Lancet Oncol 9: 419. https://doi.org/10.1016/s1470-2045(08)70120-7
29. Kuriakose A, Chirmule N, Nair P (2016) Immunogenicity of biotherapeutics: causes and association with posttranslational modifications. J Immunol Res 2016:1–18. https://doi.org/10.1155/2016/1298473
30. Scheiblhofer S, Laimer J, Machado Y, Weiss R, Thalhamer J (2017) Influence of protein fold stability on immunogenicity and its implications for vaccine design. Expert Rev Vaccines 16: 479–489. https://doi.org/10.1080/14760584.2017.1306441
31. Ryff JC (1997) Clinical investigation of the immunogenicity of interferon-alpha 2a. J Interf Cytokine Res 17(Suppl 1):S29–S33. http://www.ncbi.nlm.nih.gov/pubmed/9241613
32. Jakimovski D, Kolb C, Ramanathan M, Zivadinov R, Weinstock-Guttman B (2018) Interferon β for multiple sclerosis. Cold Spring Harb Perspect Med 8:a032003. https://doi.org/10.1101/cshperspect.a032003
33. Guo S, Bozkaya D, Ward A, O'Brien JA, Ishak K, Bennett R, Al-Sabbagh A, Meletiche DM (2009) Treating relapsing multiple sclerosis with subcutaneous versus intramuscular interferon-Beta-1a. PharmacoEconomics 27:39–53. https://doi.org/10.2165/00019053-200927010-00005
34. Narhi LO, Schmit J, Bechtold-Peters K, Sharma D (2012) Classification of protein aggregates. J Pharm Sci 101:493–498. https://doi.org/10.1002/jps.22790
35. Ross CA, Poirier MA (2004) Protein aggregation and neurodegenerative disease. Nat Med 10 (Suppl):S10–S17. https://doi.org/10.1038/nm1066
36. Baker MP, Reynolds HM, Lumicisi B, Bryson CJ (2010) Immunogenicity of protein therapeutics: the key causes, consequences and challenges. Self Nonself 1:314–322. https://doi.org/10.4161/self.1.4.13904
37. Ratanji KD, Derrick JP, Dearman RJ, Kimber I (2014) Immunogenicity of therapeutic proteins: influence of aggregation. J Immunotoxicol 11:99–109. https://doi.org/10.3109/1547691X.2013.821564

38. Alberts B, Johnson A, Lewis J (2002) Chapter 24 - the adaptive immune system. Mol Biol Cell:1363–1422
39. Ward JR, Francis SE, Marsden L, Suddason T, Lord GM, Dower SK, Crossman DC, Sabroe I (2009) A central role for monocytes in toll-like receptor-mediated activation of the vasculature. Immunology 128:58–68. https://doi.org/10.1111/j.1365-2567.2009.03071.x
40. Jawa V, Terry F, Gokemeijer J, Mitra-Kaushik S, Roberts BJ, Tourdot S, De Groot AS (2020) T-cell dependent immunogenicity of protein therapeutics pre-clinical assessment and mitigation-updated consensus and review 2020. Front Immunol 11:1301. https://doi.org/10.3389/fimmu.2020.01301
41. Rehman K, Hamid Akash MS, Akhtar B, Tariq M, Mahmood A, Ibrahim M (2016) Delivery of therapeutic proteins: challenges and strategies. Curr Drug Targets 17:1172–1188. https://doi.org/10.2174/1389450117666151209120139
42. Jarvi NL, Balu-Iyer SV (2021) Immunogenicity challenges associated with subcutaneous delivery of therapeutic proteins. BioDrugs 35:125–146. https://doi.org/10.1007/s40259-020-00465-4
43. Zaman R, Islam RA, Ibnat N, Othman I, Zaini A, Lee CY, Chowdhury EH (2019) Current strategies in extending half-lives of therapeutic proteins. J Control Release Off J Control Release Soc 301:176–189. https://doi.org/10.1016/j.jconrel.2019.02.016
44. Kromminga A, Schellekens H (2005) Antibodies against erythropoietin and other protein-based therapeutics: an overview. Ann N Y Acad Sci 1050:257–265. https://doi.org/10.1196/annals.1313.027
45. Nagata S, Pastan I (2009) Removal of B cell epitopes as a practical approach for reducing the immunogenicity of foreign protein-based therapeutics. Adv Drug Deliv Rev 61:977–985. https://doi.org/10.1016/j.addr.2009.07.014
46. Jevsevar S, Kunstelj M, Porekar VG (2010) PEGylation of therapeutic proteins. Biotechnol J 5:113–128. https://doi.org/10.1002/biot.200900218
47. Dozier JK, Distefano MD (2015) Site-specific PEGylation of therapeutic proteins. Int J Mol Sci 16:25831–25864. https://doi.org/10.3390/ijms161025831
48. Hutt M, Färber-Schwarz A, Unverdorben F, Richter F, Kontermann RE (2012) Plasma half-life extension of small recombinant antibodies by fusion to immunoglobulin-binding domains. J Biol Chem 287:4462–4469. https://doi.org/10.1074/jbc.M111.311522
49. Trüssel S, Dumelin C, Frey K, Villa A, Buller F, Neri D (2009) New strategy for the extension of the serum half-life of antibody fragments. Bioconjug Chem 20:2286–2292. https://doi.org/10.1021/bc9002772
50. Veronese FM, Mero A (2008) The impact of PEGylation on biological therapies. BioDrugs 22:315–329. https://doi.org/10.2165/00063030-200822050-00004
51. Gaberc-Porekar V, Zore I, Podobnik B, Menart V (2008) Obstacles and pitfalls in the PEGylation of therapeutic proteins. Curr Opin Drug Discov Devel 11:242–250
52. Roopenian DC, Low BE, Christianson GJ, Proetzel G, Sproule TJ, Wiles MV (2015) Albumin-deficient mouse models for studying metabolism of human albumin and pharmacokinetics of albumin-based drugs. MAbs 7:344–351. https://doi.org/10.1080/19420862.2015.1008345
53. Fathallah AM, Bankert RB, Balu-Iyer SV (2013) Immunogenicity of subcutaneously administered therapeutic proteins--a mechanistic perspective. AAPS J 15:897–900. https://doi.org/10.1208/s12248-013-9510-6
54. Tiede A, Friedrich U, Stenmo C, Allen G, Giangrande P, Goudemand J, Hay C, Holmström M, Klamroth R, Lethagen S, McKenzie S, Miesbach W, Negrier C, Yuste VJ, Berntorp E (2011) Safety and pharmacokinetics of subcutaneously administered recombinant activated factor VII (rFVIIa). J Thromb Haemost 9:1191–1199. https://doi.org/10.1111/j.1538-7836.2011.04293.x
55. Bartelds GM, Krieckaert CLM, Nurmohamed MT, van Schouwenburg PA, Lems WF, Twisk JWR, Dijkmans BAC, Aarden L, Wolbink GJ (2011) Development of antidrug antibodies

against adalimumab and association with disease activity and treatment failure during long-term follow-up. JAMA 305:1460–1468. https://doi.org/10.1001/jama.2011.406
56. Peng A, Kosloski MP, Nakamura G, Ding H, Balu-Iyer SV (2012) PEGylation of a factor VIII-phosphatidylinositol complex: pharmacokinetics and immunogenicity in hemophilia A mice. AAPS J 14:35–42. https://doi.org/10.1208/s12248-011-9309-2
57. Torosantucci R, Sharov VS, van Beers M, Brinks V, Schöneich C, Jiskoot W (2013) Identification of oxidation sites and covalent cross-links in metal catalyzed oxidized interferon Beta-1a: potential implications for protein aggregation and immunogenicity. Mol Pharm 10:2311–2322. https://doi.org/10.1021/mp300665u
58. Jenkins N (2007) Modifications of therapeutic proteins: challenges and prospects. Cytotechnology 53:121–125. https://doi.org/10.1007/s10616-007-9075-2
59. Cudna RE, Dickson AJ (2003) Endoplasmic reticulum signaling as a determinant of recombinant protein expression. Biotechnol Bioeng 81:56–65. https://doi.org/10.1002/bit.10445
60. Soto C, Pritzkow S (2018) Protein misfolding, aggregation, and conformational strains in neurodegenerative diseases. Nat Neurosci 21:1332–1340. https://doi.org/10.1038/s41593-018-0235-9
61. Demeule B, Lawrence MJ, Drake AF, Gurny R, Arvinte T (2007) Characterization of protein aggregation: the case of a therapeutic immunoglobulin. Biochim Biophys Acta 1774:146–153. https://doi.org/10.1016/j.bbapap.2006.10.010
62. Wang W, Singh S, Zeng DL, King K, Nema S (2007) Antibody structure, instability, and formulation. J Pharm Sci 96:1–26. https://doi.org/10.1002/jps.20727
63. Loladze VV, Makhatadze GI (2011) Energetics of charge-charge interactions between residues adjacent in sequence. Proteins 79:3494–3499. https://doi.org/10.1002/prot.23132
64. Walsh G, Jefferis R (2006) Post-translational modifications in the context of therapeutic proteins. Nat Biotechnol 24:1241–1252. https://doi.org/10.1038/nbt1252
65. Shishido SN, Varahan S, Yuan K, Li X, Fleming SD (2012) Humoral innate immune response and disease. Clin Immunol 144:142–158. https://doi.org/10.1016/j.clim.2012.06.002
66. Dingman R, Balu-Iyer SV (2019) Immunogenicity of protein pharmaceuticals. J Pharm Sci 108:1637–1654. https://doi.org/10.1016/j.xphs.2018.12.014
67. Ghaderi D, Taylor RE, Padler-Karavani V, Diaz S, Varki A (2010) Implications of the presence of N-glycolylneuraminic acid in recombinant therapeutic glycoproteins. Nat Biotechnol 28:863–867. https://doi.org/10.1038/nbt.1651
68. Matucci A, Nencini F, Maggi E, Vultaggio A (2019) Hypersensitivity reactions to biologics used in rheumatology. Expert Rev Clin Immunol 15:1263–1271. https://doi.org/10.1080/1744666X.2020.1684264
69. Tang L, Persky AM, Hochhaus G, Meibohm B (2004) Pharmacokinetic aspects of biotechnology products. J Pharm Sci 93:2184–2204. https://doi.org/10.1002/jps.20125
70. Shankar G, Devanarayan V, Amaravadi L, Barrett YC, Bowsher R, Finco-Kent D, Fiscella M, Gorovits B, Kirschner S, Moxness M, Parish T, Quarmby V, Smith H, Smith W, Zuckerman LA, Koren E (2008) Recommendations for the validation of immunoassays used for detection of host antibodies against biotechnology products. J Pharm Biomed Anal 48:1267–1281. https://doi.org/10.1016/j.jpba.2008.09.020
71. Barbosa MDFS, Vielmetter J, Chu S, Smith DD, Jacinto J (2006) Clinical link between MHC class II haplotype and interferon-beta (IFN-beta) immunogenicity. Clin Immunol 118:42–50. https://doi.org/10.1016/j.clim.2005.08.017
72. Carpenter J, Cherney B, Lubinecki A, Ma S, Marszal E, Mire-Sluis A, Nikolai T, Novak J, Ragheb J, Simak J (2010) Meeting report on protein particles and immunogenicity of therapeutic proteins: filling in the gaps in risk evaluation and mitigation. Biol J Int Assoc Biol Stand 38:602–611. https://doi.org/10.1016/j.biologicals.2010.07.002
73. Hansel TT, Kropshofer H, Singer T, Mitchell JA, George AJT (2010) The safety and side effects of monoclonal antibodies. Nat Rev Drug Discov 9:325–338. https://doi.org/10.1038/nrd3003
74. Raoufi E, Hemmati M, Eftekhari S, Khaksaran K, Mahmodi Z, Farajollahi MM, Mohsenzadegan M (2020) Epitope prediction by novel immunoinformatics approach: a

state-of-the-art review. Int J Pept Res Ther 26:1155–1163. https://doi.org/10.1007/s10989-019-09918-z

75. Tomar N, De RK (2014) Immunoinformatics: a brief review. Methods Mol Biol 1184:23–55. https://doi.org/10.1007/978-1-4939-1115-8_3
76. Antonets DV, Maksiutov AZ (2010) TEpredict: software for T-cell epitope prediction. Mol Biol (Mosk) 44:130–139
77. Zhou Y, Penny HL, Kroenke MA, Bautista B, Hainline K, Chea LS, Parnes J, Mytych DT (2022) Immunogenicity assessment of bispecific antibody-based immunotherapy in oncology. J Immunother Cancer 10:e004225. https://doi.org/10.1136/jitc-2021-004225
78. Brinks V, Weinbuch D, Baker M, Dean Y, Stas P, Kostense S, Rup B, Jiskoot W (2013) Preclinical models used for immunogenicity prediction of therapeutic proteins. Pharm Res 30: 1719–1728. https://doi.org/10.1007/s11095-013-1062-z
79. Strand V, Balsa A, Al-Saleh J, Barile-Fabris L, Horiuchi T, Takeuchi T, Lula S, Hawes C, Kola B, Marshall L (2017) Immunogenicity of biologics in chronic inflammatory diseases: a systematic review. BioDrugs 31:299–316. https://doi.org/10.1007/s40259-017-0231-8
80. Mufarrege EF, Giorgetti S, Etcheverrigaray M, Terry F, Martin W, De Groot AS (2017) De-immunized and functional therapeutic (DeFT) versions of a long lasting recombinant alpha interferon for antiviral therapy. Clin Immunol 176:31–41. https://doi.org/10.1016/j.clim.2017.01.003
81. De Groot AS, Terry F, Cousens L, Martin W (2013) Beyond humanization and de-immunization: tolerization as a method for reducing the immunogenicity of biologics. Expert Rev Clin Pharmacol 6:651–662. https://doi.org/10.1586/17512433.2013.835698
82. De Groot AS, Desai AK, Lelias S, Miah SMS, Terry FE, Khan S, Li C, Yi JS, Ardito M, Martin WD, Kishnani PS (2021) Immune tolerance-adjusted personalized immunogenicity prediction for Pompe disease. Front Immunol 12:636731. https://doi.org/10.3389/fimmu.2021.636731
83. Sydow JF, Lipsmeier F, Larraillet V, Hilger M, Mautz B, Mølhøj M, Kuentzer J, Klostermann S, Schoch J, Voelger HR, Regula JT, Cramer P, Papadimitriou A, Kettenberger H (2014) Structure-based prediction of asparagine and aspartate degradation sites in antibody variable regions. PLoS One 9:e100736. https://doi.org/10.1371/journal.pone.0100736
84. Irudayanathan FJ, Zarzar J, Lin J, Izadi S (2021) Divining deamidation and isomerization in therapeutic proteins: effect of neighboring residue. Bio Rxiv. 2021.07.26.453885. https://doi.org/10.1101/2021.07.26.453885
85. Wang W, Wang EQ, Balthasar JP (2008) Monoclonal antibody pharmacokinetics and pharmacodynamics. Clin Pharmacol Ther 84:548–558. https://doi.org/10.1038/clpt.2008.170
86. Craik DJ, Fairlie DP, Liras S, Price D (2013) The future of peptide-based drugs. Chem Biol Drug Des 81:136–147. https://doi.org/10.1111/cbdd.12055
87. Verma S, Goand UK, Husain A, Katekar RA, Garg R, Gayen JR (2021) Challenges of peptide and protein drug delivery by oral route: current strategies to improve the bioavailability. Drug Dev Res 82:927–944. https://doi.org/10.1002/ddr.21832
88. Krause ME, Sahin E (2019) Chemical and physical instabilities in manufacturing and storage of therapeutic proteins. Curr Opin Biotechnol 60:159–167. https://doi.org/10.1016/j.copbio.2019.01.014
89. Hsiao K, Alves J, Patel R, Adams M, Nashine V, Goueli S (2017) A high-throughput bioluminescent assay to monitor the deamidation of asparagine and isomerization of aspartate residues in therapeutic proteins and antibodies. J Pharm Sci 106:1528–1537. https://doi.org/10.1016/j.xphs.2017.02.022
90. Naik J, Kulkarni D, Kadu P, Pandya A, Kale P (2022) Use of in silico tools for screening buffers to overcome physical instability of abatacept. Transpl Immunol 71:101551. https://doi.org/10.1016/j.trim.2022.101551
91. Azevedo Reis Teixeira A, Erasmus MF, D'Angelo S, Naranjo L, Ferrara F, Leal-Lopes C, Durrant O, Galmiche C, Morelli A, Scott-Tucker A, Bradbury ARM (2021) Drug-like

antibodies with high affinity, diversity and developability directly from next-generation antibody libraries. MAbs 13:1980942. https://doi.org/10.1080/19420862.2021.1980942
92. Pandey LM (2022) Physicochemical factors of bioprocessing impact the stability of therapeutic proteins. Biotechnol Adv 55:107909. https://doi.org/10.1016/j.biotechadv.2022.107909
93. Butreddy A, Janga KY, Ajjarapu S, Sarabu S, Dudhipala N (2021) Instability of therapeutic proteins - an overview of stresses, stabilization mechanisms and analytical techniques involved in lyophilized proteins. Int J Biol Macromol 167:309–325. https://doi.org/10.1016/j.ijbiomac.2020.11.188
94. Kintzing JR, Filsinger Interrante MV, Cochran JR (2016) Emerging strategies for developing next-generation protein therapeutics for cancer treatment. Trends Pharmacol Sci 37:993–1008. https://doi.org/10.1016/j.tips.2016.10.005
95. Scott AM, Wolchok JD, Old LJ (2012) Antibody therapy of cancer. Nat Rev Cancer 12:278–287. https://doi.org/10.1038/nrc3236
96. Han TH, Zhao B (2014) Absorption, distribution, metabolism, and excretion considerations for the development of antibody-drug conjugates. Drug Metab Dispos 42:1914–1920. https://doi.org/10.1124/dmd.114.058586
97. Tibbitts J, Canter D, Graff R, Smith A, Khawli LA (2016) Key factors influencing ADME properties of therapeutic proteins: a need for ADME characterization in drug discovery and development. MAbs 8:229–245. https://doi.org/10.1080/19420862.2015.1115937
98. Kariolis MS, Wells RC, Getz JA, Kwan W, Mahon CS, Tong R, Kim DJ, Srivastava A, Bedard C, Henne KR, Giese T, Assimon VA, Chen X, Zhang Y, Solanoy H, Jenkins K, Sanchez PE, Kane L, Miyamoto T, Chew KS, Pizzo ME, Liang N, Calvert MEK, DeVos SL, Baskaran S, Hall S, Sweeney ZK, Thorne RG, Watts RJ, Dennis MS, Silverman AP, Zuchero YJY (2020) Brain delivery of therapeutic proteins using an Fc fragment blood-brain barrier transport vehicle in mice and monkeys. Sci Transl Med 12. https://doi.org/10.1126/scitranslmed.aay1359
99. Hager T, Spahr C, Xu J, Salimi-Moosavi H, Hall M (2013) Differential enzyme-linked immunosorbent assay and ligand-binding mass spectrometry for analysis of biotransformation of protein therapeutics: application to various FGF21 modalities. Anal Chem 85:2731–2738. https://doi.org/10.1021/ac303203y
100. Li F, Weng Y, Zhang G, Han X, Li D, Neubert H (2019) Characterization and quantification of an Fc-FGF21 fusion protein in rat serum using immunoaffinity LC-MS. AAPS J 21:84. https://doi.org/10.1208/s12248-019-0356-4
101. Burnett MJB, Burnett AC (2020) Therapeutic recombinant protein production in plants: challenges and opportunities. Plants People Planet 2:121–132. https://doi.org/10.1002/ppp3.10073
102. Demain AL, Vaishnav P (2009) Production of recombinant proteins by microbes and higher organisms. Biotechnol Adv 27:297–306. https://doi.org/10.1016/j.biotechadv.2009.01.008
103. Margolin E, Chapman R, Williamson A-L, Rybicki EP, Meyers AE (2018) Production of complex viral glycoproteins in plants as vaccine immunogens. Plant Biotechnol J 16:1531–1545. https://doi.org/10.1111/pbi.12963
104. Sahdev S, Khattar SK, Saini KS (2008) Production of active eukaryotic proteins through bacterial expression systems: a review of the existing biotechnology strategies. Mol Cell Biochem 307:249–264. https://doi.org/10.1007/s11010-007-9603-6
105. Montero-Morales L, Steinkellner H (2018) Advanced plant-based glycan engineering. Front Bioeng Biotechnol 6:81. https://doi.org/10.3389/fbioe.2018.00081
106. Lico C, Santi L, Twyman RM, Pezzotti M, Avesani L (2012) The use of plants for the production of therapeutic human peptides. Plant Cell Rep 31:439–451. https://doi.org/10.1007/s00299-011-1215-7
107. Fischer R, Emans N (2000) Molecular farming of pharmaceutical proteins. Transgenic Res 9:279–299. https://doi.org/10.1023/A:1008975123362
108. Biron KK (2006) Antiviral drugs for cytomegalovirus diseases. Antivir Res 71:154–163. https://doi.org/10.1016/j.antiviral.2006.05.002

109. Barnes LM, Bentley CM, Dickson AJ (2003) Stability of protein production from recombinant mammalian cells. Biotechnol Bioeng 81:631–639. https://doi.org/10.1002/bit.10517
110. Lim Y, Wong NSC, Lee YY, Ku SCY, Wong DCF, Yap MGS (2010) Engineering mammalian cells in bioprocessing – current achievements and future perspectives. Biotechnol Appl Biochem 55:175–189. https://doi.org/10.1042/BA20090363
111. Kamerzell TJ, Esfandiary R, Joshi SB, Middaugh CR, Volkin DB (2011) Protein–excipient interactions: mechanisms and biophysical characterization applied to protein formulation development. Adv Drug Deliv Rev 63:1118–1159. https://doi.org/10.1016/j.addr.2011.07.006
112. Kiran P, Khan A, Neekhra S, Pallod S, Srivastava R (2021) Nanohybrids as protein-polymer conjugate multimodal therapeutics. Front Med Technol 3. https://doi.org/10.3389/fmedt.2021.676025
113. Stevens CA, Kaur K, Klok H-A (2021) Self-assembly of protein-polymer conjugates for drug delivery. Adv Drug Deliv Rev 174:447–460. https://doi.org/10.1016/j.addr.2021.05.002
114. Baker SL, Munasinghe A, Kaupbayeva B, Rebecca Kang N, Certiat M, Murata H, Matyjaszewski K, Lin P, Colina CM, Russell AJ (2019) Transforming protein-polymer conjugate purification by tuning protein solubility. Nat Commun 10:4718. https://doi.org/10.1038/s41467-019-12612-9
115. Moncalvo F, Martinez Espinoza MI, Cellesi F (2020) Nanosized delivery systems for therapeutic proteins: clinically validated technologies and advanced development strategies. Front Bioeng Biotechnol 8. https://doi.org/10.3389/fbioe.2020.00089

Efficacy and Safety of Therapeutic Proteins 10

Pramodkumar P. Gupta, Mala M. Parab, Santosh S. Chhajed, and Virupaksha A. Bastikar

Abstract

Protein-based therapeutics have grown into a significant category of drugs involved in new treatment and personalized care. In recent times, permitted recombinant protein therapeutics are well established to treat several diseases, disease conditions, and genetic disorders. With the advent of in silico and in vitro protein-engineering technologies, drug industries can fine-tune and deeply explore the required therapeutic functional characteristics of proteins while maintaining and enhancing product safety, efficacy, or both. Therapeutic proteins-based therapy focuses on replacing a protein that is anomalous or lacking in a particular disease. These therapeutic proteins are proposed to act as a supplement to the body and fulfill the need for vital proteins that helps in minimizing the effect of disease or chemotherapy-based side effects. The genetically engineered proteins resemble closely to natural proteins, and hence, they are easily replaceable.

Keywords

Protein · Target · Drug · Therapeutics · Monoclonal antibody · Safety · Efficacy · Toxicity

P. P. Gupta (✉) · M. M. Parab
School of Biotechnology and Bioinformatics, D Y Patil Deemed to be University, Navi Mumbai, Maharashtra, India

S. S. Chhajed
Department of Pharmaceutical Chemistry, MET's Institute of Pharmacy, Nashik, India

V. A. Bastikar
Center for Computational Biology and Translational Research, Amity Institute of Biotechnology, Amity University Mumbai, Mumbai, Maharashtra, India

D. B. Singh, T. Tripathi (eds.), *Protein-based Therapeutics*,
https://doi.org/10.1007/978-981-19-8249-1_10

10.1 Introduction

Amino acids are the unique combination of the organic bases (C, H, N, and O) known as the basic building blocks of proteins and play a critical role in the metabolism of all organisms. Twenty essential amino acids make peptides and proteins with desired functions. The amino acids, peptides, and proteins are involved in diverse roles, including building proteins, enzymes, hormone synthesis, neurotransmitters, etc. Some act as a supplementary diet as well as a therapeutic drug. The peptides range from 2 to 50 amino acids, while proteins are biological macromolecules [1, 2], and both are currently used as therapeutic products in various diseases such as cancer, AIDS, hypertension, diabetes, and numerous infectious diseases [1, 3].

The discovery of insulin in 1922 laid the foundations for protein-based therapeutics in the pharmaceutical market [4, 5]. Over 250 protein-based therapeutics are clinically approved for several diseases and indications [6]. In 2020, the peptide drug market was valued at US$ 32.1 billion globally. With an increasing health hazard, rising cases of several types of cancer, metabolic diseases, and upcoming infections will increase the cost of protein-based therapy and hence a manifold upsurge of revenue in the near future [7]. Developing novel peptides and repurposing therapeutic peptides was accepted globally in the COVID-19 emergency [8–10]. In February 2020, US Food and Drug Administration (FDA) revised the definition of biologics that covered the chemically synthesized polypeptides ranging between 40 and 100 amino acid residues [11]. The category of synthetic peptides is below 40 amino acids. In the COVID-19 pandemic, peptide- and protein-based therapy emerged as an optional treatment strategy, and the pharmaceutical industries have invested massively. With the increased demand for peptide- and protein-based management, COVID-19 positively impacted the protein therapeutic market [11]. For SARS-CoV-2, 33 protein-subunit-based vaccines are in the clinical phase, and 75 are in the preclinical phase [12].

Protein is an essential class of medicines in serving patients with a specific disease or disorder that needs special. Near-approved recombinant protein and peptides as therapeutic agents have been designed to treat multiple clinical abnormalities, diseases, and genetic disorders [13], after they undergo a rigorous experimental, go-no-go trial phase, which unfortunately includes a high risk of failure. The progress in computational-based modeling, including 2D, and 3D structure modeling, molecular docking, molecular dynamics (MD) simulation, and protein engineering technologies such as synthesis and quality control, have widely supported fast production and maintenance of the product safety or efficacy or both [13]. Accurate information on the 3D structure of protein acting as a receptor/target and ligand is very important. In case of unavailability or poor structural data, the known computational methods such as homology modeling, fold-based method, threading, and ab-initio have supported the modeling of protein and peptide 2D and 3D structures.

The molecular interactions can be studied in silico by employing molecular docking tools such as HEX protein-protein docking server (http://hexserver.loria.fr/),

HADDOCK protein-protein docking (https://wenmr.science.uu.nl/haddock2.4/), etc. and these tools are widely used to predict the binding and molecular interactions such as protein-protein, protein-peptide, and peptide-peptide interactions. Detailed information about interactions such as hydrogen bonding, charge interaction, hydrophobic, etc., can be obtained. These interactions are deeply associated with the efficacy and safety of the protein therapeutics. The MD simulation helps to identify the dynamic conformational space of the protein and motions over time. Higher conformational variations in the protein and protein-protein complex lead to higher energy fluctuation and lower stability, which contributes to poor safety and efficacy. The computational approaches help in deciding the appropriate lead candidates for further study, support early screening of biological compounds, and assist in the prediction of late-phase failure of the biological candidates in the trial phases.

The recent outbreak of Omicron raised a concern to the medical fraternity about the decline in the efficacy of the vaccines and neutralizing antibodies due to the mutations [14]. Using MD simulation data of human ACE2 and wild-type SARS-CoV-2 receptor-binding domain (RBD) complex with the 15 mutations in the RBD complexed with human ACE2, Lupala et al. demonstrated that omicron RBD interacts more strongly with the human ACE2 protein as compared to the original strain [14]. The RBD mutations exhibited a high number of hydrogen bonding in the ACE2-RBD interface, enhancing the tight-binding compared to the wild-type SARS-CoV-2 [14], possibly leading to the loss of vaccine efficacy.

10.2 Classification of Therapeutic Proteins

The therapeutic proteins are classified based on their pharmacological activity, molecular types, and molecular mechanism.

10.2.1 Classification Based on Pharmacological Activity

The pharmacological activity depends on how the drugs interact with proteins, enzymes, receptors, nucleic acid, or biomembranes. The pharmacological-based activity of proteins is categorized into four subcategories.

Group I: Therapeutic proteins with either enzymatic activity or regulatory activity.

(a) These therapeutic proteins act as a replacement for the protein that is abnormal or deficient, for example, Increlex.
(b) These proteins help in the amplification of a current pathway, for example, Ovidrel, Neupogen.
(c) These proteins provide an appropriate activity or function, for example, Myoblock.

Group II: Therapeutic proteins with specific targeting activity.

(a) These therapeutic proteins interfere with an organism or a molecule with a specific targeting activity, for example, Avastin.
(b) These proteins help deliver other compounds or proteins to the desired target, for example, Ontak.

Group III: Therapeutic proteins as vaccines.

(a) These therapeutic proteins act as vaccines and help in the protection against a harmful foreign agent, for example, Engerix.
(b) These therapeutic proteins are used to treat autoimmune diseases, for example, Rophylac.

Group IV: Therapeutic proteins used for diagnostics purposes, for example, Geref [1, 15].

10.2.2 Classification Based on Molecular Types

This class of therapeutic proteins pools antibody-based drugs, anticoagulants, blood factors, interferons, growth factors, hormones, interleukins, Fc fusion protein, etc. [1].

10.2.3 Classification Based on the Molecular Mechanism

The molecular mechanism defines the interaction between the therapeutic compound and the desired target that generates the physiological response.

(a) These proteins bind to the target noncovalently, for example, monoclonal antibodies (mAbs).
(b) These proteins act by binding with the covalent bonds, for example, enzymes.
(c) These proteins are based on their activity without specific interactions, for example, serum albumin [1].

10.3 Safety

Proteins are organized in a stable structure that must be maintained to effectively preserve their biological and functional properties [13, 16]. However, it is challenging to synthesize full-length therapeutic proteins through chemical reactions and processes. Hence, it is vital to produce them in live cells or organisms, where the selection of the appropriate cell line, the origin of species, desirable culture conditions, chemicals, and quality affect the characteristics of the final product

[16–18]. Furthermost, biologically active proteins undergo posttranslational modifications (PTMs) that may result in conformational or functional changes. For this, heterologous expression systems such as yeast and mammalian cells are used, which include complex purification processes. Resins and filters are used to remove the viral contamination from the produced therapeutic protein. Low pH or detergents are used as an inactivation step in removing virus particles [19]. These methods are widely used by manufacturers to avoid serious safety issues that may arise due to viral contamination in the therapeutic proteins. Considering the complexity in the production of the therapeutic proteins and their supply to the end market, attaining the highest safety of the product is the prime concern [13].

In 2019 a human mAb–Risankizumab was globally permitted for use in the management of mild-to-severe psoriasis. The drug targets the IL-23A interleukin and exhibits a vital role in keratinocyte proliferation [20]. In clinical trial experiments, the safety profile of the Risankizumab was average compared to the protein-drug/mAb Adalimumab and Ustekinumab [20]. The mAb Adalimumab is commonly used to treat arthritis, ankylosing spondylitis, Crohn's disease, rheumatoid arthritis, psoriatic arthritis, ulcerative colitis, plaque psoriasis, hidradenitis suppurativa, uveitis, and juvenile idiopathic [21–23]. The protein-drug Ustekinumab is commonly used for Crohn's disease, ulcerative colitis, plaque psoriasis, and psoriatic arthritis [24]. In short-term general safety, a 16-week study reported the adverse events from mild to moderate severity after administration of a 150 mg dose of Risankizumab with 48.9% compared to Ustekinumab with 52.3%, Adalimumab with 56.9%, and placebo with 48.3% of subjects. Serious adverse events in the initial 16 weeks varied from 2.4% of subjects receiving the Rrisankizumab to 5% of subjects receiving Ustekinumab [20].

In the case of long-term general safety, the data were collated from two Phase 3 studies for 52-week-long clinical trial results. The observed serious adverse event in the Risankizumab subjects was 9.4% patient-year compared to Ustekinumab with 10.9% patient-year [25]. No upsurge in the serious adverse events was reported over long-term exposure, and no novel safety signals were identified in the Risankizumab-dosed subjects [20]. From the clinical trials experiments and extended use of Risankizumab, the standardized mortality ratio was approximately 15 deaths per year in the general population [26].

Drug toxicity is widely associated with the mechanism of action—on-target and off-target activity. Here the activity of the therapeutic protein can be considered on two factors: i) interactions with an intended target (on-target activity) and ii) interactions with the un-intended target (off-target activity).

10.3.1 On-Target Activity

When a therapeutic protein interacts and binds to its desired target, it may lead to inhibitory or stimulatory activity [27]. Tumor necrosis factor-α (TNF-α) is involved in the activation and metabolic signaling pathway of inflammation and inflammation-based diseases [28]. TNF-α levels are highly and selectively elevated

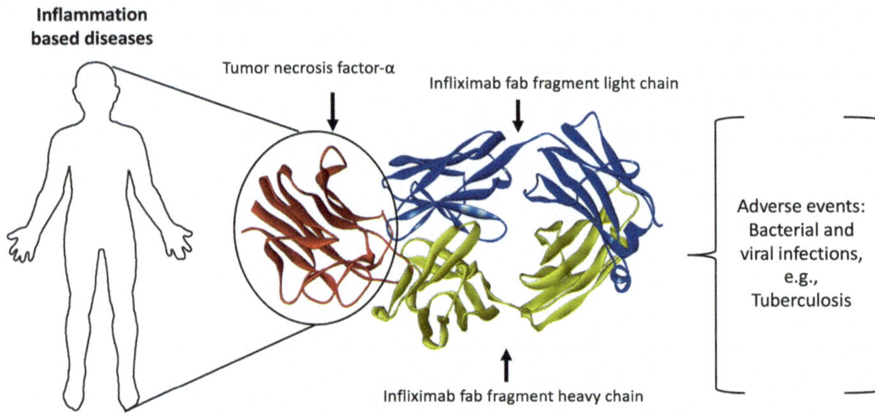

Fig. 10.1 Infliximab interacting with TNF-α (PDB code: 4G3Y) and relative adverse events. Red color represents TNF-α, blue represents Infliximab Fab fragment light chain, and yellow represents Infliximab fab fragment heavy chain

at the inflammation site in chronic inflammatory diseases like rheumatoid arthritis or Crohn's disease. The anti-TNF-α drugs were shown to reduce tissue damage and increase the quality of life of the patient. The administration of suppressor therapeutic protein may have various adverse effects, including a weakened immune response. Protein drugs such as Certolizumab pegol, Etanercept, Infliximab, and Adalimumab that target TNFα exhibit infectious complications in humans [27]. There is a possibility of unfavorable side effects of these drugs. The unwanted side effects may arise due to the immunomodulatory antibodies and could be through either suppressory or stimulatory behavior [28].

The first approved anti-TNF-α drug Infliximab, a chimeric IgG mAb, was followed by Entanercept, a dimeric, recombinant, complete two soluble TNF-α receptor protein [29], Adalimumab, a human IgG1 monoclonal antibody [30], and the recent Certolizumab pegol, a PEG-ylated anti-TNF-α Fab 2 antibody fragment. TNF-α plays a crucial role in the body's defense system against several bacteria and virus infections [31], and anti-TNF-α drug therapy such as Infliximab was associated with severe infections [28, 32–34] (Fig. 10.1). The US FDA Adverse Reporting System reported 70 cases of tuberculosis in infliximab-treated patients [28, 35]. Rituximab is a chimeric mAb used to treat nonlymphoma Hodgkin's and chronic lymphocytic leukemia. It functions as an anti-CD 20 antibody. Rituximab inhibits or stops the proliferation of cancer cells. It lowers joint discomfort and swelling and is employed to treat rheumatoid arthritis. When Rituximab was applied as monotherapy in B lymphomas or autoimmune disease, the outcome was associated with bacterial or viral infection [36] (Fig. 10.2).

In certain cases, there is a chance of life-threatening illness due to the excessive stimulation of the immune system. A study on protein therapeutic stimulatory anti-CD28 mAb TGN1412 with a single dose of administration reported the initiation of systemic inflammatory response identified by a spontaneous escalation of

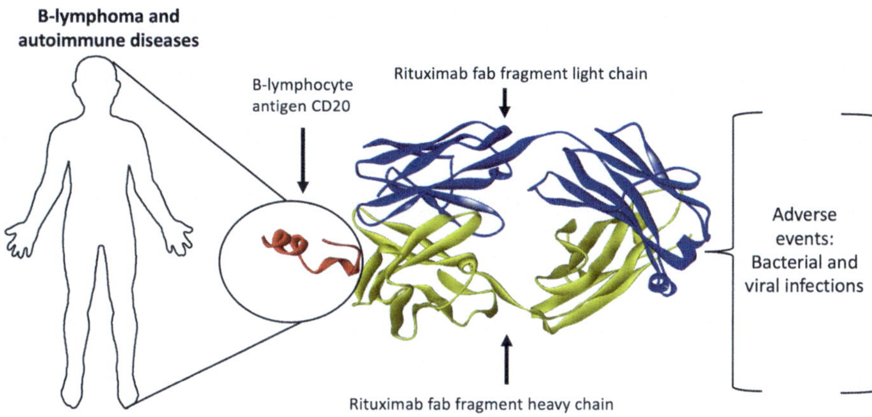

Fig. 10.2 Rituximab interacting with B-lymphocyte antigen CD20 (PDB code: 2OSL) and relative adverse events. Red color represents B-lymphocyte antigen CD20, blue represents Rituximab fab fragment light chain, and yellow represents Rituximab fab fragment heavy chains

proinflammatory cytokines in all six volunteers, leading to critical illness within 12–16 h [28, 37].

10.3.2 Off-Target Activity

In some instances, the protein therapeutics interact with unintentional/nonspecific targets and cause various adverse effects with poorly understood mechanisms. Protein therapeutics such as Alemtuzumab, Cetuximab, Infliximab, Panitumumab, Rituximab, and Trastuzumab reported adverse acute infusion reactions post administration. In the case of Rituximab, the discharge of cellular substances from disintegrated malignant B cells was reported as an infusion side effect. In certain cases, the infusion of therapeutic proteins leads to hypersensitivity reactions, anaphylactic shock (fatal allergic reaction), and serum sickness [27, 28, 38, 39]. Due to off-target or nonspecific target binding, the pharmacokinetic activities of the mAb, such as dissemination, toxicity, and efficacy, might differ or be altered.

Bumbaca et al. elucidated the effect of therapeutic antibody (FGFR4), a humanized antifibroblast growth factor 4 for hepatocellular carcinoma (HCC). An established chimeric anti-FGFR4 mAb (chLD1) has previously been shown to block ligand binding, decrease FGFR4-mediated signaling, and halt tumor growth in vivo. Analogous binding affinity and in vitro blocking activity have been confirmed for the humanized version of chLD1, hLD1.vB. The studies of humanized chLD1and hLD1.vB in HUH7 human HCC xenograft mouse model (human HCC, preclinical mouse model) have turned up fast clearance, poor target tissue distribution with limited efficacy. Comparative research showed that hLD1.vB had a particular off-target interaction with mouse complement C3, which led to higher clearance and low effectiveness [40, 41].

Furthermore, severe adverse reactions observed in the cynomolgus monkey toxicity studies of humanized mAb ABT-736 targeting amyloid-beta (Aβ) protein oligomer in induced Alzheimer's disease lead to withdrawal of molecule. Later it was identified that ABT-736 has a polyspecificity target affinity, including the unintended interaction or off-target activity with high affinity to the monkey and human plasma protein platelet factor 4 (PF-4). It is widely recognized that PF-4 has a role in serious adverse effects, including heparin-induced thrombocytopenia (HIT) in humans [42].

Due to the fear of late-phase failure in drug development and unpredictable outcome of the off-target activity, it is advised to perform an early screening of mAb for nonspecific binding via in vitro experiments such as tissue cross-reactivity assays, light-directed peptide synthesis arrays, and protein microarrays [43–45].

10.4 Efficacy

Efficacy or Effectiveness is the second most crucial factor that the US FDA and all regulatory agencies consider when deciding whether to approve a drug. Many therapeutic proteins are well known to be highly successful in vitro and in vivo, as well as in treating disease.

Protein therapeutic such as Rituximab for non-Hodgkin lymphoma [39], insulin for diabetes [46], Epoetin-alfa for anaemia [47], and Altemtuzumab for haematological malignancies [48] are well known for their pharmacological activity. The therapeutic protein Trastuzumab (Herceptin) is used to treat early-stage and advanced-stage metastatic HER2-positive breast cancer. Data collated from six clinical trial studies randomizing >14,000 women subjects with HER2-positive breast cancer were treated with Trastuzumab versus non-Trastuzumab-based adjuvant chemotherapy. The outcome reported that the inclusion of Trastuzumab decreases the repetitive frequency of disease by nearly 50% and boosts patient survival by 30% [27, 49–51] (Fig. 10.3).

Due to numerous factors, the efficacy of the protein therapeutics is not high or optimum due to their variability. Over 50% of patients with HER2-positive breast cancer who receive Trastuzumab as a therapeutic candidate experience nonresponding tumors, and most of these patients experience disease progression within a year [52]. However, when HER2-positive advanced breast cancer develops in female patients and is treated with Anthracycline-, Taxane-, and Trastuzumab-based therapy (amalgamation therapy), which maintains a dual tyrosine kinase inhibitor of EGFR and HER2, the drug Lapatinib to drug Capecitabine, a prodrug, displayed a prodigious efficacy in these patients [53].

Sotrovimab (VIR-7831), an anti-SARS-CoV-2 mAb approved by US FDA in May 2021 (manufactured by GSK and Vir Biotechnology, Inc.), binds to the highly conserved epitope of the RBD of the viral spike protein. The US FDA approval was given based on the randomized controlled trial experiment investigating the safety and efficacy of the Sotrovimab 500 mg intravenous injection given to the subjects

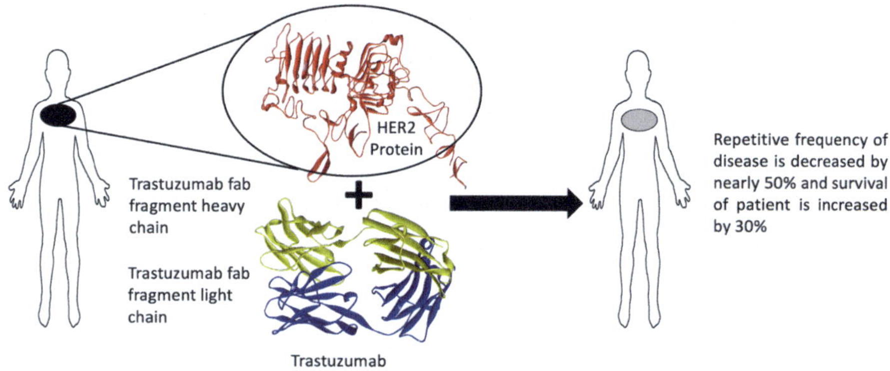

Fig. 10.3 Trastuzumab interacting with HER2 (7MN8.pdb). Trastuzumab interacting with HER2 and relative effect. Red color represents HER2, blue represents Trastuzumab fab fragment light chain, and yellow represents Trastuzumab fab fragment heavy chain

within 5 days of symptom onset. A parallel study reported the efficacy of the Sotrovimab against most circulating SARS-CoV-2 variants [54].

10.5 Consequences of Clinical Approach

As they are proposed to be safer than synthesized small molecules, therapeutic proteins are envisioned for pharmaceutical and therapeutic applications. Therapeutic proteins are produced in vivo using the same methods as their natural counterparts [28]. The treatment of patients with protein-based drugs may exhibit adverse immune response, low or poor-clinical outcomes, antibody reaction, and cross-reactivity. Throughout the design and development phase of therapeutic proteins, the disclosure, identification, and comprehensive justification of the immunologic response and adverse events are actively encouraged. The identification of adverse effects in the early phase thus facilitates the design and development of therapeutic proteins with a low risk of late phase toxicity and prevents experimental failure. This understanding of the toxic phenomena reduces the probability of fatalities and thus saves the late phase failure that impacts the timeline and cost of the project. The risk-benefit measurement depends upon the depth and degree of information and might vary among individual patients, populations, individual products, types of disease, comorbid conditions, and the immune response [55].

As a result of patient exposure to protein drugs that triggers an antibody response, the effectiveness of therapeutic proteins may be restricted or reduced. The antibodies binding to the enzymatic, catalytic, or binding site domain of the therapeutic protein/enzyme may cause the deactivation or loss of product efficacy. Counterbalancing antibodies may target the catalytic domain involved in the efficacy of the therapeutic protein, which may suppress the activity. The cross-reactivity due to the neutralizing antibodies with the therapeutic protein may also affect the safety. The pharmacokinetic properties may alter or inactivate by the neutralizing and nonneutralizing antibodies, and hence, the efficacy will be at high risk with poor absorption, distribution, metabolism, and excretion (ADME) activity [48]. Safety is one of the crucial apprehensions in drug discovery and the life span of protein therapeutics. It is challenging and sometimes unpredictable to calculate the safety parameters in a patient, considering the immunogenicity activity that varies broadly and is not readily predictable in patients dosed with the therapeutic proteins. Anaphylaxis, cytokine release syndrome, infusion responses, nonacute reactions, and cross-reactivity to endogenous proteins are among the many variables influencing safety [48].

10.6 Pharmacokinetics of Therapeutic Protein

The field of pharmacokinetics focuses on the movement of the therapeutic protein into, through, and out of the human body. Essentially, it studies the time-course of the absorption, bioavailability, distribution, metabolism, and excretion of the drug. Similar to the small molecule drugs, the pharmacokinetics of protein-based therapeutics is followed by the ADME process, that is, absorption, distribution, metabolism, and excretion.

10.6.1 Absorption

The intravenous route of therapeutic protein to the human body is highly preferred, as it enhances the bioavailability of the drug. Hence, most therapeutics are administered via the intravenous route to humans. However, there are certain limits to the intravenous route, such as (1) the intravenous route is painful and not economical for many, and (2) the rapid infusion of antibodies into the human system may cause adverse events wherein the patient may need hospitalization. The other routes considered are oral, subcutaneous, and intramuscular. Monoclonal antibodies such as Adalimumab, Efalizumab, Omalizumab, and Etanercept are administered subcutaneously, while Alefacept and Palivizumab are administered intramuscularly. The dry powders, inhalers, or aerosol formulations of Exubera can be delivered via the pulmonary route. Many therapeutic proteins cannot be absorbed by the gastrointestinal system due to factors such as molecular size, hydrophilicity, and gastric breakdown. The absorption of therapeutic proteins administered through the subcutaneous route is slower than the small molecules due to their size. Thus, the variables affecting subcutaneous absorption parameters depend on intrinsic subject characteristics (like body weight, sex, age, and level of activity), species characteristics concerning skin morphology and physiology, target interaction, charge, formulation of therapeutic protein, and mode of administration. Thus, the route of administration is highly dependent on delivering the safety and efficacy of the therapeutic protein within the defined subjects [56, 57].

10.6.2 Distribution

The distribution of the therapeutic protein within a body is constrained by the size of the molecules. Thus, the smaller the molecular size of a therapeutic protein, the higher the tissue penetration. The distribution path of a therapeutic protein in vivo is also influenced by other elements such as physical and chemical properties (charge and shape), binding qualities (molecular interaction, receptor-mediated uptake), route of administration (intravenous, subcutaneous, intramuscular-formulation, etc.), and production process (which affect PTMs, such as glycosylation). Thus, to enhance tissue penetration, one can design and model the molecular size and analyze the binding affinity of designed therapeutic molecules using in silico approaches. In recent years, a series of therapeutic proteins have been designed, modeled, and analyzed based on molecular size and binding affinity, which has further encouraged the development of therapeutic proteins for clinical use.

Despite the low tissue penetration of large-sized biotherapeutics like mAbs, which typically have effectiveness even when the site of action clings to the tissue, it is conceivable to construct a therapeutic technique so that the tissue disclosure is adequate to regulate the target at the site of action. Since delivering therapeutic proteins to specific tissues and targets is a complex and demanding endeavor, many other aspects of pharmaceutical research remain to be explored [56, 57].

10.6.3 Metabolism

Like small molecules, therapeutic proteins or antibodies may be removed via excretion or metabolism. The protein-based therapeutics, with their large molecular size, are filtered through the kidney and excreted through the urine. In general, the low-molecular-weight protein fragments are filtered through the system; however, most of the filtered protein is not eliminated but gets reabsorbed and or metabolized by the proximal tubular cells of the nephron. By a variety of metabolic routes, including proteolytic degradation, target-mediated clearance, Fc receptor-mediated clearance, nonspecific endocytosis, and creation of immunological complexes, therapeutic proteins are mostly cleared from the bloodstream or interstitial fluid. However, the detailed understanding of the complete metabolism of these small/large therapeutic molecules is yet to be elucidated. Using incubations with plasma, liver, and kidney homogenates, peptides that help select lead molecules have been explored in vitro; however, the relationships between in vitro and in vivo investigations have not yet been proven.

When the body metabolizes a therapeutic protein into a peptide or amino acid, it interacts with phagocytic cells or cells with their target antigen. One of the primary reasons for nonlinear elimination kinetics is target-mediated clearance. The binding of a therapeutic protein to the target present in cells leads to its internalization in the cell and degradation in lysosomes. Target-mediated clearance is the main channel for clearance in the case of Cetuximab, where the endothelial growth factor receptor (EGFR) is the drug target. By comparing the disposition kinetics between normal, healthy animals and animals overexpressing the target, target-mediated clearance can be understood. To characterize saturable kinetic patterns linked to the target-mediated clearance in humans, pharmacokinetic/pharmacodynamic (PK/PD) models are often developed [56, 57].

10.6.4 Excretion

In humans, the renal excretion system plays a vital role in the excretion of metabolized therapeutic proteins. Antibodies and similar structures may be expelled from the body by either excretion or metabolism. The renal route mostly excretes the protein degraded products and low-molecular-weight biological compounds of <30 kDa [58]. Studies suggest that the proteins are degraded in the liver, and the degraded materials are transported to the bile for excretion [59], such as insulin and EGFR [57]. In the excretion of oligomeric biologic therapeutics, the plasma protein binding plays an important role and results in altered excretion profiles. Understanding the relationship between structure and activity and then changing the lipophilic properties of oligomers revealed the link between biological adherence and increasing plasma protein binding, lengthening in vivo half-life, and reducing renal excretion [56, 57, 60].

10.7 Examples of Therapeutic Proteins

For the treatment of chronic diseases, modern recombinant DNA technology has been used to produce a wide range of therapeutic proteins, including vaccines, enzymes, natural and recombinant cytokines, and antibody-drug conjugates, which are readily available medications at low prices. These products have been utilized for more than just therapy—they have also been employed for diagnosis, prevention, disease management, and cure.

10.7.1 Therapeutic Proteins for Diabetes

Diabetes is a metabolic disease of unbalanced carbohydrates, fats, and proteins metabolism, triggering the hyperglycemic condition and is a major trigger of mortality and morbidity [61]. Though an extensive bag of FDA-approved parenteral and oral medicines are available, the incomplete effectiveness, adverse effects, cost, contraindications, dosage adjustments, dosing schedules, and weight gain limit their use [62]. The L cells of the small intestine generate glucagon-like peptide (GLP-1) analogues, which boost glucose-dependent insulin secretion, lower glucagon secretion, decrease the weight deficit, slow down gastric draining, reduce hunger, and promote cell regeneration. Additionally, they do not cause hypoglycemia in the absence of treatments [63]. It has been noted that glucose kinase, also known as hexokinase IV or D, significantly affects glucose homeostasis.

Glucokinase activators (GKAs) stimulate the production and release of insulin. Additionally, they have been observed to enhance glucose metabolism and associated activities in many glucokinase-expressing cells by increasing the affinity of glucokinase for glucose and its maximum catalytic rate, which is mediated by GKA. Through the widespread increase in cell activity and fasting-restricted alterations in glucose turnover, GKAs mediate their antidiabetic actions [64]. PPAR α-agonists have also been scrutinized as potential antidiabetics. The subjugation of PPAR α-agonists like fibrates can lower plasma triglycerides and very low-density lipoprotein particles. Also, PPAR γ-agonists like thiazolidinedione impact free fatty acid flux and decrease insulin resistance and blood glucose levels. The PPAR α/γ dual agonism addresses both dyslipidemia and insulin resistance. They have been shown to improve inflammation, vascular function, vascular remodelling, and increased insulin sensitivity [65]. Mudgil et al. have reported a few antidiabetic peptides from camel milk protein hydrolysates against dipeptidyl peptidase-IV (DPP-IV), porcine pancreatic α-amylase (PPA), and porcine pancreatic lipase (PPL) [66].

Bioactive proteins and peptides produced during the enzymatic breakdown of dietary proteins are being researched as potential antidiabetic treatments. They have the potential to treat diabetes by controlling the glycemic index. Food proteins can be converted into these beneficial compounds either chemically, through microbial fermentation, or by enzymatic hydrolysis [67].

10.7.2 Anticancer Proteins

Cancer is responsible for millions of fatalities worldwide. Chemotherapy and radiotherapy are contemporary cancer treatments, but they are limited due to the known and unknown adverse effects on normal cells and intermittent progression of multidrug resistance in cells. This emphasizes the demand to develop novel therapeutics and means to combat cancer [68]. In 1997, Rituximab was the first anticancer biologic authorized by FDA for non-Hodgkin's lymphoma. To treat follicular lymphoma, diffuse large B cell lymphoma, and chronic lymphocytic leukemia (CLL), the FDA approved Rituximab with hyaluronidase in 2017 [69]. A list of various therapeutic proteins and vaccines acting as anticancer agents is listed in Table 10.1.

10.7.3 Therapeutic Proteins for Cardiac Diseases

Cardiovascular diseases (CVDs) are among the leading cause of mortality globally, which include coronary heart disease, rheumatic heart disease, cerebrovascular disease, and disease of the arteries. According to WHO, out of five CVD cases, four deaths are due to heart attacks and strokes. Although extensive breakthroughs in the treatment areas of CVDs have occurred, scientists and medical practitioners have faced serious objections and claims in prevention and designing more efficient

Table 10.1 Anticancer protein-based drug/vaccines

Drug	Category	Disease	Ref.
Oncophage or Vitespen or HSPPC-96	Therapeutic vaccine	Renal cell carcinoma, melanoma	[69–71]
Cervarix	Therapeutic vaccine	Prevention of HPV 16, 18-associated cervical cancer, and other cancers of the reproductive organ	[69, 72]
Sipeuleucel-T	Therapeutic vaccine	Castration-resistant prostate cancer, hormone-refractory prostate cancer	[69, 73]
Ipilimumab	Immunotherapeutic/ biologic	Melanoma (metastatic) and other tumors	[69, 74]
Vemurafenib	Chemotherapeutic	Advanced melanoma with BRAF V600E mutation	[69, 75, 76]
Brentuximab vedotin	Drug-linked biologic	Advanced Hodgkin lymphoma, anaplastic large cell lymphoma, T cell lymphoma	[69, 77–79]
Peginterferon alfa-2b	Biologic	Melanoma, chronic hepatitis C	[69, 80, 81]
Carfilzomib	Chemotherapeutic	Relapsed or refractory multiple melanoma	[69, 82, 83]
Pomalidomide	Chemotherapeutic	Relapsed or refractory multiple melanoma	[69, 84, 85]
Blinatumomab	Biologic	B cell acute lymphoblastic leukemia	[69, 86, 87]
Tisagenlecleucel	CAR T cell immunotherapy	B cell cancer, B cell acute lymphoblastic leukemia	[69, 88–90]
Ramucirumab	Biologic / human anti-VEGFR-2 monoclonal antibody	Hepatocellular carcinoma, advanced stomach cancer, and gastro-esophageal junction adenocarcinoma	[69, 91–93]
Pembrolizumab	Humanized monoclonal anti-PD1 antibody	Melanoma, NSCLC, colorectal cancer	[69, 94–96]

therapies. Protein engineering methods have been used to develop drugs for disease management, cardiac repair, and therapy; however, none have yet received approved clinical use [97]. The TNF-α inhibitor, Etanercept, showed promise but was unsuccessful in cardiovascular therapy. Like insulin, the IGF-1 protein showed protection against the progression of heart failure in mice, but low levels of IGF-1 in humans have been linked to an increased risk of ischemic heart disease. Other side effects of IGF-1 systemic delivery include an increased risk of cancer and diabetic retinopathy. A "two-pronged" defense of the myocardium may be provided by the stromal-cell-derived factor-1 (SDF-1), which protects against acute myocardial ischemia-reperfusion damage (IRI) and energizes the healing by directing stem cells to the site of injury [97, 98].

10.7.4 Therapeutic Protein in COVID-19

The SARS-CoV-2 outbreak has cost millions of lives with a massive burden of health and medical cost to the world's economy. With a rapid disease outbreak, therapy design was a big challenge to the medical practitioners and scientific communities. To fight COVID-19, the scientific community repurposed protein therapeutics for inflammation and related diseases or common viral symptoms [3]. Several FDA-approved immunomodulatory therapeutics proteins were studied and used in interventional COVID-19 trials, such as IL6-inhibitor—Tocilizumab, Sarilumab, and Siltuximab, IL-1 inhibitor—Anakinra and Canakinumab, IFN-γ inhibitor—Emapalumab–Izsg, TNF-α inhibitor—Infliximab, VEGF inhibitor—Bevacizumab, C5 inhibitor—Eculizumab and Ravulizumab–Cwvz. PD1 inhibitor—Nivolumab and Pembrolizumab, Kallilrein inhibitor—Lanadelumab–flyo, P-selectin inhibitor—Crizanlizumab–tmca, IL-17 inhibitor—Secukinumab, IL-2 inhibitor—Aldesleukin and ILT101, and GM-CSF inhibitor—Sargramostim [3] (Fig. 10.4).

10.8 Conclusions and Future Perspectives

Therapeutic proteins substitute a protein and augment the supply of a functional protein, thereby diminishing the influence of disease. Protein therapeutics are thus exceptionally precise and offer a complex set of functions that have been predicted to have minimal potential to impede normal biological processes and cause adverse effects. These proteins are produced through microbial fermentation in transgenic plants, animals, and cell cultures. Most therapeutic proteins that are commercially available are delivered via the parenteral route, since they are delicate molecules.

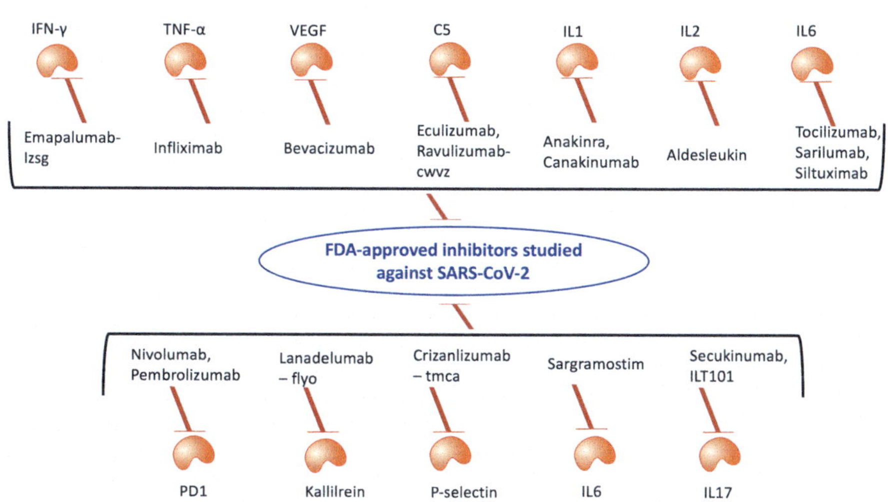

Fig. 10.4 FDA-approved protein-based drugs studied against SARS-CoV-2

Efforts have been made to improve the pharmacodynamic and pharmacokinetic properties of proteins by making changes in formulation. Many new approaches are designed to increase the half-life, decrease adverse effects, and increase patient compliance and quality of life. Studies are being conducted to solve the issue of protein distribution inside cells. Technologies that minimize risk are continually being developed to solve the problems with the manufacture of mAbs. The combined natural barriers of drug permeability, stability, pharmacodynamics, and pharmacokinetics of protein therapies still provide significant challenges. Research on new delivery systems, including nanoengineering, is ongoing. In the future, therapeutic proteins will be in high demand to treat diseases because of advantages such as quick time to market and high success rates in clinical trials compared with conventional small molecule drugs.

References

1. Chakraborty A, Singha DK, Chakraborty M, Mukherjee P (2021) Protein therapeutics: an updated review. Int J Drug Dev Res 13:6139. https://www.ijddr.in/drug-development/protein-therapeutics-an-updated-review.pdf
2. Singh DB, Tripathi T (2020) Frontiers in protein structure, function, and dynamics. Springer Nature
3. Ji P, Chen J, Golding A, Nikolov NP, Saluja B, Ren YR, Sahajwalla CG (2020) Immunomodulatory therapeutic proteins in COVID-19: current clinical development and clinical pharmacology considerations. J Clin Pharmacol 60:1275–1293. https://doi.org/10.1002/jcph.1729
4. Muttenthaler M, King GF, Adams DJ, Alewood PF (2021) Trends in peptide drug discovery. Nat Rev Drug Discov 20:309–325. https://doi.org/10.1038/s41573-020-00135-8
5. Kurtzhals P, von Scholten BJ, Kreiner FF, Gough SCL (2021) Insulin at 100: still central in protein-based therapy for chronic disease. Commun Med 1:9. https://doi.org/10.1038/s43856-021-00004-4
6. Dingman R, Balu-Iyer SV (2019) Immunogenicity of protein pharmaceuticals. J Pharm Sci 108: 1637–1654. https://doi.org/10.1016/j.xphs.2018.12.014
7. Global peptide therapeutics markets, 2021–2026 – Increase bioavailability of drugs and broaden application of peptides therapeutics – ResearchAndMarkets.com (2021). https://www.businesswire.com/news/home/20211005005869/en/Global-Peptide-Therapeutics-Markets-2021-2026%2D%2D-Increase-Bioavailability-of-Drugs-and-Broaden-Application-of-Peptides-Therapeutics%2D%2D-ResearchAndMarkets.com. Accessed 26 Jan 2022
8. Schütz D, Ruiz-Blanco YB, Münch J, Kirchhoff F, Sanchez-Garcia E, Müller JA (2020) Peptide and peptide-based inhibitors of SARS-CoV-2 entry. Adv Drug Deliv Rev 167:47–65. https://doi.org/10.1016/j.addr.2020.11.007
9. Madhavan M, AlOmair LA, Deepthi KS, Mustafa S (2021) Exploring peptide studies related to SARS-CoV to accelerate the development of novel therapeutic and prophylactic solutions against COVID-19. J Infect Public Health 14:1106–1119. https://doi.org/10.1016/j.jiph.2021.06.017
10. Shah JN, Guo G-Q, Krishnan A, Ramesh M, Katari NK, Shahbaaz M, Abdellattif MH, Singh SK, Dua K (2022) Peptides-based therapeutics: emerging potential therapeutic agents for COVID-19. Therapies 77:319–328. https://doi.org/10.1016/j.therap.2021.09.007
11. Peptide therapeutics market – growth, trends, COVID-19 impact, and forecasts (2021–2026), (2021). https://www.researchandmarkets.com/reports/5265155/peptide-therapeutics-market-growth-trends. Accessed 26 Jan 2022

12. Print RB (2022) COVID-19 vaccine tracker and landscape. https://www.who.int/publications/m/item/draft-landscape-of-covid-19-candidate-vaccines
13. Lagassé HAD, Alexaki A, Simhadri VL, Katagiri NH, Jankowski W, Sauna ZE, Kimchi-Sarfaty C (2017) Recent advances in (therapeutic protein) drug development. F1000Research 6:113. https://doi.org/10.12688/f1000research.9970.1
14. Lupala CS, Ye Y, Chen H, Su X-D, Liu H (2022) Mutations on RBD of SARS-CoV-2 omicron variant result in stronger binding to human ACE2 receptor. Biochem Biophys Res Commun 590:34–41. https://doi.org/10.1016/j.bbrc.2021.12.079
15. Creative Biomart Therapeutic Protein (n.d.) https://www.creativebiomart.net/therapeutic-proteins/targets/#:~:text=Basically%2C_therapeutic_proteins_can_be,III%3A_Therapeutic_proteins_as_vaccines. Accessed 1 Feb 2022
16. Bandaranayake AD, Almo SC (2014) Recent advances in mammalian protein production. FEBS Lett 588:253–260. https://doi.org/10.1016/j.febslet.2013.11.035
17. Butler M, Meneses-Acosta A (2012) Recent advances in technology supporting biopharmaceutical production from mammalian cells. Appl Microbiol Biotechnol 96:885–894. https://doi.org/10.1007/s00253-012-4451-z
18. Zhu J (2012) Mammalian cell protein expression for biopharmaceutical production. Biotechnol Adv 30:1158–1170. https://doi.org/10.1016/j.biotechadv.2011.08.022
19. Tobin PH, Richards DH, Callender RA, Wilson CJ (2014) Protein engineering: a new frontier for biological therapeutics. Curr Drug Metab 15:743–756. https://doi.org/10.2174/1389200216666141208151524
20. Huang Y-W, Tsai T-F (2020) A drug safety evaluation of risankizumab for psoriasis. Expert Opin Drug Saf 19:395–402. https://doi.org/10.1080/14740338.2020.1736034
21. "Adalimumab Monograph for Professionals" (2022) Drugs.com, https://www.drugs.com/monograph/adalimumab.html
22. BNF 76 (British National Formulary), 76th Revis, Pharmaceutical Press (2018)
23. Burmester GR, Gordon KB, Rosenbaum JT, Arikan D, Lau WL, Li P, Faccin F, Panaccione R (2020) Long-term safety of adalimumab in 29,967 adult patients from global clinical trials across multiple indications: an updated analysis. Adv Ther 37:364–380. https://doi.org/10.1007/s12325-019-01145-8
24. Cingoz O (2009) Ustekinumab. MAbs 1:216–221. https://doi.org/10.4161/mabs.1.3.8593
25. Gordon KB, Strober B, Lebwohl M, Augustin M, Blauvelt A, Poulin Y, Papp KA, Sofen H, Puig L, Foley P, Ohtsuki M, Flack M, Geng Z, Gu Y, Valdes JM, Thompson EHZ, Bachelez H (2018) Efficacy and safety of risankizumab in moderate-to-severe plaque psoriasis (UltIMMa-1 and UltIMMa-2): results from two double-blind, randomised, placebo-controlled and ustekinumab-controlled phase 3 trials. Lancet 392:650–661. https://doi.org/10.1016/S0140-6736(18)31713-6
26. Leaonardi C, Bachelez H, Wu J, Sinvhal R, Valdes J, Waterhouse B, Valdecantos W, Lebwohl M (2019) Safety of risankizumab in patients with moderate to severe psoriasis: analysis of pooled short-term and long-term clinical trial data. In: 24th World Congress of Dermatology Milan 2019. https://www.wcd2019milan-dl.org/abstract-book/documents/abstracts/35-psoriasis/safety-of-risankizumab-in-patients-3591.pdf
27. Dimitrov DS (2012) Therapeutic proteins. Methods Mol Biol 899:1–26. https://doi.org/10.1007/978-1-61779-921-1_1
28. Descotes J, Gouraud A (2008) Clinical immunotoxicity of therapeutic proteins. Expert Opin Drug Metab Toxicol 4:1537–1549. https://doi.org/10.1517/17425250802525496
29. Reddy SP, Shah VV, Lin EJ, Wu JJ (2016) Etanercept. In: Ther Sev Psoriasis. Elsevier, pp 83–96. https://doi.org/10.1016/B978-0-323-44797-3.00008-6
30. Scheinfeld N (2005) Adalimumab: a review of side effects. Expert Opin Drug Saf 4:637–641. https://doi.org/10.1517/14740338.4.4.637
31. Baker DE (2009) Certolizumab pegol for the treatment of Crohn's disease. Expert Rev Clin Immunol 5:683–691. https://doi.org/10.1586/eci.09.48

32. Lin J, Ziring D, Desai S, Kim S, Wong M, Korin Y, Braun J, Reed E, Gjertson D, Singh RR (2008) TNFalpha blockade in human diseases: an overview of efficacy and safety. Clin Immunol 126:13–30. https://doi.org/10.1016/j.clim.2007.08.012
33. Bongartz T, Sutton AJ, Sweeting MJ, Buchan I, Matteson EL, Montori V (2006) Anti-TNF antibody therapy in rheumatoid arthritis and the risk of serious infections and malignancies: systematic review and meta-analysis of rare harmful effects in randomized controlled trials. JAMA 295:2275–2285. https://doi.org/10.1001/jama.295.19.2275
34. Strangfeld A, Listing J (2006) Infection and musculoskeletal conditions: bacterial and opportunistic infections during anti-TNF therapy. Best Pract Res Clin Rheumatol 20:1181–1195. https://doi.org/10.1016/j.berh.2006.08.010
35. Keane J, Gershon S, Wise RP, Mirabile-Levens E, Kasznica J, Schwieterman WD, Siegel JN, Braun MM (2001) Tuberculosis associated with infliximab, a tumor necrosis factor alpha-neutralizing agent. N Engl J Med 345:1098–1104. https://doi.org/10.1056/NEJMoa011110
36. Kimby E (2005) Tolerability and safety of rituximab (MabThera). Cancer Treat Rev 31:456–473. https://doi.org/10.1016/j.ctrv.2005.05.007
37. Vial T, Descotes J (2007) Clinical adverse effects of cytokines on the immune system. In: Cytokines in human health. Humana Press, pp 319–348. https://doi.org/10.1007/978-1-59745-350-9_16
38. Chung CH (2008) Managing premedications and the risk for reactions to infusional monoclonal antibody therapy. Oncologist 13:725–732. https://doi.org/10.1634/theoncologist.2008-0012
39. Winter MC, Hancock BW (2009) Ten years of rituximab in NHL. Expert Opin Drug Saf 8:223–235. https://doi.org/10.1517/14740330902750114
40. Bumbaca D, Wong A, Drake E, Reyes AE, Lin BC, Stephan J-P, Desnoyers L, Shen B-Q, Dennis MS (2011) Highly specific off-target binding identified and eliminated during the humanization of an antibody against FGF receptor 4. MAbs 3:376–386. https://doi.org/10.4161/mabs.3.4.15786
41. Mark D, Luc D, Dorothy F (2014) Ant-FGFR4 antibodies and methods of use, US20140037624A. https://patents.google.com/patent/US20140037624A1/en
42. Loberg LI, Chhaya M, Ibraghimov A, Tarcsa E, Striebinger A, Popp A, Huang L, Oellien F, Barghorn S (2021) Off-target binding of an anti-amyloid beta monoclonal antibody to platelet factor 4 causes acute and chronic toxicity in cynomolgus monkeys. MAbs 13:1887628. https://doi.org/10.1080/19420862.2021.1887628
43. Liu L (2018) Pharmacokinetics of monoclonal antibodies and fc-fusion proteins. Protein Cell 9: 15–32. https://doi.org/10.1007/s13238-017-0408-4
44. Kelly RL, Sun T, Jain T, Caffry I, Yu Y, Cao Y, Lynaugh H, Brown M, Vásquez M, Wittrup KD, Xu Y (2015) High throughput cross-interaction measures for human IgG1 antibodies correlate with clearance rates in mice. MAbs 7:770–777. https://doi.org/10.1080/19420862.2015.1043503
45. Hötzel I, Theil F-P, Bernstein LJ, Prabhu S, Deng R, Quintana L, Lutman J, Sibia R, Chan P, Bumbaca D, Fielder P, Carter PJ, Kelley RF (2012) A strategy for risk mitigation of antibodies with fast clearance. MAbs 4:753–760. https://doi.org/10.4161/mabs.22189
46. Lee S-H, Yoon K-H (2021) A century of progress in diabetes care with insulin: a history of innovations and foundation for the future. Diabetes Metab J 45:629–640. https://doi.org/10.4093/dmj.2021.0163
47. Patel S, Patel JB (2022) Epoetin alfa. StatPearls Publishing. http://www.ncbi.nlm.nih.gov/pubmed/32119434
48. Flynn JM, Byrd JC (2000) Campath-1H monoclonal antibody therapy. Curr Opin Oncol 12: 574–581. https://doi.org/10.1097/00001622-200011000-00010
49. Castillo J, Winer E, Quesenberry P (2008) Newer monoclonal antibodies for hematological malignancies. Exp Hematol 36:755–768. https://doi.org/10.1016/j.exphem.2008.04.018
50. Mariani G, Fasolo A, De Benedictis E, Gianni L (2009) Trastuzumab as adjuvant systemic therapy for HER2-positive breast cancer. Nat Clin Pract Oncol 6:93–104. https://doi.org/10.1038/ncponc1298

51. Bedard PL, Piccart-Gebhart MJ (2008) Current paradigms for the use of HER2-targeted therapy in early-stage breast cancer. Clin Breast Cancer 8(Suppl 4):S157–S165. https://doi.org/10.3816/CBC.2008.s.012
52. Hall PS, Cameron DA (2009) Current perspective – trastuzumab. Eur J Cancer 45:12–18. https://doi.org/10.1016/j.ejca.2008.10.013
53. Cameron D, Casey M, Press M, Lindquist D, Pienkowski T, Romieu CG, Chan S, Jagiello-Gruszfeld A, Kaufman B, Crown J, Chan A, Campone M, Viens P, Davidson N, Gorbounova V, Raats JI, Skarlos D, Newstat B, Roychowdhury D, Paoletti P, Oliva C, Rubin S, Stein S, Geyer CE (2008) A phase III randomized comparison of lapatinib plus capecitabine versus capecitabine alone in women with advanced breast cancer that has progressed on trastuzumab: updated efficacy and biomarker analyses. Breast Cancer Res Treat 112:533–543. https://doi.org/10.1007/s10549-007-9885-0
54. Anti-SARS-CoV-2 Monoclonal Antibodies (2022). https://www.idsociety.org/covid-19-real-time-learning-network/therapeutics-and-interventions/monoclonal-antibodies/#Sotrovimab. Accessed 5 Apr 2022
55. Guidance for industry immunogenicity assessment for therapeutic protein products (2014) fda.gov/media/85017/download
56. Lobo ED, Hansen RJ, Balthasar JP (2004) Antibody pharmacokinetics and pharmacodynamics. J Pharm Sci 93:2645–2668. https://doi.org/10.1002/jps.20178
57. Vugmeyster Y (2012) Pharmacokinetics and toxicology of therapeutic proteins: advances and challenges. World J Biol Chem 3:73. https://doi.org/10.4331/wjbc.v3.i4.73
58. Maack T, Johnson V, Kau ST, Figueiredo J, Sigulem D (1979) Renal filtration, transport, and metabolism of low-molecular-weight proteins: a review. Kidney Int 16:251–270. https://doi.org/10.1038/ki.1979.128
59. LaRusso NF (1984) Proteins in bile: how they get there and what they do. Am J Physiol Liver Physiol 247:G199–G205. https://doi.org/10.1152/ajpgi.1984.247.3.G199
60. Geary RS (2009) Antisense oligonucleotide pharmacokinetics and metabolism. Expert Opin Drug Metab Toxicol 5:381–391. https://doi.org/10.1517/17425250902877680
61. Inzucchi SE, Bergenstal RM, Buse JB, Diamant M, Ferrannini E, Nauck M, Peters AL, Tsapas A, Wender R, Matthews DR (2012) Management of hyperglycemia in type 2 diabetes: a patient-centered approach. Diabetes Care 35:1364–1379. https://doi.org/10.2337/dc12-0413
62. DeFronzo RA, Davidson JA, Del Prato S (2012) The role of the kidneys in glucose homeostasis: a new path towards normalizing glycaemia. Diabetes Obes Metab 14:5–14. https://doi.org/10.1111/j.1463-1326.2011.01511.x
63. Sharma S, Mells JE, Fu PP, Saxena NK, Anania FA (2011) GLP-1 analogs reduce hepatocyte steatosis and improve survival by enhancing the unfolded protein response and promoting macroautophagy. PLoS One 6:e25269. https://doi.org/10.1371/journal.pone.0025269
64. Gude D (2012) Red carpeting the newer antidiabetics. J Pharmacol Pharmacother 3:127–131. https://doi.org/10.4103/0976-500X.95507
65. Liu C-H, Lee T-H, Lin Y-S, Sung P-S, Wei Y-C, Li Y-R (2020) Pioglitazone and PPAR-γ modulating treatment in hypertensive and type 2 diabetic patients after ischemic stroke: a national cohort study. Cardiovasc Diabetol 19:2. https://doi.org/10.1186/s12933-019-0979-x
66. Mudgil P, Kamal H, Yuen GC, Maqsood S (2018) Characterization and identification of novel antidiabetic and anti-obesity peptides from camel milk protein hydrolysates. Food Chem 259: 46–54. https://doi.org/10.1016/j.foodchem.2018.03.082
67. Paterson MA, Smart CEM, Lopez PE, McElduff P, Attia J, Morbey C, King BR (2016) Influence of dietary protein on postprandial blood glucose levels in individuals with type 1 diabetes mellitus using intensive insulin therapy. Diabet Med 33:592–598. https://doi.org/10.1111/dme.13011
68. Kesikli SA, Kilickap S (2013) Therapeutic proteins in tumors and targeted therapeutic agents for cancer patients. Protein Pept Lett 20:1345–1359. https://doi.org/10.2174/092986652012131112123040

69. Dembic Z (2020) Antitumor drugs and their targets. Molecules 25:5776. https://doi.org/10.3390/molecules25235776
70. di Pietro A, Tosti G, Ferrucci PF, Testori A (2008) Oncophage: step to the future for vaccine therapy in melanoma. Expert Opin Biol Ther 8:1973–1984. https://doi.org/10.1517/14712590802517970
71. Reitsma DJ, Combest AJ (2012) Challenges in the development of an autologous heat shock protein based anti-tumor vaccine. Hum Vaccin Immunother 8:1152–1155. https://doi.org/10.4161/hv.21382
72. Monie A, Hung C-F, Roden R, Wu T-C (2008) Cervarix: a vaccine for the prevention of HPV 16, 18-associated cervical cancer. Biologics 2:97–105. http://www.ncbi.nlm.nih.gov/pubmed/19707432
73. Anassi E, Ndefo UA (2011) Sipuleucel-T (provenge) injection: the first immunotherapy agent (vaccine) for hormone-refractory prostate cancer. Pharm Ther 36:197–202. http://www.ncbi.nlm.nih.gov/pubmed/21572775
74. Tarhini A, Lo E, Minor DR (2010) Releasing the brake on the immune system: ipilimumab in melanoma and other tumors. Cancer Biother Radiopharm 25:601–613. https://doi.org/10.1089/cbr.2010.0865
75. Garbe C, Abusaif S, Eigentler TK (2014) Vemurafenib. Small Mol Oncol 201:215–225. https://doi.org/10.1007/978-3-642-54490-3_13
76. Donadieu J, Larabi IA, Tardieu M, Visser J, Hutter C, Sieni E, Kabbara N, Barkaoui M, Miron J, Chalard F, Milne P, Haroche J, Cohen F, Hélias-Rodzewicz Z, Simon N, Jehanne M, Kolenova A, Pagnier A, Aladjidi N, Schneider P, Plat G, Lutun A, Sonntagbauer A, Lehrnbecher T, Ferster A, Efremova V, Ahlmann M, Blanc L, Nicholson J, Lambilliote A, Boudiaf H, Lissat A, Svojgr K, Bernard F, Elitzur S, Golan M, Evseev D, Maschan M, Idbaih A, Slater O, Minkov M, Taly V, Collin M, Alvarez J-C, Emile J-F, Héritier S (2019) Vemurafenib for refractory multisystem langerhans cell histiocytosis in children: an international observational study. J Clin Oncol 37:2857–2865. https://doi.org/10.1200/JCO.19.00456
77. Van Der Weyden C, Dickinson M, Whisstock J, Prince HM (2019) Brentuximab vedotin in T-cell lymphoma. Expert Rev Hematol 12:5–19. https://doi.org/10.1080/17474086.2019.1558399
78. Nikolaenko L, Nademanee A (2020) Brentuximab vedotin and its use in the treatment of advanced Hodgkin's lymphoma. Future Oncol 16:2273–2282. https://doi.org/10.2217/fon-2020-0026
79. Shustov A, Soma L (2019) Anaplastic large cell lymphoma: contemporary concepts and optimal management. Cancer Treat Res 176:127–144. https://doi.org/10.1007/978-3-319-99716-2_6
80. McHutchison JG, Lawitz EJ, Shiffman ML, Muir AJ, Galler GW, McCone J, Nyberg LM, Lee WM, Ghalib RH, Schiff ER, Galati JS, Bacon BR, Davis MN, Mukhopadhyay P, Koury K, Noviello S, Pedicone LD, Brass CA, Albrecht JK, Sulkowski MS (2009) Peginterferon alfa-2b or alfa-2a with ribavirin for treatment of hepatitis C infection. N Engl J Med 361:580–593. https://doi.org/10.1056/NEJMoa0808010
81. Herndon TM, Demko SG, Jiang X, He K, Gootenberg JE, Cohen MH, Keegan P, Pazdur R (2012) U.S. Food and Drug Administration approval: peginterferon-alfa-2b for the adjuvant treatment of patients with melanoma. Oncologist 17:1323–1328. https://doi.org/10.1634/theoncologist.2012-0123
82. Kortuem KM, Stewart AK (2013) Carfilzomib. Blood 121:893–897. https://doi.org/10.1182/blood-2012-10-459883
83. Lee MS, Lim SH, Yu A-R, Hwang CY, Kang I, Yeo E-J (2021) Carfilzomib in combination with bortezomib enhances apoptotic cell death in B16-F1 melanoma cells. Biology 10:153. https://doi.org/10.3390/biology10020153
84. Engelhardt M, Wäsch R, Reinhardt H, Kleber M (2014) Pomalidomide. Recent Results Cancer Res 201:359–372. https://doi.org/10.1007/978-3-642-54490-3_22

85. Attal M, Richardson PG, Rajkumar SV, San-Miguel J, Beksac M, Spicka I, Leleu X, Schjesvold F, Moreau P, Dimopoulos MA, Huang JS-Y, Minarik J, Cavo M, Prince HM, Macé S, Corzo KP, Campana F, Le-Guennec S, Dubin F, Anderson KC, Attal M, Richardson PG, Rajkumar V, San-Miguel J, Beksac M, Spicka I, Leleu X, Schjesvold F, Moreau P, Dimopoulos MA, Huang JS-Y, Minarik J, Cavo M, Prince HM, Macé S, Corzo KP, Campana F, Le-Guennec S, Dubin F, Anderson KC, Harrison S, Janowski W, Kerridge I, Spencer A, Delforge M, Fostier K, Vlummens P, Wu KL, Leblanc R, Pavic M, Sebag M, Hajek R, Maisnar V, Pour L, Gregersen H, Benbouker L, Caillot D, Escoffre-Barbe M, Facon T, Frenzel L, Hulin C, Karlin L, Kolb B, Pegourie B, Perrot A, Tiab M, Vincent L, Niederwieser D, Anagnostopoulos A, Delimpasi S, Kyrtsonis M-C, Symeonidis A, Illes A, Mikala G, Nagy Z, Bringen S, Corradini P, Fabio C, Lemoli R, Liberati A, Nozzoli C, Zambello R, Iida S, Ikeda T, Iyama S, Matsumoto M, Shimazaki C, Sunami K, Suzuki K, Uchiyama M, Koh Y, Kim K, Lee JH, Min C-K, Blacklock H, Goodman H, Neylon A, Simpson D, Grosicki S, Jurczyszyn A, Walter-Croneck A, Warzocha K, Araujo L, Moreira C, Doronin V, Mendeleeva L, Vorobyev V, Vranovsky A, Alegre A, Gironella M, Perez MSG, Montes C, Ocio E, Rodriguez P, Hardling M, Lauri B, Wang M-C, Yeh S-P, Arat M, Demirkan F, Gulbas Z, Besisik SK, Karadogan I, Tuglular T, Unal A, Vural F, Sive J, Streetly M, Yong K, Tache J (2019) Isatuximab plus pomalidomide and low-dose dexamethasone versus pomalidomide and low-dose dexamethasone in patients with relapsed and refractory multiple myeloma (ICARIA-MM): a randomised, multicentre, open-label, phase 3 study. Lancet 394:2096–2107. https://doi.org/10.1016/S0140-6736(19)32556-5
86. Kantarjian H, Stein A, Gökbuget N, Fielding AK, Schuh AC, Ribera J-M, Wei A, Dombret H, Foà R, Bassan R, Arslan Ö, Sanz MA, Bergeron J, Demirkan F, Lech-Maranda E, Rambaldi A, Thomas X, Horst H-A, Brüggemann M, Klapper W, Wood BL, Fleishman A, Nagorsen D, Holland C, Zimmerman Z, Topp MS (2017) Blinatumomab versus chemotherapy for advanced acute lymphoblastic leukemia. N Engl J Med 376:836–847. https://doi.org/10.1056/NEJMoa1609783
87. Locatelli F, Zugmaier G, Rizzari C, Morris JD, Gruhn B, Klingebiel T, Parasole R, Linderkamp C, Flotho C, Petit A, Micalizzi C, Mergen N, Mohammad A, Kormany WN, Eckert C, Möricke A, Sartor M, Hrusak O, Peters C, Saha V, Vinti L, von Stackelberg A (2021) Effect of blinatumomab vs chemotherapy on event-free survival among children with high-risk first-relapse B-cell acute lymphoblastic leukemia: a randomized clinical trial. JAMA 325:843–854. https://doi.org/10.1001/jama.2021.0987
88. No author (2020) Tisagenlecleucel for B-cell cancers. Aust Prescr 43:30–31. https://doi.org/10.18773/austprescr.2019.077
89. Stirrups R (2018) CAR T-cells for relapsed B-cell ALL in children and young adults. Lancet Oncol 19:e144. https://doi.org/10.1016/S1470-2045(18)30087-1
90. Liu Y, Chen X, Han W, Zhang Y (2017) Tisagenlecleucel, an approved anti-CD19 chimeric antigen receptor T-cell therapy for the treatment of leukemia. Drugs Today 53:597–608. https://doi.org/10.1358/dot.2017.53.11.2725754
91. Syed YY (2020) Ramucirumab: a review in hepatocellular carcinoma. Drugs 80:315–322. https://doi.org/10.1007/s40265-020-01263-6
92. Khan U, Shah MA (2019) Ramucirumab for the treatment of gastric or gastro-esophageal junction cancer. Expert Opin Biol Ther 19:1135–1141. https://doi.org/10.1080/14712598.2019.1656715
93. Turkes F, Chau I (2019) Ramucirumab and its use in the treatment of hepatocellular carcinoma. Future Oncol 15:979–988. https://doi.org/10.2217/fon-2018-0822
94. Kwok G, Yau TCC, Chiu JW, Tse E, Kwong Y-L (2016) Pembrolizumab (keytruda). Hum Vaccin Immunother 12:2777–2789. https://doi.org/10.1080/21645515.2016.1199310

95. Robert C, Schachter J, Long GV, Arance A, Grob JJ, Mortier L, Daud A, Carlino MS, McNeil C, Lotem M, Larkin J, Lorigan P, Neyns B, Blank CU, Hamid O, Mateus C, Shapira-Frommer R, Kosh M, Zhou H, Ibrahim N, Ebbinghaus S, Ribas A (2015) Pembrolizumab versus ipilimumab in advanced melanoma. N Engl J Med 372:2521–2532. https://doi.org/10.1056/NEJMoa1503093
96. Coupez D, Hulo P, Touchefeu Y, Bossard C, Bennouna J (2020) Pembrolizumab for the treatment of colorectal cancer. Expert Opin Biol Ther 20:219–226. https://doi.org/10.1080/14712598.2020.1718095
97. Jay SM, Lee RT (2013) Protein engineering for cardiovascular therapeutics. Circ Res 113:933–943. https://doi.org/10.1161/CIRCRESAHA.113.300215
98. Bromage DI, Davidson SM, Yellon DM (2014) Stromal derived factor 1α: a chemokine that delivers a two-pronged defence of the myocardium. Pharmacol Ther 143:305–315. https://doi.org/10.1016/j.pharmthera.2014.03.009

Emerging Trends, Challenges, and Opportunities in Protein Therapeutics 11

S. Silpa, X. Janet Bertilla, and S. Rupachandra

Abstract

Protein-based therapies are proving effective in the clinic, and their potential is recognized. More than 100 genuine and equivalent modified therapeutic proteins have been approved for clinical use in the European Union and the United States, with revenues of US$ 108 billion in 2010, with monoclonal antibodies(mAbs) accounting for over half of the sales (48%). Based on their pharmacological activity, the therapeutic protein-based treatments can be divided into five categories: (1) replacing a missing/abnormal protein, (2) augmenting an existing pathway, (3) providing a novel function/activity, (4) interfering with a molecule/ organism, and (5) delivering other compounds, proteins, cytotoxic drug, or effector proteins. Therapeutic proteins include antibodies, Fc fusion proteins, anticoagulants, enzymes, growth factors, hormones, interleukins, and thrombolytics. Recombinant proteins are used in most commercialized protein therapies, and several are now being investigated in clinical trials for cancer, immunological disorders, infections, and other health issues. Bispecific mAbs and multispecific fusion proteins, mAbs, and proteins with improved pharmacokinetics are currently being developed. To overcome substantial obstacles such as resistance to therapy, access to targets, biological system complexity, and individual heterogeneity, a paradigm change in techniques and knowledge of processes is necessary.

S. Silpa · X. Janet Bertilla · S. Rupachandra (✉)
Department of Biotechnology, School of Bioengineering, SRM Institute of Science and Technology, Kattankalathur, India
e-mail: rupachas@srmist.edu.in

D. B. Singh, T. Tripathi (eds.), *Protein-based Therapeutics*,
https://doi.org/10.1007/978-981-19-8249-1_11

Keywords

Therapeutics · Proteins · Antibodies · Vaccines · Cancer · Immune diseases · Immunogenicity · Safety · Efficacy

11.1 Introduction

Proteins emerged as a major new class of pharmaceuticals in the early 1980s, where treatments included a limited number of diagnostics and vaccines. Based on their pharmacological activity, a protein-based drug that targets an aberrant or anomalous protein, augments a pathway, provides a unique feature or activity, or interacts with a molecule or organism can be classified as protein therapeutics. Antibodies, anticoagulants, blood factors, bone morphogenetic proteins, engineered protein scaffolds, enzymes, Fragment crystallizable (Fc) fusion proteins, growth factors, hormones, interferons, interleukins, and thrombolytics are a few molecular categories of protein therapies. Since the early 1980s, protein therapeutics and its associated sister branch, protein engineering, have gained importance. Recombinant versions of natural proteins were the first therapeutic proteins. Clinical enhancements generated from protein modifications were quickly followed by enhancements derived from protein or glycoengineering, Fc fusion, or conjugation with polyethylene glycol (PEG). The motivation for producing better protein treatments with improved efficacy, safety, immunogenicity, or delivery is based on a convergence of clinical, scientific, technological, and commercial drives highlighting unmet needs and providing solutions. Enhanced efficacy, increased safety, lower immunogenicity, and enhanced delivery are significant prospects for protein therapies. Advances in biochemistry, protein chemistry, and molecular biology have fuelled the increased use and development of therapeutic proteins.

Proteins introduced into the circulation are exposed to a variety of cells, enzymes, and pathways that aid in their clearance and degradation. Rapid elimination of small proteins can impair therapeutic efficacy. Covalent attachment of PEG or dextran chains to proteins and protein-protein cross-linking have been used to prevent the elimination of small proteins. While any alterations may lower the biological function of the protein of interest or cause an immune response in recipient animals or humans, there is now enough knowledge in this field to design and successfully execute an appropriate extended clearance method. With the rapid expansion of genomic and proteomic data, many altered proteins are anticipated to be created, necessitating clearance-related modifications. Future protein therapies, such as antibodies and Fc fusion proteins with better effector functions or longer half-lives, will be extensively modified to improve their performance. Antibody-drug conjugates and bispecific antibodies, both old notions for enhancing antibodies, are on the verge of clinical success. This chapter focuses on protein therapeutics, current trends, limitations, and prospects [1].

11.2 Emerging Strategies for Protein-Based Therapeutics

Protein therapeutics have transformed medical science since their emergence in the market around two decades ago. They can be characterized as agents possessing pharmacologic properties such as the ability to replace an abnormal/deficient protein and transport a cytotoxic drug or radioactive isotope, etc. They are classified into several groups like blood factors, drug-based antibodies, interleukins, growth factors, etc. [2].

11.2.1 Generation of Glycosylated Proteins to Synthesize First- and Second-Generation Drugs

Human insulin, developed using recombinant DNA (rDNA) technology, was the first human protein-based therapy. Since then, the field has expanded to create therapeutics to replace defective genes or expand pathways (erythropoietin, interferon-A). The understanding of the clinical potential of proteins is extended due to the remodeling of the proteins. Recombinant insulin, for example, replicates the action of native insulin but is genetically altered to change the way it acts in the body [2]. Multiple Insulin analogues have been approved as a treatment for diabetes mellitus. These analogues provide varied benefits over recombinant drugs, including prolonged steady action, decreasing the risk of hyperglycemic, etc. Glycosylation of therapeutic proteins is a promising strategy for optimizing the in vivo efficacy of protein-based drugs. Glycoengineering of darbepoetin alfa for half-life extension is an example that, when administered, vitalizes the synthesis of RBCs in the bone marrow of anemic patients. Two glycosylation sites were added to the first-generation drug epoetin alfa in order to design darbepoetin alfa, which is a long-acting, commercially successful second-generation drug [3].

11.2.2 Fc Fusion Proteins

This entails combining genes that code for two or more proteins. Fusion proteins tend to link the functions of the constituent parts. Besides extending the half-life, Fc fusion has other advantages, such as improving stability, solubility, and efficacy by increasing valency, and enabling purification using protein A chromatography. Immunoglobulins containing the Fc region have been the most therapeutically effective fusion protein therapies to date [4]. These Fc fusions supply peptides by binding to the neonatal Fc receptor (FcRn). Proteins of relatively smaller sizes are eliminated by renal filtration. The most profitable Fc fusion protein therapeutic is etanercept. It is a Tumor Necrosis Factor (TNF) receptor 2-Fc fusion protein consumed worldwide as a therapeutic for autoimmune diseases and diseases like Rheumatoid Arthritis.

11.2.3 Assisted Design of Antibody and Protein Therapeutics (ADAPT)

The goal of the assisted design of antibody and protein therapeutics (ADAPT) is to aid in the discovery of mutants that improve the affinity of antibodies and other new treatments. It enhances the strength of mutant selection and design by combining computational estimates with experimental confirmation [5]. The first step in the working of ADAPT is the virtual mutagenesis throughout the entire complementarity-determining region (CDR) sequences. The binding affinity is estimated using a parent sequence in which each amino acid is changed to one of 18 different amino acids. A median-based affinity score is calculated using scoring functions. Around ten mutants are chosen depending on the diversity of the site and amino acids. Recently, this approach was evaluated on three Fab-antigen complexes: bH1-VEGF-A, bH1-HER2, and herceptin-HER-2. The complex crystal structures were available, which were used for virtual mutation simulations [6]. ADAPT provided mutational contributions to binding free energy are reported to have acceptable characteristics. The experimental validation is the reason for the success of ADAPT, as it has the potential to be a functional plan of action for affinity maturation of antibodies. This strategy can be expanded by providing more importance to enhancing other biophysical properties in addition to further biotherapeutic frameworks [5].

11.2.4 Injectable Implants in In Situ

These implants are constructed to deliver a wide range of molecules or proteins for an extended period of time. The structure consists of liquid or injectable semisolids, over injection becoming gel/implant. Depending upon the mechanism of implant or gel formation, these formulations can be broadly grouped into either in situ phase inversion or cross-linking mechanisms, respectively. In situ implant formations have been recently used to transport proteins and peptides for an extended period of time at a controlled rate. Implants coagulated with Polyethylene glycol (PEG) have the ability to escalate protein stability and decrease immunogenicity through steric hindrance. This mechanism provides a suitable way to deliver proteins without disturbing their stability to protect macromolecules against degradation [7].

11.3 Role of Protein Therapeutics in the Field of Medicine

Protein therapeutics have shown much success in treating various diseases, and their potential is now being acknowledged significantly [8]. Protein therapies delivered directly to target cells and tissues are considered a possible medical approach, and the potency of the created therapeutics will be greatly improved if this technique is advanced [9]. We now briefly discuss the applications of protein therapeutics in different fields of medicine.

11.3.1 Cancer-Related Disorders

Protein-based therapies have revolutionized the field of cancer for decades by targeting the cancer cells by adhering to the receptors or other indicators that are unique to the cancer cells than the healthy cells [10]. Bax, a protein that plays a crucial role in apoptosis, is used as a direct target for drug development. Numerous Bax activators have been found to be promising candidates for cancer treatment due to their selectivity and ability to overcome chemo-resistance [11]. Whey protein is also considered a good candidate in the field of cancer, since it can increase glutathione levels as the deficiency of glutathione is targeted in cancer treatment [12]. Studies show that Bovine α-lactalbumin, a key protein found in bovine milk, and human α-lactalbumin, a protein that helps regulate lactose synthesis, promote cell death in tumor cells [13]. Azurin, a bacteriocin, can induce cell death by interfering with the signaling pathways and other surface receptors in cancer cell lines and hence is a potential therapeutic for cancer treatment [14]. The use of cationic lipidoid nanoparticles in *PTEN* gene (a tumor suppressor gene) deficient prostate cell lines acts as a vehicle for the delivery of intracellular PTEN protein, which can be considered a potential treatment for cancer [15].

11.3.2 Inflammatory Diseases

Protein therapeutics are gaining much attention to produce newer, more specific treatments for inflammatory disorders [16]. DX890, a Kunitz domain peptide depelstat, is tested for the treatment of cystic fibrosis and has shown to diminish neutrophil trans-epithelial translocation and inflammatory activity [17]. Canakinumab, a monoclonal antibody (mAb), has a high affinity for blocking cytokine interleukin-1β (IL-1β) and plays a vital role in the treatment of cryopyrin-associated periodic syndrome (CAPS) [18]. Infliximab and adalimumab are mAbs that bind to tumor necrosis factor α (TNF-α) and inhibit their binding to the receptors, which are used in treating rheumatoid arthritis, ulcerative colitis, and Crohn's disease. Administration of glucose-regulated protein 78 (GRP78) antibodies stimulated microvascular endothelial cells in the cell lines of mice and thereby served as a viable target for enhancing the central nervous system transit of therapeutic antibodies [19].

11.3.3 Genetic Disorders

Proteins have emerged as a key treatment for various genetic disorders ever since the advent of the first recombinant protein therapy, insulin [20]. Emicizumab, a mAb, enhances thrombin production by imitating factor VIII activity independent of the insufficiency or the presence of factor VIII inhibitors in patients with Hemophilia [18]. Tocilizumab, an anti-IL-6R antibody that binds to different forms of IL-6R, is used to treat myocardial infarction [21]. Arimoclomal, a heat shock protein, can boost the HSP70 chaperone that aids in the folding of glucocerebrosidase. HSP70 can traverse the blood-brain barrier and improve the folding and maturation activity in the cells of patients with Gaucher's disease [22]. Phosphodiesterase 10A (PDE10A) controls a variety of critical processes in neurons, and PDE10A inhibitors have shown promising results in preclinical Huntington's disease (HD) models and are currently being tested in human trials [23]. Fabry disease is caused due to mutations in the α-galactosidase A (α-Gal A), which leads to the build-up of globotriaosylceramide (GL-3). Enzyme replacement treatment using recombinant-Gal A decreases cellular GL-3 accumulation, which is associated with a considerable decrease in albuminuria and a steady glomerular filtration rate in patients.

11.4 Process of Drug Development and Food and Drug Administration (FDA)-Approved Protein Therapeutics

11.4.1 Overall Process of Drug Development

The multistep process by which a drug is discovered and then introduced into the pharmaceutical market is called drug development. The steps involved in drug development are ("Drug Development: The Journey of a Medicine from Lab to Shelf—The Pharmaceutical Journal").

1. Preclinical research
2. Clinical research
3. FDA review
4. Postmarket safety monitoring and surveillance

The process of drug development basically starts in a laboratory where a researcher or a group of researchers undertake research ideas for a disease from its gene or molecular level (Fig. 11.1). Once the potential target is identified, a search for a suitable compound or molecule to work on the target is done. The next stage is

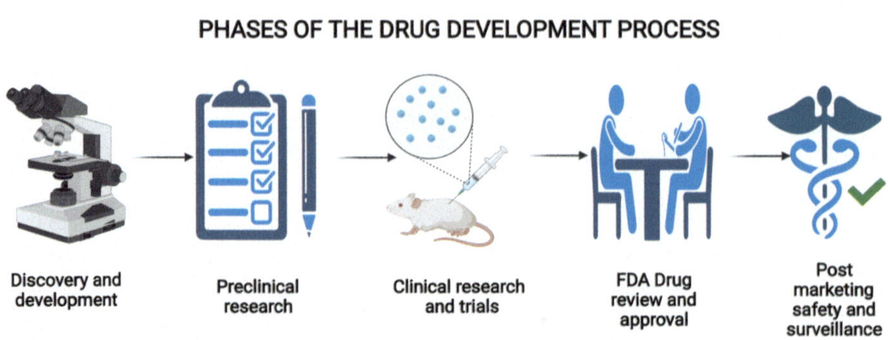

Fig. 11.1 Phases in the drug development process

to ensure that these drugs would not have any adverse effects and are safe for the human population, so preclinical tests are conducted using computerized models, simulations, cells, and laboratory animals. Once the preclinical trials are passed and the drugs are labeled safe for human use, a clinical trial application is submitted to the medical board of the country, where a panel of experts would determine if further research is required or if the drug is clear to undergo human trials.

11.4.2 Drug Discovery

Drug discovery begins with researchers finding new insights that give them potential ideas to reverse the effect of the disease. Multiple compounds are tested to see possible responses to the disease. After testing the compounds, those with favorable responses are selected for further testing. New technologies based on nanoparticles are used for enhancing drug delivery and its effects. The development process involves further testing on the selected compounds. During the discovery phase, the researchers also analyze the following:

1. Mechanism of drug absorption, action, and pathways
2. Mode of distribution, metabolism, and excretion
3. Suitable dosage and method of delivery
4. Chemical interaction with different age groups, treatment methods, and drugs
5. Comparing effectiveness with that of similar drugs in the market
6. Toxicity and side effects

11.4.3 Preclinical Research

According to the WHO guidelines, preclinical research is an integral part of the development of a new pharmaceutical intervention. Preclinical studies are performed under in vitro, in vivo, and even in silico models to obtain details about the mechanism, safety, and efficiency of the drug before releasing it to the human population. These studies must strictly follow the good laboratory practices (GLP) guidelines to ensure safe and reliable results. The FDA lays out a set of tests to be held on safety profiles, pharmacodynamics, toxicology, and pharmacokinetics. The set of tests varies depending on the type of drug, its mode of action, target, etc. [24].

11.4.4 Clinical Research

Clinical research involves research work using human volunteers to develop medical knowledge on an existing treatment method, a new method, or a new pharmaceutical drug. Clinical studies are of two types:

11.4.4.1 Interventional Studies (Clinical Trials)

In clinical trials, the volunteers are given specific interventions according to a research plan designed by the investigators. There are four phases to any clinical trial designed by the FDA.

Phase 1: First time testing of the new drug on a small group of people to test for its side effects and suitable dosage.
Phase 2: Increased scale tests for the drugs which cleared phase 1.
Phase 3: Wide range tests, that is, populations from different regions and countries. This phase is often carried out right before the treatment is approved.
Phase 4: Studies carried out after approval for long-term monitoring.

11.4.4.2 Observational Studies

Observational studies are primarily based on observing procedures of routine medical care or, in a few cases, involve a specific intervention (medical procedure or drug). If the drug passes preclinical and clinical studies and is safe and effective for intended use, the researcher or company can file a new drug application (IND). The FDA team would review all the data and decide if clearance should be issued or not. An NDA must include the complete history of the drug. The company or researcher must provide every detail from preclinical trials to phase 3 trial results. Apart from the trial details, it must also include drug abuse details, patent details, labeling, institutional review board (IRB) compliance details, data for studies conducted outside the country, and dosage/directions.

11.4.5 Food and Drug Administration (FDA)

In the USA, the FDA is a federal agency responsible for public health by ensuring the efficiency, safety, and security of drugs, bioproducts, and medical devices used [24]. For a drug to get FDA approval, the investigators working on the new drug must file an IND application. The approval of the IND application means further studies can be conducted on the drug. The FDA asks the investigators to seek early consultation through the preinvestigational new drug application consultation program before submitting the IND application. The IND application demands three categories of details that are

1. Preclinical studies with animals, toxicology tests, and previous human experience with the drug.
2. Manufacturing details that include the chemical composition of the drug, manufacturing process, stability details, and control measures.
3. Clinical study details and protocols to be followed, proof that the investigator is qualified to conduct the trials, and submission of Institutional Review Board (IRB) approval.

Once the FDA approves the IND application, further studies can be done on the drug. The studies to be conducted are mentioned at the beginning of the chapter. Once all the studies are successfully completed and the drug enters phase 3 of clinical trials, the company applies for NDA approval. The FDA demands clear evidence of all the preclinical studies, drug safety, and efficacy tests and at least two sufficient and efficiently performed phase 3 trials data with adequate proof for NDA approval. Post NDA approval, the drug is clear for circulation in the market for human use [25].

11.4.6 Protein Therapeutics

Therapeutic proteins are engineered proteins used for pharmaceutical and medicinal purposes with a wide range of advantages, such as high specificity and complex functions and low potential to interfere with the biological process, without adverse effects. In addition, they have a high tolerance, do not elicit immune responses, and are effective and easy replacement treatment, thereby cancelling the requirement for gene therapy. Protein engineering is the conception and production of unnatural polypeptides, through modifications of amino acids. Protein engineering has revolutionized protein therapeutics by providing tools to customize pre-existing proteins or mechanisms to create novel proteins depending on the requirement or specificity of the clinical application [26]. Currently, there are more than 100 FDA-approved therapeutic proteins [27]. Based on their function and therapeutic activity, they can be divided into four categories, discussed below.

11.4.6.1 Group 1: Regulatory and Enzyme Proteins

Group 1 proteins function by a classic mechanism where the patient has a deficiency, which is treated by an exogenous protein source. Group 1a proteins are used to treat conditions where a protein is either deficient or abnormal. Examples of this type of protein therapeutics include the following:

1. Insulin: It is used for the treatment of diabetes by regulating blood sugar levels and shifting potassium into cells.
2. Lipase, protease, and amylase: Cystic fibrosis patients are treated with combinations of porcine pancreatic enzymes, which include lipase, protease, and amylase.
3. Mecasermin: It is used for treating children with growth failure due to deletion of the GH gene or severe Insulin-like Growth Factor(IGF1) deficiency. The therapeutic protein is a recombinant IGF1 protein that can induce mitogenesis, chondrocyte, and organ growth, thereby helping achieve proper overall growth.

Group Ib proteins have a major effect on treating thrombosis and hemostasis. Examples include Factors 8/9, which help in hemophilia A/B respectively, and protein C concentrate used to prevent and treat venous thrombosis and patients with vitamin C deficiency. Group Ic proteins function by providing a novel activity or function. Examples include the following:

1. Collagenase: It is used in the debridement of chronic ulcers and severe burns. It functions by digesting collagen in the necrotic tissue near the injury.
2. Hyaluronidase: It is used to increase the absorption of injected drugs, mostly during injection anesthetics for optic surgery.

11.4.6.2 Group 2: Proteins with Special Targeting Activity

Group 2 therapeutic proteins work on the principle of binding to receptors of the native protein ligands or the antigen recognition sites of immunoglobulins and guide the immune system to destroy the specific molecules or cells. They trigger an immune response, and cell killing occurs through macrophages, complement fixation, or other immune cells. These groups of proteins are used in the treatment of inflammatory diseases, cancer, and infectious diseases. The group 2a proteins work on the principle of interference with a molecule or organism. Examples include the following:

1. Bevacizumab: It is used to treat non-small-cell lung cancer and colorectal cancer. It works through the binding of humanized mAb to isoforms of Vascular Endothelial Growth Factor (VEGF).
2. Alemtuzumab: It is used in the treatment of B-cell chronic lymphocytic leukemia. It works by stimulating the humanized mAb against the CD52 antigen present on T and B cells.
3. Abatacept: It is used to treat rheumatoid arthritis.
4. Enfuvirtide: It is used to treat advanced HIV infection. A 36 amino acid peptide prevents the entry of HIV into host cells by binding to the envelope gp120/gp41 protein of HIV.

Group 2b works by delivering other compounds or proteins. Examples include denileukin diftitox that is used to treat cutaneous T-cell lymphoma with its malignant cells expressing the CD25 component. It works by directing diphtheria toxin to cells that express the IL2 receptor.

11.4.6.3 Group 3: Protein Vaccines

These proteins are used as therapeutic or prophylactic vaccines. Group 3a protein therapeutics are generally used to treat infectious diseases or toxins. Examples include the following:

1. HPV vaccine: It is used to prevent human papillomavirus (HPV) infection. It is a recombinant vaccine that contains capsid proteins of four different HPV strains (strains 6, 11, 18, 16).
2. OspA: It is a vaccine for Lyme disease. The vaccine contains a noninfectious lipoprotein extracted from *Borrelia burgdorferi*.

Group 3b proteins are used to treat patients with autoimmune disorders. Examples include anti-Rhesus D antigen Ig. Group 3c therapeutic proteins can be used as anticancer vaccines; however, currently, there are no FDA-approved anti-cancer therapeutic vaccines.

11.4.6.4 Group 4: Protein Diagnostics

These proteins are not used to treat any diseases or disorders, but these can be purified, and the recombinant proteins isolated can be used for both in vivo and in vitro medical diagnosis. It plays a significant role as the diagnosis lays the path for the complete treatment plan. Examples include the following:

1. Glucagon: It is used to reduce gastrointestinal motility for radiographic studies and hypoglycemia reversal.
2. Secretin: It is used in the diagnosis of gastrinoma and the identification of ampulla of Vater during endoscopy.
3. Satumomab pendetide: It is used for ovarian and colon cancer detection. The imaging agent used is indium-111-labeled mAb specific for Tumor-associated glycoprotein (TAG-72).
4. Nofetumomab: It is used for detecting and staging small-cell lung cancer. A technetium-labeled antibody is the imaging agent that is specific for this type of cancer.
5. Hepatitis C antigens: It is used for the diagnosis of hepatitis C exposure. The diagnosis is made through the presence of antibodies to hepatitis C.
6. Apcitide: It is used for imaging acute venous thrombosis. A technetium-labeled synthetic peptide (imaging agent) is used, which functions by binding GPIIb/IIIa receptors to activated platelets.

11.5 Engineered Protein Scaffold as Emerging Therapeutic Proteins

Non-Ig scaffolds are smaller than antibodies (usually up to 100 amino acid residues or 11 kDa), more stable, and frequently lack disulfide linkages. Over 50 other nonantibody protein scaffolds have been offered as alternative mAb creation platforms over the last decade [28]. The non-Ig scaffold proteins such as Affibody, Anticalins, Adnectins/Monobody, and DARPins are discussed below.

11.5.1 Affibody

In Affibody libraries, amino acid residues on the Fc-binding surface are randomized, eliminating the Fc binding [27] (Fig. 11.2). Affibodies are chosen from combinatorial libraries in which 13 residues on the Fc-binding surface of helices 1 and 2 are randomly assigned. Bio panning the phage-displayed library against desired targets identifies specific binders to target proteins [28]. The B-domain in the Ig-binding

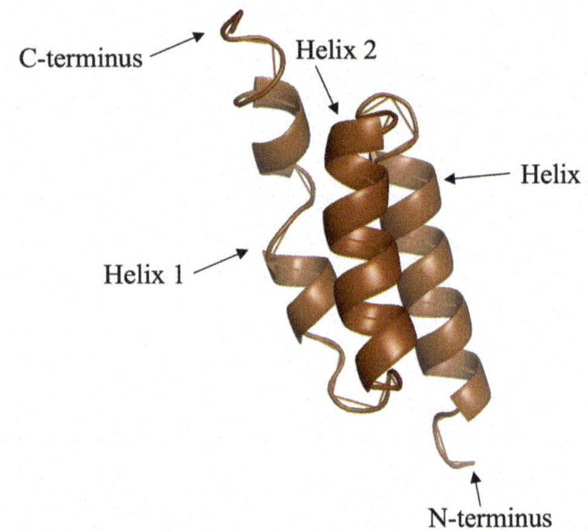

Fig. 11.2 Affibody (Z-domain-dark region)

region of staphylococcal protein A has been used to synthesize affibody molecules [29]. The B-domain is a 58 amino acid cysteine-free peptide that folds into a three-helical bundle structure with one of the fastest folding rates. The designed variation was named the Z-domain. The engineered Z-domain maintained its affinity for the Fc portion of the antibody but lost practically all of its affinity for the Fab region. The Z-domain scaffold basically terminated the Fab binding. Selections against various targets have yielded HER2, EGFR, insulin, IgA, Alzheimer's amyloid-beta peptides, HIV-1 gp120, and other targets are recognized by affibody molecules. Affibodies have been used for in vivo tumor detection and, more recently, therapeutic applications, particularly those directed against medically important protein targets. For example, the Fc part of mAbs, which mediates antibody-dependent cell-mediated cytotoxicity (ADCC) and complement activation, is not necessary for therapeutic applications that rely entirely on preventing the function of disease-related target proteins as a mechanism of action [30].

11.5.2 Adnectin

Adnectin is synthesized from the tenth domain of fibronectin type III (10Fn3), a human extracellular matrix protein and a 94 amino acid thermostable binding protein [31]. Along with its original structure as a polyfunctional multidomain species, this fibronectin-based protein is highly suitable for developing multispecific therapeutic targets. This domain, resembling the VH domain of IgG, has an overall beta-sandwich topology and loop structure but folds reliably with no disulfide bonds. Adnectins with a high affinity for target proteins of interest may be engineered efficiently [32]. The first generation of Adnectins was composed of monobodies focused on a library of small FG loops and selected by phage display. As a result, various monobody derivatives of FN3 scaffolds optimized for loops that bind IκBα caused a change in expression levels; therefore, the e10-FN3s were assembled with an altered AB loop ("AAPTSOL" changed to "EASPTSLIQ") and liberated expression levels. While the thermostability and solubility of 10Fn3 can be significantly reduced when its loops are altered to facilitate target binding. The exceptionally high stability of the wild type suggests that even destabilized variations can be appropriately stable with therapeutic properties [33].

11.5.3 Anticalins

Anticalins are non-Ig binding proteins that have been engineered via combinatorial design approaches based on a human lipocalin scaffold and have the potential to emerge as therapeutic reagents [33]. They are small in size (~20 kDa) (Fig. 11.3) [34]. Although anticalins have a high sequence variation, they share an intact barrel of eight antiparallel β-strands with an α-helix as their scaffold, and four loops connect each pair of strands at the barrel's end; these loops form the binding pocket, which controls the specificity of the ligand [35] binding, enclosing a ligand binding site with a repeating +1 topology. Researchers have demonstrated that anticalin

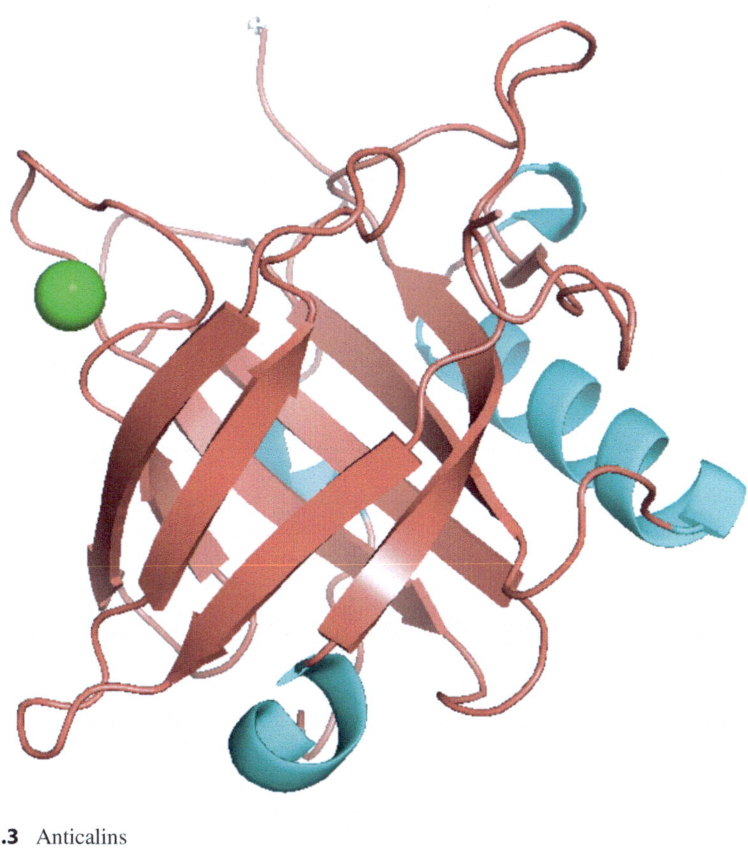

Fig. 11.3 Anticalins

libraries can be displayed well on *E. coli*, allowing for FACS and presenting a viable future option to manufacture unique specialized anticalins. Several anticalins have been produced with affinity in the nanomolar to picomolar range against many therapeutically important targets, including cytotoxic T-lymphocyte antigen 4 (CTLA-4, 10 CD152), VEGF, oncogene c-Met, PRS-343 (an anticalin–antibody fusion protein that binds both 4-1BB), and the well-known breast cancer antigen human epidermal growth factor receptor 2 (HER2). Individuals with HER2-positive malignancies such as breast, bladder, and stomach tumors may benefit from PRS-343 as a therapy option [36].

11.5.4 DARPins

Ankyrin repeat proteins (DARPins) are made up of closely linked repetitions of 33 amino acid residues. Each repeat consists of a β-turn followed by two antiparallel α-helices to form a structural unit [37]. The interaction with the target is mediated by the library modules in their randomized placements. These residues are found on the concave molecular surface of the DARPin, producing a prolonged contact interface that binds structural epitopes on the target protein surface (Fig. 11.4) [38]. The significant display of DARPins on LAB surfaces widened the spectrum of binding proteins and opened up various therapeutic and other possibilities [39].

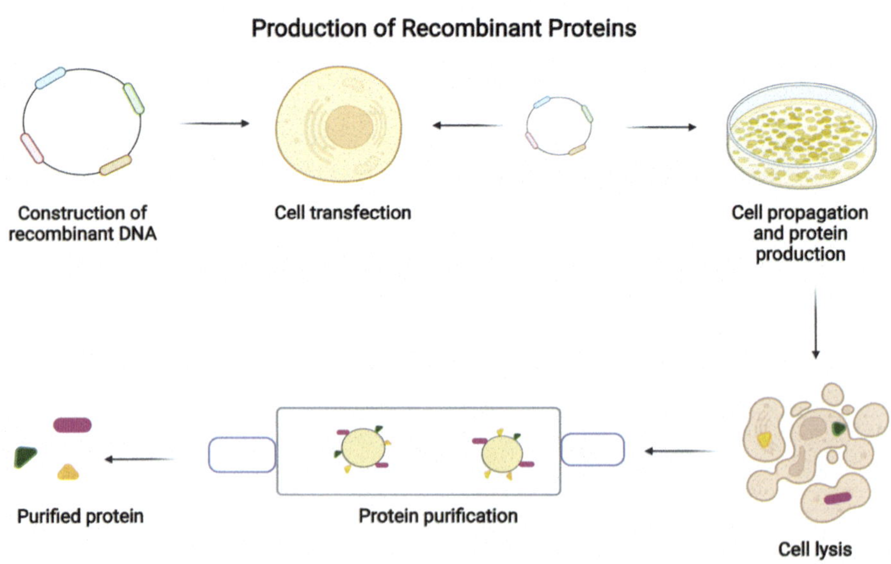

Fig. 11.4 Cloning, expression, and production of recombinant proteins

11.6 Production of Recombinant Protein Therapeutics and their Applications

Recombinant proteins are generated by the transfection of foreign genes in the desired host cell. Pharmaceutical products, protein-based polymers for drug administration, antibodies and enzymes for disease therapy, protein scaffolds for tissue engineering, and a variety of other applications are all accomplished by recombinant proteins [40]. Recombinant proteins are utilized not only in biological research but also as medications in therapy. In 1982, recombinant human insulin became the first recombinant protein used in medicine. The US FDA has authorized more than 130 recombinant proteins for clinical use [41].

11.6.1 Cloning and Initial Preparation to Produce Recombinant Proteins

To produce heterologous proteins, a variety of microbial hosts are used, including *Aspergillus niger*, *Bacillus* sp., *E. coli*, *Pichia pastoris*, *S. cerevisiae*, *T. reesei*, etc. [42]. Complete intracellular organelles and membrane-bound compartments are seen in *S. cerevisiae*; as a result, a yeast cell may generate and fold a variety of proteins. The translated proteins in a yeast cell are exposed to posttranslational modifications (PTMs) such as signal peptide proteolysis, disulfide bond formation, subunit assembly, acylation, glycosylation, etc. [43].

A well-established plan is frequently followed when developing a manufacturing process of a recombinant protein in mammalian cells. The cells are first transfected with the recombinant gene with the requisite transcriptional regulatory components. A second gene is also transfected, which gives recipient cells a selection advantage. Only cells expressing the selector gene survive in the presence of the selection agent, which is administered a few days after gene transfer. Dihydrofolate reductase (DHFR), a nucleotide metabolic enzyme, and glutamine synthetase (GS) are the most common genes for selection. In both situations, nontransformed cells are prevented from growing in the absence of the relevant metabolite (hypoxanthine and thymidine in the case of DHFR, glutamine in the case of GS) [44].

To produce recombinant proteins in plants, foreign DNA must be introduced into the plant cell. Gene transfer can be accomplished in one of three ways. The first is *Agrobacterium tumefaciens*, a soil pathogen that can transfer a short fragment of DNA from a resident plasmid (Ti plasmid) into the plant genome. By removing the bacterial genes from the transferred DNA segment (T-DNA) and replacing them with foreign DNA, this natural system can be exploited. Direct DNA transfer is the second method, which includes a variety of procedures that drive DNA over the cell wall and/or membrane. Particle bombardment, electroporation, PEG-mediated protoplast transformation, and transformation of intact cells after perforation of the cell wall with silicon carbide whiskers are examples of such techniques. The third approach takes advantage of the innate capacity of plant viruses to infiltrate cells and transport nucleic acids directly inside. The genomes of the plant viruses can be altered to include foreign genes, resulting in the production of recombinant proteins and viral gene products [45].

The recommended method for generating stable cell lines for industrial applications is nonviral gene transfer. Calcium phosphate transfection, electroporation, lipofection, and biolistic and polymer-mediated gene transfer are all standard methods of gene transfer that are effective and reliable [46].

11.6.2 Expression of Recombinant Proteins

Transfecting a cell with an rDNA vector carrying the gene of interest is one method of producing recombinant protein. Cells with rDNA are then grown to transcribe and translate the desired protein. The cells are then lysed or broken to release the produced protein, which can then be purified using various procedures. Both prokaryotic and eukaryotic systems are employed in the expression of protein from the rDNA [45] (Fig. 11.4).

11.6.2.1 Mammalian Systems

Proteins required for functional and structural analysis, antibody synthesis, interaction analysis, virus production, etc., are produced in mammalian systems. The ability to produce proteins in either a transient or stable manner is the major benefit of this approach. Furthermore, methods have been designed to produce proteins quickly and with a high yield. However, the necessity for big yields in suspension cultures is one of the downsides. In this situation, the cell culture conditions are also more difficult to sustain [47].

11.6.2.2 Insect Systems

This system is utilized for producing proteins that are used for functional and structural analysis, intracellular expression of protein and protein complexes, virus production, etc. Here, the processing procedures are comparable to those used in mammalian systems to process proteins. However, the cell culture conditions in this system are more demanding than those for cultivating prokaryotic cells.

11.6.2.3 Yeast Systems

The yeast system is utilized for producing proteins for structural and functional analysis, antibody production, and protein interaction research. It is scalable up to fermentation and can handle eukaryotic proteins. In this situation, the media requirements are simple. Fermentation, on the other hand, is essential for producing extremely high yields, and growth conditions must often be optimized.

11.6.2.4 Bacterial Systems

Structural and functional experiments can also be performed using proteins expressed in the bacterial systems. It is scalable, low-cost, and has simple media requirements to work. However, some mammalian proteins may be difficult to express in bacterial systems [48].

11.6.2.5 Algal Systems

This method is used to produce proteins used to investigate photosynthesis, plant biology, and metabolism. It may help genetically modify and express photosynthetic microalgae systems, which is one of the key advantages. It can also be used to provide effective experimental controls for producing biofuels and nutraceuticals. However, this platform is still in its infancy and is less developed than others.

11.6.2.6 Cell-Free Systems

This system is used to investigate toxic proteins, incorporate nonnatural amino acids, perform translational inhibitor screening, etc. However, scaling up recombinant protein production beyond multimilligram volumes might be costly and time-consuming.

11.6.3 Applications of Recombinant Proteins

11.6.3.1 Medicine

The dysfunction of particular proteins causes or contributes to most human disorders. Diabetes, cancer, viral disorders, hemophilia, and anemia are a few conditions for which therapeutic proteins are helpful. Antibodies, Fc-fusion proteins, hormones, interleukins, enzymes, and anticoagulants are common examples of therapeutic proteins (Fig. 11.5). Therapeutic proteins can be categorized into four groups:

Group I: These therapeutic proteins have enzymatic or regulatory actions. They either replace a missing or defective protein, upregulate an existing route, or provide a new function or activity to an existing pathway.

Group II: These therapeutic proteins have a unique ability to target specific organs. They either interfere with or transport other molecules to a molecule or organism.

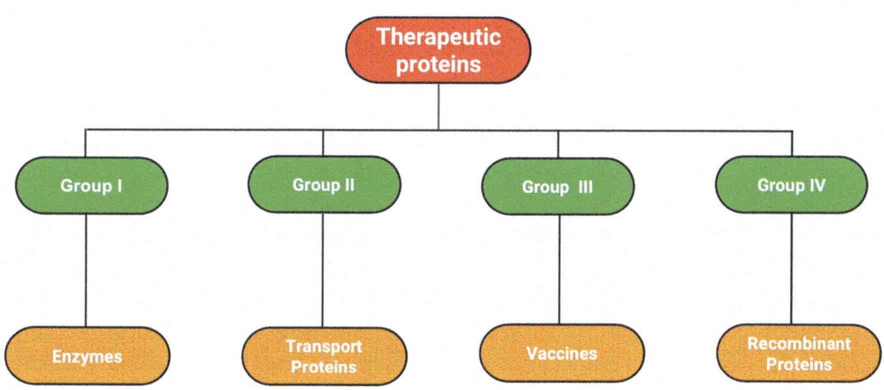

Fig. 11.5 Classification of therapeutic proteins

Group III: These are vaccines based on therapeutic proteins. They are used to treat foreign agents, autoimmune disorders, cancer, etc.

Group IV: These are therapeutic proteins used for diagnostic purposes. Purified and recombinant proteins are the most common diagnostics.

11.6.3.2 Research

Recombinant proteins help in understanding the fundamental principles of biology. They may be used to discover the function of genes in cellular processes such as cell signaling, metabolism, growth, replication, death, transcription, translation, and protein modification, as well as to identify and analyze the protein encoded by any gene. As a result, recombinant proteins are widely employed in molecular biology, cell biology, biochemistry, structural and biophysical research, etc. Understanding protein-protein interactions require the use of recombinant proteins. Recombinant proteins are used in various laboratory techniques, including ELISA, Western Blot, and immunohistochemistry. Enzymatic assays can be performed using recombinant proteins. They are also utilized as standards, such as ELISA standards, when combined with a matching antibody pair. Furthermore, in Western blots, recombinant proteins are used as positive controls.

11.6.3.3 Industrial Applications

Various industrial fields use recombinant proteins, including food production, agriculture, bioengineering, etc. Enzymes can be added to animal feed to augment the nutritional content of feed ingredients, reduce feed and waste management costs, support animal gastrointestinal health, boost overall performance, and improve the environment. Lactic acid bacteria have long been used to make fermented foods, and more recently, they have been engineered to produce recombinant proteins, which could have a wide range of applications, including improving human/animal digestion and nutrition.

11.6.4 Recombinant Proteins in Diseases and Vaccines

The liver produces alpha-1-antitrypsin, which is released and circulated throughout the body to protect the lungs. Patients who are unable to make this protein are given an infusion of alpha-1-antitrypsin protein isolated from donor blood on a regular and quantitative basis. The discoverers of PD-L2, one of the key players in immune checkpoint therapy, were awarded the 2018 Nobel Prize for Physiology & Medicine. The discovery of PD-L2 led to treatments of several cancers, such as melanoma, lung cancer, and others, based on immune checkpoint therapy [49].

The human papillomavirus (HPV) causes cervical cancer in women and has been related to head and neck malignancies in males in a few cases. Currently, the virus is prevented by the Gardasil-9 vaccination. Gardasil 9 is a combination of nine vaccines that can protect against numerous HPV types. The HPV virus-like particles (VLPs) were created by attaching the L2 protein to a pre-existing hepatitis B virus-like particle and then combining the L2 protein with an antibody to create a recombinant immune complex.

11.7 Developments in Recombinant Protein-Based Vaccines

The recombinant protein subunit vaccine comprises one or a few microbiological components produced in heterologous expression systems. A general property in developing recombinant protein-based vaccines is that they require adjuvants and can take advantage of recent advances in the formulation and stabilization of recombinant therapeutic proteins [50]. Protein-based vaccines are T cell dependent and are expected to be immunogenic in infants and the elderly. More importantly, coverage of protein-based vaccines is expected to be broader, because they contain protein antigens common to all or most disease serotypes for which the vaccine is being prepared. The expression and folding of proteins may be incorrect in prokaryotic cells for different reasons. Factors such as signal sequences and disulfide bonds are considered when designing protein-based vaccines. These factors lead to toxicity, insolubility, and rapid degeneration within bacteria. With a significant focus on antigen-presenting cells (APCs), T cells, and B cells, better variants of subunit vaccines can be designed. Dendritic cells are considered necessary as they are designed to capture and process antigens into small fragments with a subsequent presentation at the cell surface when in association with MHC molecules. The occurrence of an immune cascade is necessary to enhance the delivery of subunit antigens. These antigens target the dendritic cells and prevent their degradation in vivo. For example, adjuvants used in vaccines for diphtheria and hepatitis B vaccines like aluminum hydroxide, when combined with surfactants such as polyoxyethylene sorbitan monooleate, and sorbitan trioleate, have the ability to increase the immunogenicity of vaccines. These adjuvants exhibited a decent safety profile and provided successful results in the implementation of successful vaccine programs.

Production of recombinant proteins can be done using various microbial systems, plant cells, mammalian cells, etc. the most predominantly used recombinant proteins are derived from *Escherichia coli* as it is cost-effective and proliferates rapidly. Various vaccine-based antigens are manufactured using *E. coli*, such as the Lyme disease vaccine VLA 15 [51]. Other recombinant protein-based vaccines include vaccines against meningococcal serogroup B infections like Trumenba by Pfizer. This vaccine uses two variants of meningococcal factor H-binding proteins called fHBP as antigens, Bexsero® by GSK, which requires three meningococcal antigens (fHbp, NadA, and NHBA), which are immunogenic. The FDA authorized these two vaccines in 2014 and 2017, respectively. Examples of yeast-based vaccines include Recombivax HB and HEPLISAV-B, where surface antigen like recombinant hepatitis B (HBsAg) is synthesized from yeast. Another example is Gardasil, where the capsid protein L1 is derived from four types of human papillomaviruses (HPVL1) as its antigens. Cervarix®, an HPV vaccine that employs the recombinant HPV L1 antigen, and Flublok®, an influenza vaccine that uses a recombinant trivalent hemagglutinin antigen, are the only two licensed vaccines in the US that use insect cell-expressed antigens. Insect-cell expressed recombinant proteins are well-folded and contain desired PTMs [52]. Recombinant antigens can also be made up of viruses where they are assembled spontaneously into multiprotein VLP structures without the incorporation of the viral genome [53]. VLPs instigate immune responses similar to the ones prompted by natural viral infection. Several VLP-based vaccines have been commercialized recently. One such example is GlaxoSmithKline's Engerix-B® [54].

11.8 Challenges in Protein Therapeutics

11.8.1 Problems Related to In Vivo Administration

Several studies have found considerable discrepancies between therapeutic protein shelf-life stability and in vivo stability. Although in vivo degradation may directly impact the biological performance of therapeutic proteins, there are no reliable methodologies or models for determining the in vivo stability [55]. The pharmaceutical sector is particularly interested in IV, SC, IVT, and IT administration. Human body compartments have physical and chemical properties that differ significantly from those of routinely used dispersible powder (DP) formulations. Various physiological properties, including flow rate, temperature, macromolecules, metabolites, and endogenous degradation products, exist in these body compartments/fluids, which might lead to different physical/chemical degradation profiles [56].

11.8.2 Mechanism Involved in Clearance

Advances in biochemistry, protein chemistry, and molecular biology have fuelled the increased usage and development of proteins as injectable medicinal agents. Proteins introduced into the circulation are exposed to a variety of cells, enzymes, and pathways that aid in their clearance and degradation [57]. Notably, the rapid elimination of small proteins can impair their therapeutic efficacy. Proteins are fused to long-lived plasma proteins like albumin or immunoglobulins, or sections of these proteins, through gene modifications [58]. With the rapid expansion of genomic and proteomic data, an ever-increasing number of deliberately engineered proteins are anticipated to be created, necessitating clearance-related modifications [59].

11.8.3 Strategies for Increasing the Half-Life

Therapeutics based on macromolecular proteins and peptides have been shown to be successful in treating severe human ailments. One of the most significant obstacles to protein therapies in clinical use is their rapid breakdown in the blood and elimination due to enzymatic degradation, renal clearance, hepatic metabolism, and immunogenicity, all of which contribute to their short half-lives [60]. Protein molecules are susceptible to renal filtration and liver metabolism due to their size and hydrophobicity. Artificially increasing the therapeutic half-life of a protein by attaching a molecule that increases its overall size (e.g., PEG) or aids receptor-mediated recycling (e.g., albumin), or manipulating the amino acid chain in a way that makes it more prone to aggregate formation are some of the revolutionary approaches to avoid rapid degradation in vivo [61]. Half-life extension technologies have been successfully applied to many protein therapeutics, including hormones and enzymes, growth regulators, clotting factors, and interferon, to massively increase the half-lives of proteins in circulation (2–100-folds) and thus improve their overall pharmacokinetic parameters.

11.8.4 Developmental Challenges

Pharmaceutical companies have started relying on protein-based research for new and effective treatments. The focus of researchers has led to substantial improvements in the manufacture and use of therapeutic proteins in recent years [62]. Despite the fact that protein-based treatments have taken center stage in pharmaceutical research and have improved human safety, many obstacles remain, including safety and immunogenicity issues, protein stability, and degradation issues.

11.8.5 Safety and Immunogenicity Issues

The intrinsic complexity of proteins, as well as the complexities of the manufacturing process, plays a role in biopharmaceutical immunogenicity. Biopharmaceuticals are made through a series of complex processing stages that result in a product that is very pure, stable, safe, and effective. The host cell development, master cell bank establishment, protein production, purification, analysis, formulation, and storage and handling are the seven stages of the process, which might take months. Even modest alterations at any of these stages can result in clinically significant differences in the efficacy and/or safety of the final product. The way forward for dealing with immunogenicity issues in protein therapeutics is likely to involve a combined approach in which new molecules are generated through rational sequence design (via methods such as deimmunization), and the lead proteins are tested in the appropriate animal model and/or in vitro assay during preclinical development from which the least immunogenic leads are selected. Other product-specific factors such as physicochemical analysis, aggregation, PTMs, and the impacts of excipients should also be tested [63].

11.9 Strategies to Overcome Challenges in Protein Therapeutics

Chemical and physical modifications in protein therapeutics have been performed for clinical benefit. Such modifications are essential since the drug must pass through various barriers before reaching the target. The active targeting of the drug is typically achieved by conjugating it to a target entity, thereby improving the bioavailability. These modifications also solve the issue of protein aggregation. Insulin has a tendency to self-associate to form hexamers whose absorption is lower than that of the monomeric insulin analog. Mutation of amino acids involved in the self-association results in the formation of a monomeric insulin analog called insulin lispro, which is characterized by rapid onset of action following subcutaneous injection. Chemical modification can also be used to produce prodrugs, which can be helpful in the protection against enzymatic degradation [64, 65]. Gene modification can be used to create therapeutic proteins with altered isoelectric points and dynamics. These modifications can modulate both enzyme selectivity and the intrinsic activity of the protein [66]. One such example is the modification of Neprilysin, a protease that has its possible application in the treatment of Alzheimer's disease. The genetically engineered Neprilysin mutant showed a 20-fold increase in activity on amyloid-beta 1–40 [67].

Protein PEGylation using PEG is an approach that prolongs the circulation time of protein therapeutics. PEGylation of proteins can be performed by chemically reacting specific chemical groups within a protein (e.g., the side chains of lysine, histidine, arginine, cysteine, aspartic acid, glutamic acid, threonine, tyrosine, and serine as well as the N-terminal amino and the C-terminal carboxylic acid groups) with a suitable PEGylation reagent [68]. In a study conducted to optimize Extendin-4, PEGylation was carried out at C-terminus (C40-tPEG50K-Ex4-Cys) using Ex4-Cys and activated trimeric PEG. The resulting derivative showed a 7.53-fold increase in the circulation time and its AUC_{inf} (measure of total exposure to the drug) by 45-fold. Its pharmacologic activity was also increased eightfold relative to the native Ex4-Cys [69].

Antibody-directed enzyme prodrug therapy (ADEPT) aims to restrict the action of a high concentration of a cytotoxic drug to target sites by using an antibody (or antibody fragment) to deliver a nonhuman enzyme to those sites [70]. The modified therapeutic proteins aim to produce enzymatic variants with good catalytic efficacy, high levels of stability, and reduced immunogenicity. These features can increase a protein's half-life, which gradually leads to a decrease in the number of doses required. This also limits the possibility of developing resistance against the particular protein drug.

Another approach is to increase the degree of interaction of therapeutic proteins or peptides with serum components (e.g., albumins) to increase the half-life of circulating protein [68, 69]. Serum components like albumin and immunoglobulins have long half-lives in humans [70].

The development of albumin-based drug carriers is another approach for transporting endogenous and exogenous compounds. Human serum albumin is widely used as an excipient for biological products [71]. Therapeutic compounds have been pharmaceutically enhanced by multiple techniques using albumin to improve their distribution, bioavailability, and half-life. For example, noncovalent interactions allow the binding of the albumin to a broad range of endogenous and exogenous ligands. Albumin dimerization, in particular, has significant potential and advantages for clinical applications as both a plasma expander and a drug carrier. Such dimers are present at elevated levels in the circulating blood of patients with chronic renal disease and also result from oxidative damage in the blood [72]. Many molecules of therapeutic interest bind to endogenous albumin in the blood through their fatty acid binding sites, thereby prolonging their half-life and bioavailability. For example, the human insulin analogue, Detemir (marketed by Novo Nordisk as Levemir), is long-acting due to the myristic acid moiety bound to the Lys residue at position B29 of insulin. The attached fatty acid facilitates binding to albumin, thereby prolonging the circulatory half-life of this insulin derivative in blood [73, 74]. Various domains of the HSA molecule have also been used to make bioconjugates with increased stability, better targeting properties, and/or extended half-lives in blood. For example, domain I of HSA has been used to prepare antibody conjugates. This was achieved through the use of a cyclohexene sulfonamide compound that site-selectively labels Lys64 in this HSA domain [75]. Similarly, the half-life of the granulocyte colony stimulating factor (G-CSF) was prolonged by genetic fusion to domain III of I to its N-terminus [76].

Encapsulation of therapeutic proteins into micro or nanoparticles is a promising approach to enhance the stability of therapeutic proteins [77]. Nanoparticles can transport therapeutic molecules to the sites of interest, providing access to molecular interest and modulating interactions [78]. It can also protect the therapeutic proteins from degradation by biological agents. Hyaluronic acid-Fe_2O_3 hybrid magnetic nanoparticles have been designed to deliver proteins to HEK293 and A529 cells at 100% level [79].

11.10 Opportunities in Protein Therapeutics

Protein treatments evolve in tandem with protein engineering technology and regulatory frameworks. Better drug targeting, increased potency, and functionality can be used to create improved versions of existing medicines. Understanding the mode of action and structure-function relationship allows to alter the activity of a protein or introduce new activities, resulting in the customization of current proteins and the development of novel medicines for particular clinical applications. There are examples of sensible alterations to current protein medications that resulted in the approval of revolutionary second-generation therapies thanks to protein engineering [77]. Pharmacogenetics can help identify responders and nonresponders to drugs, prevent side effects, and optimize drug dosage. Medication labels may include genetic biomarker information, as well as information on pharmacological exposure, clinical response variability, risk of adverse events, genotype-specific dosage, drug mechanisms of action, and polymorphism drug target and disposition genes. Drug candidates have significantly varied safety, effectiveness, pharmacokinetics, and pharmacogenetic profiles. Drug makers can use premarket evaluation in early-phase clinical studies and recommendations for labeling to help them perform exploratory pharmacogenomic studies, enrichment techniques for clinical trials, adaptive trial designs, or companion diagnostics [78]. However, as a result, the potential dangers of immunogenicity will rise, necessitating the development of novel tools for risk assessment and reduction [79]. The introduction and implementation of innovative plasma-derived medicines, rise in the frequency of chronic and life-threatening illnesses, and increased public awareness of high efficacy are the primary drivers driving the worldwide protein therapeutics market. However, the market expansion is hampered by the high cost of protein treatments and the complicated reimbursement landscape. In contrast, corporates are expected to benefit from a rise in fresh indications for existing protein therapies and undiscovered developing markets.

11.11 Conclusions

Healthcare is undergoing dramatic changes in which disease management decisions are determined at the genomic and protein levels, supporting all biological and protein therapies. Protein therapies have remained in focus for decades in research and development. Over the last few decades, protein-based treatments have enabled novel, tailored strategies for treating various diseases. These approaches have had a significant impact on diagnostics and industry, and the demand to develop protein-based medicine continues to grow. Folding, consistency, efficiency, and PTMs of proteins intended as biopharmaceuticals have been scientifically proven to benefit from the genetic engineering of traditional recombinant proteins. However, the physiological constraints of the current protein production platforms provide little room for improving product quality and output and challenge to overcome regulatory restraints. Many innovative uses of protein therapies incorporate targeting and delivery fuelled by the precise binding characteristic and expanding on previous triumphs. Inflammatory diseases can be cured or prevented by using different protein therapies as they can reduce the clinical symptoms and also have the potential to avoid the relapse of the disease. Recently, repositories like THPdb have been developed that provide detailed information about US FDA-approved therapeutic peptides and their drug variants, along with the structure and sequence of the respective proteins. Proteins have emerged to be a notable therapeutic by garnering attention as an alternative source of traditional treatments and small molecules in drug discovery. The need for cost-effective treatments makes protein therapy an essential requirement in the future. Because transport and stability are two main concerns with protein and peptide treatments, solving these problems will inevitably lead to even more peptide and protein treatments with significant therapeutic promise.

References

1. Nicolaides N, Sass P, Grasso L (2010) Advances in targeted therapeutic agents. Expert Opin Drug Discovery 5(11):1123–1114
2. Vigneri R, Squatrito S, Sciacca L (2010) Insulin and its analogs: actions via insulin and IGF receptors. Acta Diabetol 47(4):271–278
3. Kiss Z, Elliott S, Jedynasty K, Tesar V, Szegedi J (2010) Discovery and basic pharmacology of erythropoiesis-stimulating agents (ESAs), including the hyperglycosylated ESA, darbepoetin alfa: an update of the rationale and clinical impact. Eur J Clin Pharmacol 66(4): 331–340
4. Walsh G (2010) Biopharmaceutical benchmarks 2010. Nat Biotechnol 28(9):917–924
5. Agarwal P, Rupenthal ID (2013) Injectable implants for the sustained release of protein and peptide drugs. Drug Discov Today 18(7–8):337–349
6. Ovchinnikova L, Terekhov S, Ziganshin R, Bagrov D, Filimonova I, Zalevsky A, Lomakin Y (2021) Reprogramming extracellular vesicles for protein therapeutics delivery. Pharmaceutics 13(6):768
7. Kintzing J, Filsinger Interrante M, Cochran J (2016) Emerging strategies for developing next-generation protein therapeutics for cancer treatment. Trends Pharmacol Sci 37(12):993–1008
8. Liu Z, Ding Y, Ye N, Wild C, Chen H, Zhou J (2015) Direct activation of bax protein for cancer therapy. Med Res Rev 36(2):313–341
9. Teixeira F, Santos H, Howell S, Pimentel G (2019) Whey protein in cancer therapy: a narrative review. Pharmacol Res 144:245–256
10. Svensson M, Fast J, Mossberg AK, Düringer C, Gustafsson L, Hallgren O, Brooks CL, Berliner L, Linse S, Svanborg C (2003) Alpha-lactalbumin unfolding is not sufficient to cause apoptosis, but is required for the conversion to HAMLET (human alpha-lactalbumin made lethal to tumor cells). Protein Sci 12(12):2794–2804
11. Gao M, Zhou J, Su Z, Huang Y (2017) Bacterial cupredoxin azurin hijacks cellular signaling networks: protein-protein interactions and cancer therapy. Protein Sci 26(12):2334–2341
12. Altınoğlu S, Wang M, Li K, Li Y, Xu Q (2016) Intracellular delivery of the PTEN protein using cationic lipidoids for cancer therapy. Biomater Sci 4(12):1773–1780
13. La Manna S, Di Natale C, Florio D, Marasco D (2018) Peptides as therapeutic agents for inflammatory-related diseases. Int J Mol Sci 19(9):2714
14. Simeon R, Chen Z (2017) In vitro-engineered non-antibody protein therapeutics. Protein Cell 9(1):3–14
15. Kuemmerle-Deschner J (2015) CAPS—pathogenesis, presentation and treatment of an autoinflammatory disease. Semin Immunopathol 37(4):377–385
16. Shimizu F, Schaller K, Owens G, Cotleur A, Kellner D, Takeshita Y et al (2017) Glucose-regulated protein 78 autoantibody associates with blood-brain barrier disruption in neuromyelitis optica. Sci Transl Med 9(397):eaai9111
17. Usmani S, Bedi G, Samuel J, Singh S, Kalra S, Kumar P et al (2017) THPdb: database of FDA-approved peptide and protein therapeutics. PLoS One 12(7):e0181748
18. Mannucci PM (2020) Hemophilia therapy: the future has begun. Haematologica 105(3): 545–553
19. Kleveland O, Kunszt G, Bratlie M, Ueland T, Broch K, Holte E et al (2016) Effect of a single dose of the interleukin-6 receptor antagonist tocilizumab on inflammation and troponin T release in patients with non-ST-elevation myocardial infarction: a double-blind, randomized, placebo-controlled phase 2 trial. Eur Heart J 37(30):2406–2413
20. Sam R, Ryan E, Daykin E, Sidransky E (2021) Current and emerging pharmacotherapy for Gaucher disease in pediatric populations. Expert Opin Pharmacother 22(11):1489–1503
21. Shannon K, Fraint A (2015) Therapeutic advances in Huntington's disease. Mov Disord 30(11): 1539–1546
22. Ciociola AA, Cohen LB, Kulkarni P (2014) How drugs are developed and approved by the FDA: current process and future directions. Am J Gastroenterol 109(5):620–623

23. Carter P (2011) Introduction to current and future protein therapeutics: a protein engineering perspective. Exp Cell Res 317(9):1261–1269
24. Löfblom J, Feldwisch J, Tolmachev V, Carlsson J, Ståhl S, Frejd FY (2010) Affibody molecules: engineered proteins for therapeutic, diagnostic and biotechnological applications. FEBS Lett 584(12):2670–2680
25. Feldwisch J, Tolmachev V, Lendel C, Herne N, Sjöberg A, Larsson B, Rosik D, Lindqvist E, Fant G, Höidén-Guthenberg I, Galli J (2010) Design of an optimized scaffold for affibody molecules. J Mol Biol 398(2):232–247
26. Ståhl S, Gräslund T, Karlström AE et al (2014) Affibody molecules in biotechnological and medical applications. Trends Biotechnol 36(8):691–712
27. Frejd FY, Kim KT (2017) Affibody molecules as engineered protein drugs. Exp Mol Med 49(3):306–e306
28. Simeon R, Chen Z (2018) *In vitro*-engineered non-antibody protein therapeutics. Protein Cell 9(1):3–14
29. Gunneriusson E, Samuelson P, Ringdahl J, Grönlund H, Nygren PÅ, Ståhl S (1999) Staphylococcal surface display of immunoglobulin A (IgA)-and IgE-specific in vitro-selected binding proteins (affibodies) based on Staphylococcus aureus protein A. Appl Environ Microbiol 65(9): 4134–4140
30. Emanuel SL, Engle LJ, Chao G, Zhu RR, Cao C, Lin Z, Yamniuk AP, Hosbach J, Brown J, Fitzpatrick E, Gokemeijer J (2011) A fibronectin scaffold approach to bispecific inhibitors of epidermal growth factor receptor and insulin-like growth factor-I receptor. MAbs 3(1):38–48
31. Gross GG, Junge JA, Mora RJ, Kwon HB, Olson CA, Takahashi TT, Liman ER, Ellis-Davies GC, McGee AW, Sabatini BL, Roberts RW (2013) Recombinant probes for visualizing endogenous synaptic proteins in living neurons. Neuron 78(6):971–985
32. Chandler PG, Buckle AM (2020) Development and differentiation in monobodies based on the fibronectin type 3 domain. Cell 9(3):610
33. Lipovšek D (2011) Adnectins: engineered target-binding protein therapeutics. Protein Eng Des Sel 24(1–2):3–9
34. Bandaranayake AD, Almo SC (2014) Recent advances in mammalian protein production. FEBS Lett 588(2):253–260
35. Rothe C, Skerra A (2018) Anticalin® proteins as therapeutic agents in human diseases. BioDrugs 32(3):233–243
36. Sennhauser G, Grütter MG (2008) Chaperone-assisted crystallography with DARPins. Structure 16(10):1443–1453
37. Baneyx F (1999) Recombinant protein expression in Escherichia coli. Curr Opin Biotechnol 10(5):411–421
38. Rising A, Widhe M, Johansson J, Hedhammar M (2011) Spider silk proteins: recent advances in recombinant production, structure-function relationships and biomedical applications. Cell Mol Life Sci 68(2):169–184
39. Demain AL, Vaishnav P (2009) Production of recombinant proteins by microbes and higher organisms. Biotechnol Adv 27(3):297–306
40. Mitsui R, Yamada R (2021) Chapter 16 – Saccharomyces cerevisiae as a microbial cell factory. In: Singh V (ed) Microbial cell factories engineering for production of biomolecules. Academic Press, pp 319–333
41. Li Z, Rinas U (2020) Recombinant protein production associated growth inhibition results mainly from transcription and not from translation. Microb Cell Factories 19(1):83
42. Twyman RM, Stoger E, Christou P (2003) GENETIC MODIFICATION, APPLICATIONS | Molecular Farming. In: Thomas B (ed) Encyclopedia of applied plant sciences. Elsevier, Oxford, pp 436–442
43. Norton PA, Pachuk CJ (2003) Methods for DNA introduction into mammalian cells. In: New comprehensive biochemistry, vol 38. Elsevier, pp 265–277
44. Tripathi NK, Shrivastava A (2019) Recent developments in bioprocessing of recombinant proteins: expression hosts and process development. Front Bioeng Biotechnol 7:420

45. Lai T, Yang Y, Ng SK (2013) Advances in mammalian cell line development technologies for recombinant protein production. Pharmaceuticals 6(5):579–603
46. Rosano GL, Ceccarelli EA (2014) Recombinant protein expression in Escherichia coli: advances and challenges. Front Microbiol 5:172
47. Yearley JH, Gibson C, Yu N, Moon C, Murphy E, Juco J, Lunceford J, Cheng J, Chow LQM, Seiwert TY, Handa M, Tomassini JE, McClanahan T (2017) PD-L2 expression in human tumors: relevance to anti-PD-1 therapy in cancer. Clin Cancer Res 23(12):3158–3167
48. Jain NK, Sahni N, Kumru OS, Joshi SB, Volkin DB, Russell Middaugh C (2015) Formulation and stabilization of recombinant protein based virus-like particle vaccines. Adv Drug Deliv Rev 93:42–55
49. Roldao A, Mellado MC, Castilho LR, Carrondo MJ, Alves PM (2010) Virus-like particles in vaccine development. Expert Rev Vaccines 9:1149–1176
50. Keating GM, Noble S (2003) Recombinant hepatitis B vaccine (Engerix-B): a review of its immunogenicity and protective efficacy against hepatitis B. Drugs 63:1021–1051
51. Schuster J, Koulov A, Mahler H-C, Detampel P, Huwyler J, Singh S, Mathaes R (2020) In vivo stability of therapeutic proteins. Pharm Res 37(2):23
52. Geraldes DC, Beraldo-de-Araújo VL, Pardo BOP, Junior AP, Stephano MA, de Oliveira-Nascimento L (2020) Protein drug delivery: current dosage form profile and formulation strategies. J Drug Target 28(4):339–355
53. Sheffield WP (2001) Modification of clearance of therapeutic and potentially therapeutic proteins. Curr Drug Targets Cardiovasc Haematol Disord 1(1):1–22
54. Muzykantov VR (2010) Drug delivery by red blood cells: vascular carriers designed by mother nature. Expert Opin Drug Deliv 7(4):403–427
55. Kraynov E, Martin SW, Hurst S, Fahmi OA, Dowty M, Cronenberger C, Loi C-M et al (2011) How current understanding of clearance mechanisms and pharmacodynamics of therapeutic proteins can be applied for evaluation of their drug-drug interaction potential. Drug Metab Dispos 39(10):1779–1783
56. Zaman R, Islam RA, Ibnat N, Othman I, Zaini A, Lee CY, Chowdhury EH (2019) Current strategies in extending half-lives of therapeutic proteins. J Control Release 301:176–189
57. Binder U, Skerra A (2012) Half-life extension of therapeutic proteins via genetic fusion to recombinant PEG mimetics. In: Therapeutic proteins: strategies to modulate their plasma half-lives. Wiley, pp 63–80
58. Lagassé HAD, Alexaki A, Simhadri VL, Katagiri NH, Jankowski W, Sauna ZE, Kimchi-Sarfaty C (2017) Recent advances in (therapeutic protein) drug development. F1000Research 6 (February):113
59. Atsmon J, Kate-Ilovitz E, Shaikevich D, Singer Y, Volokhov I, Haim KY, Ben-Yedidia T (2012) Safety and immunogenicity of multimeric-001--a novel universal influenza vaccine. J Clin Immunol 32(3):595–603
60. Baker MP, Reynolds HM, Lumicisi B, Bryson CJ (2010) Immunogenicity of protein therapeutics: the key causes, consequences and challenges. Self/nonself 1(4):314–322
61. Frokjaer S, Otzen DE (2005) Protein drug stability: a formulation challenge. Nat Rev Drug Discov 4(4):298–306
62. Morishita M, Peppas NA (2006) Is the oral route possible for peptide and protein drug delivery? Drug Discov Today 11(19–20):905–910
63. Ishida T (2010) Effects of point mutation on enzymatic activity: correlation between protein electronic structure and motion in chorismate mutase reaction. J Am Chem Soc 132(20): 7104–7118
64. Webster CI, Burrell M, Olsson LL, Fowler SB, Digby S, Sandercock A, Snijder A, Tebbe J, Haupts U, Grudzinska J, Jermutus L (2014) Engineering neprilysin activity and specificity to create a novel therapeutic for Alzheimer's disease. PLoS One 9(8):e104001
65. Veronese FM, Pasut G (2005) PEGylation, successful approach to drug delivery. Drug Discov Today 10(21):1451–1458

66. Kim TH, Jiang HH, Lim SM, Youn YS, Choi KY, Lee S, Chen X, Byun Y, Lee KC (2012) Site-specific PEGylated Exendin-4 modified with a high molecular weight trimeric PEG reduces steric hindrance and increases type 2 antidiabetic therapeutic effects. Bioconjug Chem 23(11): 2214–2220
67. Bagshawe KD (2006) Antibody-directed enzyme prodrug therapy (ADEPT) for cancer. Expert Rev Anticancer Ther 6(10):1421–1431
68. Schellenberger V, Wang CW, Geething NC, Spink BJ, Campbell A, To, W, Scholle MD, Yin Y, Yao Y, Bogin O, Cleland JL (2009) A recombinant polypeptide extends the in vivo half-life of peptides and proteins in a tunable manner. Nat Biotechnol 27(12):1186–1190
69. Andersen JT, Dalhus B, Viuff D, Ravn BT, Gunnarsen KS, Plumridge A, Bunting K, Antunes F, Williamson R, Athwal S, Allan E (2014) Extending serum half-life of albumin by engineering neonatal Fc receptor (FcRn) binding. J Biol Chem 289(19):13492–13502
70. Park K (2012) Albumin: a versatile carrier for drug delivery. J Control Release 1(157):3
71. Ogasawara Y, Namai T, Togawa T, Ishii K (2006) Formation of albumin dimers induced by exposure to peroxides in human plasma: a possible biomarker for oxidative stress. Biochem Biophys Res Commun 340(2):353–358
72. Patterson JT, Wilson HD, Asano S, Nilchan N, Fuller RP, Roush WR, Rader C, Barbas CF III (2016) Human serum albumin domain I fusion protein for antibody conjugation. Bioconjug Chem 27(10):2271–2275
73. Zhao S, Zhang Y, Tian H, Chen X, Cai D, Yao W, Gao X (2013) Extending the serum half-life of G-CSF via fusion with the domain III of human serum albumin. Biomed Res Int 2013: 107238
74. Niven RW, Lott FD, Ip AY, Cribbs JM (1994) Pulmonary delivery of powders and solutions containing recombinant human granulocyte colony-stimulating factor (rhG-CSF) to the rabbit. Pharm Res 11(8):1101–1109
75. Kim BYS, Rutka JT, Chan WCW (2010) Nanomedicine. N Engl J Med 363(25):2434–2443
76. Cavadas M, González-Fernández Á, Franco R (2011) Pathogen-mimetic stealth nanocarriers for drug delivery: a future possibility. Nanomedicine 7(6):30–743
77. Food and Drug Administration (2016) Guidance for industry: pharmacogenomic data submissions 2005. FDA
78. Food and Drug Administration (2016) Guidance for industry: clinical pharmacogenomics: premarket evaluation in early-phase clinical studies and recommendations for labeling 2013. FDA
79. Shankar G, Pendley C, Stein KE (2007) A risk-based bioanalytical strategy for the assessment of antibody immune responses against biological drugs. Nat Biotechnol 25(5):555–561

Biosimilar, Biobetter, and Biosuperior Therapeutic Proteins

12

Abhishek Pathak, Satya Pal Singh, Apoorv Tiwari, Atul Mani Tripathi, Tarannum Jahan, and Dev Bukhsh Singh

Abstract

Therapeutic proteins are therapeutic agents that help patients who desperately need new treatments. Recently, recombinant proteins have been produced and licensed for therapeutic use against a wide range of clinical conditions, including the treatment of autoimmune diseases/inflammation, malignancies, infection-related illnesses, hereditary diseases, etc. Biosimilars, biobetters, and biologics can benefit from the patent expiration of drugs or biopharmaceuticals. Biobetters, for instance, are novel medications derived from current drugs and have superior qualities such as increased stability, selectivity, half-life, reduced immunogenicity, and toxicity. Glycosylation is a common modification of proteins, while PEGylation is the favoured option in most clinical trials. Currently, protein-based treatments have taken center stage in drug research and development, but several challenges must be addressed, including safety, immunogenicity, protein stability, and degradation. We discuss the emerging trends and approaches in different therapeutic protein drug development with the conceivable elements and issues confronted during the development.

A. Pathak (✉) · S. P. Singh (✉) · T. Jahan
Department of Veterinary Pharmacology and Toxicology, G. B. Pant University of Agriculture and Technology, Pantnagar, Uttarakhand, India

A. Tiwari
Department of Computational Biology and Bioinformatics, Jacob School of Biotechnology and Bio-Engineering, Sam Higginbottom University of Agriculture, Technology and Sciences, Prayagraj, Uttar Pradesh, India

A. M. Tripathi
Pt. Ram Charitra Mishra P. G. College, Kadipur, Sultanpur, Uttar Pradesh, India

D. B. Singh
Department of Biotechnology, Siddharth University, Kapilvastu, Siddharth Nagar, Uttar Pradesh, India

D. B. Singh, T. Tripathi (eds.), *Protein-based Therapeutics*,
https://doi.org/10.1007/978-981-19-8249-1_12

Keywords

Therapeutic proteins · Biologics · Biosimilar · Biobetter · Biosuperior · Recombinant DNA · Protein-based therapeutics · Pegylation · Antibody-drug conjugates · Protein production technology

12.1 Introduction

The global burden of chronic diseases continues to rise, making it critical for patients to access safe and effective treatment. In the field of medicine, biologics are a powerful family of pharmaceuticals that have transformed the way doctors treat diseases such as diabetes, autoimmune disorders, cancer, and other ailments. The positive results for patients have been astounding. Currently, over 200 biologics and vaccines are available in the market, with therapeutic proteins accounting for the vast bulk of these medicines. Therapeutic proteins are proteins produced in the laboratory for medicinal application. Several proteins that are absent or produced in low quantities in patients suffering from hepatitis B/C, hemophilia, infectious disease, anemia, cancer, sclerosis, and other diseases are artificially manufactured on a massive scale using genetically engineered cells. Protein therapy is a therapeutic strategy that uses proteins and peptides to cure ailments. The concept of protein therapy is similar to gene therapy; however, in contrast to gene therapy, protein therapy provides the body with specific amounts of protein to aid in repairing illness, treating pain, and restoring structures.

Human insulin, introduced in the 1920s, is widely regarded as the first medicinal therapeutic protein. It is probably not possible for the pharmaceutical sector to function without using proteins and peptides [1]. Recent advancements in pharmaceutical biotechnology have enhanced the market value and availability of protein- and peptide-based therapeutics. Over 100 proteins have been approved by the US Food and Drug Administration (FDA) to be used in medicine, and many more are in clinical trials. The FDA has approved several naturally occurring/recombinant cytokines, biotechnological pharmaceuticals, vaccines, antibody-drug conjugates, enzymes, and interferons as recombinant proteins [2]. In addition, these recombinant therapeutic proteins can also be used to diagnose, prevent, and manage and treat diseases. Furthermore, protein therapies account for one-third of all novel biologic drugs introduced to the market [3]. Various protein therapeutics have been produced using recombinant DNA (rDNA) technology and are now readily available as treatments for treating chronic disorders at reasonable costs [4]. Therapeutic proteins are becoming increasingly popular, since they have been shown to be beneficial in treating a wide range of potentially fatal conditions such as heart disease, diabetes, and cancer. Interestingly, proteins have now been shown to be efficient as vaccines, as they aid in the stimulation of the body's natural defense mechanism, which is responsible for the immunogenic response [5].

Therapeutic proteins are validated in the pharmaceutical industry through cloning and expression of cDNA encoding heterologous proteins. However, a critical

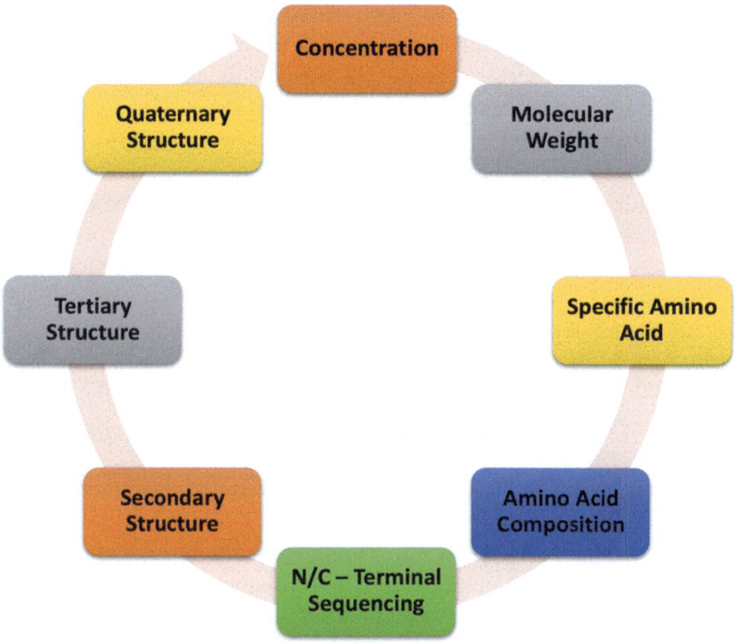

Fig. 12.1 Protein therapeutics analysis: an overview

formulation is needed for their safe and effective action. For this, a full grasp of the fundamentals for formulating and distributing protein-based therapeutic agents is required. Although therapeutic proteins have risen to the forefront of drug development and discovery, there are still several obstacles that must be addressed in the future. Some protein-based treatments cause immunological responses and other undesirable side effects. However, in recent years, researchers have paid close attention to therapeutic proteins, resulting in substantial advancements in the manufacture and application of therapeutic proteins (Fig. 12.1) [6].

12.2 A Brief History of Therapeutic Proteins

Proteins, such as wheat gluten, serum albumin, fibrin, and albumin, were identified as biological entities with specific characteristics in the eighteenth century, owing to their capacity to coagulate when exposed to heat or acid. Jons Jakob Berzelius introduced the term "protein" to characterize these molecules in 1838. Although it is possible to point out which class of biomolecules are the most crucial for life, proteins are perhaps the most critical biologicals in terms of therapeutic value. In the European Union (EU) and the United States, around 100 authentic, unmodified therapeutic proteins have been licensed for clinical use, while we are still waiting for the first approved DNA-based therapy. When it comes to the production of therapeutic antibodies, there have been two major paradigm shifts [7]. Von Behring produced an antitoxin in the 1880s that didn't kill bacteria but neutralized their toxin and discovered diphtheria serum therapy. He received the Nobel Prize in Physiology or Medicine for his contributions that saved many lives. The second significant paradigm shift occurred in the 1970s with the development of hybridoma technology that allowed the production of an infinite number of monoclonal antibodies (mAbs) with preset specificity [8]. The development of chimeric and humanized mAbs was facilitated by applying various molecular biology techniques, primarily rDNA technology, and a better knowledge of antibody structure and function. It was only via phage display and other molecular-biology-based methods (such as those involving transgenic animals) that it was possible to create fully human antibodies, effectively capping off the paradigm shift that began in the 1970s and continued through the 1990s. The scientific revolution that occurred in the discovery of antibody therapies had far-reaching consequences for the field of protein therapeutics [9].

In 1922, patients with diabetes mellitus were treated with insulin, the first protein therapy used. This marked the beginning of a new age marked by the first paradigm shift. However, the price, supply, and immunogenicity of animal-derived insulin limited its application. The second paradigm shift happened in the 1970s when the first recombinant protein therapy, humulin (human insulin), was developed [10]. In the 1970s, the invention of rDNA technologies, polymerase chain reaction (PCR,), and other developments triggered research in molecular biology recombinant production of therapeutic proteins. Monoclonal antibodies were developed that are distinguished from those other protein therapeutics as they were produced using hybridoma technology using single-cell without rDNA or any other molecular biology approaches. In 1986, the first therapeutic mAb was discovered and approved for clinical use. Currently, therapeutic proteins are being incrementally improved in terms of safety, quality, efficacy, and cost, but there are no new ideas comparable to those that resulted in the development of recombinant proteins [11].

12.3 Classification of Therapeutic Proteins

Both proteins and peptides possess many beneficial qualities, yet their chemical and physical properties are distinct. Proteins are defined as long, linear chains of amino acids containing more than 100 amino acid residues; On the other hand, peptides are classified as smaller polymers with fewer than 50 amino acids [12]. Clinically useful therapeutic proteins have been classified by Leader based on the functions and therapeutic applications of the particular protein/peptide [2]. They classified them based on whether they had been certified by the FDA (Group I and II) or whether they had been examined in vivo (Group III) and in vitro (Group IV). Proteins used to cure metabolic diseases or neuroendocrine dysfunctions, like IL-1Ra for type 2 diabetes and/or diseases requiring clotting factor VIII, such as haemophilia A, have been classified in Group Ia (Fig. 12.2) [13]. Group Ib includes proteins that promote different hematological and immunological processes; for example, interferon-alpha

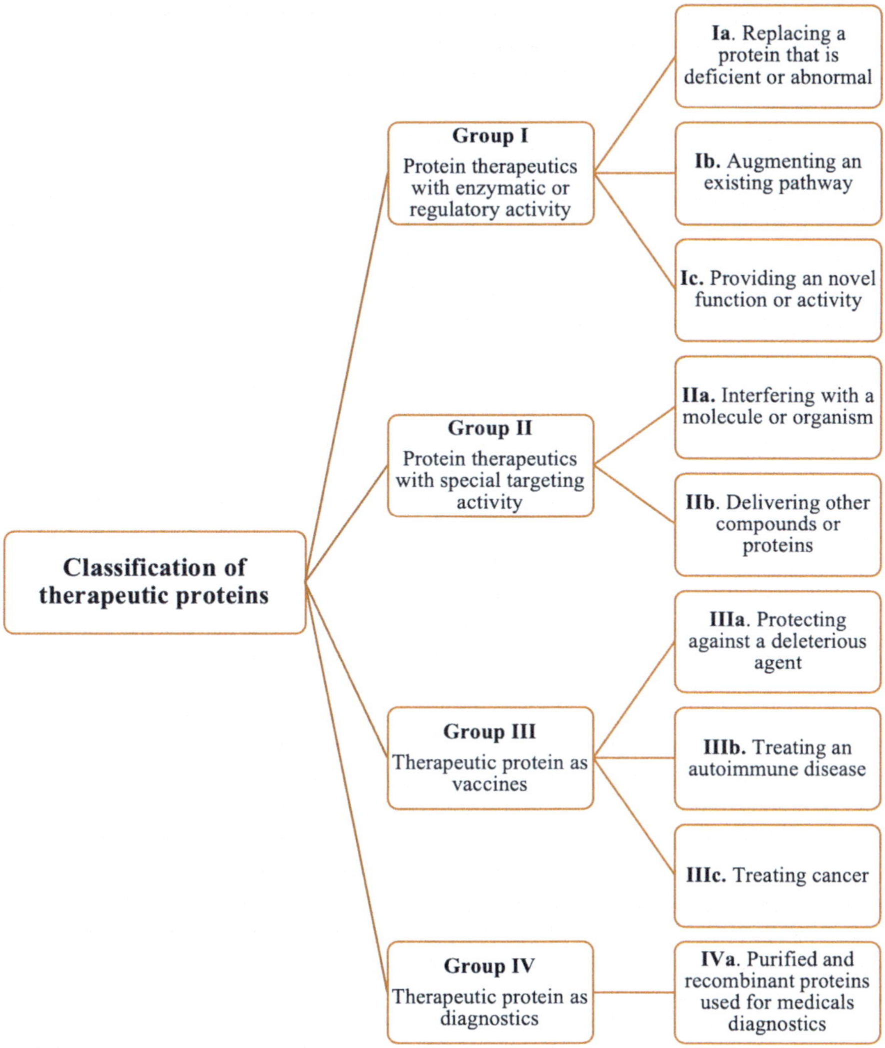

Fig. 12.2 Classification of therapeutic proteins

is used to treat hepatitis C, while erythropoietin is used to treat anaemia [14]. Group Ic proteins treat disease by altering its pathophysiology; for example, botulinum toxin subtypes A and B are used to treat multiple dystonias, and lepirudin is used to treat heparin-induced thrombocytopenia [15]. The proteins that have been categorized into Group IIa either enhance or obstruct the operation of molecules or organisms by directly interacting with their targets; for example, colorectal cancers are treated with cetuximab, a human mAb that interacts with an epidermal growth factor receptor [16] and anakinra (IL-1Ra), which is used to treat moderate-to-severe rheumatoid arthritis [17]. Compounds that can potentially distribute proteins in a targeted manner have been classified as Group IIb drugs, for example, denileukin diftitox and Ibritumomab tiuxetan used to treat persistent cutaneous T-cell lymphoma and transformed non-lymphoma Hodgkin's lymphoma, respectively [18]. Group III contains proteins that are employed in vaccines (both preventative and therapeutic) like antirhesus (Rh) immunoglobulin G (IgG) vaccine for immunization in Rh(D)-negative women and HBs Ag against hepatitis B infection [19]. Finally, proteins from Group IV are used for diagnostic applications, such as cancer imaging and infectious disease diagnostics [20]. To summarize, all proteins, particularly recombinant human proteins with FDA permission, occupy a unique niche among biomedical commodities with potential applications in every field involving biologics. In addition to eukaryotic-derived protein therapeutics, several prokaryotic-derived protein therapeutics have also been produced; for example, the most widely used therapeutic proteins in the treatment of many diseases include methionine gamma-lyase (a potential anticancer agent), L-asparaginase (a chemotherapeutic enzyme), and L-glutaminase (an antileukemia enzyme) [21].

12.4 Overview of Approved Therapeutic Proteins

Protein therapies recently authorized by the US FDA's Center for Biologics Evaluation and Research (CBER) and Center for Drug Evaluation and Research (CBER) are 62 recombinant therapeutic proteins (from 01 January 2011 to 31 August 2016) listed in the "Purple Book" of licensed biological products, including biosimilars and interchangeable biological products [21]. Monoclonal antibodies made up over half of the approved therapeutic proteins (48%), followed by coagulation factors (19%) and replacement enzymes (11%). Growth factors, hormones, fusion proteins, and plasma proteins received the remaining approvals (22%). Therapeutic proteins that have recently been approved help a vast spectrum of populations for their health benefits [11]. The FDA has authorized these therapeutic proteins for their use in a wide variety of medical disorders. Over half of the approved therapeutic proteins were indicated for oncology (26%) and hematology (29%), whereas the remaining 45% had primary indications in cardiology/vascular disease (5%), dermatology (3%), endocrinology (6%), gastroenterology (2%), genetic disease (2%), immunology (6%), infectious diseases (3%), musculoskeletal (8%), nephrology (2%), ophthalmology (3%), pulmonary/respiratory disease (3%), and rheumatology (2%). A complete list of approved medications is provided in Table 12.1.

12.5 Scope of Therapeutic Proteins

Therapeutic proteins are proteins that are not present in sufficient quantities in the body to perform specific functions but are used in the form of a treatment procedure. For example, some studies have explored certain proteins in treating cardiovascular disease, particularly when capillaries or arteries become obstructed. The correct combinations of proteins could assist by forming new blood flow pathways. Some

Table 12.1 Protein pharmaceuticals approved by the US FDA (2011–2016)[a]

S. no.	Drug	Market name	Approval date (dd/mm/yyyy)	Description
CDER-approved therapeutic proteins (2011–2016[a])				
1.	Asparaginase erwinia chrysanthemi	Erwinazze	18/11/2011	Enzyme [*Asparaginase erwinia chrysanthemi*]
2.	Brentuximabvedotin	Adcetris	19/08/2011	Antibody-drug conjugate [mouse/Human chimeric anti CD30]
3.	Belimumab	Benlysta	09/03/2011	mAb [Human anti-B-cell-activating factor (BAFF)]
4.	Belatacept	Nulojix	15/06/2011	Fc fusion [CTLA-4 Fc-fusion]
5.	Ipilimumab	Yervoy	25/03/2011	mAb [Human anti-CTLA-4]
6.	Pertuzumab	Perjeta	08/06/2012	mAb [(HER2)]
7.	Tbo-filgrastim	Granix	29/08/2012	Growth factor [G-CSF]
8.	Glucarpidase	Voraxaze	17/01/2012	Enzyme [glucarpidase]
9.	Ocriplasmin	Jetrea	17/10/2012	Enzyme [ocriplasmin]
10.	Olimumab injection for IV use	Simponi Aria	18/07/2013	mAb [Human anti-TNFα]
11.	Ocriplasmin	Jetrea	17/10/2012	Enzyme [ocriplasmin]
12.	Idarucizumab	Praxbind	16/10/2015	Fab [Humanized antidabigatran]
13.	Aflibercept	Eylea	18/01/2011	Fc fusion [VEGFR Fc-fusion]
14.	Tocilizumab	Actemra	21/10/2013	mAb [Humanized anti-IL-6 receptor]
15.	Raxibacumab	Abthrax	14/12/2012	mAb [Human antianthrax protective antigen (PA)]
16.	Evolocumab	Repatha	27/08/2015	mAb (PCSK9)
17.	Asfotase-alfa	Strensiq	23/10/2015	Fc fusion/enzyme
18.	Taliglucerase alfa	Elelyso	01/05/2012	Enzyme [β-glucocerebrosidase]
19.	Ado-trastuzumab emtansine	Kadcyla	22/02/2013	Antibody-drug conjugate
20.	Ziv-aflibercept	Zaltrap	03/08/2012	Fc fusion [VEGFR Fc fusion]
21.	Obiltoxaximab	Anthim	18/03/2016	mAb [mouse/human chimeric anti-*Bacillus anthracis*]
22.	Siltuximab	Sylvant	23/04/2014	mAb [mouse/human chimeric anti-IL-6]
23.	Elosulfase alfa	Vimizim	14/02/2014	Enzyme [elosulfase alfa]
24.	Albiglutide	Tanzeum	15/04/2014	Albumin fusion/hormone [GLP-1 dimer.albumin fusion]
25.	Peginterferon beta-1	Plegridy	15/08/2014	Cytokine [PEGylated IFNb-1b]
26.	Vedolizumab	Entyvio	20/05/2014	mAb [Humanized.antiintegrin a4b7 (LPAM-1)]
27.	Obinutuzumab	Gazyva	01/11/2013	mAb [Humanized anti-CD20]
28.	Nivolumab	Opdivo	22/12/2014	mAb [Human anti-PD-1]
29.	Asfotase-alfa	Strensiq	23/10/2015	Fc fusion/enzyme

(continued)

Table 12.1 (continued)

S. no.	Drug	Market name	Approval date (dd/mm/yyyy)	Description
30.	Etanercept-szzs	Erelzi	30/08/2016	Fc fusion [TNFR Fc-fusion]
31.	Daclizumab	Zinbryta	27/05/2016	mAb [Humanized anti-CD25]
32.	Eslizumab	Cinqair	23/03/2016	mAb [Humanized anti-IL-5]
33.	Parathyroid hormone	Natpara	23/01/2015	Hormone [parathyroid hormone]
34.	Infliximab-dyyb	Inflectra	05/04/2016	mAb [mouse/Human chimeric anti-TNFα]
35.	Ramucirumab	Cyramza	21/04/2014	mAb [Human anti-VEGFR2 (KDR)]
36.	Elotuzumab	Empliciti	30/11/2015	mAb [Humanized anti CD319 (SLAMF7)]
37.	Metreleptin	Myalept	24/02/2014	Hormone [metreleptin]
38.	Pembrolizumab	Keytruda	04/09/2014	mAb [Humanized anti-PD-1]
39.	Dinutuximab	Unituxi	10/03/2015	mAb [mouse/human chimeric anti GD2]
40.	Ixekizumab	Taltz	22/03/2016	mAb [Humanized anti-IL-17a]
41.	Dulaglutide	Trulicity	18/09/2014	Fc fusion [Glucagon-like peptide-1 receptor agonist]
42.	Alirocumab	Praluent	24/07/2015	mAb [PCSK9]
43.	Filgrastim-sndz	Zarxio	06/03/2015	Growth factor [G-CSF]
44.	Secukinumab	Cosentyx	21/01/2015	mAb [Human anti-IL-17A]
45.	Blinatumomab	Blincyto	03/12/2014	Ab [mouse bispecific anti-CD19/anti-CD3]
46.	Antihemophilic factor	Afstyla	25/05/2016	Coagulation factor (hemophilia A); [recombinant factor VIII]
CBER-approved therapeutic proteins (2011–2016)				
47.	Antihemophilic factor	Kovaltry	16/03/2016	Coagulation factor (hemophilia A); (recombinant factor VIII full length)
48.	Coagulation factor IX	Idelvion	04/03/2016	Coagulation factor (hemophilia B); (recombinant factor IX albumin fusion)
49.	Antihemophilic factor, PEGylated	Adynovate	13/11/2015	Coagulation factor (hemophilia A); (recombinant factor VIII PEGylated)
50.	Von Willebrand factor	Vonvendi	08/02/2015	Plasma protein (recombinant VWF); Von Willebrand disease
51.	Antihemophilic factor	Nuwiq	04/09/2015	Coagulation factor (hemophilia A); (recombinant factor VIII)
52.	Coagulation factor IX	Ixinity	29/04/2015	Coagulation factor (hemophilia B); (recombinant factor IX)
53.	Antihemophilic factor porcine, B-domain truncated	Obizur	23/10/2014	Coagulation factor (hemophilia A); (recombinant factor VIII)

(continued)

Table 12.1 (continued)

S. no.	Drug	Market name	Approval date (dd/mm/yyyy)	Description
54.	C1 esterase inhibitor	Ruconest	16/07/2014	Plasma protein [recombinant C1 esterase inhibitor; hereditary angioedema]
55.	Antihemophilic factor, Fc fusion protein	Eloctate	06/06/2014	Fc fusion/coagulation factor (hemophilia A); (recombinant factor VIII Fcfusion)
56.	Coagulation factor IX, Fc fusion protein	Alprolix	28/03/2014	Fc fusion/coagulation factor (hemophilia B); (recombinant factor IXFc fusion)
57.	Coagulation factor XIII A-subunit	Tretten	23/02/2013	Coagulation factor (recombinant factor XIIIA subunit); congenital factor XIII deficiency
58.	Antihemophilic factor	Novoeight	15/10/2013	Coagulation factor (hemophilia A); [recombinant factor VIII]
59.	Coagulation factor IX	Rixubis	26/06/2013	Coagulation factor (hemophilia B); (recombinant factor IX)

[a] Comprehensive listing of all FDA-approved therapeutic proteins granted orphan designation upon original submission from 01 January 2011 to 31 August 2016. In addition, the class of protein and a brief description are included

experts believe that such protein therapy will potentially be so effective that it will prevent the necessity for complex surgeries such as bypass surgery.

Protein-based medicines have come a long way in the last two decades. Biological medications are currently frequently employed in disease prevention and therapy. However, protein-based treatments, such as peptides, recombinant proteins, mAbs, and vaccinations, are proving to be highly effective in clinical trials, and their promise is being appreciated on an incredible scale. Following are some of the benefits of protein therapeutics:

1. They can perform specific and complex functions that small-molecule medicines cannot do.
2. They have a low risk of interfering with normal biological processes and causing negative consequences.
3. Protein-based therapeutics are frequently well tolerated and lower the risk of triggering immunological responses.
4. There are no risks associated with gene therapy.
5. The clinical development time for the FDA approval is reduced to half.
6. Extensive patent protection. The quality of biopharmaceuticals must be assessed to ensure that protein treatments are accurately produced and changed and that old reference structures are maintained. Additionally, the biological activity or potency of the biological medication must be established before clinical trials.

7. Proteomics-based methods provide a platform for high-resolution analysis that enable the complete characterization of therapeutic proteins.

12.6 Biosimilar, Biobetter, and Biosuperior Therapeutic Proteins

12.6.1 Biosimilars

Biosimilar medications or simply biosimilars, are replicas of approved biologics whose patent duration has completed or expired [22]. Additionally, they are referred to as following entry biologics or follow-on biologics. To be considered a biosimilar, a drug must be safe, pure, and potent with no clinically significant differences from its reference biological product. Although the biosimilar and the reference drug typically have the same core structure and other molecular properties, the therapeutically inactive components of the two drugs may differ slightly in their composition. The misconception that "biosimilar are generic biologics" must be avoided. Generic drugs are exact reproductions of brand-name pharmaceuticals, contain the same active ingredient, and are identical to their brand-name equivalents in dosage form, safety, efficacy, route of administration, purity, performance attributes, and use. In other words, the brand name and the generic are both called "bioequivalent". Compared to the reference product, biosimilars are "quite comparable"; yet, they are allowed variances, because they are made from live organisms. Biosimilars have the potential to provide more cost-effective options for patients while also encouraging a competitive environment for future biologics development and marketing [23]. Reference products, originator products, and innovator products are commonly used terms to refer to the original branded and approved biologic products in the context of biosimilar drugs. Following the demonstration of pharmacological equivalence (i.e., the use of similar active ingredients) and bioequivalence (i.e., the use of comparable pharmacokinetics), conventional generics are considered therapeutically equivalent to a reference and do not require extensive clinical effectiveness and safety studies. As implied by the name "biogeneric", it is possible to simply identify and show that a biosimilar's active component is structurally and functionally identical to the active ingredient in the reference product. In the case of biosimilars, this is not the case at all. The active ingredient in a biopharmaceutical is a collection of large protein isoforms rather than a single molecular entity, as in the case of conventional small-molecule drugs. So, it is very rare for two products to have the same active ingredients, and there are no methods for determining biopharmaceutical equivalency [22]. Due to the complexity of biopharmaceuticals, it is challenging to prevent heterogeneity across batches from the same manufacturing process and between identical proteins from different producers [22]. Additionally, without clinical studies, it is difficult to prove the therapeutic equivalence between biosimilars and reference drugs [24].

The safety of biosimilars is also an important consideration. Regarding safety, the potential to elicit an immune response (immunogenicity) distinguishes biopharmaceuticals from traditional drugs [25]. The European Medicines Agency (EMA) has published documents outlining the criteria for biosimilar market clearance, which provide critical information [26]. In addition, the European Medicines Agency (EMA) produced a public assessment report for biopharmaceutical items, which summarizes the product's qualities and justifies why it should be authorized. The EMA proposed biosimilar analogues of recombinant human EPO (rHuEPO), recombinant somatropin, and filgrastim, among other drugs, for licensing in January 2008. A single somatropin product has been approved in the USA, but additional approvals of follow-on protein products will necessitate formal legal and regulatory procedures [27]. It is imperative that physicians are informed of the fundamental issues of the use of biosimilars. It is critical for the safety of patients that doctors are well versed in the potential differences between different medications.

12.6.2 Biobetters

G.V. Prasad, executive chairman of Dr. Reddy's Laboratory in India [28], introduced the term "biobetter" in 2007. It relates to a biological therapy that excels or outperforms the parent molecule in at least one aspect while still achieving the therapeutic goal. This can be achieved by altering the amino acid sequence, protein folding, altering a chemical compound, altering humanization, or implementing a more effective purification approach. These modifications can impact the pharmacokinetics of biological drugs, such as increasing the half-life or decreasing the clearance. If one is looking for a recombinant protein drug that is better than the original, they are looking for a "biobetter", sometimes known as "biosuperior". Biobetters are better than their competitors at one or more product features. Biobetter applications look for characteristics such as an extended biological half-life, reduced aggregation risk, better efficacy, purity, or lower undesirable effects. In contrast to creating an entirely new biologic class, biobetters build on the accomplishments of previously approved biologics and are therefore considered to carry a reduced commercial risk [29]. Biobetters and Biosuperiors are upgraded versions of original biological products that are safer, more effective, or have a more convenient dose regimen. A Biobetter is a biological that achieves the same aim as the conventional biological but does so for a prolonged duration and with minimal side effects, sometimes at lower doses than the original biological [28].

12.6.3 Biosuperiors

Rather than being a carbon clone, a biosuperior is meant to have features superior to the first-generation product. The overall pipeline portfolio risk of a biosuperior is higher than that of a biosimilar but lower than that of an innovator product, because the goal is to demonstrate an enhancement over the existing treatment based on previously proven targets. Biobetters are approved when a thorough biologic license application (BLA) is completed and belongs to the same pharmacological class as the original drug. They allow pharmaceutical firms to concentrate their efforts on a mechanism of action that has been clinically and commercially proven beneficial. Improved impact on the target, low dose for effectiveness, and reduced unwanted effects are some modifications that have been made to the medicine. These breakthroughs have been made possible by technological advancements such as glycosylation, chemical modification, protein fusion, altered amino acid sequence, humanization, PEGylation, etc. As biobetters are modified to improve upon reference products, they represent novel molecular entities, and alterations in molecular structure can dramatically affect the mechanism of action. "Biobetters" and "Biosuperiors" are two new types of follow-on biologics that go beyond "similar". They make changes in chemistry, change the formulation, and use new delivery methods to make the original biologic better. Like licensed biologics, these drugs have distinct advantages, such as a longer duration of therapeutic efficacy and less severe adverse event profiles [30]. Current biosimilar products are developed and approved through the traditional 301(a) Biologics License Application (BLA) pathway for biologics and are required to demonstrate efficacy and safety without the need for comparability studies to prove their similarity to the originator molecule, thereby relieving the need to conduct large Phase III comparative trials. Of course, the biosuperior product's effectiveness and safety data must show a benefit/risk ratio compared to the innovator product, but not a head-to-head comparison. The findings achieved by the innovator in their existing product description will be compared to the biosuperior developer's outcomes. As a result, the work required for approval of a biosuperior would be more similar to the 505(b) new drug approval (NDA) regulatory pathway for "improved" approved drugs, which would leverage the FDA's knowledge of previously approved innovator products, rather than the 505 (j) NDA regulatory pathway for generic drugs, which would be based on the expectation of interchangeability (Fig. 12.3) [31].

12.7 Similarities and Differences Between Biosimilar, Biobetter, and Biosuperior Therapeutic Proteins

Both biosimilars and biobetters are modifications of an original biologic molecule. Biosimilars are almost identical replicas of biotherapeutics that have already been approved for use. In contrast to small-molecule generics, where structural identity is required, the term biosimilar implies that there may be some variances from the reference molecule. In some cases, the biosimilar seems so similar to the originator that it is referred to as "interchangeable". Biosimilars should fulfill the same safety, purity, and efficacy standards as the original molecule [32]. On the other hand, a biobetter has increased bioavailability or lower side effects than the parent molecule. In contrast to the original product, biobetters have a distinct active ingredient. Biobetters can be copyrighted [33]; however, since the active substance is highly similar to a reference product, patenting all biobetters may be difficult. This is because patents can only be issued to things that demonstrate a significant improvement over previously available technology. Many biobetters may be unable to obtain

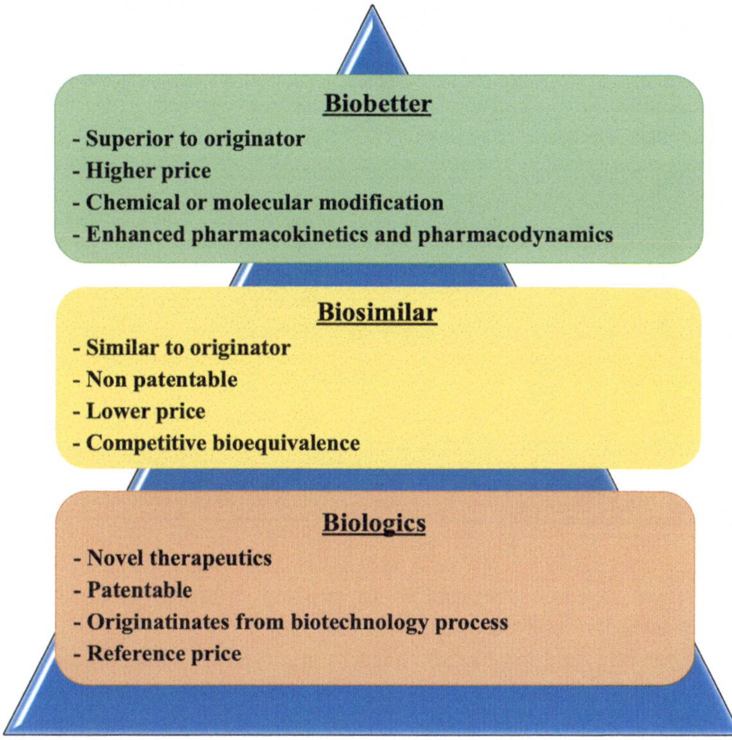

Fig. 12.3 Biological reference drugs, biosimilars, and biobetters

patent protection. Regulatory authorities in the EU and the USA can stimulate the development of biobetters rather than biosimilars, even if patent protection is not provided. Instead of the shorter process used for biosimilars or generic pharmaceuticals, biobetters would follow the traditional scientific approval route. As novel medications, biobetters could profit from market exclusivity rights, even if they are not sufficiently distinct to warrant patent protection (Table 12.2) [32].

12.8 Advantages of Biobetters over Biosimilars

Biobetters have several advantages over biosimilars; a few of them are listed below:

1. Biologics are costly pharmaceuticals, and their high costs strain the public health system's budget and potentially limit patients' access to these increasingly vital drugs [34]. The emergence of biosimilar pharmaceuticals has lowered the average purchasing cost of biologic treatments, making them more accessible [35]. Biosimilars still face challenges, such as regulatory restrictions and intellectual property rights held by third parties [36]. The advancements in

Table 12.2 Differences between biosimilar and biobetter therapeutic proteins

S. no.	Biosimilar	Biobetter
1.	Biosimilars and innovator products are quite similar.	They are a type of innovative product that has been changed.
2.	Biosimilars should also have similar effectiveness and safety characteristics.	Improved safety and/or effectiveness profiles are desirable in biobetters.
3.	Biosimilars are limited in composition, as they must contain an active component equivalent to that of reference products.	Biobetters are structurally unrestricted and may incorporate molecular/chemical alterations; hence, they are deemed to have a unique "active compound" than the original product.
4.	Patent rights or information exclusivity do not apply to biosimilars.	Biobetters may be granted patent or database exclusivity depending on their inventiveness.
5.	Biosimilars can be authorized after showing that the biosimilar and original product are similar, which is done through equivalence data.	Biobetters are novel medications that must go through a fresh updated product application or a composite product application that includes all clinical and nonclinical trial data.

pharmaceutical research are applied to existing products to developing new biopharmaceuticals, referred to as biobetters [36].

2. Biobetters can alleviate some of the hazards involved with the commercialization of biosimilars. Since biobetters, unlike biosimilars, are expected to be more efficient and effective than reference products, they can take advantage of existing tactics for circumventing patent and regulatory hurdles, avoiding the need to wait for third-party patents to expire [37].
3. Biosimilars are expected to have the same potency as the original drug at a lower cost. They are designed to be an upgraded version of the original with structural or chemical changes and lower side-effect profiles. Due to a verified biologic target, biobetters are likely to have a better chance of success than original biologics, although an improved biologic is far from certain and may necessitate significant testing. Biobetters are generated by exploring the protein folding process and how it influences medicine, while biosimilars are identical replicas of the source [35].
4. Biobetters are based on a well-known target concept, which means they have cheaper R&D expenses in the early stages.
5. Unlike biosimilars, biobetters do not have a distinct regulatory pathway. A biobetter is a medication already established to be a therapeutic and financial success, lowering the risk of failure compared to any new therapeutics. A biobetter does not have to wait for the originator product's patent to expire before launching the product to the market. Because biobetter does not claim similarities to the originating product, it has a better chance of avoiding patent infringement or, at the very least, lowering litigation expenses [36].
6. Although biobetters need discovery and an original Biologics License Application (BLA) with all preclinical and clinical data in order to be approved for

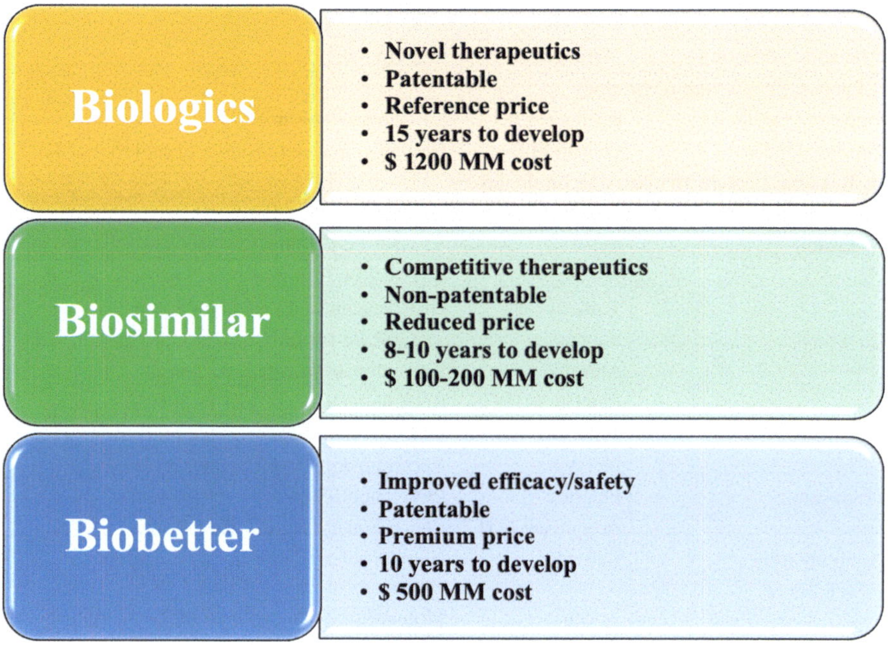

Fig. 12.4 Biobetters: advantages over biosimilars

marketing, they also provide several benefits. Biobetters have a competitive edge over biosimilars, since they offer a better or longer-acting drug.

7. Biobetter drugs can be used instead of biosimilars to improve the quality, safety, and innovation of health-care products while lowering prices, increasing access to biologic drugs, and overcoming market barriers such as third-party IP rights, competitor products, and doctor and patient acceptance (Fig. 12.4) [37].

12.9 Technologies to Produce Biobetter and Biosuperior Therapeutic Proteins

The development and manufacture of biological therapeutic products employing genetically altered bacteria, yeast, fungi, cells, or even animals and plants is a significant industrial application of biotechnology. In some cases, these biopharmaceuticals were originally derived from human tissues and fluids, often extracted in relatively modest volumes. Recombinant DNA technology made it possible to produce large amounts of highly purified products, such as proteins that had been PEGylated (treated with polyethylene glycol to make them more stable), posttranslationally modified, or with modified DNA sequences. Biological processes have been tinkered with or altered with the aid of biotechnology. The

Table 12.3 Protein-engineering platform and technologies

Production technology	Market name	Description
I. Protein production technology		
Transgenic animals	Ruconest [38]	C1 esterase inhibitor
Transgenic plant	Elelyso [39]	Human glucocerebrosidase (carrot root cells)
II. Rational protein structure technologies		
Glyco-engineering	Gazyva [40]	Humanized anti-CD20 mAb
Fc-Fusion	Eylea	VEGFR Fc-fusion
	Nulojix	CTLA-4 Fc-fusion
	Trulicity	GLP-1 receptor agonist Fc-fusion
	Zaltrap	VEGFR Fc-fusion
	Alprolix	Recombinant factor IX Fc fusion
	Eloctate [41]	Recombinant factor VIII Fc-fusion
Albumin fusion	Tanzeum	GLP-1 receptor agonist-albumin fusion
	Idelvion	Recombinant factor IX albumin fusion [42]
PEGylation	Plegridy	PEGylated IFNβ-1a
	Adynovate [43]	Recombinant factor VIII PEGylated
Antibody-drug conjugates	Kadcyla	Humanized anti-HER2/neu conjugated to emtansine
	Adcetris	Mouse/human chimeric anti-CD30 [44]
Mab humanization/ chimerism	Perjeta	Antihuman epidermal growth factor receptor 2 (HER2)
	Entyvio	Anti-HER2/neu conjugated antiintegrin a4b7 (LPAM-1)
	Actemra	Anti-IL-6 receptor
	Gazyva	Anti-CD20
	Keytruda	Anti-PD-1
	Cinqair	Anti-IL-5
	Zinbryta [45]	Anti-CD25

techniques listed in Table 12.3 are used to modify the original biological molecule to improve one or more of its attributes.

12.9.1 Protein Production Technology

12.9.1.1 Transgenic Animals

The need for safe and efficacious therapeutic proteins intensified research and resulted in the development of transgenic animals as potential bioreactors. Due to certain safety issues, the human-origin plasma protein inspired the development of recombinant mammalian cells. However, the constraints in development and production capacity for the culture of mammalian cells encouraged the use of livestock as transgenic animals as a source of therapeutic proteins [46]. Transgenic animals are preferred as they have the most common properties of animal cells in culture. Animal cells are in an ideal metabolic situation and can efficiently synthesize proteins. Biological fluids like the blood of transgenic animals are an excellent source to extract recombinant proteins [47].

12.9.1.2 Transgenic Plants

Humans have been using plants to prevent and cure diseases since prehistoric times, but it is in the past three decades that the genome of plants has been manipulated to produce therapeutic molecules. Plants as bioreactors are very economical as they can be produced on a large scale. Transgenic plants are preferred over transgenic animals or recombinant microbes, because the plant proteins are safer as plants are not human pathogens, the cost of production is low, plants have reduced capitalization cost, and the plants can produce multimeric proteins in the correct assembly form [48]. Human serum albumin was the first therapeutic protein produced in potato and tobacco leaves in 1990 [49].

12.9.2 Rational Protein Structure Technologies

12.9.2.1 Glycosylation

Glycosylation is the process of enzymatically linking sugar moieties, usually oligosaccharides, onto a protein. It happens naturally in cells as part of protein posttranslational alteration. The insertion of glycans to the original molecule adds extra sites, giving stability by limiting aggregation, degradation, or denaturation and contributing to a longer half-life for therapeutic proteins [50]. For example, Amgen's ARANESP® (Darbepoetin-alfa) was developed utilizing this technology that resulted in a longer half-life of the therapeutic protein. Glycosylation also reduced the frequency of erythropoietin administration to once every 2 weeks in clinical practice [51].

12.9.2.2 PEGylation

PEG (polyethylene glycol) is a hydrophilic polyether. PEGylation is the process of attaching a PEG molecule to a protein covalently. Because the hydrophilic characteristic of PEG polymers causes an increase in the molecular size of the protein, renal filtration is reduced because of the pore size in the nephrons. As a result, the half-life become longer. Furthermore, the PEGylation acts as a big mask to protect the protein from proteolytic destruction. PEGylation technology is often used to modify a variety of products, including PEGINTRON® (PEGylated interferon), PEGASYS® (PEGylated interferon), NEULASTA® (PEGylated granulocyte colony-stimulating factor), and MIRCERA® (PEGylated granulocyte colony-stimulating factor) and MIRCERA® (PEGylated granulocyte colony-stimulating factor; methoxy-polyethylene glycol-epoetin beta) (Fig. 12.5) [52].

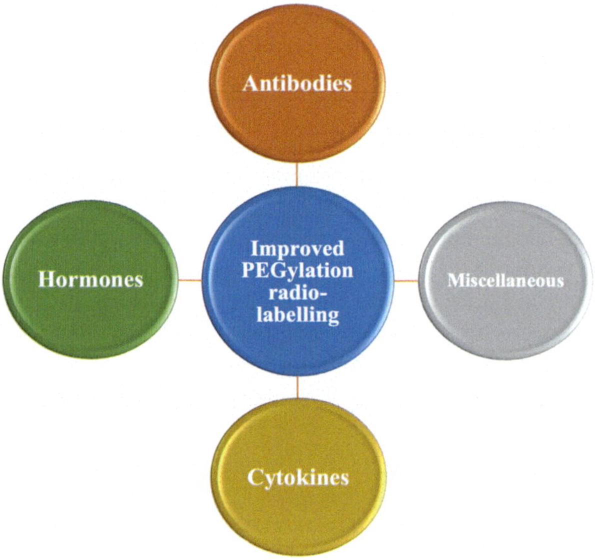

Fig. 12.5 Different approaches used for designer proteins

12.9.3 Modification of Therapeutic Proteins Via Fusion

12.9.3.1 Fusion with the Fc Domain of Immunoglobulins

During fusion, a recombinant protein is linked to a companion protein with a lengthy half-life, enhancing its pharmacokinetics. Albumin and the Fc portion of immunoglobulins are two examples of naturally occurring partner proteins used for fusion. ELOCTATE™ is a recombinant factor VIII fusion protein (BDDrFVIIIFc) that increases the half-life of the partner protein by 1.5–2 times [53].

Various techniques have been developed to extend the plasma half-life of the therapeutic proteins, like genetic modifications, sustained delivery, chemical modification, and fusion with carrier proteins. The Fc domain of immunoglobulins is directly linked to another peptide to create an Fc-Fusion protein. This Fc domain is fused with binding moiety to extend their plasma half-life. Biologically active proteins and peptides have a fast renal clearance, resulting in a short half-life. This decreases the therapeutic efficacy of the therapeutic protein drug as the time of exposure to the organ is less. Fc domain prolongs the half-life of Fc-Fusion protein because of the pH-dependent binding to neonatal Fc receptor. This prevents proteins from degradation by endosomes [9].

12.9.3.2 Albumin Fusion

Fusion with albumin also prolongs the plasma half-life of therapeutic proteins as the proteins confer albumin-like pharmacokinetic properties. Albumin also benefits the therapeutic protein by providing easier cellular penetration due to its plasma protein properties. However, the immunogenicity of albumin fusion proteins is low. Notably, they are easy to construct and purify as albumin can be easily obtained

[54]. Studies conducted on activated recombinant factor VII fusion protein with albumin found that it resulted in better pharmacokinetics. A long-acting insulin analog was developed using the albumin fusion technology [55].

12.9.3.3 Antibody-Drug Conjugates

Antibody-drug conjugates (ADCs) have two components, the recombinant mAb and cytotoxic chemicals, bound via synthetic linkers. The cytotoxic chemicals are known as warheads. These are high cytotoxic immune-conjugates that are antitumor, highly selective, stable, and have improved pharmacokinetic properties of mAbs. ADCs were developed for anticancer therapeutic proteins, and the warheads efficiently perform this activity. The linkers play an essential role, and the optimization of linkers is important. The linkers should be stable while circulating in the blood and should be able to release the drug once it is inside the target cancer cells. The concept of ADCs is not new. In the 1960s, first use of ADCs in laboratory animals was documented, and then in the 1980s, mouse IgG was used for ADC preparation. However, the major problem in developing ADCs is identifying and proving the accuracy of antigenic targets for the protein drugs [56].

12.9.4 Humanization

Nonhuman mAbs have a significant risk of causing immune-mediated adverse effects. As a result, chimeric mAbs have been created by substituting nonhuman Fc domains with human Fc regions. Humanized mAbs have also been generated by transforming major portions of the Fab regions into human counterparts. Although immunogenicity remains an issue, advances in transgenic mouse technology and the invention of phage display techniques have made fully human mAbs possible. GAZYVA® (Obinutuzumab) is a fully humanized mAb that binds to an epitope on CD20 cells and partially overlaps the epitope recognized by Rituximab, making it a biobetter to Rituximab. Another example is Golimumab, a human mAb used as an immunosuppressive drug and marketed under the brand name SIMPONI® (Janssen Biotech and Merck) [57].

12.9.5 Altering Amino Acid Sequences

In this process, attaching or modifying a peptide sequence (C-terminal peptide, CTP) to the original proteins stabilizes them and prolongs their shelf life without extra toxicity or loss of favored biological activity [58].

12.9.6 Sustained Release

Biobetters are occasionally created to minimize dose frequency by producing a new sustained-release formulation. Dong-A Pharmaceutical Co., Ltd. made DA-3091, a sustained-release, subcutaneous version of exenatide to treat type 2 diabetes mellitus [59].

12.9.7 New Routes of Administration

Oral, dermatological, topical, subcutaneous injections or inhaled formulations may be developed for easier administration. An excellent example is FLUMIST®(AstraZeneca), an intranasal influenza vaccine for easier administration than complicated injectable routes [60]. Also, AFREZZA® (Mankind), the only inhalable insulin to control adult diabetic patients' blood sugar during mealtime, is approved [61]. A new method involves changes such as the use of a new cell platform (the introduction of a newer cell culture method to replace the egg-based manufacturing system results in improved vaccine efficacy), the shift from live-attenuated or inactivated vaccines to recombinant products, and the use of a greater number of serotypes [28].

12.10 Delivery System of Therapeutic Proteins

Therapeutic proteins can be covalently modified with a peptide sequence capable of quickly translocating membranes, referred to as protein transduction domain (PTD) or cell-piercing peptide (CPP), using the techniques listed below:

1. Direct expression of recombinant fusion proteins derived from a vector carrying the DNA sequence of CPP.
2. For the chemical conjugation of CPP to the protein, a cleavable disulfide linkage in a reductive environment is utilized.
3. Synthetically simple, easier to characterize, less toxic, and more immunogenic than large polymers, peptide-based biomaterials are used for delivery. These biomaterials can also shield proteins from protease degradation and increase delivery efficiency.
4. Synthetic peptides have also been developed to protect proteins from protease degradation and develop a technique for improving efficiency, delivery, and noncovalent encapsulation. Such peptides may self-assemble quickly with protein cargos due to their amphipathic nature, presumably through noncovalent hydrophobic contacts.
5. Protein polymers are coupled with PEG and poly(*N*-isopropyl acrylamide, PNIPAAMm), the most prevalent polymers, for targeted delivery. When subjected to pH or temperature variations, these polymers vary their solubility or proclivity for self-assembly, imparting sensitivity to the protein to which they are linked. Because it can be functionalized with active esters, it can also be combined with protein amines [62].

12.11 Challenges in the Production of Therapeutic Proteins

Despite the tremendous technological advancements in the last two decades, there are multiple limitations during the production of therapeutic proteins. The major challenges are listed below:

1. Protein solubility, delivery method, distribution, and stability limit the proper implementation of a protein treatment.
2. Proteins are large macromolecules with both hydrophilic and hydrophobic properties, which make their entry into cells and subcellular compartments challenging. Additionally, proteases, protein-modifying compounds, and other clearance processes can substantially decrease the half-life of therapeutic proteins. PEGylated versions of therapeutic proteins are being developed to overcome these problems.
3. PEG-interferon is a modified version of interferon in which the polymer PEG is added to extend absorption, delay the enzymatic breakdown, reduce renal clearance, enhance the elimination half-life, and diminish interferon immunogenicity.
4. The body may produce an immunological reaction against the therapeutic protein, which is another significant hurdle. This immune response can sometimes neutralize the protein and, at times, induce an adverse reaction in the patient. As evidenced by the production of antifactor VIII antibodies (inhibitors) in individuals with severe hemophilia A, administrated with recombinant human factor VIII, immune responses can be developed against group Ia therapeutic proteins. However, advances in rDNA technologies and related fields have led to the production of several antibody products that are less likely to elicit an immunological response than unmodified murine antibodies. Humanized antibodies have portions of the antibody that are not critical for antigen-binding specificity and replaced with human Ig sequences to provide additional protein stability and biological activity but do not trigger an antiantibody response. Also, fully human antibodies can be made using transgenic animals.
5. Post-translational changes such as glycosylation, phosphorylation, and proteolytic cleavage are frequently necessary for a therapeutic protein to be physiologically active. These modifications require the use of certain cells capable of adequately expressing and modifying the protein. Furthermore, for large-scale production, recombinant proteins must be generated in genetically modified cell types.
6. In addition to producing physiologically active therapeutic protein, the host cells must also express it in a large quantity to fulfill the demand. The technology must also allow for the purification and long-term preservation of the protein in a therapeutically active state.
7. Developing systems in which entire cascades of genes involved in protein folding are induced along with the therapeutic protein could be a potential solution. The impetus for this work comes from observing that in plasma cells, which are natural protein production factories, such gene cascades produce large amounts of mAbs.

8. While bacteria and yeast are simple to grow, certain mammalian cell types can be challenging and expensive to maintain. Other manufacturing methods, such as genetically altered animals and plants, might offer a competitive edge.
9. Recombinant protein-producing transgenic sheep, cows, and goats have been developed, and transgenic hens that lay eggs containing recombinant proteins are expected in the future. Transgenic plants can create large amounts of protein without any waste or the requirement of bioreactors; for example, potatoes may be genetically modified to express recombinant proteins, allowing for the production of edible vaccinations.
10. The cost of protein-based treatments is a significant limitation. Switching from tedious purification of placentally derived protein to rDNA methods has enabled the manufacture of enough glucocerebrosidase to treat Gaucher's disease in many individuals.
11. Considering the hundreds of proteins produced by the human body and the thousands of proteins produced by other species, the future of protein therapy has a massive opportunity. However, as protein therapy's role in medicine develops in the future, we cannot overlook its problems and the associated ethical issues. The medical community must weigh the risks of protein treatment against the possible therapeutic advantages for patients. And as with every new treatment breakthrough, moral and ethical questions arise. Protein therapies that can improve conditions previously thought to be deviations from the normal may also question the notion of sickness or disease. With the prospect of utilizing growth hormone to improve a child's growth, the concept of small height may be rethought [63].

12.12 Next-Generation Biosimilar and Bio-Betters: Success and Difficulties

The success of protein-based therapeutics is due to the use of models and concepts developed during the second paradigm shift several decades ago, which led to dramatic advancements in three main characteristics of candidate therapies (efficacy, safety, and quality), all of which are required for regulatory approval. They are critical to the effectiveness of any treatment, but especially for antibody-based therapies, which confront similar challenges as other protein therapeutics.

12.12.1 Safety

The side effects of therapeutic proteins can be divided into two categories based on the interactions with desired targets and undesired targets. Binding the desired target might have unfavorable side effects, such as suppressory or stimulatory immunomodulatory antibodies. The use of suppressor therapeutic proteins might result in various adverse effects associated with immune system dysfunction: for example, the most popular antibody-based protein treatments targeting TNF (etanercept, infliximab, certolizumabpegol, and adalimumab) can cause infections [64]. The overstimulation of the immune system can potentially result in life-threatening diseases. Data indicates that a single dose of the anti-CD28 mAb (TGN1412) elicited an inflammatory reaction in all six volunteers, characterized by rapid activation of pro-inflammatory cytokines, leading to severe sickness after 12–16 h [65]. One crucial difference between antibody-based treatments incorporating Fc and other bioactive molecules such as proteins is that antibody effector actions such as ADCC and complement-dependent cytotoxicity (CDC) might induce toxicities in organs other than those targeted. When antibodies are administered simultaneously or consecutively with an anthracycline, the cardiotoxicity caused by trastuzumab increases [66].

Associations with undesired targets can have a wide variety of side effects, some of which are produced by incompletely understood mechanisms. Panitumumab, trastuzumab, cetuximab, rituximab, alemtuzumab and infliximab, insulin, and interferon are some examples of hazardous acute infusion events after protein delivery, where cytokine release plays a key role; however, unidentified pathways may also be implicated [67]. The discharge of cellular components from lysed malignant B cells might cause rituximab infusion adverse effects [68]. Protein supplementation can potentially cause hypersensitivity events such as anaphylactic shock and serum sickness [64]. Pre-existing IgEs that cross-react with protein therapies can increase the number and magnitude of these responses, which can happen even during the initial administration of the protein. The immunogenicity is usually linked to hypersensitivity, and the administration of cetuximab is a good illustration of this [67].

12.12.2 Immunogenicity

The discovery of fewer immunogenic proteins was crucial to the success of mAb-based treatments [64]. Murine mAbs were investigated as treatment options in the 1980s, but their strong immunogenicity leads to high antibody titer of human antimouse antibodies (HAMAs) and toxicities with poor efficacy. The successful treatment of the products required the development of less immunogenic chimeric mAbs with human Fc segments and humanized mAbs with mouse complements defining regions (CDRs) grafted into human antibody frameworks. Human antibodies are now the most popular form of antibodies in development, even though the majority of therapeutic antibodies authorized for clinical use are still chimeric and humanized mAbs.

The immunogenicity of a therapeutic protein is dependent on the structure, formulation, contaminants, proportions, post-translational alterations, heterogeneity, aggregation, degradation, storage conditions, characteristics of its interacting partner, route, disease condition, patient's immune response, concurrent treatments, time and dosing frequency, particularly when offered in multiple dosages over a significant duration of time [69]. Human proteins can trigger antihuman antibodies in humans. The antiepidermal growth factor receptor (EGFR) mAbs cetuximab and panitumumab have shown clinical activity against metastatic colorectal cancer when delivered alone or combined with chemotherapy [70]. In solid tumors, mAbs and other proteins have demonstrated significant clinical effects in a single therapy or combination [71]. The processes driving the limited efficacy of some therapeutic proteins and the diversity of treatment responses are unknown. However, numerous variables are likely involved in resistance, whether pre-existing or developing, which is a major issue for any therapy. EGFR-targeted treatments use a variety of strategies, including mutations, multidrug transport stimulation, and signaling protein overexpression or stimulation [72]. Poor penetration into tissues, such as solid tumors, is another challenge. Inadequate or missing binding to the surface of certain molecules (in the presence of steric barriers) is a relevant issue for full-size mAbs, as seen in the case of HIV envelope (Env)glycoprotein. Improvements in signaling pathways, half-life, tumor and tissue permeability, and reliability are among the new approaches being developed to improve the efficacy of mAbs and other therapeutic proteins. Here, the methods used include both protein and glycol-engineering, and so far, the results have been promising [73]. Various small designed antibody domains (~10-fold smaller than IgG) are being created to enable greater tissue penetration and concealed epitope access [74]. The present state of specific antibody treatments indicates that features of currently used therapeutic proteins are gradually improving, and the emergence of new proteins and targets is expected to continue in the coming years. One of the most challenging problems is dramatically improving therapeutic antibody efficacy and using them for a larger spectrum of illnesses. Creating successful tailored antibody-based treatments and predicting toxicity or potentially poor efficacy in vivo are key obstacles.

12.12.3 Quality

The FDA considers quality to be a critical factor during the approval of any drug. The heterogeneity of mAbs and other biologics is a crucial trait that separates them from small molecule medications. Alterations like N-terminal pyroglutamine cyclization, glycosylation, formation of partial disulfide bonds, C-terminal lysine preparation, isomerization, deamidation, oxidation, amidation of the C-terminal amino acids, and maleuric acid revision of the N-terminal amino acids, noncovalent connections with other molecules, conformational diversity, and aggregation, all contribute to the heterogeneity of therapeutic proteins. There might be tens of thousands of variations with the same sequence. For safety and FDA clearance, high-quality protein therapies with minimum heterogeneity and contamination are required. The process of manufacturing therapeutic proteins includes rDNA techniques, gene amplification, use of a potent expression system, classification of a stable host cell expression system, improvement and design of mammalian cell culture fermentation system, and innovation of an efficacious recovery process leading to greater yields and quality of products. To maintain steady long-term production and delivery of therapeutic proteins, three techniques can be applied in vivo: (1) implanting a capsule antibody that generates heterologous or autologous cells, (2) transplanting genetically manipulated autologous cells ex vivo, and (3) direct in vivo injection of gene-carrying integrating vectors. Another exciting potential is to use molecular techniques to manufacture low-cost therapeutic proteins in plants, such as in genetically altered tobacco leaves [75].

12.13 Concluding Remarks

In recent decades, much progress has been made to produce potent therapeutic proteins [76]. This raises several questions about where this field will go next. A significant concern is whether there are signs of a third paradigm shift that could result in radically new therapies, as occurred a few decades ago, resulting in a flood of clinically approved protein therapies. Meanwhile, incremental improvements in the attributes of current protein therapeutics, the exploration of novel protein-based drugs and targets, and the evolution of innovative protein-based therapeutic agents by conjugating therapeutics using holistic techniques based on molecular biology, biotechnology, and genetic engineering research. Therapeutic proteins have emerged as a crucial treatment procedure for a variety of disorders in the last few decades. Furthermore, in addition to the better therapeutic characteristics of protein therapeutics, the production and purification techniques have developed exponentially, helping the pharmaceutical industry to accelerate research in this area. However, several parameters must be considered during the manufacturing, purifying, and formulating protein therapeutics to improve their quality. Furthermore, a greater comprehension of their pharmacokinetic characteristics and the relationship between their pharmacokinetic and pharmacodynamic impacts will aid in developing and administering better protein-based therapeutics. Biotechnological advancements have boosted and accelerated the synthesis of therapeutically important proteins. There is a need to better understand the therapeutic protein delivery routes and their absorption mechanisms. Biobetters are on the verge of becoming the next generation of biologics with clearly visible benefits. Since they have a validated target and a reference molecule with verified safety and efficacy, their exploration should be fostered by relaxing the IP norms. They have the potential to alleviate the economic burden that these diseases impose on society due to their improved formulation and dosing schedule. There is an urgent need for biobetter drugs to improve the quality, safety, and innovation of health-care products while lowering prices. This will make biologic drugs more accessible by breaking down market barriers such as third-party IP rights and competitor products and the acceptance of doctors and patients.

Acknowledgments The authors acknowledge the Department of Pharmacology and Toxicology, College of Veterinary and Animal Sciences, GBPUA&T, Pantnagar, for providing a state-of-the-art facility for their research.

References

1. Craik DJ, Fairlie DP, Liras S, Price D (2013) The future of peptide-based drugs. Chem Biol Drug Des 81:136–147
2. Leader B, Baca QJ, Golan DE (2008) Protein therapeutics: a summary and pharmacological classification. Nat Rev Drug Discov 7:21–39
3. Tang L, Persky AM, Hochhaus G, Meibohm B (2004) Pharmacokinetic aspects of biotechnology products. J Pharm Sci 93:2184–2204
4. Karvar S (2014) The role of ABC transporters in anticancer drug transport. Turk J Biol 38:800–805
5. Karacali S, İzzetoğlu S, Deveci R (2014) Glycosylation changes leading to the increase in size on the common core of N-glycans, required enzymes and related cancer-associated proteins. Turk J Biol 38:754–771
6. Regan L, Jackson SE (2003) Engineering and design: protein design: theory and practice. Curr Opin Struct Biol 13:479–481
7. Dimitrov DS, Marks JD (2009) Therapeutic antibodies: current state and future trends—Is a paradigm change coming soon? Methods Mol Biol 525:1–27
8. Kohler G, Milstein C (1975) Continuous cultures of fused cells secreting antibody of predefined specificity. Nature 256:495–497
9. Beck A, Reichert JM (2011) Therapeutic Fc-fusion proteins and peptides as successful alternatives to antibodies. MAbs 3(5):415–416
10. Goeddel DV, Kleid DG, Bolivar F, Heyneker HL, Yansura DG, Crea R, Hirose T, Kraszewski A, Itakura K, Riggs AD (1979) Expression in Escherichia coli of chemically synthesized genes for human insulin. Proc Natl Acad Sci U S A 76:106–110
11. Dimitrov DS (2012) Therapeutic proteins. Methods Mol Biol 899:1–26
12. Sato AK, Viswanathan M, Kent RB, Wood CR (2006) Therapeutic peptides: technological advances driving peptides into development. Curr Opin Biotechnol 17:638–642
13. Akash MSH, Rehman K, Chen S (2013) IL-1Ra and its delivery strategies: inserting the association in perspective. Pharm Res 30:2951–2966
14. Van Zonneveld M, Honkoop P, Hansen BE, Niesters HG, Darwish Murad S, de Man RA, Schalm SW, Janssen HL (2004) Longterm follow-up of alpha-interferon treatment of patients with chronic hepatitis B. Hepatology 39:804–810
15. Eriksson BI, Wille-Jorgensen P, Kalebo P, Mouret P, Rosencher N, Bosch P, Baur M, Ekman S, Bach D, Lindbratt S, Close P (1997) A comparison of recombinant hirudin with a low-molecular weight heparin to prevent thromboembolic complications after total hip replacement. N Engl J Med 337:1329–1335
16. Saltz LB, Meropol NJ, Loehrer PJ Sr, Needle MN, Kopit J, Mayer RJ (2004) Phase II trial of cetuximab in patients with refractory colorectal cancer that expresses the epidermal growth factor receptor. J Clin Oncol 22:1201–1208
17. Tesser J, Fleischmann R, Dore R, Bennett R, Solinger A, Joh T, Modafferi D, Schechtman J (2004) Concomitant medication use in a large, international, multicenter, placebo controlled trial of anakinra, a recombinant interleukin 1 receptor antagonist, in patients with rheumatoid arthritis. J Rheumatol 31:649–654
18. Ohya S, Matsuda T (2005) Poly (N-isopropylacrylamide)(PNIPAM)-grafted gelatin as thermoresponsive three-dimensional artificial extracellular matrix: molecular and formulation parameters vs. cell proliferation potential. J Biomater Sci Polym 16:809–827
19. MacKenzie IZ, Bichler J, Mason GC, Lunan CB, Stewart P, Al-Azzawi F, De Bono M, Watson N, Andresen I (2004) Efficacy and safety of a new, chromatographically purified rhesus (D) immunoglobulin. Eur J Obstet Gynecol Reprod Biol 117:154–161
20. Campos-Neto A, Rodrigues-Junior V, Pedral-Sampaio DB, Netto EM, Ovendale PJ, Coler RN, Skeiky YA, Badaro R, Reed SG (2001) Evaluation of DPPD, a single recombinant Mycobacterium tuberculosis protein as an alternative antigen for the Mantoux test. Tuberculosis (Edinb) 81:353–358

21. Sharma B, Singh S, Kanwar SS (2014) L-methionase: a therapeutic enzyme to treat malignancies. Biomed Res Int 2014:506287
22. Crommelin DJ, Bermejo T, Bissig M et al (2005) Pharmaceutical evaluation of biosimilars: important differences from generic low-molecular weight pharm. Eur J Hosp Pharm Sci 1: 11–17
23. Roger SD (2006) Biosimilars: how similar or dissimilar are they? Nephrology (Carlton) 11: 341–346
24. Locatelli F, Roger S (2006) Comparative testing and pharmacovigilance of biosimilars. Nephrol Dial Transplant 21(5):13–16
25. Kessler M, Goldsmith D, Schellekens H (2006) Immunogenicity of biopharmaceuticals. Nephrol Dial Transplant 21(5):9–12
26. Wiecek A, Mikhail A (2006) European regulatory guidelines for biosimilars. Nephrol Dial Transplant 21(5):17–20
27. National Health Service (2008) European CHMP issues positive opinion on four biosimilar filgrastim products. Biosimilar therapeutics—what do we need to consider? Clin Kidney J. Oxford Academic (oup.com). Accessed 26 May 2022
28. Dinwoodie N (2006) Biobetters and the future biologics market. BioPharm Int 24(11):31–35. Available at: http://www.biopharminternational.com/biobetters-andfuture-biologics-market. Accessed 19 May 2022
29. Stone K (2016) Biobetter—definition and examples, 26 Feb 2016. About Money, US. Available at: http://ijdra.com/index.php/journal/article/view/177. Accessed 23 May 2022
30. Konski AF (2011) Generic biologics. BioProcess Int 9(8):34–41
31. Sashidhar KS (2011) Biobetters—advent of a new concept. Frost, North America. Available at: https://www.frost.com/sublib/display-marketinsight.do?id=247671489. Accessed 26 May 2022
32. US Code, Title 42, The Public Health and Welfare, Chap. 6A, Public Health Service, Subchapter II, General Powers and Duties, Part F, Licensing of Biological Products and Clinical Laboratories, Subpart 1, Biological Products, Sec. 262, Regulation of Biological Products
33. European Medicines Agency (2005) Guideline on similar biological medicinal products. Evaluation of medicines for human use. European Medicines Agency, London. Available at: https://www.ema.europa.eu/en. Accessed 26 May 2022
34. Silva EP, da Silva Guerra LD (2019) Revisão integrativa: intercambialidade e aspectos farmacoeconômicos dos medicamentos. JMPHC. ISSN: 2179-6750. https://doi.org/10.14295/jmphc.v11iSup.747
35. de Oliveira Taveira RMS (2019). Biossimilares: Desafios no acesso ao mercado. Dissertação de Mestrado-Portugal, Universidade de Coimbra, set
36. Verma S, Jain P, Chauhan P (2018) Biobetters: the better biologics and their regulatory overview. Int J Drug Regul Affairs 4(9):13–20
37. Royzman I, Cohen AD (2014) Biobetters: the advantages and challenges of being better. Nat Rev Drug Discov 13:99–100
38. Van Veen HA, Koiter J, Vogelezang CJ et al (2012) Characterization of recombinant human C1 inhibitor secreted in milk of transgenic rabbits. J Biotechnol 162(2):319–326
39. Grabowski GA, Golembo M, Shaaltiel Y (2014) Taliglucerase alfa: an enzyme replacement therapy using plant cell expression technology. Mol Genet Metab 112(1):1–8
40. Mössner E, Brünker P, Moser S et al (2010) Increasing the efficacy of CD20 antibody therapy through the engineering of a new type II anti-CD20 antibody with enhanced direct and immune effector cell-mediated B-cell cytotoxicity. Blood 115(22):4393–4402
41. Powell JS, Josephson NC, Quon D et al (2012) Safety and prolonged activity of recombinant factor VIII Fc fusion protein in hemophilia A patients. Blood 119(13):3031–3037
42. Peters RT, Low SC, Kamphaus GD et al (2010) Prolonged activity of factor IX as a monomeric Fc fusion protein. Blood 115(10):2057–2064
43. Pelegri-O'Day EM, Lin EW, Maynard HD (2014) Therapeutic protein-polymer conjugates: advancing beyond PEGylation. J Am Chem Soc 136(41):14323–14332

44. Dumont JA, Liu T, Low SC et al (2012) Prolonged activity of a recombinant factor VIII-Fc fusion protein in hemophilia A mice and dogs. Blood 119(13):3024–3030
45. Pharmacoeconomic Review Report: Daclizumab (Zinbryta) [Internet]. Canadian Agency for Drugs and Technologies in Health, Ottawa (ON), 2017 July
46. Cott V, Kevin E, Velander WH (1998) Transgenic animals as drug factories: a new source of recombinant protein therapeutics. Expert Opin Investig Drugs 7(10):1683–1690
47. Houdebine L-M (1994) Production of pharmaceutical proteins from transgenic animals. J Biotechnol 34(3):269–287
48. Larrick JW, David WT (2001) Producing proteins in transgenic plants and animals. Curr Opin Biotechnol 12(4):411–418
49. Twyman RM, Stefan S, Rainer F (2005) Transgenic plants in the biopharmaceutical market. Expert Opin Emerg Drugs 10(1):185–218
50. Sola R, Griebenow K (2009) Effects of glycosylation on the stability of protein pharmaceuticals. J Pharm Sci 98:1223–1245
51. Aranesp® prescribing information. Amgen Inc., Thousand Oaks, CA (2012). Available at: http://pi.amgen.com/united_states/aranesp/ckd/aranesp_pi_hcp_english.pdf. Accessed 18 May 2022
52. Malkin BJ (2012) Biobetters and extending into the marketplace beyond patent expiration. FDA Lawyer's Blog. Available at: http://www.fdalawyersblog.com. Accessed 20 May 2022
53. Peters RT, Toby G, Lu Q, Liu T, Kulman JD, Low SC et al (2013) Biochemical and functional characterization of a recombinant monomeric factor VIII-Fc fusion protein. J Thromb Haemost 11:132–141
54. Rogers B, Daoyuan D, Li Z, Li Z (2015) Recombinant human serum albumin fusion proteins and novel applications in drug delivery and therapy. Curr Pharm Des 14(1):1899–1907
55. Duttaroy A, Kanakaraj P, Osborn BL, Schneider H, Pickeral OK, Chen C, Zhang G, Kaithamana S, Singh M, Schulingkamp R, Crossan D (2005) Development of a long-acting insulin analog using albumin fusion technology. Diabetes 54(1):251–258
56. Beck A, Liliane G, Charles D, Nathalie C (2017) Strategies and challenges for the next generation of antibody-drug conjugates. Nat Rev Drug Discov 16(5):315–337
57. Schuster M, Jost W, Mudde G, Wiederkum S, Schwager C, Janzek E et al (2007) In vivo glyco-engineered antibody with improved lytic potential produced by an innovative non-mammalian expression system. Biotechnol J 2(6):700–708
58. Selvaraju R. Prolor Biotech, Inc. Biotechnology and Specialty Pharmaceutics. Available at: http://d494288.u26.1host.co.il/prolor_13.7.11.pdf. Accessed 18 May 2022
59. Exenatide sustained release—Dong-A ST. Available at: http://adisinsight.springer.com/drugs/800028637. Accessed 21 May 2022
60. FluMist Quadrivalent Influenza Vaccine Live, Intranasal. Astrazeneca. Available at: https://www.flumistquadrivalent.com. Accessed 26 May 2022
61. Afrezza. Mannkind. Available at: https://www.afrezza.com/. Accessed 19 May 2022
62. Tariq M, Akash MSH, Rehman K, Shuqing C (2015) Development of therapeutic proteins: advances and challenges. Turk J Biol 39:343–358
63. Dixit R, Coats S (2009) Preclinical efficacy and safety models for mAbs: the challenge of developing effective model systems. IDrugs 12:103–108
64. Descotes J, Gouraud A (2008) Clinical immunotoxicity of therapeutic proteins. Expert Opin Drug Metab Toxicol 4:1537–1549
65. Suntharalingam G, Perry MR, Ward S, Brett SJ, Castello-Cortes A, Brunner MD, Panoskaltsis N (2006) Cytokine storm in a phase 1 trial of the anti-CD28 monoclonal antibody TGN1412. N Engl J Med 355:1018–1028
66. Ewer SM, Ewer MS (2008) Cardiotoxicity profile of trastuzumab. Drug Saf 31:459–467
67. Chung CH (2008) Managing premedications and the risk for reactions to infusional monoclonal antibody therapy. Oncologist 13:725–732
68. Winter MC, Hancock BW (2009) Ten years of rituximab in NHL. Expert Opin Drug Saf 8:223–235

69. Schellekens H (2008) How to predict and prevent the immunogenicity of therapeutic proteins. Biotechnol Annu Rev 14:191–202
70. Tassev DV, Cheung NK (2009) Monoclonal antibody therapies for solid tumors. Expert Opin Biol Ther 9:341–353
71. Hopper-Borge EA, Nasto RE, Ratushny V, Weiner LM, Golemis EA, Astsaturov I (2009) Mechanisms of tumor resistance to EGFR-targeted therapies. Expert Opin Ther Targets 13:339–362
72. Presta LG (2008) Molecular engineering and design of therapeutic antibodies. Curr Opin Immunol 20:460–470
73. Jefferis R (2009) Glycosylation as a strategy to improve antibody-based therapeutics. Nat Rev Drug Discov 8:226–234
74. Chen W, Zhu Z, Feng Y, Xiao X, Dimitrov DS (2008) Construction of a large phage-displayed human antibody domain library with a scaffold based on a newly identified highly soluble, stable heavy chain variable domain. J Mol Biol 382:779–789
75. Zhang MY, Dimitrov DS (2007) Novel approaches for identification of broadly cross-reactive HIV-1 neutralizing human monoclonal antibodies and improvement of their potency. Curr Pharm Des 13:203–212
76. Singh DB, Tripathi T (2020) Frontiers in protein structure, function, and dynamics. Springer Nature, Singapore

Therapeutic Protein-Based Vaccines 13

Shaban Ahmad, Vandana Dahiya, Arpana Vibhuti, Ramendra Pati Pandey, Manish Kumar Tripathi, and Manoj Kumar Yadav

Abstract

Infectious diseases are reported worldwide, and the emergence of highly mutated antibiotic-resistant strains is a major concern globally. Developing efficient vaccines is the only way to prevent and treat diseases effectively. Though developing conventional vaccines is an intricate and time-consuming process due to several rate-limiting steps, these vaccines help treat an array of existing diseases. There is a dire need for new forms of vaccines as many incidents of resistance are reported, and the efficacy of newly developed vaccines must be enhanced to treat the infections well in time. The human immune system fights against several infections utilizing antibody and the non–antibody-based immune mechanism, providing significant protection against identified pathogens. Nowadays, much effort is being made to develop vaccines focussing on the role of cellular responses to clear several complicated infections. This chapter concentrates on strategies for designing therapeutic protein-based vaccines and their diverse clinical and nonclinical applications.

S. Ahmad
Department of Biomedical Engineering, SRM University, Delhi-NCR, Sonepat, Haryana, India

Department of Computer Science, Jamia Millia Islamia, New Delhi, India

V. Dahiya · M. K. Yadav (✉)
Department of Biomedical Engineering, SRM University, Delhi-NCR, Sonepat, Haryana, India
e-mail: manoj.yadav@srmuniversity.ac.in

A. Vibhuti · R. P. Pandey
Department of Biotechnology, SRM University, Delhi-NCR, Sonepat, Haryana, India

M. K. Tripathi
Department of Biophysics, All India Institute of Medical Sciences, New Delhi, India

D. B. Singh, T. Tripathi (eds.), *Protein-based Therapeutics*,
https://doi.org/10.1007/978-981-19-8249-1_13

Keywords

Protein-based vaccines · Antibodies · Immune mechanism · SARS-CoV-2 · Cancer · HIV

13.1 Introduction

Over 17 million individuals die annually from preventable infectious or communicable diseases globally. The high mutation rates in pathogens and the development of antibiotic-resistant strains are the key barriers to the prevention and treatment of such diseases. Vaccines offer a safer and faster method to overcome infectious diseases over a large population scale. However, the high cost of conventional vaccines in developing nations is a major concern, which is attributed to a lack of proper storage and supply facilities [1–3]. Hence, there is a vital need to develop new methods of vaccine development with low cost and high specificity. Vaccines are usually an inactivated pathogen or its component (DNA, RNA, protein) that stimulates a benign effect in an immune reaction to generate defense against an infection/disease on subsequent exposure to the pathogen when introduced to the host [2, 4–6]. Vaccines usually provide neutralizing activity in the body by generating protective antibodies against infections, and these antibodies develop in a few weeks to several months. Therefore, vaccines deliver the antigens to induce specific protecting antibodies to control, eliminate, and protect humans from pathogens and associated diseases [6–8]. Vaccinology is the conjunction of epidemiology, microbiology, immunology, and pharmacy principles. The ever-changing high mutation rates in pathogens are the vital challenges associated with safe and effective vaccination. Moreover, the efficacy and immune response of vaccination depends on numerous factors such as disease complexity, host immunity (cell- and antibody-mediated responses), gender, age, genetic variations, medical conditions of the host, etc. [2].

13.2 Types of Vaccines

The desired properties of any vaccine comprise safety, efficacy, specificity, long-lasting neutralizing activity against pathogens, lack of autoimmunity, storage, and ease of administration to the host. The use of different kinds of vaccines (DNA-based, RNA-based, protein-based) resulted in successful and long-lasting immunity; however, in many cases, the immunity is not long-lasting enough to cure the disease [9]. The advantages and disadvantages of various vaccines, based on their development, are summarized in Table 13.1.

Table 13.1 Summary of mechanisms, advantages, and limitations of different types of vaccines

Types of vaccine	Mechanism of action	Advantages	Limitation	References
DNA-based vaccine	Also called the third-generation vaccines. Induces an immunologic response in the individual against bacteria, parasites, viruses, and potential cancer by using engineered DNA.	It can induce both cellular and humoral responses at the same time. Non–live-cell approach. DNA molecules are more stable over time. Removes the need for protein purification and increases safety and efficacy.	The immune effect is very low as only a small amount can enter intracellular space.	[6, 10, 11]
mRNA-based vaccine	To achieve the expression of target antigens, they need to enter the cytoplasm.	Theoretically safe.	Some immune responses like headache, muscle pain, and fatigue can be there.	[5, 12]
Protein-based vaccine	Includes an inactivated bacterial toxoid protein, and induces an immune response, e.g., human papillomavirus (HPV) vaccines.	Cost-effective, easy to administer, and stable.	Cannot be produced through MHC-I.	[4, 13–15]
Pure polysaccharide vaccine	Includes polysaccharide molecules (sugar/ carbohydrate) found on the outside of some bacteria, e.g., some vaccines to protect against *Pneumococci.*	Less expensive.	Not able to offer long-lasting herd immunity due to hyporesponsiveness.	[16]
Live attenuated vaccine	Functional/alive and weakened virus or bacteria is used. Also, it can replicate in the body to produce	Produces a strong and lasting cell-mediated immune response.	In some cases, the disease can develop due to the multiplication of weakened viruses or bacteria. Production and	[2, 5, 12]

(continued)

Table 13.1 (continued)

Types of vaccine	Mechanism of action	Advantages	Limitation	References
	an immune response without causing the disease, e.g., chickenpox.		maintenance are complex.	
Dead/ inactivated vaccines	Viruses or bacteria in these vaccines are inactivated/ dead, e.g., polio.	Cause a humoral immune response. Easy to prepare.	Always require repeated doses for immunity.	[5, 14, 17]

13.2.1 Protein-Based Vaccines

DNA- and RNA-based vaccines are better choices in terms of effectiveness and long-term immunity; however, the limitation with using these vaccines is the presence of gene coding materials, which can induce health issues in the host. It is well known that antigens are solely responsible factors for generating the adaptive immune response. Most antigens are either proteins, polysaccharides, or peptides. The variations in the structure of different proteins lead to distinct immune responses in individuals. Recent studies against various infections provided detailed knowledge about viral envelop, protein conformation, and epitope information. This can be incredibly advantageous for designing specific vaccines against these harmful viruses. Due to these facts, it is essential to focus on protein-based vaccines (PBVs) [4, 7, 8, 14]. The advancement in genetic sequencing, microbiology, X-ray crystallography, nuclear magnetic resonance (NMR), spectroscopy, and genetic engineering provides a better and more detailed understanding of the structure of proteins with explicit knowledge of why some proteins are more immunogenic than others proteins [8, 18, 19]. PBVs are designed using weakened or inactivated proteins that can trigger immune responses inside the host. These protein antigens are obtained from the pathogen by isolation and purification. Further, the advantage of this method is that it confiscates the after side effects of the dose. At the same time, this method requires multiple doses to enhance a more potent and durable effect. The antigen-presenting cells (APCs) are responsible for producing adaptive immune responses in the host [10, 12, 20]. The first PBV vaccine was a bacterial toxin vaccine that was made from antitoxin isolated from an animal immunized with the unmodified toxin in a small amount. Later it was realized that the success rate of this active immunization could be increased if, before administration, the toxin was chemically or thermally treated or coadministered with proper antitoxin. The human trials of PBV are in progress against SARS-COV-2 [21]. The safety and efficacy were the primary reason for PBV use over live attenuated and inactivated vaccines, as the immune response produced by PBV is usually based on the antigen used. However, the safety can be influenced due to genetic modification or mutations in antigen structure, as in the case of SARS-CoV-2, both positively and negatively (autoimmunity). To avoid the problem related to autoimmunity or efficacy, the conjugated vaccine is a better option. The conjugated vaccine is designed using the unnatural amino acids (p-nitrophenylalanine) incorporated with PBV structure, for example, vaccines against RANKL and TNFα.

13.3 Design and Development of Protein-Based Vaccines

Initially, when protein-based vaccines were designed, they relied on natural sources for the antigens [22, 23]. In recent decades, the technical approaches for developing and producing new vaccines have grown exponentially, especially during the COVID-19 pandemic. The vaccine design methodology combines the various interrelated fields like genetics/reverse vaccinology, molecular biology, polysaccharide chemistry, protein biology, virology, immunology, bacteriology, fermentation technology, macromolecular purification, formulation of the complexes, etc. [24]. A significant fraction of the previously developed vaccines is from the preventive category (prophylaxis) of infectious disease rather than therapy of infections [23]. Modern technological developments have facilitated the development of vaccines for noninfectious diseases, such as autoimmune disease, cancer allergy, drug addiction, and therapeutic vaccines for a specific group of infectious and noninfectious diseases. The significant development in this area in the last decade redefined the vaccine development process. Due to the recent achievements in vaccine design, several vaccines could be created in just a few months against SARS-CoV-2. Vaccines can be classified into active and inactive vaccines. Active vaccines stimulate the immune system and produce either specific antibodies or cellular immune responses [25]. In some cases, both the responses are activated concurrently and help to treat the disease condition. While in the passive vaccination, preformed antibodies can bind to a human cellular antigen and, thus, completely neutralize a pathogen. An inactive vaccine is administered before or around the time of exposure to a pathogen or a subject showing initial symptoms of infection. The vaccine design strategies can be divided into several categories that are discussed below.

13.3.1 Glycoconjugate Vaccines

In the early and late twentieth century, polysaccharide vaccines were prepared that protect against *Haemophilus influenzae*, pneumococcal, and meningococcal infections. These vaccines are derived using capsular polysaccharides (CPS) from the surface of these bacteria. The high abundance and surface exposure of CPS provoke immune responses and, thus, result in bactericidal activities. A significant improvement is required in the formulation of these polysaccharide vaccines, because they are only effective in adults compared to infants and young children. Only a single serotype caused most *H. influenzae* type b (Hib) diseases, while several immunologically distinct serogroups circulate during infections; thus, more complex epidemiology exists for other pathogens. A broadly protective glycoconjugate vaccine that can improve the immune response to polysaccharide antigens can be designed for pathogens by including multiple CPS serogroups in a multivalent formulation [22, 23, 25]. The glycoconjugate vaccine (7-valent) against *Streptococcus pneumoniae* showed a significant reduction in pneumococcal disease across all age groups. Though such multivalent vaccines broadly offer protection, due to the discovery of more than 90 distinct disease-causing pneumococcal serotypes, there is a dire need to develop alternative pneumococcal vaccines based on one or a few highly conserved protein antigens.

13.3.2 Protein Subunit Vaccines and Structure-Based Antigen Design

The initial success of glycoconjugate vaccines paved the way for scientists to develop alternate methods of vaccine design using modern techniques. In the early twentieth century, toxoid protein-based vaccines were developed against diphtheria, tetanus, influenza, etc. Vaccines were also created using hemagglutinin as the primary antigen. Hemagglutinin is the glycoprotein that plays a crucial role in the early stage of the infection in the influenza virus [22, 25]. Similarly, various protein-based vaccines for different disease targets started developing. Many vaccines exist for different serogroups against *Neisseria meningitides I*, but the existing arsenal of vaccines could not provide a universal solution for serogroup B (MenB) patients. A multicomponent vaccine, 4CMenB, was developed against MenB by applying principles of reverse vaccinology. This initiative greatly appreciated and accelerated the vaccine formulation using computational identification and reversed vaccinology techniques. These initial developments resulted in the first genome-derived recombinant protein-based vaccine, Bexsero®, against MenB. Initially, the Bexsero® vaccine was approved by the European Medicines Agency in 2013, and later on, it received approval in more than 35 countries worldwide.

In an attempt to provide proper antigen characterization, advances in structural biology methods such as X-ray crystallography, cryogenic electron microscopy (cryo-EM), NMR spectroscopy, and computational studies are making an immense contribution to designing and optimizing new vaccine antigens. Recently, various approaches to vaccine antigen design in combination with structural biology techniques have been reported. This multidisciplinary approach is also termed "structural vaccinology." There are three ways in which structural vaccinology is helping in vaccine research. Firstly, poor biochemical behaviour is resolved using structural biology approaches where potential weakness can be highlighted in an antigen. Secondly, these structural studies can identify conformational heterogeneity in an antigen, leading to the formulation of different mutated antigen forms. Thirdly, when this approach is combined with the epitope mapping, the antigens' multiple regions can be identified, which is necessary for elevating protection or neutralizing antibody responses [26]. Structural vaccinology is being used in many vaccine research against the human immunodeficiency virus (HIV), where they focus on designing immunogens capable of producing protective antibody responses against gp120 or gp41 segments of HIV envelope glycoprotein trimer [27]. Structural vaccinology has also collaborated with nanobiotechnology to create self-assembling protein nanoparticles that present many copies of an antigen in an ordered array. A large antigen nanoparticle is more immunogenic than the recombinant proteins [8, 25]. This multidisciplinary combination of structural vaccinology and nanobiotechnology has shown multiple benefits. The technological advances in human B-cell cloning and antibody production have made it feasible to develop an effective structure-based antigen design.

13.4 Delivery and Mechanism of Action of Protein-Based Vaccines

13.4.1 B-Cell Repertoires, Antibody Discovery, and the Human Immune Response

Antibody-mediated immune responses play a decisive role in preventing infection, and T cell-mediated responses are crucial in killing the cells infected with a virus. Adaptive immunity mainly relies on the production of specific antibodies. The maintenance of protective levels of antibodies is critical for eliciting an adequate immune response after vaccination. The B-cell receptors (BCRs) present on the surface of B-cells process the antigen, and interactions between the antigen-specific T-cells and B-cells are mainly responsible for initiating specific B-cell responses. This cognate T-cell and B-cell interaction leads to the expansion of antigen-specific B-cells and their differentiation into short-lived plasma cells. This results in the production of unmutated antibodies, mostly of IgM isotype, and provides the first-line defense against the infection. These responses are followed by the formation of the germinal centre (GC) in the lymphoid organs. Plasma cells with a higher affinity for the antigen come from the GC of the bone marrow. This pool of plasma cells has a long life that will continuously release antibodies, and they are thus responsible for nourishing antibody levels even if the antigen is absent. Memory B-cells generated through a GC reaction start recirculating in the secondary lymphoid organs and peripheral blood. Thus, their affinity for BCRs makes them highly competent in capturing antigens. Usually, the production of new plasma cells attains its highest level in the blood on the 7th day of antigenic boost. This results in a continuous rise in antibody titer in serum; however, not all antibody titers are equally competent. T-cell-independent antibody responses to free polysaccharides have a short life, whereas T-cell-dependent antigens can evoke immunity for decades or a lifetime. Almost all licensed vaccines provide protection against disease by producing antibodies by B-cell. However, the underlying nature of promising antibody response has been challenging to generate. Antibodies of heavy (m, a, g, d, ε) and light (k, l) chains are linked by disulfide bonds and contain variable and constant domains. One of the most remarkable developments in understanding antibody responses is the development of technologies to produce human monoclonal antibodies (mAbs) by using Epstein-Barr (EB) virus transformation by performing phage displays in genetically modified mice. Since 2008, next-generation sequencing (NGS) technologies have provided a way of amplification and cloning vectors of heavy and light chain immunoglobin genes from B-cells. This advancement is mainly used to identify high-affinity influenza-specific antibodies and segregate broadly neutralizing antibodies (bnAbs) against HIV [8]. The advances in NGS technologies have led to sequencing antibody genes from millions of cells, thus providing a detailed characterization of the antibody sequence repertoire and reactions that occur after vaccinations. The antibody repertoires are examined after immunization with influenza and tetanus using NGS technologies. Various analyses of the antibody repertoire allowed tracing the evolutionary paths, resulting in bnAbs. The presence of new methods to separate human mAbs and study the atomic details of their protein structures has helped in describing the antigen-antibody interactions comprehensively. Various developments in this field focus on the vaccine against HIV; however, it should also be done for other pathogens, including influenza and respiratory syncytial viruses (RSVs), which are responsible for high morbidity and mortality in children. In recent studies, the immunoglobin gene repertoire information is combined with the antigen-specific repertoire information that consists of human serum polyclonal response [28]. Recent studies also suggest that many peripheral B-cell-encoded antibodies are not present in the blood or secretions, so they cannot contribute to humoral immunity. Overall, these studies are taking us toward an era in which antigen-specific immunological research on various antigens can increase the pace of vaccine development and design more effective vaccine antigens.

13.4.2 Nucleic Acid Vector Vaccine Delivery Systems

The human immune system is a redundant, non–antibody-based immune mechanism that can provide significant protection against pathogens alone or with antibodies. Considering this, many efforts are made to design vaccines focussing on cellular responses to clear challenging pathogens such as HIV, hepatitis C, Ebola, etc. The CD8+ responses can be increased using DNA vectors that harbor the genes responsible for encoding intracellular antigen expressions. Many attempts are being made to achieve this, such as using naked DNA fragments or systems based on alphavirus, poxvirus, vaccinia virus, or lentivirus. The human adenovirus 5 (Ad5) has been used for gene delivery in many vaccine development studies. The use of Ad5 showed favourable results in preclinical and clinical trials. Since most humans are already exposed to Ad5, this affects the immunological potency of these vector delivery systems. Recent studies in the viral-based delivery of genetic vaccines have a prime-boost strategy that combines the Chimpanzee adenoviral (ChAd) vector with the modified vaccinia virus Ankara (MVA) vector. Favourable results come from the high levels of both CD8+ and CD4+ T-cells, especially for the antigens delivered genetically for HCV. The clinical efficacy of the ChAd vectors is yet to be fully established. After the outbreak of the West African Ebola virus, which resulted in more than 8500 deaths, a vaccine development program started a clinical trial to study the working of the monovalent ChAd3 vaccine encoding the peripheral glycoprotein of the Zaire Ebola virus. RNA-based vaccines have several advantages compared to DNA vaccines. Using DNA increases the possibility of integrating plasmid DNA into the genome of the immunized host, which can then be directly translated into the cytoplasm [23, 29]. The RNA-based vaccines have better antigen expression during acute infection and can generate more robust antigen-specific immune responses. The effectiveness of the RNA vaccine is also dependent on the fact that RNA is a rich stimulator of innate immunity, and the results in animal models have been highly favourable. Earlier, RNA vaccines were not preferred due to the presence of unstable naked RNA in the tissue fluids. Several studies have been performed to increase the efficiency and stability of RNA-based vaccines. Clinical works in metastatic melanoma and renal cell carcinoma have shown a rise in antigen-specific immune responses (both antibodies and T cells). Currently, RNA vaccines against prostate cancer, melanoma, rabies, influenza, HIV, tuberculosis, etc., are in clinical trials [29], and RNA vaccines against infectious diseases are under assessment. However, the future of RNA vaccines relies on new and synthetic delivery systems.

13.4.3 Synthetic Viral Seeds for Rapid Generation of Influenza Vaccines

With the global emergence and rapid spread of new SARS-CoV-2 variants in the human population, health organizations and pharma companies rapidly developed responses to provide well-matched vaccines against the variants. In a pandemic, there is little hope that any pre-existing vaccines will boost the immune responses of human populations worldwide. Nowadays, scientists are trying to improve the vaccine responses against the emergence of new influenza variants. Multiple influenza strains are used to design universal influenza vaccines and develop new methods to speed up vaccine production. The advances in synthetic biology enable rapidly identifying genes responsible for encoding new influenza variants. Recently, scientists constructed a synthetic seed virus with hemagglutinin (HA) and neuraminidase (NA) genes taken from influenza (H7N9) virus sequence, using Madin-Darby canine kidney (MDCK) cell lines. The combined approaches significantly improved vaccine production rates compared to existing methods. The cell-culture-derived H7N9 vaccine was found to be safe and immunogenic in the phase I trial. After two doses, the vaccine shows potentially significant immune responses in most subjects with no pre-existing immunity against the H7N9 virus. These observations have provided a strong rationale for further clinical development of synthetic vaccine reagents.

13.5 Advantages and Limitations of Protein-Based Vaccines

The protein-based vaccines display several advantages over the other vaccine platforms. Nonetheless, there are associated limitations too. Both the advantages and limitations are detailed below:

1. One of the most important advantages of protein-based vaccines is that they can be easily accessible to low- and middle-income countries. Also, protein-based vaccines against some diseases such as hepatitis B are made locally in Brazil, Indonesia, and India.
2. Researchers from the University of Liverpool and the MRC Laboratory of Molecular Biology in Cambridge stated that protein-based subunit vaccines are good alternatives to mRNA-based vaccines [30]. Currently, few protein-based vaccines are approved for COVID-19 disease and are reported better in terms of ease and cost production, transportation, administration, and effectiveness of protection.
3. Unlike inactivated whole-cell vaccines, protein-based vaccines do not contain live parts of the pathogens. They only have antigenic characteristics of the pathogen, so they are considered safe comparatively.
4. The major limitation of protein-based vaccines is that they require adjuvants and booster shots to generate an effective immune response. Adjuvants are ingredients being used for decades in vaccines to enhance their immunogenicity. Some adjuvants can lead to more local reactions (like redness, swelling, itching, and pain at the injection site) and systematic reactions (like fever and body ache) in the patients.
5. In addition, the design of protein-based vaccines may also take time to determine the perfect antigen combination.

13.6 Recombinant Production of Protein-Based Vaccines

13.6.1 Bacterial Systems

E. coli bacteria was the first recombinant expression system. It helps understand molecular biochemistry, offers a large yield of defined proteins, has a fast growth rate, and requires a short production time with low cost, simple process scale-up, upstream processing, and high productivity [17, 31]. However, the *E. coli* system lacks machinery for posttranslational modifications (PTMs) such as glycosylation and multimer assembly. It is essential to focus on the PTMs, as it is the primary reason for protein misfolding, low solubility, and nonfunctionality. To resolve the issue of PTMs, engineered bacteria is a better choice [32]. Furthermore, the development of molecular biology, biopharmaceutical applications, and bioinformatics tools helps predict potential expression issues. Leucogen® (Virbac, Carros, France), a purified recombinant p45 FeLV-envelope antigen, was the first recombinant veterinary vaccine successfully produced in *E. coli*. For recombinant protein production, the cytoplasm and periplasm are the possible targets in the *E. coli* cells. The success rate of recombinant protein production usually depends on the total metabolic load imposed on cells and the ability of the host cell to produce proteins [21]. Three pathways are exploited for recombinant protein production: (1) the twin-arginine translocation (TAT) pathway, (2) SecB-dependent pathway, and (3) SRP-mediated pathway. Among these, SecB-dependent pathway is the most popular method [33]. However, the limitation of using the standard Sec pathway is its incapability to transport folded proteins. In such cases, the TAT pathway is a better choice as it can export fully folded proteins and cofactor substrates with a limit of size up to 150 kDa [34]. However, the limitation of the TAT pathway is its low product yield due to the low abundance of TAT apparatus. Therefore, this system has not been used for industrial production [35], and further technological advancement is required for its successful use.

13.6.2 Yeast System

For routine expression of proteins with PTMs, yeast has emerged as a preferred choice for clinical or veterinary use. The use of yeast offers an opportunity for an extensive range of substrates, advanced genome analysis, and specific responses against genetic manipulations [36]. With all these factors, yeast displays a straightforward method for developing a nontoxic vaccine. There are an array of applications of yeast systems (Fig. 13.1), and various techniques can be used to

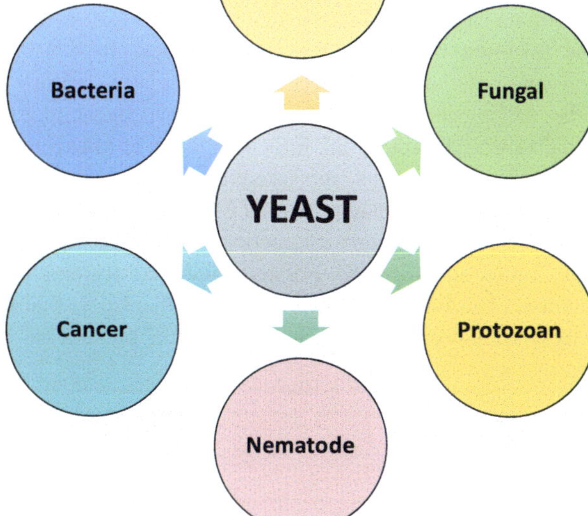

Fig. 13.1 Schematic diagram showing different applications of yeast system

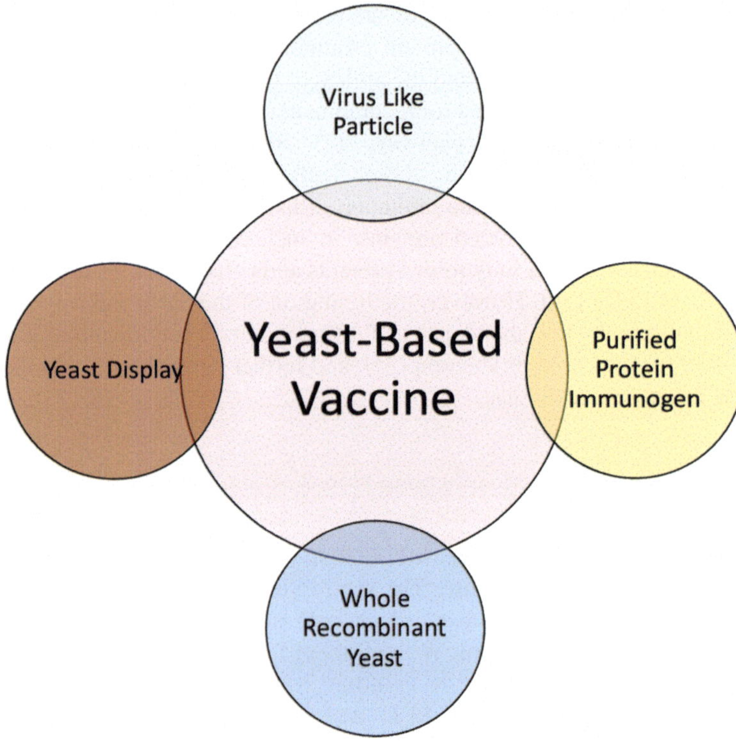

Fig. 13.2 Various methods to design yeast-based vaccines

design yeast-based vaccines (Fig. 13.2). The nonpathogenic nature of yeast is already known, but in recent studies, yeast has also shown an immunologic response in animals and is taken up via dendritic cells (DCs) and macrophages [37]. Earlier it was assumed that due to polysaccharides, such as beta-1,3-D-glucan (BG) and mannan, yeast cells possess an immunogenic nature through antigen-presenting cells (APCs), including DCs, accompanied by employing the technology of threat signals through microbial infection. Due to the robust adjuvant nature of these carbohydrates moieties, the infection can be detected utilizing sample popularity receptors like toll-like receptors (TLRs) and mannan receptors on APCs, which help activate T-cells through interaction and recognition of antigen peptides through MHC molecules [36–38]. For cell-mediated immune response, T-cells activation is essential. The major advantage of yeast display is that the soluble antibodies in the blood can directly recognize the antigen present in yeast cells and produce an immune response [39].

13.6.3 Mammalian Cells

Protein-based therapeutics are rapidly growing due to their advance and specific applications. For recombinant protein-based vaccine production, mammalian cell lines are dominantly used to generate safe and human-like glycoproteins. Mammalian cells host mAbs, enzymes, hormones, and cytokines [40, 41] and are a better choice over the other systems for recombinant protein production due to their ability to generate complex PTMs, stability over deviations in oxygen, temperature, and pH in the production stage, high productivity, and heterologous secretion of protein molecules in the site of extraction via cell lysis. However, with all these advantages, there is a challenge with mammalian cells associated with low production speed, very high cost, the requirement of supplementation of growth factors, amino acids, vitamins, and the risk of contamination during the production process. The risk of virus contamination can be reduced by following regulatory guidelines, selecting low-risk raw materials, and in-process manufacturing control to prevent contamination in the final product. An appropriate method and cell lines are needed to transport the gene of interest in the host cells; mAbs are used in more than 60% of cases. Continuous cell lines (CCLs) are used for virus propagation to develop virus-based vaccines. For this, Vero (African monkey kidney epithelial) cell line is practiced worldwide and used to produce polio and rabies virus vaccines [42]. The cell lines derived from mammalian cells can synthesize large and complex protein molecules. Mouse myeloma, human embryonic kidney 293 (HEK293) cells, and Chinese hamster ovary (CHO) cell systems are standard cell lines used for recombinant vaccine production. The human cell lines offer a greater advantage, as they could also have PTMs characteristics of human proteins [14, 40–42]. These cell lines are developed in adherent cultures or suspension cultures. Suspension culture has greater application as it is easier to scale up and is adaptable to automated processes. To express the foreign genes over these cell lines, stable or transient expression processes (a large amount of protein) can be used. CHO cells are primarily used for stable cell line expression, whereas HEK-293 cells are used for transient expression due to their high transfection efficiency. The transient method provides rapid protein production in a short period making it suitable for recombinant vaccine production. The optimization of vaccine development using mammalian cells continues; a human vaccine produced with CHO cells has been approved for use [7, 8, 14, 43].

13.6.4 Insect Cells

Insect cells are another alternative host platform for recombinant protein-based vaccine production due to the high cost of mammalian cell lines. The baculovirus expression vector system (BEVS) has emerged as a better choice. It provides a high yield of recombinant protein as it has a strong late viral polyhedrin (polh) promoter, less production time, bypassing the requirement of developing stable cell lines, no contamination by prions, and oncogenic DNA. Figure 13.3 summarizes the characteristics of the BEVS system that makes it a better option for recombinant

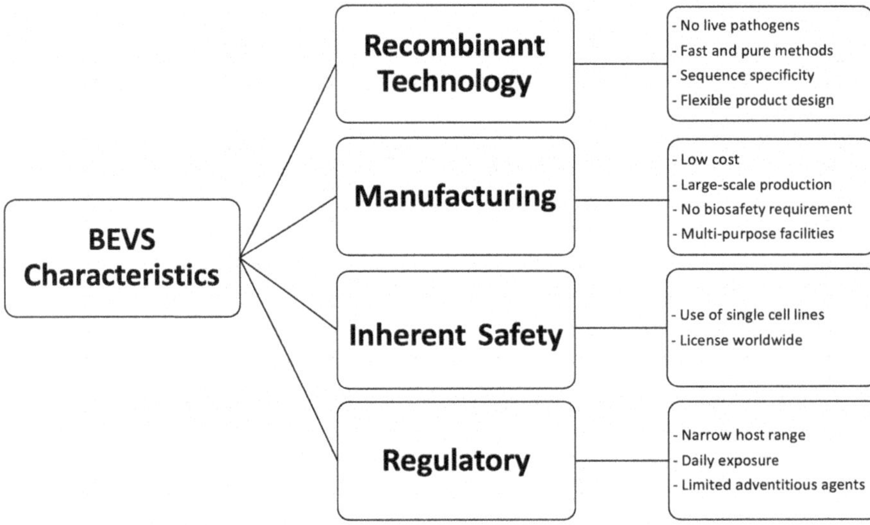

Fig. 13.3 The characteristics of the baculovirus expression vector system (BEVS)

protein-based vaccine development [41, 44]. The major limitations of the insect baculovirus expression are the lack of homogenous human-like glycosylation and cell lysis [15, 38, 45]. The insect cell growth contains two phases: (1) the insect cells are multiplied to desired cell density, and (2) infected with suitably modified baculovirus containing the gene of interest [18]. Another issue related to the insect cells is their inability to carry out N-glycosylation. However, to solve this problem, two steps are followed: (1) the mammalian glycosyltransferases can be introduced into insect cells, or (2) the coexpression of these enzymes with the gene of interest in baculoviruses.

Sf9 is the most popular cell line for the baculovirus expression system [44]. Other cell lines commonly used are S2, Sf21, Tn-368, and High-Five™ cells. The first commercially available veterinary vaccine produced in insect cells was the classical swine fever virus (CSFV) vaccine based on the E2 antigen [45]. Overall, the BEVS possesses flexibility, efficacy, safety, specificity, and single-cell line use in manufacturing multiple products.

13.6.5 Plant-Based System

The advancements in promoter selection, plasmid transformation, codon optimization, transgenic and transformation approaches, and recombinant protein-based vaccine designing using plant sources have become easy and more cost-effective than eukaryotic systems [14, 46, 47]. The expression studies of vaccine antigens in plants include whole plants, roots, moss, suspension cells, microalgae, and duckweeds. The plant-based vaccine offers high protein stability, low cost, safety, stability over pH or temperature, the capability of producing N-glycosylated proteins, and easier and more economical storage of engineered drugs. Plant-based systems have minor differences in glycosylation patterns compared to mammalian cells. The terminal galactose and sialic acid residues are common in animals, whereas plant-based systems are deficient, and instead, the plant proteins contain α-(1,3) fucose and β-(1,2) xylose, which animal proteins lack. Therefore, glycoproteins produced in plants can affect the pharmacokinetic properties and generate immune reactions. Also, controlling transgene expression levels in plants is difficult, and the purification stage is more complex, posing a greater challenge in eliminating the secondary metabolites and pesticide residues from plant sources. To avoid this issue, glycoengineering approaches are used. Currently, two major strategies are used for obtaining the desired therapeutic protein where the protein is first extracted from a plant source and then purified and examined to check its immunogenic activity [47]. These are (1) *Agrobacterium*-mediated transformation, where a stable transgene expression is acquired, and (2) via plant viral vectors, where a transient expression of the foreign gene is obtained. The stable transgene expression is advantageous but time consuming and results in low expression yields. At the same time, transient expression is easy to manipulate and quick but less stable. After evaluating and examining the production of the functional protein level at the laboratory stage, large-scale industrial production in a plant-based system is preferred. An example of such an industrial scale is the production of therapeutic protein in carrot cell cultures (ProCellEx™) to treat Gaucher disease using the human recombinant β-glucocerebrosidase (taliglucerase alfa) [14, 41, 47]. The eligible dose and combination requirements to target plant and transgenic protocol, and proper and safe procedure for cultivation, manufacturing, and processing are essential points that must be taken care in order to design a effective plant-based vaccine.

13.7 Current Status of Protein-Based Vaccines

The design, expansion, and delivery of protein-based vaccines are still a challenge in the fields of vaccine development. However, an array of protein antigens capable of inducing adequate immune responses against specific pathogens have been discovered. The development of protein-based vaccines is still in its naive phase because of existing delivery problems. There is an absence of a complete understanding of the basic requirements for formulating and delivering protein-based therapeutics. Yet proteins have recently proven to be very effective as vaccines as they can mount immunogenic responses owing to stimulation of the body's natural mechanism [48]. Several protein-based vaccines have been developed against diseases like influenza, cancer, COVID-19, etc. [49].

13.7.1 Influenza Vaccine

The development of influenza vaccines aims to elicit a broader immunity, because the seasonal influenza vaccines lack efficacy against pandemic influenza strains [50]. Though seasonal influenza vaccines have always been promising and saved countless lives, the continuous genetic mutation and immune escape mechanism in this virus need regular upgradation of vaccines. Recombinant protein vaccines are among the universal vaccine approaches that utilize innovative technologies [51]. Immunological and virological advances, along with knowledge of structural biology and bioinformatics, are boosting the development of novel vaccine approaches [52–54]. The influenza virus membrane contains two critical proteins: haemagglutinin (HA) and neuraminidase (NA). They are crucial for the entry and release of the virus from infected cells [55]. Apart from these two proteins, other structural components, such as the RNA-binding matrix protein M1, the nucleoprotein (NP) that coats the viral RNA, and the ion channel M2 protein, can be recognized by our immune systems. However, HA and NA are more accessible antibody targets than other components owing to their increased prevalence and accessibility on the viral envelope.

Protein-based vaccines contain viral haemagglutinin and neuraminidase proteins. The viruses used for vaccine production are typically grown in chicken eggs, which makes the reliability of vaccine production on a steady supply of embryonated eggs [56]. To avoid this need, a newer technology that employs cell culture for virus growth has been used. A recent report showed the increased efficacy in healthy adults and improved protection in elderly subjects upon administering recombinant HA-subunit vaccine produced from insect cells [57]. Recently vaccines based on neuraminidase, matrix protein 2 ectodomain (M2e), and nucleoprotein have proven effective [58]. The next-generation subunit protein vaccines open new avenues for meeting the escalating demand for safe, affordable, and effective influenza vaccines.

13.7.2 Cancer

Though protein-based vaccines for cancer treatment have not been that successful so far, delivering these proteins within caged protein nanoparticles has shown promise in improving the vaccine efficacy [59]. The protein nanoparticles are required to increase the immunogenicity of the tumor microenvironment. Since immune escape is the hallmark of cancer, it becomes essential to elicit better immune responses while administering any vaccine for cancer. For a cancer vaccine to be effective, it must also impart long-term immune memory to prevent tumor recurrence [60]. The vaccine must also recognize the tumor-associated antigens present specifically on the surface of cancer cells. Hence, the vaccines used for cancer treatment should recognize these antigens and destroy the cancer cells. Protein vaccines are made from tumor-associated antigens in cancer cells that can elicit immune responses quickly. For example, cervical cancer cells express the human papillomavirus HPV E7 oncoprotein (E7), which plays a crucial role in cellular transformation and maintaining the transformed phenotype. The E7 protein is a potent target for developing therapeutic subunit vaccines against cervical cancer. However, it has a limitation of having low antigenicity, so there is a need to add suitable adjuvants to increase its efficacy. A novel chimeric form of the 4-1BBL costimulatory molecule engineered with core streptavidin (SA-4-1BBL) has been developed [61, 62]. The utility of SA-4-1BBL as the immunomodulatory component of HPV-16 E7 recombinant protein-based therapeutic vaccine in the E7-expressing TC-1 tumor as a model of cervical cancer in mice showed that the results are encouraging and offer 70% efficacy in eradicating established tumors in the mice model.

13.7.3 COVID-19

Despite the administration of safe and effective COVID-19 vaccines worldwide, researchers are working to develop different vaccine strategies that could provide longer immunity. The administration of COVID-19 vaccines aims to generate neutralizing antibodies against SARS-CoV-2, particularly the antibodies against the receptor-binding domain (RBD) of the SARS-CoV-2 spike protein [63]. The spike protein is responsible for facilitating viral entry through its interaction with the epithelial angiotensin-converting enzyme 2 (ACE2) receptors. The titer of antibodies reactive to the RBD/spike protein and neutralization of viral infectivity is the primary measures of response to COVID-19 vaccines. Although the initially approved vaccines were based on mRNA, they targeted only the SARS-CoV-2 spike protein. Moreover, protein-based vaccines offer advantages over mRNA vaccines in terms of the ease and cost of production, the robustness of the material, and potency. A recent report showed that an archaeon-based ferritin-like protein coupled with different antigens from SARS-CoV-2 was highly effective in generating a stable immune response [30]. These highly stable vaccine nanoparticles completely protected the mice from SARS-CoV-2-associated pneumonia in just a single immunization. Richmond et al. tested a stabilized trimeric spike subunit protein vaccine (SCB-2019) [64]. This vaccine is unique from those approved as it uses a stabilized protein trimer as the antigen. Another group used Trimer-Tag, a protein derived from the C-terminus of human type I procollagen, which preserves the trimeric conformation of the SARS-CoV-2 spike protein [65]. The Trimer-Tag technology provides an alternative trimer stabilization strategy to the molecular clamp derived from HIV proteins [66]. This technology can be used for scalable production and rapid development of safe and effective protein-based vaccines.

13.7.4 Other Diseases

Various protein-based drugs produced by recombinant technologies are now readily available therapeutics at reasonable prices for treating chronic diseases. Therapeutic proteins are increasingly prominent because they have been effective in treating many potentially fatal diseases like diabetes, heart disorders, and cancer [48, 67,

68]. Moreover, proteins have been proven effective even as vaccines to help stimulate the body's natural defense mechanism for an immunogenic response. Therapeutic proteins are booming in the pharmaceutical industry through the cloning and expression of cDNA that encodes heterologous proteins [69]. Protein-based vaccines have been developed for Hepatitis B surface antigen (HBsAg) for hepatitis B infection and antirhesus (Rh) IgG vaccine for routine postpartum prevention of Rh (D) immunization in Rh(D)-negative women [70].

13.8 Assistance of Artificial intelligence in Vaccine Development

Artificial intelligence (AI) has revolutionized and transformed the field of medicine. The power of AI's automatic feature learning, combined with the massive volume of data, contributed to its role in its wide applications. In the medical field, the two most crucial areas, drug discovery and vaccine discovery, are immensely benefited by AI technology [41, 71–73]. In recent decades, machine learning (ML), the subfield of AI technology, also helped to improve vaccine design. The AI/ML employs an algorithm structure to interpret and learn the features from the data given in the input form. It makes independent decisions for completing specific objectives. The prominent role of AI technology is basically to analyze the existing data and use it for prediction purposes [74]. Apart from the prediction, it also helps understand and suggest the paradigm for developing future vaccines based on the case studies against a disease. The essential feature of AI is speed and accuracy, which impacts diagnosis and vaccine development processes [24].

VaxiJen is the first server that implemented ML in reverse vaccinology approaches and showed promising results for antigen prediction using its physicochemical properties [75, 76]. The recent web-based reverse vaccinology program, Vaxign-ML, is used to predict bacterial protective antigens. These pipelines, which consist of feature extraction, feature selection, data augmentation, and cross-validation, are implemented to predict the vaccine candidates against various infectious diseases [77]. Other pipelines, such as the immune epitope database (IEDB) and BlastP, use the recurrent neural network (RNN) approaches to study different pathogenic viruses [78, 79]. Recently, some pipelines have been developed that work based on the graph theory method and represent antibodies with expert-designed features. A subset of AI, namely, deep learning (DL), is also widely used on graph-based features to speed up accurate vaccine development [80]. Thus, DL-based approaches revolutionized the field of vaccinology through improved prediction methods [81–83]. Autoencoders of the DL approach have shown promising enhancement in mining the features from data, which could be utilized in vaccine discovery [84]. The critical aspects of the development of vaccine therapy are safety and reliability. The vaccine adverse event reporting system (VAERS) and vaccine safety databank (VSD) are the most popular immunization strategy for tracking, recording, and predicting the safety of vaccines. Earlier, computational simulation and mathematical modelling techniques were significantly used to

Table 13.2 Some prediction tools used for vaccine design

Vaccine types	Prediction tools
B-cell epitope	ABCpred, ElliPro, Pep-3D-Search, MimoPro, BEPro (PEPITO), BEST, SVMTriP, Pep-3DSearch
T-helper epitope	IL4pred, IFNepitope
MHC Class I binders	ProPred1, RANKPEP, nHLAPred, MMBPred
MHC Class II binders	Propred, MARIA, EpiDOCK, HLA DR4Pred, MHC2Pred, Consensus, HLA DR4Pred
Endogenous antigen processing	NetChop, CTLPred, FRED, TAPreg, TAPhunter, NetTepi, Pcleavage
Allergenicity of peptides	AllerHunter, Allertop, Hemolytik, Toxinpred, AHTPDB
Antigenicity	SVMTriP, ANTIGENpro, VaxiJen

improve the trade-off between the assessment of safety and efficacy [84, 85]. Natural language Processing (NLP) technology is now widely used to identify adverse events related to vaccine development [86]. Many prediction tools are available for vaccine design that are listed in Table 13.2.

In summary, AI has been applied in the drug discovery and vaccine development subfields. The advances in DL algorithms are significant for the rapid discovery of vaccines and drugs. The DL-based models can extract important features from the dataset with high accuracy without any manual intervention. The generative ability of DL-based models is exploited for better epitope prediction, which may lead to lower chances of failure of the designed vaccine in the trial phases. Thus, AI is a novel approach to vaccine development that uses transfer learning and leverages the learned knowledge from existing data.

13.9 Challenges and New Approaches for Protein-Based Therapeutics

Protein-based therapeutics are exceptionally effective in the clinic. Computational methods to analyze small molecule drug development use mathematical calculations to scan the underlying information and integrate it into the target molecules [87, 88] with the assumption that they will regulate its action [89]. This is an essential first step toward high-throughput screening and a suitable therapeutic approach. In general, small molecules that are not naturally occurring can be significantly more dangerous than human proteins. Despite the limitations associated with their pharmacokinetic features, therapeutic proteins are increasingly being used for a wide range of treatments [90]. The success of protein-based therapeutics is mainly due to the application of ideas and techniques developed, which resulted in significant improvements in three critical aspects of competitor therapeutics that are required for FDA approval: safety, efficacy, and quality [91]. These three are vital to the success of any treatment and are discussed in detail below.

13.9.1 Safety

Therapeutic protein-induced side effects could be associated with either interaction with expected targets or interaction with accidental targets. The organization of suppressor therapeutic proteins could have a variety of unintended consequences. Overstimulating the immune system can lead to more severe diseases [92]. Restricting to a specific goal can result in unintended consequences, such as immunomodulatory antibodies, which can either inhibit or stimulate the immune system [93]. One significant distinction between counteracting agent-based restoratives containing Fc and other helpful proteins is antibody-dependent cellular cytotoxicity (ADCC) and complement-dependent cytotoxicity (CDC) [94]. The cardiotoxicity associated with trastuzumab is amplified when the antibody is taken simultaneously or sequentially with an anthracycline [95]. A common example is the adverse acute infusion reactions after protein administration, where cytokine discharge plays a critical role, but other unidentified players could also be involved; such responses were observed for some protein therapeutics such as infliximab, rituximab, trastuzumab, and panitumumab, insulin, and interferon [96]. The leakage of cell debris from lysed harmful B cells can result in cumulative effects of rituximab [92]. Protein structure can also cause sensitivity responses such as anaphylactic shock and serum ailment. Previous IgEs that cross-respond with proteins can increase the intensity of such reactions [95].

Immunogenicity of therapeutic proteins can be a big issue [92, 97, 98]. For example, the discovery of less immunogenic proteins was crucial to the success of mAb-based treatments. In the 1980s, murine mAbs were tested as prospective treatments, but their high immunogenicity resulted in large titers of human antimouse antibodies (HAMAs), toxicities, and limited effectiveness. The development of the less immunogenic new mAbs, which contain human Fc sections, and humanized mAbs, which have mouse complementarity-determining regions (CDRs) joined into the human antibody system, showed clinical success. Fully humanized antibodies, on average, have low immunogenicity and are the most widely used form of antibody in development, despite the fact that the majority of the therapeutic antibodies approved for clinical usage are still artificial and humanized mAbs.

The protein structure, composition, PTMs, contaminations, and heterogeneity can all affect immunogenicity, along with the patient's susceptibility and disease status, following medication, course, period, and recurrence of the disease, mainly when controlled as varied dosages for a long time [98]. Human proteins can trigger antihuman antibodies in humans. Treatment with the human mAb adalimumab resulted in antibodies against the therapy ranging from 1% to 87% for different cohorts of patients, procedures, diseases, and measurement methods in one of the most researched cases of anti-TNFa mAbs [99]. The antibody sequences that contribute to antigen binding and specificity but may appear foreign are a potential explanation for human mAb immunogenicity. Human therapeutic proteins can also disrupt immunological tolerance, and antibody elicitation can be influenced by aggregation [98]. Aggregation can lead to structures that do not necessarily require T cell assistance. Protein immunogenicity may potentially influence efficacy via the pharmacokinetic or neutralizing effects of antibody responses, which are controlled by several parameters, including the affinity, specificity, and concentration of the produced antibodies [97]. Because immunogenicity affects both safety and efficacy, researchers are working hard to predict and reduce immunogenicity in therapeutic proteins [100–102].

Individual safe reactions to therapeutic proteins fluctuate generally. Despite rigorous efforts to identify critical factors associated with immunogenicity, it is difficult to predict the immunogenicity of therapeutic proteins in human subjects. However, less is known about the individual antibodies that compensate for the polyclonal reaction to therapeutic proteins. Because the germline antibody repertoire, at any given time, could be a major determinant of individual differences, knowledge of a large pool of antibodies produced by the human immune system, preferably the entire set, that is, the antibodyome, is essential [103], and could ultimately assist in predicting individual insusceptible reactions to therapeutic proteins. Therapeutic proteins have a significant benefit over small molecule therapies, which are often less selective and can attach to many molecules nonspecifically. However, there are major adverse effects in some circumstances, and safety concerns can result in therapeutic proteins being withdrawn from the market [104].

13.9.2 Efficacy

Besides safety, the FDA considers efficacy the essential factor in granting approval. Many therapeutic proteins, including insulin for diabetes, epoetin for anaemia, and rituximab for non-Hodgkin's lymphoma [96], are very effective in vivo and have changed the landscape of disease therapy. Additional examples are alemtuzumab, a drug used to treat hematological cancers [105], and trastuzumab, the human epidermal growth factor receptor type 2 (HER2) positive breast cancer adjuvant systemic medication [106]. The addition of trastuzumab to non–trastuzumab-based adjuvant treatment lowers recurrence by roughly 50% and improves overall survival by 30%, according to results from six studies involving over 14,000 women with HER2-positive early breast cancer [107]. Therapeutic mAbs and other therapeutic proteins have low overall effectiveness, and there is a lot of individual heterogeneity. Trastuzumab (Herceptin) has completely transformed the management of HER2-positive patients; most patients still have nonreacting cancers, and infection movement occurs in most cases within a year [108]. Antiangiogenic treatments, such as bevacizumab, that target the vascular endothelial growth factor (VEGF) and the VEGF receptor (VEGFR), are useful adjuncts in treating solid tumours and are usually regulated in blend with cytotoxic chemotherapy. Regardless, many patients fail to respond to angiogenic treatment of gliomas, and the response term is brief and variable [109].

New techniques, such as improved effector functions, are being explored to enhance the efficacy of mAb and other therapeutic proteins by working on the half-life, expanded cancer and tissue availability, and, more importantly, stability. The improvement in efficacy involves both protein engineering and glycoengineering fields [109–111]. The mAbs that do not interact with innate immunity are being created [112]. Antibodies with many targets are being produced and evaluated in clinical studies. Modulation of immune responses by mAbs targeting T cell immune response regulators is also a viable approach. The inhibitory regulator of such responses is the cytotoxic T lymphocyte antigen 4 (CTLA-4) expressed on activated T cells. Human antibodies and Fc fusion proteins that block CTLA-4 function have been evaluated in the clinic and proven to have antimelanoma activity [113, 114].

Second- and third-generation mAbs against already established targets, including HER2, CD20, and tumor necrosis factor-α (TNF-α), are now in clinical trials or have already been authorized. Several methodologies have been employed to find new, relevant targets, but progress has been slow. An enhanced selection of cross-reactive antibodies by sequential antigen panning [115] and competitive antigen panning for focused selection of antibodies targeting a specific protein domain or subunit have been described as modifications to normal panning processes [116, 117]. To enable greater tissue penetration and concealed epitope access, a variety of modestly designed antibody domains (approximately tenfold smaller than IgG) are being developed [118, 119]. Antibodyome information might be utilized to create semi-synthetic libraries for selecting high-affinity binders with small sizes and low immunogenicity [103]. The development of antibody-based therapeutics means that existing antibodies are gradually improving in characteristics and being designed. A continuous upgradation is required in the properties of existing therapeutic proteins and in identifying novel therapeutic protein targets. The future challenge is how to increase the efficacy of therapeutic antibodies and how to go for their mass production without compromising standard protocols. Developing successful personalized antibody-based treatments and predicting toxicity or potentially poor efficacy in vivo are other key obstacles [91].

13.9.3 Quality

The FDA considers quality to be a critical factor in approving any pharmaceutical. The heterogeneity of mAbs and other biologics is a key feature that separates them from small-molecule medications. Modifications, such as incomplete disulfide bond formation, glycosylation, N-terminal pyroglutamine cyclization, C-terminal lysine processing, deamidation, isomerization, oxidation, amidation of the C-terminal amino acid, modification of the N-terminal amino acids by maleuric acid, as well as noncovalent associations with other molecules, conformational diversity, and aggregation, cause heterogeneity [120–122]. A vast number of variations with a similar arrangement might exist together. Improving excellent protein therapeutics with negligible heterogeneity and defilement is fundamental for their security and endorsement by the FDA [123]. The possibility of using molecular cloning and genetically engineered approaches for manufacturing low-cost therapeutic proteins in plants and delivering therapeutic proteins by in vivo methods are other methods to improve quality and reduce the treatment cost [91].

13.10 Conclusion and Future Perspectives

DNA and RNA vaccines are generally preferred in terms of effectiveness and long-term immunity. To avoid problems related to autoimmunity or efficacy, the conjugated vaccine is a better option. The conjugated vaccine is designed using unnatural amino acids (p-nitrophenylalanine) incorporated into the protein-based vaccine structure, for example, vaccines against activator of NF-κB ligand (RANKL) and TNF-α. However, the design and development of protein-based vaccines remain challenging. The advancement in technical approaches leads to the identification of new protein antigens that can induce immunity to infectious pathogens. The availability of new methods would allow investigators to focus on best-suited resources for different applications in the field of vaccine research. A lot of quick progress made in recent decades toward developing effective therapeutic proteins gives hope for the future. Antibody treatments will benefit immensely from studies evaluating the synergistic effects of antibodies with chemotherapeutic drugs, radiation, or other biologic agents in the future. Furthermore, the discovery of new biomarkers can potentially increase the efficacy and specificity of antibody-based therapies for human diseases.

References

1. Davies J (1994) Inactivation of antibiotics and the dissemination of resistance genes. Science 264:375–382. https://doi.org/10.1126/science.8153624
2. Lahariya C (2016) Vaccine epidemiology: a review. J Family Med Prim Care 5:7–15. https://doi.org/10.4103/2249-4863.184616
3. Singh DB, Tripathi T (2020) Frontiers in protein structure, function, and dynamics. Springer Nature, Singapore
4. Anasir MI, Poh CL (2019) Structural vaccinology for viral vaccine design. Front Microbiol 10. https://doi.org/10.3389/fmicb.2019.00738
5. Brisse M, Vrba SM, Kirk N, Liang Y, Ly H (2020) Emerging concepts and technologies in vaccine development. Front Immunol 11. https://doi.org/10.3389/fimmu.2020.583077
6. Zimmermann P, Curtis N (2019) Factors that influence the immune response to vaccination. Clin Microbiol Rev 32. https://doi.org/10.1128/cmr.00084-18
7. Graham BS, Gilman MSA, McLellan JS (2019) Structure-based vaccine antigen design. Annu Rev Med 70:91–104. https://doi.org/10.1146/annurev-med-121217-094234
8. Saylor K, Gillam F, Lohneis T, Zhang C (2020) Designs of antigen structure and composition for improved protein-based vaccine efficacy. Front Immunol 11. https://doi.org/10.3389/fimmu.2020.00283
9. Vartak A, Sucheck SJ (2016) Recent advances in subunit vaccine carriers. Vaccines (Basel) 4. https://doi.org/10.3390/vaccines4020012
10. Kaur SP, Gupta V (2020) COVID-19 vaccine: a comprehensive status report. Virus Res 288: 198114. https://doi.org/10.1016/j.virusres.2020.198114
11. Suschak JJ, Williams JA, Schmaljohn CS (2017) Advancements in DNA vaccine vectors, non-mechanical delivery methods, and molecular adjuvants to increase immunogenicity. Hum Vaccin Immunother 13:2837–2848. https://doi.org/10.1080/21645515.2017.1330236
12. Speiser DE, Bachmann MF (2020) COVID-19: mechanisms of vaccination and immunity. Vaccine 8:404

13. Blakney AK, McKay PF (2021) Next-generation COVID-19 vaccines: here come the proteins. Lancet 397:643–645. https://doi.org/10.1016/s0140-6736(21)00258-0
14. Cid R, Bolívar J (2021) Platforms for production of protein-based vaccines: from classical to next-generation strategies. Biomolecules 11:1072
15. Contreras-Gómez A, Sánchez-Mirón A, García-Camacho F, Molina-Grima E, Chisti Y (2014) Protein production using the baculovirus-insect cell expression system. Biotechnol Prog 30:1–18. https://doi.org/10.1002/btpr.1842
16. Weber C, Drogoz A, David L, Domard A, Charles MH, Verrier B, Delair T (2010) Polysaccharide-based vaccine delivery systems: macromolecular assembly, interactions with antigen presenting cells, and in vivo immunomonitoring. J Biomed Mater Res A 93:1322–1334. https://doi.org/10.1002/jbm.a.32605
17. Francis MJ (2018) Recent advances in vaccine technologies. Vet Clin North Am Small Anim Pract 48:231–241. https://doi.org/10.1016/j.cvsm.2017.10.002
18. Owczarek B, Gerszberg A, Hnatuszko-Konka K (2019) A brief reminder of systems of production and chromatography-based recovery of recombinant protein biopharmaceuticals. Biomed Res Int 2019:4216060. https://doi.org/10.1155/2019/4216060
19. Tripathi T, Dubey VK (2022) Advances in protein molecular and structural biology methods, 1st edn. Academic Press, Cambridge, MA
20. Zhu FC, Guan XH, Li YH, Huang JY, Jiang T, Hou LH, Li JX, Yang BF, Wang L, Wang WJ, Wu SP, Wang Z, Wu XH, Xu JJ, Zhang Z, Jia SY, Wang BS, Hu Y, Liu JJ, Zhang J, Qian XA, Li Q, Pan HX, Jiang HD, Deng P, Gou JB, Wang XW, Wang XH, Chen W (2020) Immunogenicity and safety of a recombinant adenovirus type-5-vectored COVID-19 vaccine in healthy adults aged 18 years or older: a randomised, double-blind, placebo-controlled, phase 2 trial. Lancet 396:479–488. https://doi.org/10.1016/s0140-6736(20)31605-6
21. Ma C, Li Y, Wang L, Zhao G, Tao X, Tseng CT, Zhou Y, Du L, Jiang S (2014) Intranasal vaccination with recombinant receptor-binding domain of MERS-CoV spike protein induces much stronger local mucosal immune responses than subcutaneous immunization: implication for designing novel mucosal MERS vaccines. Vaccine 32:2100–2108. https://doi.org/10.1016/j.vaccine.2014.02.004
22. Effio CL, Hubbuch J (2015) Next generation vaccines and vectors: designing downstream processes for recombinant protein-based virus-like particles. Biotechnol J 10:715–727. https://doi.org/10.1002/biot.201400392
23. Murthy N, Xu M, Schuck S, Kunisawa J, Shastri N, Fréchet JMJ (2003) A macromolecular delivery vehicle for protein-based vaccines: acid-degradable protein-loaded microgels. Proc Natl Acad Sci 100:4995–5000. https://doi.org/10.1073/pnas.0930644100
24. Kalita P, Tripathi T (2022) Methodological advances in the design of peptide-based vaccines. Drug Discov Today 27:1367–1380. https://doi.org/10.1016/j.drudis.2022.03.004
25. Dale JB, Smeesters PR, Courtney HS, Penfound TA, Hohn CM, Smith JC, Baudry JY (2017) Structure-based design of broadly protective group a streptococcal M protein-based vaccines. Vaccine 35:19–26. https://doi.org/10.1016/j.vaccine.2016.11.065
26. Oyarzún P, Kobe B (2016) Recombinant and epitope-based vaccines on the road to the market and implications for vaccine design and production. Hum Vaccin Immunother 12:763–767. https://doi.org/10.1080/21645515.2015.1094595
27. Kovacs JM, Nkolola JP, Peng H, Cheung A, Perry J, Miller CA, Seaman MS, Barouch DH, Chen B (2012) HIV-1 envelope trimer elicits more potent neutralizing antibody responses than monomeric gp120. Proc Natl Acad Sci U S A 109:12111–12116. https://doi.org/10.1073/pnas.1204533109
28. Rees AR (2020) Understanding the human antibody repertoire. MAbs 12:1729683. https://doi.org/10.1080/19420862.2020.1729683
29. Pollard C, De Koker S, Saelens X, Vanham G, Grooten J (2013) Challenges and advances towards the rational design of mRNA vaccines. Trends Mol Med 19:705–713. https://doi.org/10.1016/j.molmed.2013.09.002

30. Salzer R, Clark JJ, Vaysburd M, Chang VT, Albecka A, Kiss L, Sharma P, Gonzalez Llamazares A, Kipar A, Hiscox JA, Owen A, Aricescu AR, Stewart JP, James LC, Löwe J (2021) Single-dose immunisation with a multimerised SARS-CoV-2 receptor binding domain (RBD) induces an enhanced and protective response in mice. FEBS Lett 595:2323–2340. https://doi.org/10.1002/1873-3468.14171
31. Nag N, Khan H, Tripathi T (2022) Strategies to improve the expression and solubility of recombinant proteins in E. coli. In: Tripathi T, Dubey VK (eds) Advances in protein molecular and structural biology methods. Academic Press, San Diego, pp 1–12
32. Selas Castiñeiras T, Williams SG, Hitchcock AG, Smith DC (2018) E. coli strain engineering for the production of advanced biopharmaceutical products. FEMS Microbiol Lett 365. https://doi.org/10.1093/femsle/fny162
33. Mergulhão FJM, Summers DK, Monteiro GA (2005) Recombinant protein secretion in Escherichia coli. Biotechnol Adv 23:177–202. https://doi.org/10.1016/j.biotechadv.2004.11.003
34. Paschke M (2006) Phage display systems and their applications. Appl Microbiol Biotechnol 70:2–11. https://doi.org/10.1007/s00253-005-0270-9
35. Tullman-Ercek D, DeLisa MP, Kawarasaki Y, Iranpour P, Ribnicky B, Palmer T, Georgiou G (2007) Export pathway selectivity of Escherichia coli twin arginine translocation signal peptides. J Biol Chem 282:8309–8316. https://doi.org/10.1074/jbc.M610507200
36. Malak A, Baronian K, Kunze G (2016) Blastobotrys (Arxula) adeninivorans: a promising alternative yeast for biotechnology and basic research. Yeast 33:535–547. https://doi.org/10.1002/yea.3180
37. Stubbs AC, Martin KS, Coeshott C, Skaates SV, Kuritzkes DR, Bellgrau D, Franzusoff A, Duke RC, Wilson CC (2001) Whole recombinant yeast vaccine activates dendritic cells and elicits protective cell-mediated immunity. Nat Med 7:625–629. https://doi.org/10.1038/87974
38. Kost TA, Kemp CW (2016) Fundamentals of baculovirus expression and applications. In: Vega MC (ed) Advanced technologies for protein complex production and characterization. Springer International Publishing, Cham, pp 187–197
39. Kumar R, Kumar P (2019) Yeast-based vaccines: new perspective in vaccine development and application. FEMS Yeast Res 19. https://doi.org/10.1093/femsyr/foz007
40. Ecker JW, Kirchenbaum GA, Pierce SR, Skarlupka AL, Abreu RB, Cooper RE, Taylor-Mulneix D, Ross TM, Sautto GA (2020) High-yield expression and purification of recombinant influenza virus proteins from stably-transfected mammalian cell lines. Vaccines (Basel) 8. https://doi.org/10.3390/vaccines8030462
41. Tripathi NK, Shrivastava A (2019) Recent developments in bioprocessing of recombinant proteins: expression hosts and process development. Front Bioeng Biotechnol 7. https://doi.org/10.3389/fbioe.2019.00420
42. Berlec A, Strukelj B (2013) Current state and recent advances in biopharmaceutical production in Escherichia coli, yeasts and mammalian cells. J Ind Microbiol Biotechnol 40:257–274. https://doi.org/10.1007/s10295-013-1235-0
43. Gaillet B, Gilbert R, Broussau S, Pilotte A, Malenfant F, Mullick A, Garnier A, Massie B (2010) High-level recombinant protein production in CHO cells using lentiviral vectors and the cumate gene-switch. Biotechnol Bioeng 106:203–215. https://doi.org/10.1002/bit.22698
44. Yee CM, Zak AJ, Hill BD, Wen F (2018) The coming age of insect cells for manufacturing and development of protein therapeutics. Ind Eng Chem Res 57:10061–10070. https://doi.org/10.1021/acs.iecr.8b00985
45. Felberbaum RS (2015) The baculovirus expression vector system: a commercial manufacturing platform for viral vaccines and gene therapy vectors. Biotechnol J 10:702–714. https://doi.org/10.1002/biot.201400438
46. Margolin E, Chapman R, Williamson AL, Rybicki EP, Meyers AE (2018) Production of complex viral glycoproteins in plants as vaccine immunogens. Plant Biotechnol J 16:1531–1545. https://doi.org/10.1111/pbi.12963

47. Rybicki EP (2009) Plant-produced vaccines: promise and reality. Drug Discov Today 14:16–24. https://doi.org/10.1016/j.drudis.2008.10.002
48. Hamid Akash MS, Rehman K, Chen S (2015) Natural and synthetic polymers as drug carriers for delivery of therapeutic proteins. Polym Rev 55:371–406. https://doi.org/10.1080/15583724.2014.995806
49. Padhi AK, Rath SL, Tripathi T (2021) Accelerating COVID-19 research using molecular dynamics simulation. J Phys Chem B 125:9078–9091. https://doi.org/10.1021/acs.jpcb.1c04556
50. Wei C-J, Crank MC, Shiver J, Graham BS, Mascola JR, Nabel GJ (2020) Author correction: next-generation influenza vaccines: opportunities and challenges. Nat Rev Drug Discov 19:427. https://doi.org/10.1038/s41573-020-0066-8
51. Sebastian S, Lambe T (2018) Clinical advances in viral-vectored influenza vaccines. Vaccines (Basel) 6. https://doi.org/10.3390/vaccines6020029
52. Graham BS (2013) Advances in antiviral vaccine development. Immunol Rev 255:230–242. https://doi.org/10.1111/imr.12098
53. Koff WC, Burton DR, Johnson PR, Walker BD, King CR, Nabel GJ, Ahmed R, Bhan MK, Plotkin SA (2013) Accelerating next-generation vaccine development for global disease prevention. Science 340:1232910. https://doi.org/10.1126/science.1232910
54. Malonis RJ, Lai JR, Vergnolle O (2020) Peptide-based vaccines: current progress and future challenges. Chem Rev 120:3210–3229. https://doi.org/10.1021/acs.chemrev.9b00472
55. Neirynck S, Deroo T, Saelens X, Vanlandschoot P, Jou WM, Fiers W (1999) A universal influenza A vaccine based on the extracellular domain of the M2 protein. Nat Med 5:1157–1163. https://doi.org/10.1038/13484
56. Chen J-R, Liu Y-M, Tseng Y-C, Ma C (2020) Better influenza vaccines: an industry perspective. J Biomed Sci 27:33. https://doi.org/10.1186/s12929-020-0626-6
57. Dunkle LM, Izikson R, Patriarca P, Goldenthal KL, Muse D, Callahan J, Cox MMJ (2017) Efficacy of recombinant influenza vaccine in adults 50 years of age or older. N Engl J Med 376:2427–2436. https://doi.org/10.1056/NEJMoa1608862
58. Tan MP, Tan WS, Mohamed Alitheen NB, Yap WB (2021) M2e-based influenza vaccines with nucleoprotein: a review. Vaccines 9:739. https://doi.org/10.3390/vaccines9070739
59. Neek M, Kim TI, Wang SW (2019) Protein-based nanoparticles in cancer vaccine development. Nanomedicine 15:164–174. https://doi.org/10.1016/j.nano.2018.09.004
60. Tay BQ, Wright Q, Ladwa R, Perry C, Leggatt G, Simpson F, Wells JW, Panizza BJ, Frazer IH, Cruz JLG (2021) Evolution of cancer vaccines-challenges, achievements, and future directions. Vaccines (Basel) 9. https://doi.org/10.3390/vaccines9050535
61. Sharma RK, Yolcu ES, Shirwan H (2014) SA-4-1BBL as a novel adjuvant for the development of therapeutic cancer vaccines. Expert Rev Vaccines 13:387–398. https://doi.org/10.1586/14760584.2014.880340
62. Melief CJM, Welters MJP, Vergote I, Kroep JR, Kenter GG, Ottevanger PB, Tjalma WAA, Denys H, van Poelgeest MIE, Nijman HW, Reyners AKL, Velu T, Goffin F, Lalisang RI, Loof NM, Boekestijn S, Krebber WJ, Hooftman L, Visscher S, Blumenstein BA, Stead RB, Gerritsen W, van der Burg SH (2020) Strong vaccine responses during chemotherapy are associated with prolonged cancer survival. Sci Transl Med 12. https://doi.org/10.1126/scitranslmed.aaz8235
63. Kyriakidis NC, López-Cortés A, González EV, Grimaldos AB, Prado EO (2021) SARS-CoV-2 vaccines strategies: a comprehensive review of phase 3 candidates. npj Vaccines 6:28. https://doi.org/10.1038/s41541-021-00292-w
64. Richmond P, Hatchuel L, Dong M, Ma B, Hu B, Smolenov I, Li P, Liang P, Han HH, Liang J, Clemens R (2021) Safety and immunogenicity of S-Trimer (SCB-2019), a protein subunit vaccine candidate for COVID-19 in healthy adults: a phase 1, randomised, double-blind, placebo-controlled trial. Lancet 397:682–694. https://doi.org/10.1016/s0140-6736(21)00241-5

65. Liu H, Su D, Zhang J, Ge S, Li Y, Wang F, Gravel M, Roulston A, Song Q, Xu W, Liang JG, Shore G, Wang X, Liang P (2017) Improvement of pharmacokinetic profile of TRAIL via trimer-tag enhances its antitumor activity in vivo. Sci Rep 7:8953. https://doi.org/10.1038/s41598-017-09518-1
66. Watterson D, Wijesundara DK, Modhiran N, Mordant FL, Li Z, Avumegah MS, McMillan CL, Lackenby J, Guilfoyle K, van Amerongen G, Stittelaar K, Cheung ST, Bibby S, Daleris M, Hoger K, Gillard M, Radunz E, Jones ML, Hughes K, Hughes B, Goh J, Edwards D, Scoble J, Pearce L, Kowalczyk L, Phan T, La M, Lu L, Pham T, Zhou Q, Brockman DA, Morgan SJ, Lau C, Tran MH, Tapley P, Villalón-Letelier F, Barnes J, Young A, Jaberolansar N, Scott CA, Isaacs A, Amarilla AA, Khromykh AA, van den Brand JM, Reading PC, Ranasinghe C, Subbarao K, Munro TP, Young PR, Chappell KJ (2021) Preclinical development of a molecular clamp-stabilised subunit vaccine for severe acute respiratory syndrome coronavirus 2. Clin Transl Immunol 10:e1269. https://doi.org/10.1002/cti2.1269
67. Hermeling S, Crommelin DJ, Schellekens H, Jiskoot W (2004) Structure-immunogenicity relationships of therapeutic proteins. Pharm Res 21:897–903. https://doi.org/10.1023/b:pham.0000029275.41323.a6
68. Ibrahim M, Farooq T, Hussain N, Hussain A, Gulzar T, Hussain I, Akash MS, Rehmani FS (2013) Acetyl and butyryl cholinesterase inhibitory sesquiterpene lactones from Amberboa ramosa. Chem Cent J 7:116. https://doi.org/10.1186/1752-153X-7-116
69. Setiawan D, Brender J, Zhang Y (2018) Recent advances in automated protein design and its future challenges. Expert Opin Drug Discov 13:587–604. https://doi.org/10.1080/17460441.2018.1465922
70. Das S, Ramakrishnan K, Behera SK, Ganesapandian M, Xavier AS, Selvarajan S (2019) Hepatitis B vaccine and immunoglobulin: key concepts. J Clin Transl Hepatol 7:165–171. https://doi.org/10.14218/JCTH.2018.00037
71. Tripathi MK, Nath A, Singh TP, Ethayathulla AS, Kaur P (2021) Evolving scenario of big data and artificial Intelligence (AI) in drug discovery. Mol Divers 25:1439–1460. https://doi.org/10.1007/s11030-021-10256-w
72. Shukla R, Tripathi T (2021) Molecular dynamics simulation in drug discovery: opportunities and challenges. In: Singh SK (ed) Innovations and implementations of drug discovery strategies in rational drug design. Springer Nature, Singapore, pp 295–316
73. Shukla R, Tripathi T (2020) Molecular dynamics simulation of protein and protein-ligand complexes. In: Singh DB (ed) Computer-aided drug design. Springer Nature, Singapore, pp 133–161
74. Keshavarzi Arshadi A, Webb J, Salem M, Cruz E, Calad-Thomson S, Ghadirian N, Collins J, Diez-Cecilia E, Kelly B, Goodarzi H, Yuan JS (2020) Artificial intelligence for COVID-19 drug discovery and vaccine development. Front Artif Intell 3:65
75. Doytchinova IA, Flower DR (2007) VaxiJen: a server for prediction of protective antigens, tumour antigens and subunit vaccines. BMC Bioinformatics 8:4. https://doi.org/10.1186/1471-2105-8-4
76. Heinson AI, Woelk CH, Newell M-L (2015) The promise of reverse vaccinology. Int Health 7:85–89. https://doi.org/10.1093/inthealth/ihv002
77. He Y, Xiang Z, Mobley HLT (2010) Vaxign: the first web-based vaccine design program for reverse vaccinology and applications for vaccine development. J Biomed Biotechnol 2010:297505. https://doi.org/10.1155/2010/297505
78. Flower DR, Macdonald IK, Ramakrishnan K, Davies MN, Doytchinova IA (2010) Computer aided selection of candidate vaccine antigens. Immunome Res 6(Suppl 2):S1. https://doi.org/10.1186/1745-7580-6-S2-S1
79. He L, Zhu J (2015) Computational tools for epitope vaccine design and evaluation. Curr Opin Virol 11:103–112. https://doi.org/10.1016/j.coviro.2015.03.013
80. Magar R, Yadav P, Farimani AB (2020) Potential neutralizing antibodies discovered for novel corona virus using machine learning. bioRxiv. https://doi.org/10.1101/2020.03.14.992156

81. Sher G, Zhi D, Zhang S (2017) DRREP: deep ridge regressed epitope predictor. BMC Genomics 18:676. https://doi.org/10.1186/s12864-017-4024-8
82. Tran NH, Qiao R, Xin L, Chen X, Shan B, Li M (2019) Personalized deep learning of individual immunopeptidomes to identify neoantigens for cancer vaccines. bioRxiv:620468. https://doi.org/10.1101/620468
83. Wu J, Wang W, Zhang J, Zhou B, Zhao W, Su Z, Gu X, Wu J, Zhou Z, Chen S (2019) DeepHLApan: a deep learning approach for neoantigen prediction considering both HLA-peptide binding and immunogenicity. Front Immunol 10:2559
84. Miyake J, Kaneshita Y, Asatani S, Tagawa S, Niioka H, Hirano T (2018) Graphical classification of DNA sequences of HLA alleles by deep learning. Hum Cell 31:102–105. https://doi.org/10.1007/s13577-017-0194-6
85. Vaishnav N, Gupta A, Paul S, John GJ (2015) Overview of computational vaccinology: vaccine development through information technology. J Appl Genet 56:381–391. https://doi.org/10.1007/s13353-014-0265-2
86. Zheng C, Yu W, Xie F, Chen W, Mercado C, Sy LS, Qian L, Glenn S, Lee G, Tseng HF, Duffy J, Jackson LA, Daley MF, Crane B, McLean HQ, Jacobsen SJ (2019) The use of natural language processing to identify Tdap-related local reactions at five health care systems in the Vaccine Safety Datalink. Int J Med Inform 127:27–34. https://doi.org/10.1016/j.ijmedinf.2019.04.009
87. Gane PJ, Dean PM (2000) Recent advances in structure-based rational drug design. Curr Opin Struct Biol 10:401–404. https://doi.org/10.1016/s0959-440x(00)00105-6
88. Zeng J (2000) Mini-review: computational structure-based design of inhibitors that target protein surfaces. Comb Chem High Throughput Screen 3:355–362. https://doi.org/10.2174/1386207003331490
89. Gschwend DA, Sirawaraporn W, Santi DV, Kuntz ID (1997) Specificity in structure-based drug design: identification of a novel, selective inhibitor of Pneumocystis carinii dihydrofolate reductase. Proteins 29:59–67
90. Cho MJ, Juliano R (1996) Macromolecular versus small-molecule therapeutics: drug discovery, development and clinical considerations. Trends Biotechnol 14:153–158. https://doi.org/10.1016/0167-7799(96)10024-x
91. Dimitrov DS (2012) Therapeutic proteins. Methods Mol Biol 899:1–26. https://doi.org/10.1007/978-1-61779-921-1_1
92. Descotes J, Gouraud A (2008) Clinical immunotoxicity of therapeutic proteins. Expert Opin Drug Metab Toxicol 4:1537–1549. https://doi.org/10.1517/17425250802525496
93. Suntharalingam G, Perry MR, Ward S, Brett SJ, Castello-Cortes A, Brunner MD, Panoskaltsis N (2006) Cytokine storm in a phase 1 trial of the anti-CD28 monoclonal antibody TGN1412. N Engl J Med 355:1018–1028. https://doi.org/10.1056/NEJMoa063842
94. Ewer SM, Ewer MS (2008) Cardiotoxicity profile of trastuzumab. Drug Saf 31:459–467. https://doi.org/10.2165/00002018-200831060-00002
95. Chung CH (2008) Managing premedications and the risk for reactions to infusional monoclonal antibody therapy. Oncologist 13:725–732. https://doi.org/10.1634/theoncologist.2008-0012
96. Winter MC, Hancock BW (2009) Ten years of rituximab in NHL. Expert Opin Drug Saf 8:223–235. https://doi.org/10.1517/14740330902750114
97. Pendley C, Schantz A, Wagner C (2003) Immunogenicity of therapeutic monoclonal antibodies. Curr Opin Mol Ther 5:172–179
98. Schellekens H (2008) How to predict and prevent the immunogenicity of therapeutic proteins. Biotechnol Annu Rev 14:191–202. https://doi.org/10.1016/s1387-2656(08)00007-0
99. Emi Aikawa N, de Carvalho JF, Silva CAA, Bonfá E (2010) Immunogenicity of Anti-TNF-alpha agents in autoimmune diseases. Clin Rev Allergy Immunol 38:82–89. https://doi.org/10.1007/s12016-009-8140-3
100. Baker MP, Jones TD (2007) Identification and removal of immunogenicity in therapeutic proteins. Curr Opin Drug Discov Devel 10:219–227

101. Onda M (2009) Reducing the immunogenicity of protein therapeutics. Curr Drug Targets 10: 131–139. https://doi.org/10.2174/138945009787354511
102. Stas P, Lasters I (2009) Strategies for preclinical immunogenicity assessment of protein therapeutics. IDrugs 12:169–173
103. Dimitrov DS (2010) Therapeutic antibodies, vaccines and antibodyomes. MAbs 2:347–356. https://doi.org/10.4161/mabs.2.3.11779
104. Dixit R, Coats S (2009) Preclinical efficacy and safety models for mAbs: the challenge of developing effective model systems. IDrugs 12:103–108
105. Castillo J, Winer E, Quesenberry P (2008) Newer monoclonal antibodies for hematological malignancies. Exp Hematol 36:755–768. https://doi.org/10.1016/j.exphem.2008.04.018
106. Mariani G, Fasolo A, De Benedictis E, Gianni L (2009) Trastuzumab as adjuvant systemic therapy for HER2-positive breast cancer. Nat Clin Pract Oncol 6:93–104. https://doi.org/10.1038/ncponc1298
107. Bedard PL, Piccart-Gebhart MJ (2008) Current paradigms for the use of HER2-targeted therapy in early-stage breast cancer. Clin Breast Cancer 8(Suppl 4):S157–S165. https://doi.org/10.3816/CBC.2008.s.012
108. Hall PS, Cameron DA (2009) Current perspective—trastuzumab. Eur J Cancer 45:12–18. https://doi.org/10.1016/j.ejca.2008.10.013
109. Norden AD, Drappatz J, Wen PY (2008) Novel anti-angiogenic therapies for malignant gliomas. Lancet Neurol 7:1152–1160. https://doi.org/10.1016/s1474-4422(08)70260-6
110. Jefferis R (2009) Glycosylation as a strategy to improve antibody-based therapeutics. Nat Rev Drug Discov 8:226–234. https://doi.org/10.1038/nrd2804
111. Presta LG (2008) Molecular engineering and design of therapeutic antibodies. Curr Opin Immunol 20:460–470. https://doi.org/10.1016/j.coi.2008.06.012
112. Labrijn AF, Aalberse RC, Schuurman J (2008) When binding is enough: nonactivating antibody formats. Curr Opin Immunol 20:479–485. https://doi.org/10.1016/j.coi.2008.05.010
113. Weber J (2009) Ipilimumab: controversies in its development, utility and autoimmune adverse events. Cancer Immunol Immunother 58:823–830. https://doi.org/10.1007/s00262-008-0653-8
114. Yuan J, Gnjatic S, Li H, Powel S, Gallardo HF, Ritter E, Ku GY, Jungbluth AA, Segal NH, Rasalan TS, Manukian G, Xu Y, Roman RA, Terzulli SL, Heywood M, Pogoriler E, Ritter G, Old LJ, Allison JP, Wolchok JD (2008) CTLA-4 blockade enhances polyfunctional NY-ESO-1 specific T cell responses in metastatic melanoma patients with clinical benefit. Proc Natl Acad Sci U S A 105:20410–20415. https://doi.org/10.1073/pnas.0810114105
115. Zhang MY, Shu Y, Phogat S, Xiao X, Cham F, Bouma P, Choudhary A, Feng YR, Sanz I, Rybak S, Broder CC, Quinnan GV, Evans T, Dimitrov DS (2003) Broadly cross-reactive HIV neutralizing human monoclonal antibody Fab selected by sequential antigen panning of a phage display library. J Immunol Methods 283:17–25. https://doi.org/10.1016/j.jim.2003.07.003
116. Choudhry V, Zhang MY, Sidorov IA, Louis JM, Harris I, Dimitrov AS, Bouma P, Cham F, Choudhary A, Rybak SM, Fouts T, Montefiori DC, Broder CC, Quinnan GV Jr, Dimitrov DS (2007) Cross-reactive HIV-1 neutralizing monoclonal antibodies selected by screening of an immune human phage library against an envelope glycoprotein (gp140) isolated from a patient (R2) with broadly HIV-1 neutralizing antibodies. Virology 363:79–90. https://doi.org/10.1016/j.virol.2007.01.015
117. Zhang MY, Dimitrov DS (2007) Novel approaches for identification of broadly cross-reactive HIV-1 neutralizing human monoclonal antibodies and improvement of their potency. Curr Pharm Des 13:203–212. https://doi.org/10.2174/138161207779313669
118. Chen W, Dimitrov DS (2009) Human monoclonal antibodies and engineered antibody domains as HIV-1 entry inhibitors. Curr Opin HIV AIDS 4:112–117. https://doi.org/10.1097/COH.0b013e328322f95e
119. Chen W, Zhu Z, Feng Y, Xiao X, Dimitrov DS (2008) Construction of a large phage-displayed human antibody domain library with a scaffold based on a newly identified highly soluble,

stable heavy chain variable domain. J Mol Biol 382:779–789. https://doi.org/10.1016/j.jmb.2008.07.054
120. Liu H, Gaza-Bulseco G, Faldu D, Chumsae C, Sun J (2008) Heterogeneity of monoclonal antibodies. J Pharm Sci 97:2426–2447. https://doi.org/10.1002/jps.21180
121. Nag N, Chetri PB, Uversky VN, Giri R, Tripathi T (2022) Experimental methods to study intrinsically disordered proteins. In: Tripathi T, Dubey VK (eds) Advances in protein molecular and structural biology methods. Academic Press, San Diego, pp 505–533
122. Chetri PB, Khan H, Tripathi T (2022) Methods to determine the oligomeric structure of proteins. In: Tripathi T, Dubey VK (eds) Advances in protein molecular and structural biology methods. Academic Press, San Diego, pp 49–76
123. Birch JR, Racher AJ (2006) Antibody production. Adv Drug Deliv Rev 58:671–685. https://doi.org/10.1016/j.addr.2005.12.006

The manufacturer's authorised representative in the EU is Springer Nature Customer Service Centre GmbH, Europaplatz 3, 69115 Heidelberg, Germany. If you have any concerns regarding our products, please contact ProductSafety@springernature.com

Printed and bound by CPI Group (UK) Ltd, Croydon, CR0 4YY
30/09/2026
02225253-0014